Mechanics of Auxetic Materials and Structures

Mechanics of Auxetic Materials and Structures offers a wide range of application-based and practical considerations of smart materials and auxetic materials in engineering structures. Exploring the analytical and numerical solution procedures, the book discusses crucial characteristics of metamaterials and their response to external factors.

Covering the effect of different parameters and external factors on the mechanics of auxetic materials and structures, the book considers the benefits leading to better fracture resistance, toughness, shear modulus, and acoustic response.

The book serves as a reference for senior undergraduate and graduate students studying civil engineering, mechanical engineering, and materials science and taking courses in smart materials, metamaterials, and mechanics of materials.

Emerging Materials and Technologies

Series Editor
Boris I. Kharissov

The *Emerging Materials and Technologies* series is devoted to highlighting publications centered on emerging advanced materials and novel technologies. Attention is paid to those newly discovered or applied materials with potential to solve pressing societal problems and improve quality of life, corresponding to environmental protection, medicine, communications, energy, transportation, advanced manufacturing, and related areas.

The series takes into account that, under present strong demands for energy, material, and cost savings, as well as heavy contamination problems and worldwide pandemic conditions, the area of emerging materials and related scalable technologies is a highly interdisciplinary field, with the need for researchers, professionals, and academics across the spectrum of engineering and technological disciplines. The main objective of this book series is to attract more attention to these materials and technologies and invite conversation among the international R&D community.

Chemistry of Dehydrogenation Reactions and its Applications
Edited by Syed Shahabuddin, Rama Gaur and Nandini Mukherjee

Biosorbents
Diversity, Bioprocessing, and Applications
Edited by Pramod Kumar Mahish, Dakeshwar Kumar Verma, and Shailesh Kumar Jadhav

Principles and Applications of Nanotherapeutics
Imalka Munaweera and Piumika Yapa

Energy Materials
A Circular Economy Approach
Edited by Surinder Singh, Suresh Sundaramuthy, Alex Ibhadon, Faisal Khan, Sushil Kansal, and S.K. Mehta

Tribological Aspects of Additive Manufacturing
Edited by Rashi Tyagi, Ranvijay Kumar, and Nishant Ranjan

Emerging Materials and Technologies for Bone Repair and Regeneration
Edited by Ashok Kumar, Sneha Singh, and Prerna Singh

Mechanics of Auxetic Materials and Structures
Farzad Ebrahimi

For more information about this series, please visit: www.routledge.com/Emerging-Materials-and-Technologies/book-series/CRCEMT

Mechanics of Auxetic Materials and Structures

Farzad Ebrahimi

CRC Press
Taylor & Francis Group
Boca Raton London New York

CRC Press is an imprint of the
Taylor & Francis Group, an **informa** business

Designed cover image: Shutterstock

First edition published 2024
by CRC Press
2385 NW Executive Center Drive, Suite 320, Boca Raton FL 33431

and by CRC Press
4 Park Square, Milton Park, Abingdon, Oxon, OX14 4RN

CRC Press is an imprint of Taylor & Francis Group, LLC

© 2024 Farzad Ebrahimi

ISBN: 978-1-032-26659-6 (hbk)
ISBN: 978-1-032-26660-2 (pbk)
ISBN: 978-1-003-28929-6 (ebk)

DOI: 10.1201/9781003289296

Typeset in Times
by Apex CoVantage, LLC

Contents

About the Author

Farzad Ebrahimi is an Associate Professor in the Department of Mechanical Engineering, IKIU, Qazvin, Iran. His research interests include mechanical behaviors of nano-engineered systems, mechanics of composites and nanocomposites, functionally graded materials, viscoelasticity, and smart materials and structures. Dr. Ebrahimi has authored more than 400 high-quality peer-reviewed research articles in his fields of interest. He has also edited and authored multiple books for well-known publishers. He is an Associate Editor of the journal *Shock and Vibration*, an Editorial Board member of the *Journal of Computational Applied Mechanics*, and a distinguished reviewer whose expertise helps the editors of prestigious journals judge research articles.

AFFILIATIONS AND EXPERTISE

Associate Professor, Department of Mechanical Engineering, Faculty of Engineering, Imam Khomeini International University, Qazvin, Iran

Preface

The concept of metamaterials, which has just been introduced, underscores the influence of a material's deliberate microstructure on its overall characteristics. The recent discoveries highlight the importance of exploring the mechanical properties of materials, allowing enhanced regulation and comprehension of the "bottom-up" mechanism inherent in their formation. This suggests that the manipulation of constituent components and microstructural characteristics of engineering materials has the potential to provide novel mechanical capabilities, hence enabling the realization of conceptual designs. The emerging idea has potential for deployment over a range of sizes, depending upon the specific applications and manufacturing techniques involved. This technological progress has the potential to establish a new area of academic inquiry that combines industrial technology, materials science, and mechanics. A crucial aspect of this study is the development of metamaterials exhibiting a negative Poisson's ratio. In the last twenty years, there has been notable advancement in the development, production, and synthesis of materials and structures associated with negative Poisson's ratio.

This literary work delves into the present condition of metamaterials and structures from the perspective of mechanical analysis, elucidating the advancements that have been achieved in this domain. The work is organized in a systematic way to provide a thorough exploration of the subject area. Metamaterials and structures exhibit captivating characteristics, making them very attractive contenders for a diverse range of applications. In the preceding thirty years, significant progress has been achieved in the field of metamaterials, covering theoretical analysis, finite element simulations, and experimental investigations. The advancement in metamaterials research in recent times has presented novel opportunities for future investigation. Nevertheless, there exist a multitude of captivating inquiries that need deeper examination, hence highlighting the continuous advancement and imperative for supplementary inquiry in the realm of metamaterials.

The main purpose of this book is to tackle the problem of inadequate analytical investigations regarding the mechanical analysis of metamaterials and structures. The introductory chapter presents a comprehensive review of the essential ideas of metamaterials and auxetic materials and structures, aiming to give a succinct understanding of these topics. Chapter 2 introduces the topic of static analysis in the context of auxetic structures. In this chapter, the focus is on understanding the behavior and properties of these structures under static conditions. Moving on to Chapters 3 and 4, the investigation shifts towards the vibration analysis of different types of metamaterial structures, specifically auxetic rods, auxetic plates, and auxetic shells. The aim is to analyze and comprehend the vibrational characteristics exhibited by these structures. Lastly, the final chapter is dedicated to the examination of wave dispersion analysis in auxetic beams, plates, and shells. The objective here is to study and comprehend the dispersion of waves in these specific types of auxetic structures.

The current work aims to offer mathematical concepts in a comprehensive way, eliminating the need for readers to go to additional resources. In addition, this study will use a variational-based methodology to express the kinematic equations that control the statics and dynamics of metamaterials with beam, plate, and shell geometries. The study of metamaterial structures is facilitated with the use of classical and shear deformable kinematic theories. Furthermore, it is possible for any person with a basic comprehension of engineering mathematics to deduce all of the equations that have been given. This literary composition has the potential to serve as a significant scholarly resource for those seeking higher levels of education, particularly those with a keen interest in investigating the mechanical characteristics of continuous mechanical systems.

The author made an effort to provide a thorough and all-encompassing work in order to assist readers in understanding the intricacies of metamaterials and structures. The author respectfully invites all readers to share their opinions and comments without hesitation. The author may benefit from receiving constructive input from a discriminating reader to improve the quality of this written work in future versions.

F. Ebrahimi
Imam Khomeini International University, Qazvin, Iran
October 2023

Abbreviations

Acronym	Definition
CF	Carbon fiber
CFRP	Carbon fiber-reinforced polymer
GFR	Glass fiber-reinforced
MMCs	Metal-matrix composites
FRPs	Fiber-reinforced plastics
CNTs	Carbon nanotubes
PNC	Polymer matrix nanocomposites
MD	Molecular dynamics
PV	Photovoltaic
SWCNT	Single-walled carbon nanotubes
BNNTs	Boron nitride nanotubes
NSGT	Nonlocal strain gradient theory
EBB	Euler-Bernoulli beam
TB	Timoshenko beam
CS	Cylindrical shell
DLGSs	Double-layered graphene sheets
MHCR	Multi-hybrid nanocomposite reinforcement
ER	Electro-rheological
MR	Magneto-rheological
MEE	Magneto-electro-elastic
NET	Nonlocal elasticity theory
TWBNNT	Triple-walled boron nitride nanotubes
GPLR	Graphene platelets-reinforced
GDQM	Generalized differential quadrature method
GNPRC	Graphene nanoplatelet-reinforced composite
GF	Graphene foam
GOP	Graphene oxide powder
PGF	Porous graphene foam
FSDT	First-order shear deformation theory
HSDT	Higher-order shear deformation theory
FGM	Functionally graded material
CNCs	Carbon nanocones
CFR	Carbon fiber reinforced
CPT	Classical plate theory
NPR	Negative Poisson's ratio
2D	Two-dimensional
3D	Three-dimensional
UHMWPE	Ultra-high molecular weight polyethylene
DTM	Dilating tetrahedral model
CTM	Contemporaneous tetrahedral model
SMA	Shape memory alloys

PNCs	Polymer nanocomposites
GPLs	Graphene platelets
GDQM	Generalized differential quadrature method
FG-GOA	FG graphene origami auxetic
HPM	Homotopy perturbation method
PVDF	Polyvinylidene fluoride polymer
TSDT	Third-order shear deformation theory
ENET	Eringen's nonlocal elasticity theory
C-C	Clamped-clamped
NSGT	Nonlocal strain gradient theory
MCST	Modified couple stress theory
FG	Functionally graded
ROM	Rule of mixing

Symbols

Symbol	Definition	Symbol	Definition
U	Displacement vector in the x direction	I_i	Moment of inertia
V	Displacement vector in the y direction	E	Young's modulus
W	Displacement vector in the z direction	P	Critical buckling load
ϕ_x	Rotation in the x direction	ρ_f	Density of FG materials
ϕ_y	Rotation in the y direction	ρ_m	Density of magnetostrictive materials
u_0	Initial displacement in the x direction	ω	Frequency
v_0	Initial displacement in the y direction	Ω	Non-dimensional frequency
w_0	Initial displacement in the z direction	∇^2	Laplacian operator
a	Length	X_m	Admissible Glerkin functions along the x direction
b	Width	Y_n	Admissible Glerkin functions along the y direction
N_{xx}	Normal forces	∂T	Variation of Strain energy
M_{xx}	Bending moment	∂U	Variation of potential energy
ξ	The amount of porosity	∂W	Variation of external work
t	Time	z_0	Neutral location
k	Gradient index	k_s	Shear correction factor
ε_{ij}	Strain vector	δ	Variation
σ_{ij}	Stress vector	ε	Strain
μ	Non-local parameter	σ	Stress
λ	Length scale parameter	K_w	Winkler coefficient
[C]	Damping matrix	K_p	Pasternak coefficient
[K]	Stiffness matrix	I_i	Mass moment of interia
[M]	Mass matrix	C_{ijkl}	Elastic coefficient

1 An Introduction to Metamaterials and Auxetic Materials and Structures

1.1 BACKGROUND

The demands placed on engineering materials are tightening and getting more specific as modern engineering advances. Extensive endeavors have been undertaken for millennia to enhance the mechanical properties of engineering materials. One fascinating example of this is the craftsmanship of ancient blacksmiths, who skillfully utilized a range of steel types to achieve a delicate equilibrium between hardness and ductility in the central region and cutting edge of their swords. Moreover, their ingenuity extended to employing diverse lamination techniques for the blades' cross-section, yielding distinct levels of mechanical prowess across the entirety of the weapon. This example demonstrates the importance of components and microstructure in influencing the mechanical characteristics of an engineering material. The mechanical behavior of composite materials is widely understood in contemporary engineering. When creating composites, designers use a variety of materials and fine-tune their work to attain certain features. The idea of mechanical metamaterials, which was just recently presented, emphasizes the effect of a material's intentional microstructure on its whole qualities. These latest findings show the significance of investigating the mechanical attributes of materials, enabling better control and understanding of the "bottom-up" process involved in their development. It implies that by modifying their elements and microstructures, engineering materials with innovative mechanical properties could be created from an idea. The concept under development holds promise for implementation across various scales, depending on potential applications and production methods. This advancement may pave the way for a novel field of research integrating industrial technologies, materials science, and mechanics. An important pathway in this chapter involves the creation of metamaterials with a negative Poisson's ratio (NPR). Over the past two decades, significant progress has been made in designing, fabricating, or synthesizing NPR-related materials and structures. These encompass a wide range of materials such as polymers, metals, ceramics, composites, laminates, and fibers, covering nearly all major material categories.

In an effort to improve the understanding of this particular material, ongoing research focuses on thoroughly examining the development of NPR materials by exploring their fundamental principles. The distinct mechanical characteristics of such materials are then summarized based on their distinctive structural characteristics.

DOI: 10.1201/9781003289296-1

This chapter explores the current state of auxetic materials and structures, discussing the developments that have been made in this field. The text is structured in a methodical manner to provide comprehensive coverage of the study field. The intriguing properties that auxetic materials and structures possess make them promising candidates for a wide variety of applications. Over the past three decades, significant advancements have been made in auxetics, encompassing theoretical analysis, finite element simulations, and experimental work. Recent progress in auxetics research has opened up new prospects for further exploration. However, there are still numerous intriguing questions that require further investigation, indicating the ongoing development and need for additional research in auxetic materials.

While many published works have proposed auxetic cellular structures, only a few have successfully implemented their practical application through experimentation. Furthermore, different materials or construction techniques can result in varied properties for the same auxetic structure. This aspect necessitates further inquiry.

Previous studies have primarily focused on symmetrical and regular structural patterns, but the potential for auxetic behavior also exists in asymmetric, relatively less symmetric, or irregular systems. Tailored to specific applications, research on auxetic materials and structures demands substantial development. By utilizing advanced optimization and modeling approaches, the creation of distinctive and optimal auxetic structures with the intended range of Poisson's ratio becomes achievable.

The investigation of NPR sources brings attention to the careful analysis of classical mechanics. Traditional elasticity principles can still be applied to develop new ideas, such as negative Poisson's ratio, negative stiffness, and negative bulk modulus. Most NPR materials exhibit complex microstructures at a microscopic level and undergo significant volume changes under external loading, which may challenge specific presumptions of classical continuum mechanics. Because of this, additional theories may be required to describe particular kinds of mechanical behavior, taking into account changes in volume and the blurred boundary that separates materials and structures in metamaterials.

Looking ahead, future research in this field is expected to focus on the concept of the unit cell, incorporating hierarchical structures and homogenization of both periodic/ordered and disordered cells. This approach, from both a mechanical and computational perspective, is likely to dominate future investigations.

1.2 DEFINITION OF POISSON'S RATIO

It is widely known and evident that when a material undergoes stretching in one direction, it will contract in the opposite direction of the load that is being applied. This phenomenon is illustrated in the upper portion of Figure 1.1a (top). Consequently, if the load is changed from stretching to compression, as seen in Figure 1.1b (bottom), the materials expand perpendicularly to the applied load. As defined below, the Poisson's ratio v is [1,2]:

$$v = -\frac{\varepsilon_{\text{Trans}}}{\varepsilon_{\text{Load}}} \qquad (1.1)$$

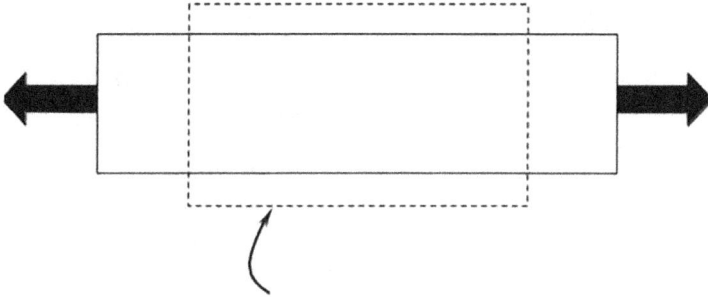

FIGURE 1.1A Original or unreformed shape.

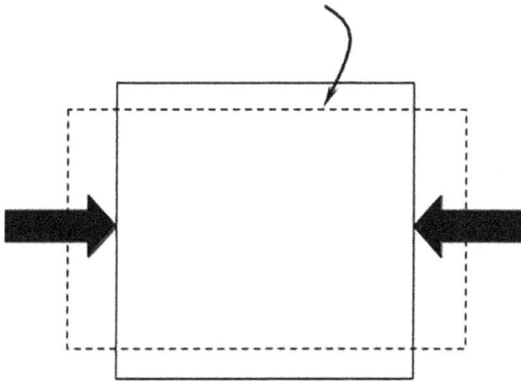

FIGURE 1.1B Schematic illustrating the positive Poisson's ratio effect in 2D deformation.

Here, ε_{Load} represents the strain occurring in the direction of the applied load, while ε_{Trans} represents the strain occurring transverse to the loading direction. In order to account for the typical observation and intuitive understanding that ε_{Load} and ε_{Trans} have opposite signs, resulting in a negative ratio of $\varepsilon_{Load}/\varepsilon_{Trans}$, equation (1.1) incorporates a negative sign. This negative sign is included to ensure that positive values are obtained for Poisson's ratios.

Tensile deformation, a positive characteristic, describes the stretching or elongation of a material, while compressive deformation, considered negative, represents the material's compression. The inclusion of a negative sign in Poisson's ratio's definition confirms the prevalence of a positive value for this ratio in most conventional materials. From a continuum standpoint, this can be elucidated by the fact that the majority of materials exhibit a stronger resistance to changes in volume compared to changes in shape [1,2]. Examining this behavior on a microstructural level reveals that it stems from the tendency of inter-atomic bonds to realign themselves as external deformation intensifies. These atomic realignments contribute to the material's ability to accommodate and distribute the applied force throughout its structure, resulting in its observed deformation characteristics.

It was widely believed, in the early stages of advancements in elasticity theory, that Poisson's ratio had a constant value of −0.25 for all isotropic materials. This belief was founded on the *uniconstant theory*. However, the development of classical elasticity theory has shown that the elasticity of isotropic materials is actually determined by two different parameters: the elastic modulus and Poisson's ratio. It has come to light that the value of Poisson's ratio can change depending on the type of material being utilized.

In modern times, the classical elasticity theory imposes limitations on the range of Poisson's ratio for isotropic materials, requiring that the ratio must fall between −1 and 0.5. These limits were established due to the requirement that the elastic moduli of an unbound block of isotropic material must be positive for stability. Under the conditions of isotropy, linearity, and elasticity, it is possible to derive a positive value for the bulk modulus B using common interrelations [1].

$$B = \frac{E}{3(1-2v)} \tag{1.2}$$

indicates a Poisson's ratio that is less than 0.5, while a positive shear modulus (G):

$$G = \frac{E}{2(1+v)} \tag{1.3}$$

implies a Poisson's ratio exceeding −1.

Hence, these two quantities are commonly utilized to represent Poisson's ratio, as shown in Figure 1.2, which provides an overview of the range of Poisson's ratio for various materials. For the vast majority of ordinary solid materials such as metals, isotropic composites, and polymers, the value of v typically falls in the interval of $0.25 < v < 0.35$. When the shear modulus G is less than or equal to the bulk modulus B, the Poisson's ratio approaches 0.5. Such materials demonstrate a propensity for shear deformations while resisting volumetric deformation. Rubber is a widely recognized illustration of such behavior. Additionally, the plot serves as a reminder that elasticity theory allows for materials with negative Poisson's ratios, with a lower limit of −1. Based on the definitions of shear modulus (G) and bulk modulus (B), this indicates that such materials exhibit exceptional compressibility when their Poisson's ratio is negative, indicating a tendency for volumetric deformation rather than shear deformation (G ≥ B). This novel and counterintuitive mechanical behavior has garnered significant attention over the past two decades.

As shown in Figure 1.2, the shear modulus (G) and bulk modulus (B) are frequently used to demonstrate the range of Poisson's ratios for different materials. The Poisson's ratio falls within the range of $0.25 < v < 0.35$ for the majority of typical solid materials, including metals, polymers, and isotropic composites. The Poisson's ratio v gets closer to 0.5 when the shear modulus is lower than or equal to the bulk modulus (G ≤ B). These materials exhibit a propensity for shear deformations while exhibiting resistance to volumetric deformation. A well-known example of one of these materials is rubber. Furthermore, the plot also illustrates that elasticity theory permits materials to possess negative Poisson's ratios, with a lower limit of −1. This

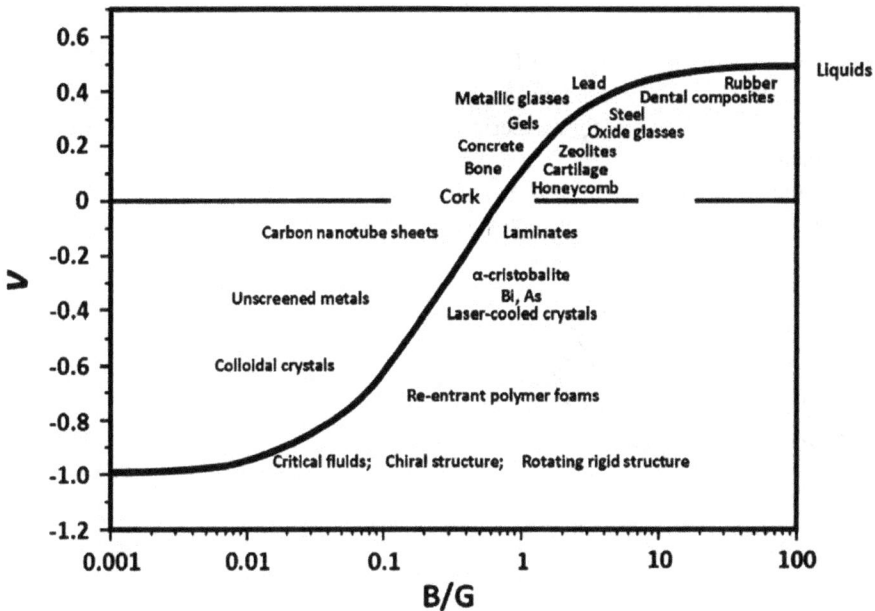

FIGURE 1.2 The relationship between Poisson's ratio and the B/G ratio for different materials.

shows that these materials are quite compressible, as defined by G and B. They typically experience volumetric deformation rather than shear deformation when the Poisson's ratio is negative (G ≥ B). Over the past two decades, this unique and unexpected mechanical behavior has garnered a lot of interest [1].

Young made an early and significant discovery in his renowned "Lectures on Natural Philosophy and the Mechanical Arts" when he delved into the effects of stretching and subsequent contraction in a sideways direction. Poisson took Young's work further and proposed a fixed value of $v = 1/4$ based on the theory of molecular interaction. This estimation gained credibility through Cagniard de la Tour's indirect measurement, which yielded a value close to $v \approx 0.357$ for a brass rod, reinforcing Poisson's proposition. Building upon this methodology, Wertheim conducted experiments and determined the Poisson's ratio as $v = 1/3$ for both brass and glass. Expanding on the research, Kirchhoff adopted a means to calculate the Poisson's ratio of numerous metals. He achieved this by employing measurements of their Young's and shear moduli, providing valuable insights into the behavior of various materials. These measurements, combined with subsequent studies, unearthed empirical evidence suggesting that the Poisson's ratio is not a fixed value but instead displays variation across different materials. Cauchy's theoretical work further solidified the understanding of the elastic behavior of solids exhibiting isotropy. He demonstrated the necessity for two distinct elasticity moduli to describe such behavior accurately. This revelation hinted at the inherent variability of Poisson's ratio among different materials, lending additional support to the notion of its dynamic nature.

1.3 DEFINITION OF AUXETIC MATERIALS

Materials that exhibit a negative Poisson's ratio are commonly referred to as auxetic materials. These unique substances possess a remarkable characteristic: when stretched in one direction, they experience expansion in the perpendicular direction, as depicted in the top image of Figure 1.1a. Conversely, if the load is switched from stretching to compression, these materials contract in the transverse direction, as shown in the bottom image of Figure 1.1b. Figure 1.3 provides several illustrations

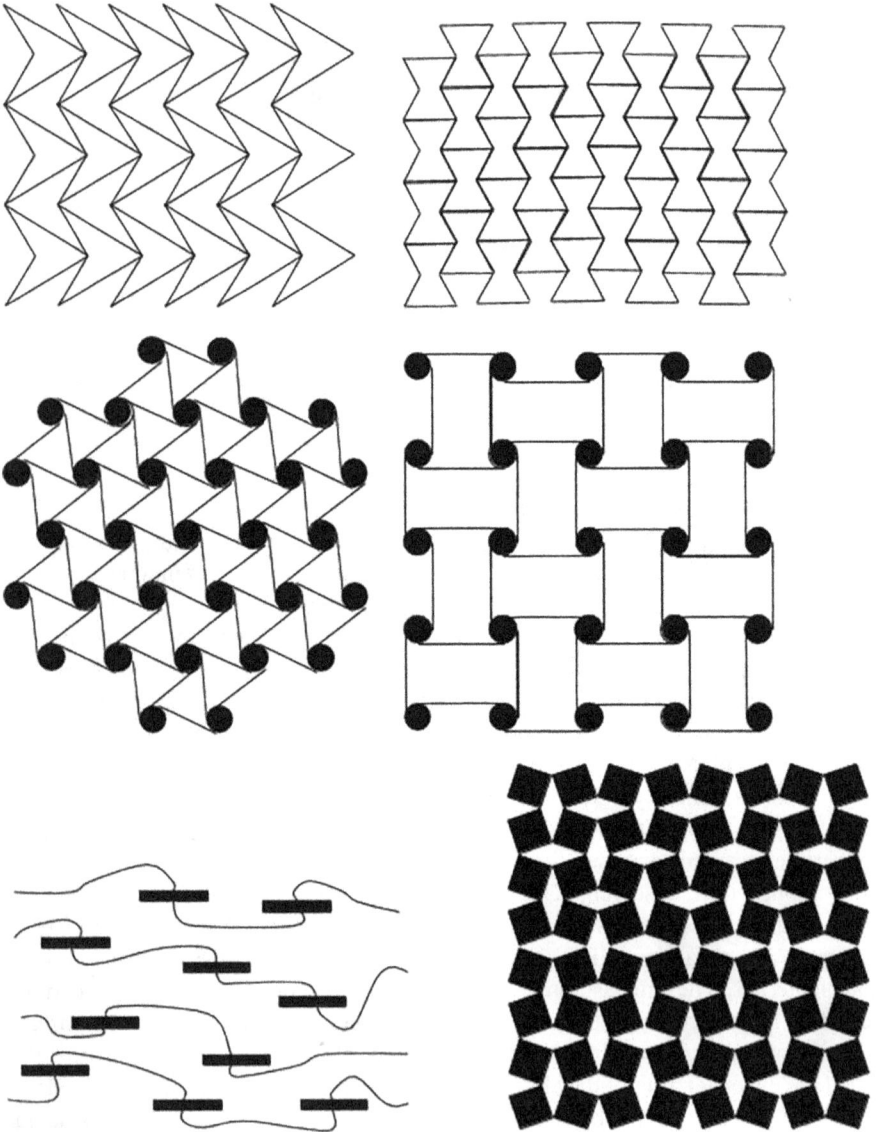

FIGURE 1.3 Several extremely simplified shapes that exhibit auxetic behavior.

TABLE 1.1

Five Indicators of Poisson's Ratio for Isotropic Solids

Poisson's ratio	Physical significance
$v = 1$(for 2D)	Preservation of area
$v = 0.5$	Preservation of volume
$v = 0$	Preservation of cross-section
$v = -0.5$	Preservation of moduli, $E = G$
$v = -1$	Preservation of shape

of highly idealized microstructural geometries that result in two-dimensional (2D) auxetic behavior. In order for auxetic phenomena to manifest in these geometries and other instances, some form of rotation must occur within the material's structure. When considering isotropic solids, the Poisson's ratio holds five key indicators that are of significant interest. These indicators, which can be found in Table 1.1, are discussed following the definitions of the Poisson's ratio and auxetic materials. Among these indicators, the most widely recognized is the concept of incompressibility or volume preservation, represented by a Poisson's ratio value of $v = 0.5$. The concept of "preservation of cross-section" encompasses the characteristic of a solid with a Poisson's ratio of 0, where the application of an axial load does not alter the shape or area of the cross-section that is perpendicular to the loading direction. This property highlights the stability and consistency of the material's cross-sectional dimensions under loading conditions. Conversely, the idea of "preservation of moduli" pertains to the observation that when a Poisson's ratio of -0.5 (for isotropic solids) is substituted into equation (1–3), it results in $E = G$. This equality indicates that the Young's modulus (E) and shear modulus (G) remain unchanged, emphasizing the preservation of these two important engineering moduli. The preservation of moduli is significant in various practical applications. The phrase "preservation of shape" refers to the scenario where a Poisson's ratio of -1 is applied. In this particular case, when a specific strain is imposed in one direction, it induces a corresponding strain in the perpendicular direction, causing a phenomenon known as dilatational deformation. This concept emphasizes the interdependence between elongation in one direction and the resulting strain in the opposite direction within the same plane, where Poisson's ratio comes into play. It is worth noting that in the context of 2D deformation, the maximum value of Poisson's ratio is 1, indicating that the area within that plane remains constant and unaffected by the deformation [2].

1.4 EFFECTIVE STRUCTURES AND DEFORMATION MECHANISMS OF AUXETIC MATERIALS

Despite the significant development of NPR materials over the past decade, a few well-established basic structures have emerged as leading candidates for explaining the deformation mechanisms observed in the majority of currently available NPR materials. These structures exhibit unique properties and include variations in geometric configuration that effectively illustrate the diverse characteristics exhibited by cellular auxetic materials and structures. The subsequent sections present a comprehensive review of six distinct models that extensively explore the properties of cellular auxetic materials and structures, highlighting their varying geometric configurations and the range of characteristics they demonstrate [1].

The first geometric model in Figure 1.4 illustrates the process by which the re-entrant structure experiences deformation. This mechanism is primarily characterized by the realignment of the ribs, assuming that the ribs are both rigid and have free pivots. The mechanism really involves axial deformation and rib deflection, making it more complex. Almost all re-entrant structures can be deformed using this process (Figure 1.4). When subjected to tension and compression, the two-arrowhead structure demonstrates a substantial NPR effect in one main direction.

The presence of a star-shaped re-entrant structure enables the construction of an in-plane isotropic structure, as it exhibits an NPR effect in both main directions. The structure depicted in Figure 1.4c, known as the missing rib structure, is a characteristic of auxetic foams. This structure, shown in the figure, induces an overall NPR

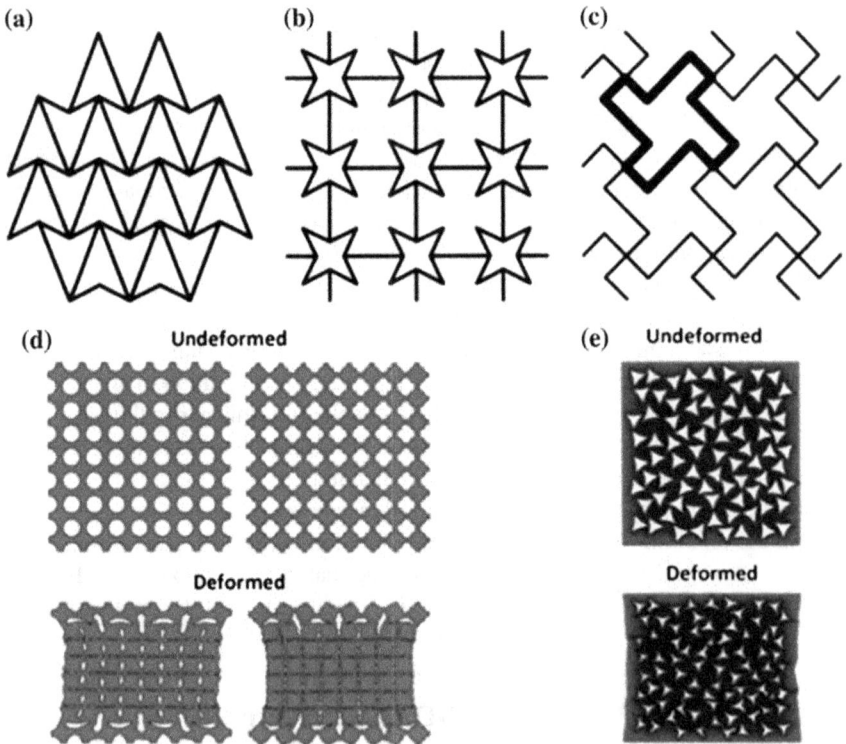

FIGURE 1.4 Illustration of common re-entrant negative Poisson's ratio (NPR) structures: (a) arrowhead shape; (b) star shape; (c) missing rib structure; (d) 2D porous soft material; (e) 2D composite featuring re-entrant hollow inclusions.

effect due to the extension of ribs in the typical unit, illustrated by bold lines. This effect occurs in both in-plane orthotropic orientations. Another example of an in-plane orthotropic NPR structure is presented in Figure 1.4d. The investigation of this phenomenon is crucial as it extends the concept of re-entrant structures by incorporating the elastic instability effect (buckling) into the study of NPR materials. The design of the re-entrant structure introduces geometric imperfections under compressive loads, potentially leading to structural instability. Research has been done to obtain structures with an isotropic NPR effect in addition to those having the NPR in specific directions. For instance, the overall Poisson's ratio demonstrates statistical isotropy when hollow re-entrant shell inclusions are introduced at random into a composite material (Figure 1.4e). The re-entrant structure theory can be applied to 3D spaces in addition to 2D ones. First, a foam-based 3D re-entrant structure was created [1].

The idealized tetrakaidecahedra model seen in Figure 1.5 was then used to illustrate its deformation mechanism. When folded ribs are stretched under tensile pressure, three orthotropic directions of expansion result, and the total structure exhibits the NPR effect. It is simple to infer that the flexure of folding ribs could cause the structure to constrict under compressive force.

A detailed investigation into the three-dimensional (3D) structure of NPR is commonly referred to as "Bucklicrystals" (Figure 1.6). Initially, it was attributed to the number of holes in a crystal, but the current analytical parameter used to examine its effect on the deformation mechanism of polyhedrons is the symmetry order. Rather than folding re-entrant structures, a geometrical imperfection in the shape of struts with variable thickness is incorporated. This flaw allows the structure to remain stable. When the structure is loaded in a compressive manner, the buckling of these struts induces a contraction of the whole structure (seen in Figure 1.6b). This research is noteworthy because it provides an understanding of rotation, which is a crucial mechanism of deformation for these kinds of structures.

The concept of a re-entrant structure may be applied to shell constructions in addition to those built using ribs and struts. A 3D hollow re-entrant shell tetrahedron

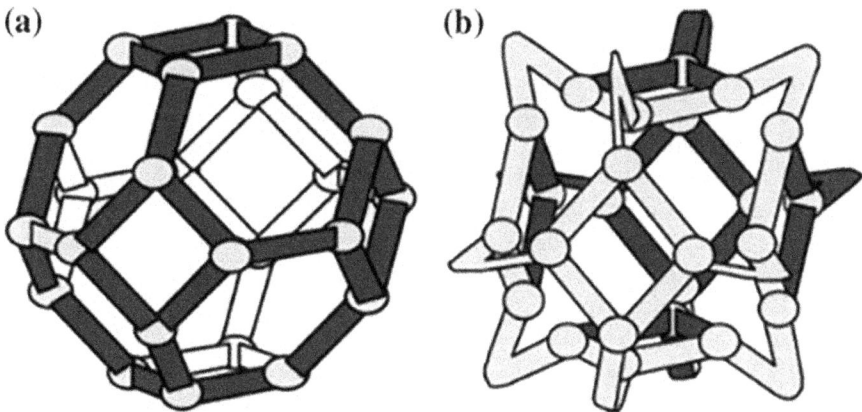

(a)   (b)

FIGURE 1.5   Tetrakaidecahedron: (a) conventional cell model; (b) re-entrant cell model.

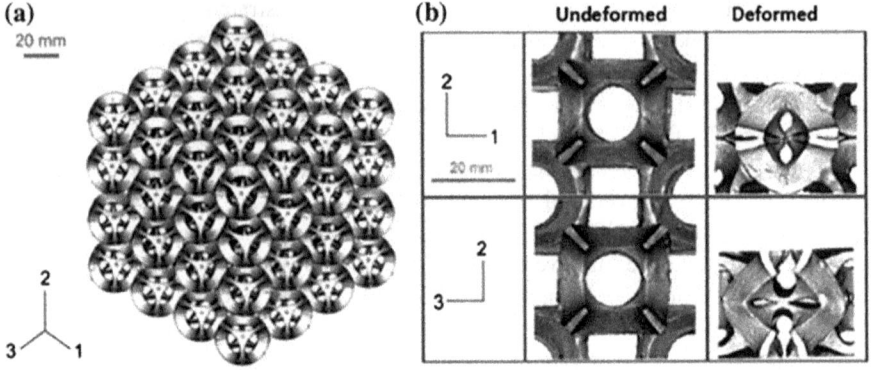

FIGURE 1.6 NPR structure (a) based on six-hole Bucklicrystal (b).

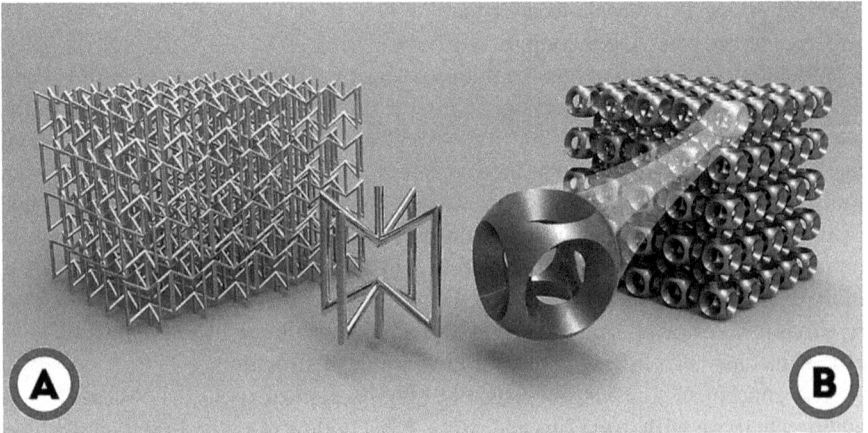

FIGURE 1.7 (a) A typical example of a re-entrant structure consisting of 3D re-entrant hexagonal unit cells, and (b) a six-hole Bucklicrystal.

is the subject of one of the research studies, as depicted in Figure 1.7b. The structure was dealt with as a composite structure inclusion. The embedded inclusion contracts when the composite structure as a whole undergoes compression because its surfaces buckle. This is referred to as the closure effect (Figure 1.8). The inclusion drags the matrix materials into its internal void, resulting in its closure, based on this deformation mechanism. As a result, the overall composite's Poisson's ratio is decreased. Additionally, it is possible to use the inclusion density as a regulating factor for the overall Poisson's ratio. If these types of inclusions are present in higher density, it is even possible for the overall Poisson's ratio of the composite to become negative [3].

Figure 1.9c depicts an NPR construction with missing ribs. It's interesting to note that its deformation process can be explained by both the rotation of its representative unit and the flexure of re-entrant ribs. Chiral structures are those that display the NPR effect and have rotational reflection as their dominant way of deformation.

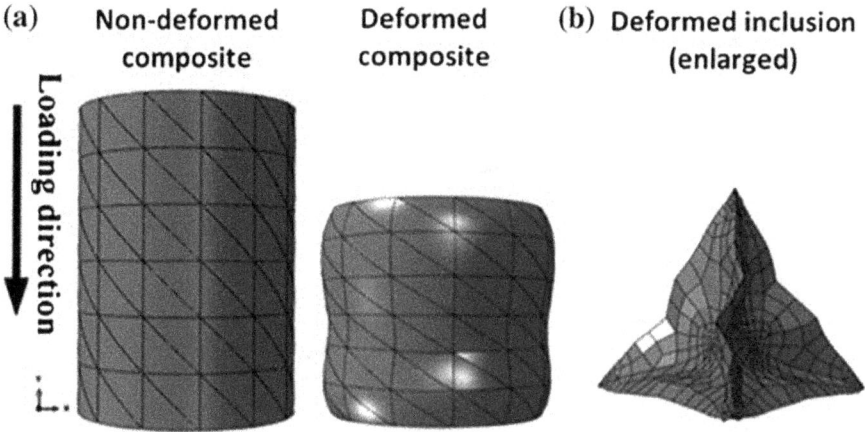

(a) Non-deformed composite   Deformed composite   **(b)** Deformed inclusion (enlarged)

Loading direction

FIGURE 1.8 Deformation mechanism of a 3D NPR composite (a) and its inclusion (b).

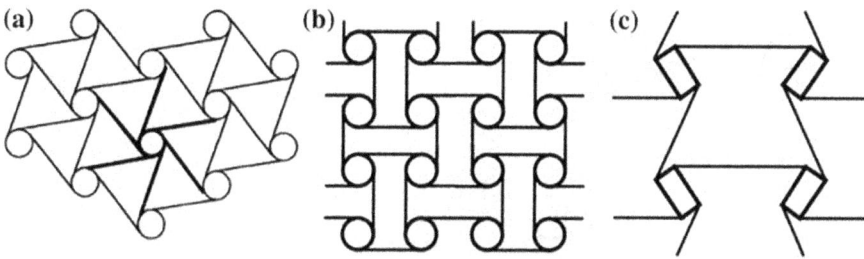

(a)   **(b)**   **(c)**

FIGURE 1.9 Common chiral structures with: (a) mono chirality accompanied by restricted rotational symmetry; (b) absence of chirality and restricted rotational symmetry; and (c) non-chirality block with relaxed rotational symmetry.

A lattice consisting of hexagonal molecules was the first structure of this type to be studied. Later, the idea was put into practice as a 2D periodic structure, widely recognized as the fundamental basis for all chiral NPR structures (Figure 1.9a). The fundamental chiral unit of this structure consists of a central disc (or node) connected to six ligaments, which are tangentially linked and highlighted in Figure 1.10a. The arrangement of these ligaments results in the chiral unit exhibiting rational symmetry of order six, indicating mechanical in-plane isotropy. As the structure consists of basic units with identical chirality, it attains global chirality, and the basic rotation unit can be either "left-handed" or "right-handed." When a uniaxial load is applied in the axial direction, the center disc will rotate, and as a direct consequence, the ligaments will bend. Because of the rotational mechanism, the ligaments that surround the disc can either fold or unfold depending on the type of loading they are subjected to, with tensile loading causing folding and compressive loading causing unfolding. This results in a lower value for the global Poisson's ratio. When the appropriate geometrical elements are added to the structure, the Poisson's ratio can approach −1,

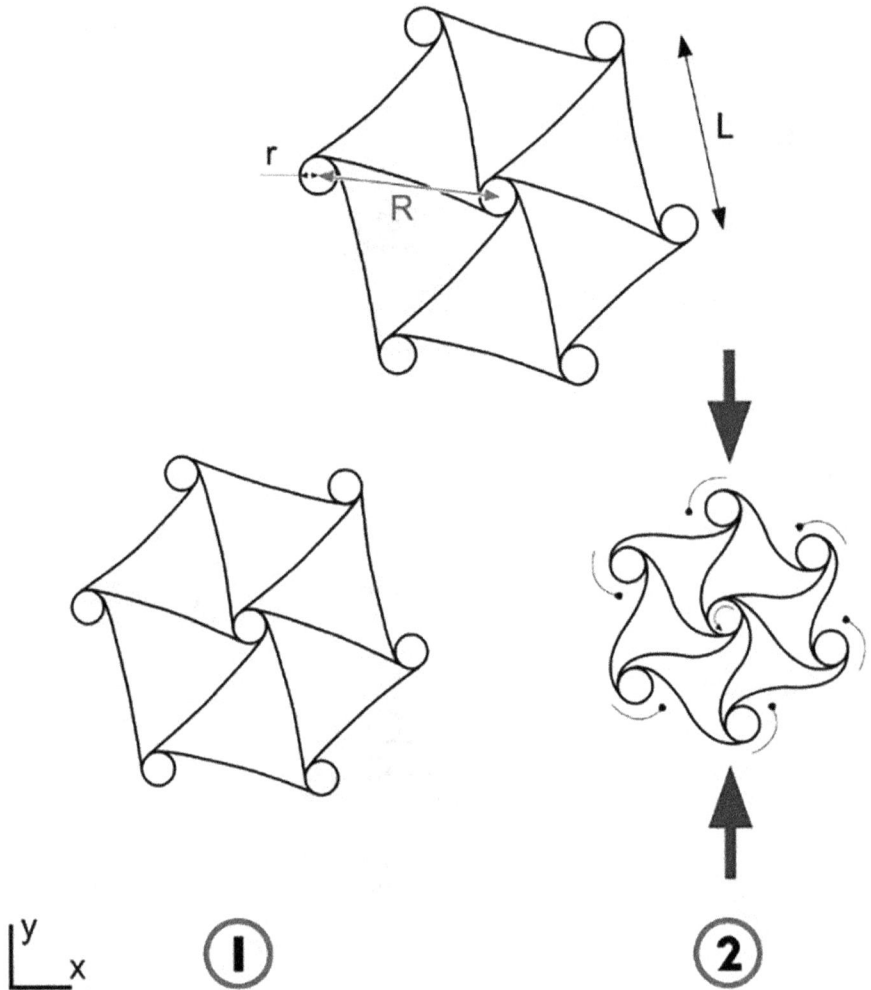

FIGURE 1.10 Hexachiral unit deformation: (1) initial configuration at rest; (2) complete deformation resulting from ligament folding and cylinder rotation in response to compression along the y-axis.

and the NPR effect can be maintained across a wide range of strain. This is possible because the NPR effect is not affected by the strain. NPR structures with both "left-handed" and "right-handed" chiral units were also constructed, in contrast to the structures that had the same chirality. Because of this, the total structures are classified as non-chiral structures, and they exhibit reflecting symmetry.

In addition to studying the in-plane behavior of chiral structures, their response to flatwise compression was investigated, considering both elastic and anelastic buckling modes. The construction of chiral structures based on chiral units presents a significant challenge due to the limitation imposed by rotational symmetry. For each basic chiral unit, the number of ligaments connected to each node must match the rotational symmetry of order "n." Specifically, when it comes to constructing space-filling periodic structures, only chiral units with rotational symmetry orders of 3, 4, and 6 are viable. Maintaining the symmetry requirement limits the number of such structures to five. By relaxing the constraint of rotational symmetry, it becomes possible to create meta-chiral structures. As illustrated in Figure 1.10c, the basic chiral unit features a rectangular node with four ligaments connected, deviating from the rotational symmetry of the unit. This concept opens up the possibility of constructing a range of structures using chiral units to achieve the NPR effect [1].

The initial investigation focused on a rotating structure to understand the NPR effect observed in various zeolites. The structure, as depicted in Figure 1.11a, consists of rigid squares (shown in grey) interconnected by simple hinges at their vertices, as illustrated in Figure 1.11b. Under the application of a load, the squares rotate at their vertices, leading to an overall increase or decrease in size under tension or compression forces. The ideal system has a Poisson's ratio of −1, independent of the loading direction or the dimensions of the squares, and it is isotropic. This characteristic holds true regardless of the loading direction. Building upon this idea, researchers have conducted extensive studies using similar principles with congruent rectangles, equilateral triangles, rhombi, and parallelograms [1].

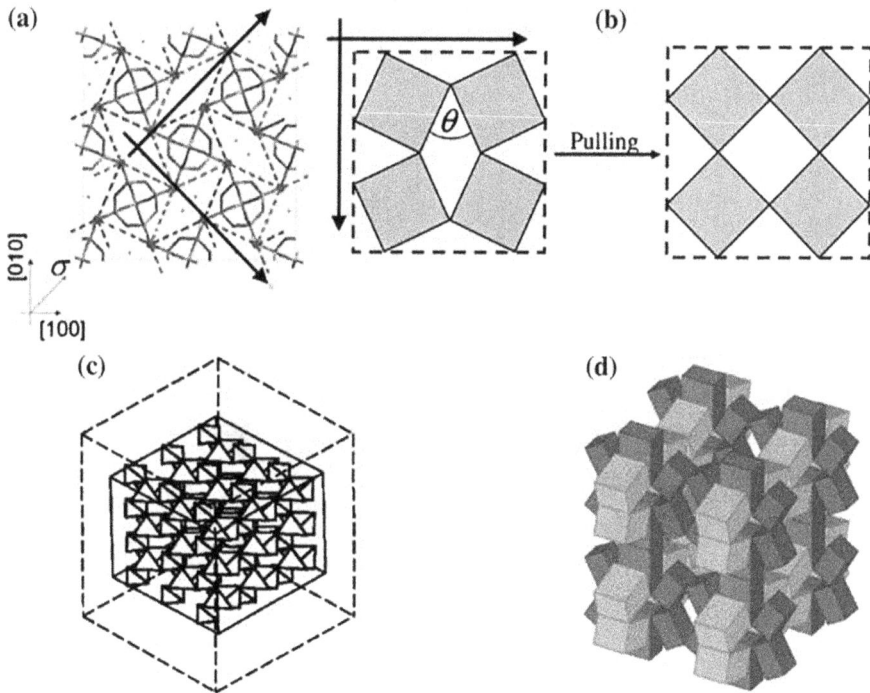

FIGURE 1.11 A representation of common NPR structures utilizing rotational units: (a) zeolite structure as well as (b) its idealized model; (c) 3D tetrahedral and (d) hexahedron structures.

Extensive study has been carried out to investigate the effects that rotational unit shape, size, connection, and arrangement have on the general behavior of the mechanical system as a whole. Furthermore, the concept has also been validated in a 3D space, as depicted in Figure 1.12. The rotational structures are primarily based on two main assumptions: the presence of rigid units and hinge connections. Some studies have also taken into account the deformation of certain rotation units, referred to as semi-rigid units. Yet, the analysis of constraints in the unit connections is hardly explored. The deformation of such structures heavily relies on the rotational freedom of these connections, which can lead to significant strain generation at the connections. As a result, the production of NPR materials with rotation structures becomes challenging and complex [1].

Two distinct configurations of rhombi can be identified (Figure 1.13): the rotating rigid rhombi of Type α and the rotating rigid rhombi of Type β. In the case of Type α, a rhombus with an obtuse angle is connected to a neighboring rhombus with an acute angle, while in the case of β, the rhombi are joined together by angles that are either both acute or both obtuse. In either case, as per the mathematical frameworks, there is the possibility for auxetic behavior to be exhibited. In the case of Type α, a

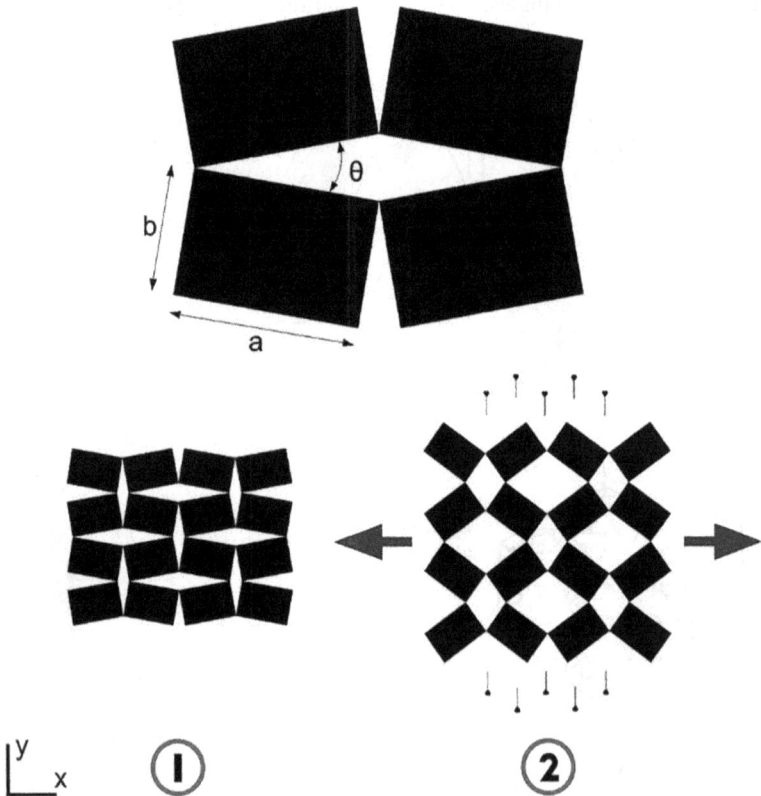

FIGURE 1.12 The deformation of a rotating rigid rectangular structure under two conditions: (1) when it is at rest, and (2) when it is subjected to tensile loading in the x-direction.

notable level of anisotropy is observed, where the Poisson's ratio depends not only on ϕ and θ but also on the direction of the applied load (Figure 1.13). This system exhibits positive and negative Poisson's ratios, which are determined by the value of θ, in a similar way to the rotating rectangles structure of Type I. The transition region displays significant Poisson's ratios, along with symmetrical distributions of v_{12} and v_{21} approaching the value of $\theta = \pi/2$. On the contrary, Type β demonstrates isotropy within the plane, maintaining a Poisson's ratio of -1 regardless of strain (or the values of angles ϕ and θ), as well as the direction in which the loading is applied. In contrast to the Type α, it is essential to keep in mind that the fully closed conformation of the Type β does not result in the occupation of space within the structure. When ϕ equals 90 degrees, the Type α simplifies into the isotropic rotating squares possessing a Poisson's ratio of -1, unless both ϕ and θ are equal to 90 degrees; however, when θ is equal to 90 degrees, the vacant spaces undergo a transformation into squares, resulting in a contrasting behavior. In this specific case, the structure demonstrates an isotropic Poisson's ratio of $+1$, with the exception of the case in which both ϕ and θ are equal to 90 degrees [3].

The combination of rotating rectangles and rotating rhombi connectivity schemes can be employed to describe the interconnection pattern of rigid parallelograms. This combination results in the emergence of four distinct systems: type I α, type I β, type

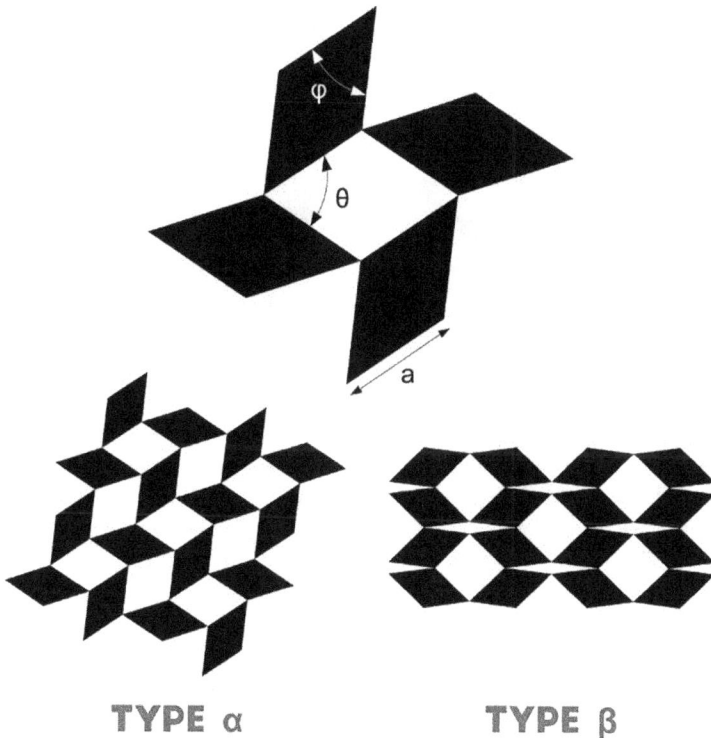

TYPE α **TYPE** β

FIGURE 1.13   Configurations of rotating rigid rhombi, classifying Type α and Type β.

II α, and type II β (see Figure 1.13). In order to establish the mechanical characteristics possessed by every mechanism, mathematical equations were developed. Akin
to the Type β rhombi, the Type II β parallelograms exhibit in-plane isotropy and have
a Poisson's ratio that remains constant at −1. This distinction is particularly evident
when comparing the Type I α, Type I β, and Type II α systems. In these cases, the
Poisson's ratios along the axis are significantly influenced by variables such as φ
and θ (as illustrated in Figure 1.14), as well as the loading direction. Additionally, the
Poisson's ratio of Type I systems is influenced by the aspect ratio a/b. These demonstrate a broad range of Poisson's ratios relying to the value of θ while transitioning
from a totally closed state to an entirely open state, with as many as four transitions
being identified in the process. This occurs when the state changes from entirely
closed to fully open. The consistent presence of auxetic behavior in the fully closed

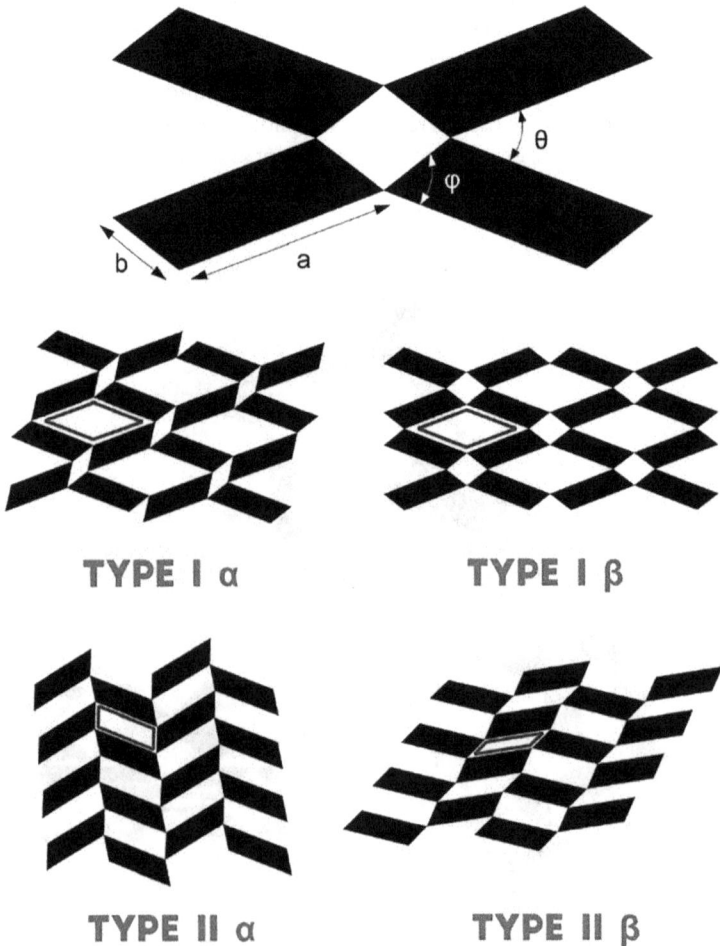

TYPE I α   **TYPE I β**

TYPE II α   **TYPE II β**

FIGURE 1.14 Configurations of rotating rigid parallelogram systems, categorized as Type
I α, Type I β, Type II α, and Type II β.

position is observed in both the Type I α and Type II α systems. However, the early performance of the Type I β shows a strong dependence on the relative magnitude of both the aspect ratio and ϕ. When $\phi = 90°$, similar to rhombi, something unique occurs. When a = b, the Type I systems are reduced to being equivalent to the Type I rectangles, and from there, they can be further simplified into the isotropic rotating squares system. Despite its structural similarity to the Type II rotating rectangles when $\phi = \pi/2$, the Type II acts more similarly to the Type I rotating rectangles than it does to the Type II rotating rectangles. Only in the exceptional circumstance where $\theta \neq \phi = \pi / 2$ is present does the Poisson's ratio of this system decrease to −1, which is the result that one would expect based on the Type II rotating rectangles. Similarly, the rotating rhombi of Type α follow the same principle, because their mathematical models have similarities with regards to Poisson's ratio and differ solely by a factor of multiplication in their moduli [3].

The rotating rhombi of Type α share a similar mathematical model with regards to Poisson's ratio and differ solely by a multiplication factor in their moduli, which leads to the same conclusion.

The NPR effect is typically governed by specific structure mechanism of deformations, and these structures can be typically categorized as such. However, because of how quickly these materials are developing, certain structures cannot be classified utilizing the concepts of re-entrant, chiral, and rotation structures. Take a look at Figure 1.10 for an illustration of a triangular cell unit that comprises an elastic unit and a truss. Such units could show a worldwide isotopic NPR effect in a network. Despite being an idealized structure and being quite complex, it offers another perspective regarding the characteristics of all NPR systems. This structure's trusses generate a triangular region in 2D that shifts as the trusses axially deform. The space begins to contract (close) when the trusses are compressed, and it opens up when they are stretched. The material's overall NPR effect is determined by the closure and expansion mechanisms.

All current 2D NPR architectures can be explained by this principle. Figure 1.15 depicts typical chiral, rotational, and 2D re-entrant structures. Regardless of the reason for the change in internal areas (as shown in Figure 1.15), the Poisson's ratio of those structures is always governed by it. The expansion effect produces an overall negative Poisson's ratio when the entire structure is under tension. The closing of the microstructure dominates the NPR effect under compressive force. Therefore, disordered 2D NPR structures can also be explained in the same way.

The principle could be explained regarding the variation in volume in a 3D space. A traditional 3D re-entrant structure, for instance, could be thought of as a polyhedral cell with a re-entrant surface (Figure 1.16). When subjected to an external force, the re-entrant surface's deformation could open or close the cell's interior volume, leading to an overall negative Poisson's ratio. A composite containing shell inclusions (Figure 1.16) is a more obvious example, where the compression-induced closure of the embedded inclusion results in a decrease in the overall volume. The global Poisson's ratio is affected by such volume changes, which is consistent with the conventional hypothesis. Under external forcing, the volume of a typical material exhibiting a Poisson's ratio below 0.5 varies. Bulk modulus B serves as a signal. When a material having a closure and expansion element is deformed, its volume

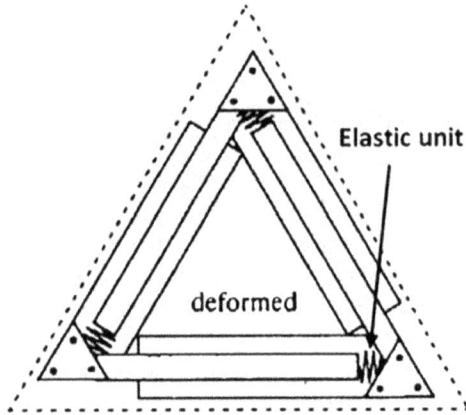

FIGURE 1.15   Deformation of triangular elastic unit.

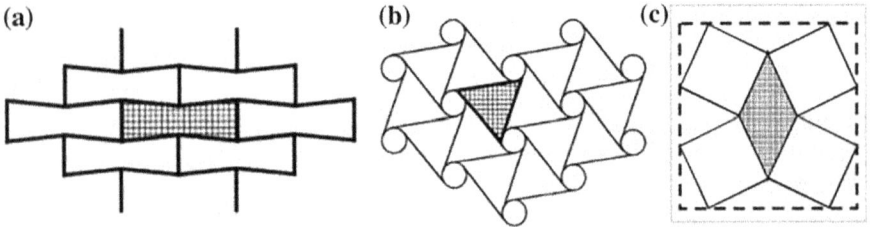

FIGURE 1.16 Illustration depicting deformation mechanisms in NPR structures: (a) re-entrant, (b) chiral, and (c) rotating.

also changes. As demonstrated in the example (Figure 1.16), the inclusion's enhanced volume change can be regarded as a separate change from the change caused by the matrix material, represented by the lower B, in terms of the overall volume. The material's Poisson's ratio diminishes and can take on negative values when the material undergoes volumetric deformation as opposed to shear ($G \geq B$). Evidently, the extent of the volume change influences the change in the Poisson's ratio; a greater extent corresponds to a larger modification in the Poisson's ratio. Additionally, it was shown that the linearity of the magnitude of volume changes reflected the linearity of the relationship between the Poisson's ratio change and strain.

If we relate the influence of the negative Poisson's ratio to changes in either the surface area (2D) or the overall volume (3D), we may think of the NPR material as a multi-phase structure. Figure 1.16 illustrates the fundamental structural elements of re-entrant, chiral, and rotational NPR structures, which consist of ribs (ligaments), stiff (semi-rigid) units, and voids, respectively [1].

Two other sorts of constituents are far more stiff than voids, whereas voids have diminishing levels of stiffness. This difference in stiffness enables the structure's stiffer phases to freely deform and considerably alter the area of the low-stiff phase, producing the NPR effect. In the case of skeleton structures, this method might not matter much, but it's crucial for creating composite materials with the NPR. A composite material is created when an NPR structure is incorporated into a matrix. The matrix, which is believed to be a compressible material with low stiffness, is shown in shaded areas in Figure 1.17, along with two examples of these materials. It is believed that the remaining components, which are referred to as inclusions, are rigid (or extremely stiff). Because of this, the inclusions will still be able to deform despite the soft matrix's resistance, which will result in a global NPR effect. In addition to being applicable to two-phase systems, this theory is also valid for multi-phase systems of higher order. For example, the in-plane isotropic NPR material depicted in Figure 1.17e may be described by a three-phase composite system that consists of the following components: (1) inclusions with a significant degree of rigidity, (2) a matrix with a low degree of stiffness, and (3) voids with no degree of rigidity at all. The distribution of inclusions can be arbitrary and discontinuous because of the three-phase composite structure. In NPR composites, the effective components are the inclusions, which exhibit closure and expansion behaviors under external loadings. Based on this research, the effects of the components' contrasting stiffnesses was also investigated. The NPR effect is typically increased by greater contrast. This idea holds true for 3D composite structures, as illustrated in Figure 1.17.

Due to the deformation experienced by the more rigid shell inclusion under external compressive pressure, the less rigid matrix material is drawn into the void. This mechanism of deformation leads to a significantly lower global Poisson's ratio than would have been observed otherwise.

The study of innovative NPR materials is not done for the pleasure of the researchers; in reality, the peculiar deformation behaviors of the materials might provide them with special mechanical features that are uncommon, useful, and tunable. The properties often oppose those of conventional materials, which demonstrates significant promise for a variety of technical applications across a wide range of fields.

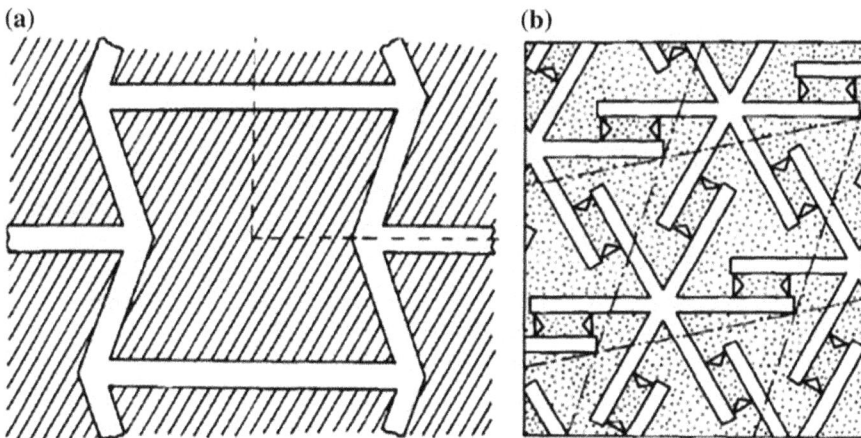

FIGURE 1.17   (a) Re-entrant honeycomb and (b) hexagonal inclusion in a 2D NPR composite.

NPR materials are characterized by noticeable alterations in their Young's modulus and shear modulus. Equation 1.3 demonstrates that the change from a positive to a negative Poisson's ratio can immediately result in an increase in the material's resistance to shear stress. This increase occurs due to the application of tearing or twisting forces. Theoretically, as the Poisson's ratio gets closer to −1, there is a possibility for the shear resistance to become infinitely large. This fascinating property of NPR materials offers intriguing possibilities for various applications in fields such as engineering and material science. The equation $E = 2G(1 + v)$ illustrates that in generally isotropic materials, the Young's modulus (E) is typically at least twice the magnitude of the shear modulus. As the Poisson's ratio decreases, the two moduli gradually approach each other in value. Notably, when the Poisson's ratio drops below −0.5, the Young's modulus and shear modulus are deemed equal. This relationship indicates a significant shift in the material's mechanical behavior, reflecting a balance between its resistance to tensile and shear stresses.

When the Poisson's ratio gets close to −1, it is indeed accurate to state that the shear modulus surpasses the elastic modulus. As a consequence of this, the material will be very compressible, but it will be resistant to deformation as a result of shearing. It is important to keep in mind that in the case of NPR materials, Young's modulus can vary and may not necessarily remain constant. Both the density ratio and the volumetric change ratio have the potential to have an influence on it. When a material is subjected to tension, the volumetric compression ratio often increases, which results in the Young's modulus decreasing. During compression, there is an increase in the volumetric compression ratio, which results in an increase in the Young's modulus. An explanation in simpler terms: materials that have a higher density typically have a higher level of stiffness. As a result, it is possible to establish a correlation between the bulk modulus, the shear modulus, and the Poisson's ratio.

The material's hardness (H) could be expressed employing the equation presented below, which is derived from the principles of elasticity:

$$H \propto \left[\frac{E}{\left(1 - v^2\right)} \right]^{\gamma}, \tag{1.4}$$

where the value of γ depends on the applied load. When the load is uniformly distributed, γ equals 1. In the case of Hertzian indentation, γ is equal to 2/3.

According to the equation, it becomes evident that an increase in the magnitude of the negative Poisson's ratio directly corresponds to a higher level of resistance to indentation in a material. According to the classical theory of elasticity, the range of v for isotropic materials typically extends from −1 up to 0.5. As the Poisson's ratio becomes closer to −1, the value of the term in the equation, that is $(1 + v)$, gets closer and closer to 0. As a consequence of this, the indentation resistance experiences a significant surge towards infinity whenever there is a preservation of E, despite an increase in the value of the negative Poisson's ratio. In Figure 1.18b, a schematic explanation is presented. When a material possessing an NPR is subjected to an impact force, it exhibits movement in the direction of the impact due to the lateral contraction induced by its NPR. The force of the impact leads to the material becoming denser along both the longitudinal and transverse axes, thereby enhancing the

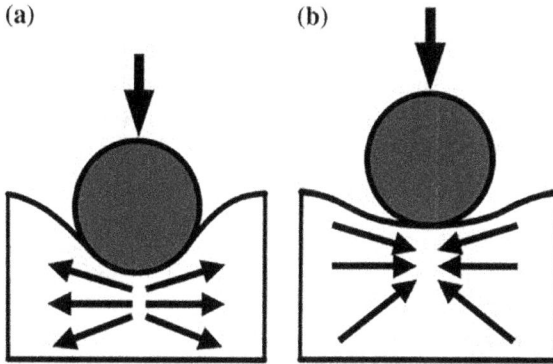

FIGURE 1.18 Comparison of indentation resistance between conventional (a) and NPR (b) materials.

substance's resistance to indentation. The phenomenon is obviously distinct from that of conventional materials (Figure 1.18a), in which the material frequently moves away from the direction of loading when an impact loading is applied, which results in a lower hardness than that of NPR materials. When an impact loading is applied, the NPR materials, on the other hand, do not deviate away from the direction in which the loading is being applied. The validity of this phenomenon has been confirmed through the utilization of a diverse range of synthetic materials with NPRs, including polymeric and metallic foams, laminates with fiber-reinforced composites, and various types of polymers.

The following equation was used to characterize the toughness of NPR materials:

$$\sigma = \frac{\pi E T}{2r\left(1 - v^2\right)} \tag{1.5}$$

where r is the diameter of the circular crack, T denotes the surface tension, and σ is the critical tensile stress. Based on this equation, it can be deduced that the material will exhibit a significant increase in its Poisson's ratio as it approaches the value -1. Numerous attempts have been undertaken to evaluate the toughness of a variety of synthetic NPR materials, such as composite laminates and re-entrant foams, because these materials may be manufactured using a wide variety of processes. The behavior of foams was studied by applying volumetric compression ratios through compressive and tensile loadings. This approach led to an observed improvement in the toughness of the investigated materials. However, due to the influence of factors such as production methods and post-processing, such as annealing, the outcomes of various research projects may not always offer conclusive answers. In contrast, NPR foams exhibit superior toughness compared to conventional foams. Additionally, they possess a greater ability to absorb energy and experience less loss of rigidity.

Research on composite materials extensively explores the influence of reinforcement types and orientations, commonly involving fibers, to comprehend their effects. The connections are challenging, though, because inter-laminate shearing and delamination also have an impact on the deformation mechanisms. In addition to investigating the response of NPR materials under straightforward quasi-static loading, extensive research was conducted to examine their behavior under cyclic loading and loadings with variable strain rates. This comprehensive study employed a combination of numerical simulations and experimental methods. The conclusive findings demonstrate that NPR composite materials exhibit superior toughness compared to conventional alternatives.

In addition, NPR materials have greater fracture resistance than their traditional equivalents. They require more energy to expand than traditional materials, despite having a low tendency to propagate cracks. The mechanisms underlying such unique features were assessed using both theoretical and experimental techniques. However, given the complex and uneven microstructure of NPR materials, particularly foams, even the slightest variations in the fabrication process and environment might result in varying mechanical properties. As a result, utilizing idealized models to describe their behaviors is difficult, but the phenomenon may be simply explained using the idea of NPR. The volume of NPR content increases both locally and globally when it is under tension. Consequently, as a crack grows, the surrounding materials have a tendency to expand and enclose it.

NPR materials exhibit additional intriguing features as a result of their unique microstructures. For instance, NPR structures exhibit excellent ultrasonic and acoustic damping in addition to having superior energy absorption capabilities when subjected to mechanical loads. When bent out of plane, a conventional material will have an anticlastic curvature, also known as a saddle shape; an NPR material, on the other hand, may have a synclastic curvature, also known as a double-curvature shape. It offers a method for creating intricate dome-shaped structures without the need for advanced procedures or further machining. Additionally, NPR materials have changeable permeability, shape memory, and abrasion resistance, among other properties.

Thus far, we have presented a comprehensive overview of the most prevalent types of auxetic mechanical metamaterials. Our discussions have primarily focused on differentiating between various auxetic structure geometries and recognizing their fundamental structures. We have discovered fundamental formations that enable us to differentiate the deformation mechanisms witnessed in various auxetic structural geometries. However, it is crucial to note that we have yet to explore other types of structures. These structures hold significant importance in the construction and development of auxetic metamaterials. While we acknowledge that their exploration is yet to be carried out, we understand the importance of conducting further investigations and gaining a deeper understanding of these structures in order to broaden our knowledge and explore the potential applications of auxetic metamaterials.

One notable category among these alternative structures is origami and kirigami, which are traditional Japanese techniques involving the folding and cutting of paper to create intricate patterns. Origami and kirigami techniques have garnered substantial attention in contemporary times as effective methods for constructing mechanical metamaterials. The term *ori* in Japanese signifies "fold," while *gami* denotes "paper." Similarly, *kiri* means "cut." The ancient arts of origami and kirigami provide invaluable design principles that are applicable even today. Originating from

Japan, these arts enable the conversion of thin planar materials into intricate and detailed 3D structures. These structures exhibit extraordinary mechanical properties that can be tailored and adjusted to meet specific requirements. The combination of folding and cutting in origami and kirigami unlocks a realm of possibilities for developing innovative and versatile metamaterials with diverse applications in engineering and beyond. These structures can be created by transforming simple 2D thin-film materials. These features include the ability to preserve various stable configurations, shape transformation, flexibility, adjustable Poisson's ratio, and adjustable stiffness, and the ability to change its shape. Notable examples of origami and kirigami include not just paper-based creations, but also extend to those crafted from polymers, metals hydrogels, and even graphene. These structures encompass a wide range of scales, from the macroscopic to the microscopic and even down to the nanoscale. The use of origami and kirigami in mechanical metamaterials has found applications in a variety of industries, such as electronics, robotics, and medical technology, among others. Origami and kirigami are methods employed to partition thin materials into pliable parts, like folds in origami, and rigid sections, including thin panels in both origami as well as kirigami. These techniques originated in 17th-century Japan and give rise to similar mechanical properties in both practices. The mechanical attributes of origami and kirigami structures crucially depend on achieving an optimal equilibrium between inherent flexibility and rigidity within the chosen pattern. Therefore, the mechanical behavior of these structures hinges on the delicate interplay between their flexible and rigid components. However, there are notable distinctions between origami and kirigami. Origami structures undergo folding from a flat state to form a condensed volume, while kirigami structures are stretched from their initial state to achieve an expanded arrangement. Furthermore, there is an increasing tendency to merge origami and kirigami in hybrid designs to leverage the benefits provided by both methods.

In the realm of mechanical metamaterial design, a fundamental principle revolves around the mechanical energy landscape. This concept encompasses the alterations in strain energy across diverse geometrical and deformation parameters within the deformed configuration space of metamaterials. The mechanical attributes of a mechanical metamaterial, including its deployability, stability, and stiffness, are profoundly affected by the distribution of mechanical energy within its structure. One notable example is bistability, which occurs due to the presence of two local energy minima, leading to elastic energy being stored in the material. Similarly, self-deployment happens when the mechanical energy decreases as the material is deployed. Both origami and kirigami techniques provide appealing approaches to create specific energy landscapes through cutting and folding patterns, enabling the development of desired mechanical properties.

Under specific conditions depicted in Figure 1.18, certain origami and kirigami patterns exhibit panels that maintain their rigidity while undergoing deformation, resulting in the storage of energy solely within the crease or connection regions. This form of rigid origami or kirigami offers more predictable mechanical behavior and kinetics, making it particularly suitable for applications involving shape morphing. Alternatively, when the panels also undergo deformation, elastic energy is stored in both the crease or linkage areas and the panel regions. This intricate energy landscape, referred to as deformable origami or kirigami, enables enhanced programmability in terms of forces, stiffness, and stability.

In order to conduct a comprehensive examination, mechanical metamaterials derived from origami or kirigami can be categorized into three distinct groups, as illustrated in Figure 1.19 The first group comprises mechanical metamaterials that are primarily based on origami and/or kirigami, further subdivided into three categories based on the manipulation of thin-film materials. These categories are as follows: (1) origami-based metamaterials, which involve exclusively folding techniques; (2) kirigami-based metamaterials, which entail solely cutting techniques; and (3) hybrid origami-kirigami-based metamaterials, which combine both folding and cutting techniques to create unique structures and functionalities.

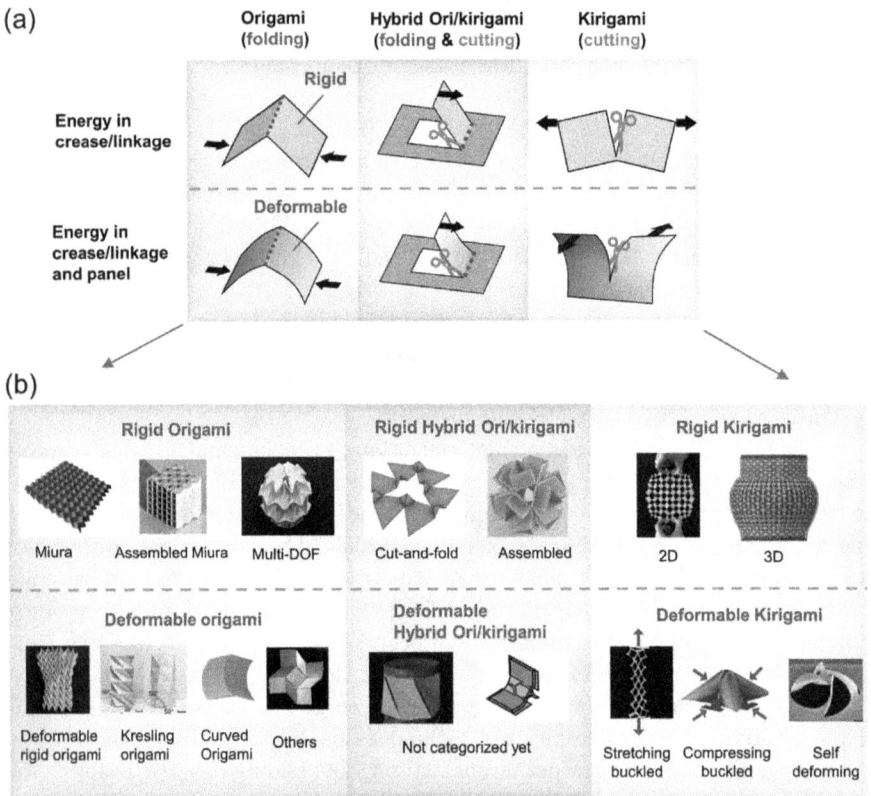

FIGURE 1.19   Classification of mechanical metamaterials based on origami and kirigami techniques: (a) graphical representation illustrating the classification of mechanical metamaterials based on origami and kirigami. These metamaterials might be divided into two main categories, rigid and deformable, depending on their energy distribution Additionally, the fabrication method employed leads to further classifications: origami-based, kirigami-based, and hybrid varieties. (b) Sub-categories within each group of mechanical metamaterials based on origami and kirigami.

In the process of designing mechanical metamaterials inspired by origami, an essential step is to introduce folding creases into thin-sheet materials. The incorporation of the folding creases enables origami to demonstrate adjustable dynamics and mechanical characteristics. The presence of creases allows neighboring panels to rotate, influencing the distribution of elastic energy throughout both the folding and unfolding phases. Origami-inspired mechanical metamaterials might be classified as either rigid or deformable, according to the distribution of elastic energy during folding and unfolding.

Miura-ori configurations are among the origami structures that have received significant attention from researchers. Several recent studies have investigated the mechanical characteristics of Miura-ori origami structures, including the original designs and various derived variants. For example, it has been demonstrated that the Poisson's ratio of Miura-ori origami structures can exhibit either positive or negative values, depending on the design parameters of the structures (Figure 1.20).

Through the application of origami principles, it becomes possible to fabricate auxetic materials, like the Miura-ori origami structures exhibiting a negative Poisson's ratio. This serves as a mere illustration of the wide-ranging applications of auxetic materials, which involve techniques of folding and unfolding materials. As for kirigami, it utilizes materials made of thin sheets that incorporate cuts and connections between panels, and the mechanical characteristics of these connections have a significant impact on the overall behavior of kirigami. When the connections are thick and robust, they restrict panel rotation and distribute strain more evenly across both the linkages and panel areas, resulting in deformable kirigami panels. On the other hand, in situations where the connections are thin and fragile, the kirigami

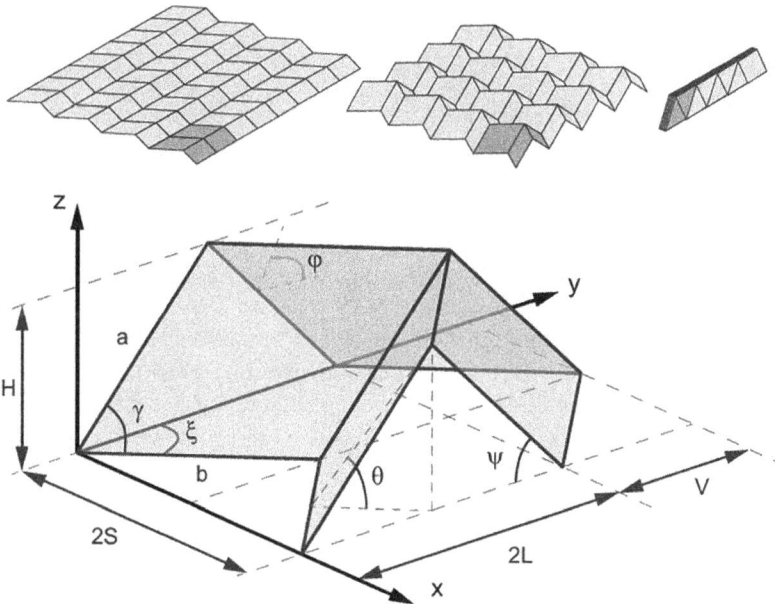

FIGURE 1.20 The structural configuration of Miura-ori origami patterns.

panels tend to retain rigidity and primarily rotate around the linkages. In contrast to origami, which is commonly classified according to specific designs, kirigami structures are typically grouped based on their distinctive deformation characteristics. As an instance, when considering rigid kirigami, it can be classified as either 2D or 3D, depending on the type of deformation. Similarly, deformation kirigami can be categorized as tension-induced, compression-induced, or self-deforming, based on the nature of their deformation. The combination of origami and kirigami is possible due to their shared utilization of thin sheets. By incorporating cutting and folding techniques, researchers have created hybrid metamaterials that fully exploit the benefits of both origami and kirigami. These hybrid forms can be further classified as either rigid or flexible, based on the manner in which energy is distributed within them [4].

- Rigid origami encompasses Miura origami, assembled Miura origami, and multi-degree-of-freedom (DOF) origami.
- Deformable origami encompasses deformable rigid origami (Kresling), curved origami, and other variations.
- Rigid kirigami might be further classified into 2D kirigami and 3D kirigami.
- Deformable kirigami includes stretch-buckled kirigami, compression-buckled kirigami, and self-deforming kirigami.
- Rigid hybrid origami-kirigami might be categorized into cut-and-fold and assembled types.

However, a comprehensive categorization for deformable hybrid origami-kirigami is yet to be established.

1.5   MECHANICAL MODELING OF AUXETIC MATERIALS AND STRUCTURES

A few common models that have been used to describe auxetic behavior have been gathered in Figure 1.21. Micromechanics underlying these models are examined in more detail in this chapter, emphasizing the geometric models employed and the analytical solutions utilized.

The foundational concept for the proposed auxetic behavior is founded on the classic 2D re-entrant structure illustrated in Figure 1.22a. This structure is made up of cells arranged in a honeycomb pattern in a hexagonal grid. Referring to Figure 1.22b, the honeycomb structure exhibits auxetic behavior when the angle θ is positive, indicating a re-entrant shape within the unit cell. Conversely, when the angle θ has a negative value, the unit cell assumes a conventional hexagonal shape. These assessments of the Young's modulus and Poisson's ratio in the loading direction determine whether the honeycomb structure is auxetic or conventional [2].

$$\nu_{12} = \frac{\sin\theta\left(h/l + \sin\theta\right)}{\cos^2\theta}$$

$$E_1 = k\frac{\left(h/l\right) + \sin\theta}{b\cos^3\theta}$$

$$(1.6)$$

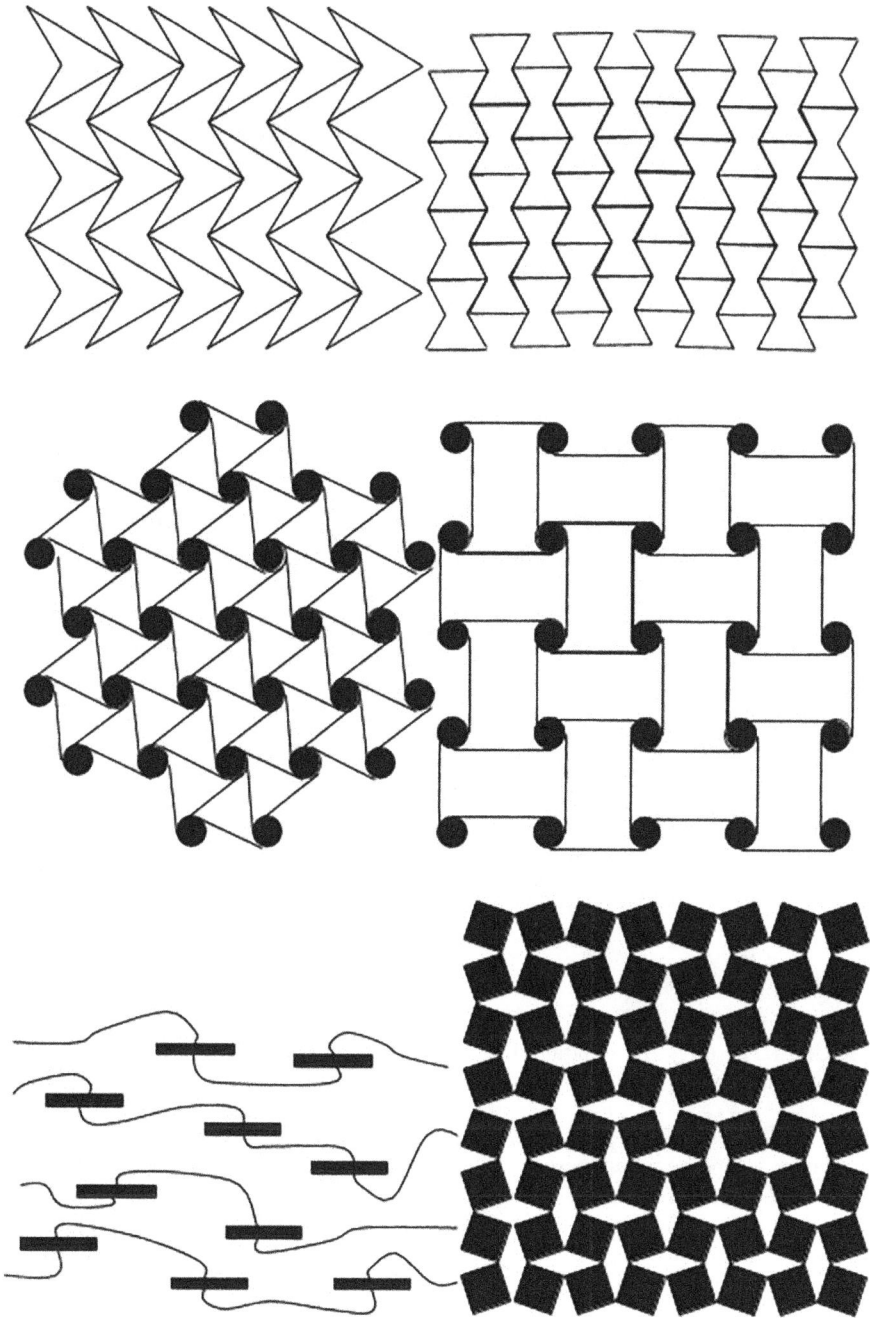

FIGURE 1.21   Some highly simplified geometries exhibiting auxetic behavior.

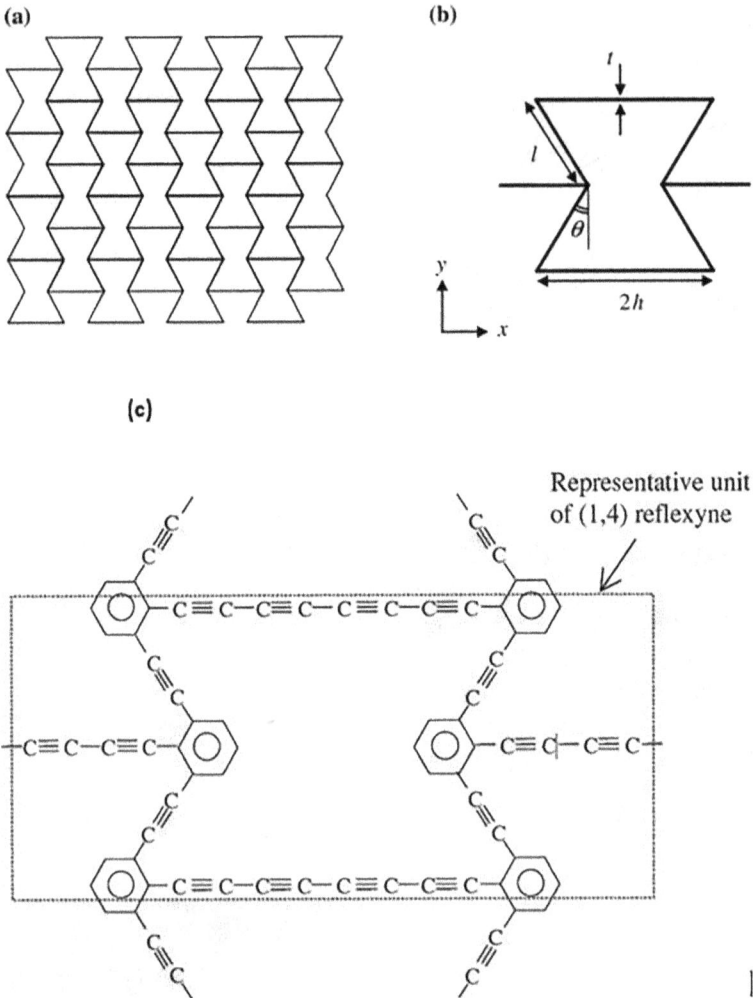

FIGURE 1.22 (a) A 2D re-entrant structure, (b) geometry, and (c) a re-entrant structure at the molecular level.

In Figure 1.22b, the parameters h and l each denote half the length of the horizontal rib and the inclined rib, respectively. The parameter t corresponds to the rib's thickness, while b represents the depth of the cell ribs, which are not depicted in Figure 1.22b. The value of the parameter k is determined by

$$k = E_s b \left(\frac{t}{l} \right)^3 \qquad (1.7)$$

where the rib material's Young's modulus, denoted as E_s, plays a significant role in the analysis, as depicted in Figure 1.22c. Additionally, another proposed similar shape, as illustrated in Figure 1.23b, is utilizing an idealized re-entrant unit cell portrayed in Figure 1.23a. Consequently, the resulting Poisson's ratio under infinitesimal strain is influenced by these factors and can be further explored.

$$v_{elastic} = -\frac{\sin(\varphi - \pi/4)}{\cos(\varphi - \pi/4)} \tag{1.8}$$

where the angle φ is shown in Figure 1.22, and the Poisson's ratio under significant strain

following the formation of a plastic hinge is provided as

$$v_{plastic} = \frac{\cos(\varphi - \pi/4) - \cos(\varphi - \pi/4 - \theta)}{\sin(\varphi - \pi/4) - \sin(\varphi - \pi/4 - \theta)} \tag{1.9}$$

The variable θ denotes the amount of angular rotation in a clockwise direction for the cell rib BC. The relationship between Poisson's ratio and strain in elastic-plastic deformation is demonstrated with the following formulation:

$$v_{elasto\text{-}plastic} = v_y - \varepsilon_{ex}\frac{1 - \cos\eta}{2\varepsilon_x} \tag{1.10}$$

in which

$$\varepsilon_x = \frac{1}{\sqrt{2}}\frac{\sin(\varphi - \pi/4) - \sin(\varphi - \pi/4 - \theta)}{1 + \sin(\pi/2 - \varphi)} = \varepsilon_{ex}\frac{\delta}{\delta_e} \tag{1.11}$$

and

$$\eta = \frac{P_e}{P}\left(3 - \frac{P}{P_e} - 2\sqrt{3 - 2\frac{P}{P_e}}\right) \tag{1.12}$$

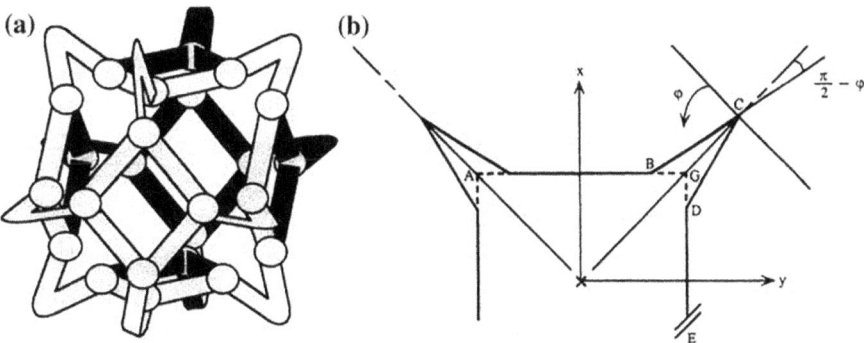

FIGURE 1.23   (a) An idealized re-entrant unit cell, and (b) a closer look at the geometry used.

where v_y represents the Poisson's ratio at the point of initial yielding, ε_x denotes the strain along the x-direction, while ε_{ex} represents the x-component of the strain measured at the beginning of yielding. δ represents the deflection, and δ_e corresponds to the deflection observed at initial yielding. P denotes the applied load, and P_e represents the load that is being applied at the point when initial yielding occurs, by assuming that $\varepsilon_{ex} = 1\%$.

Figure 1.24 presents a visual representation that compares the estimated and measured Poisson's ratio values in relation to strain for copper foam.

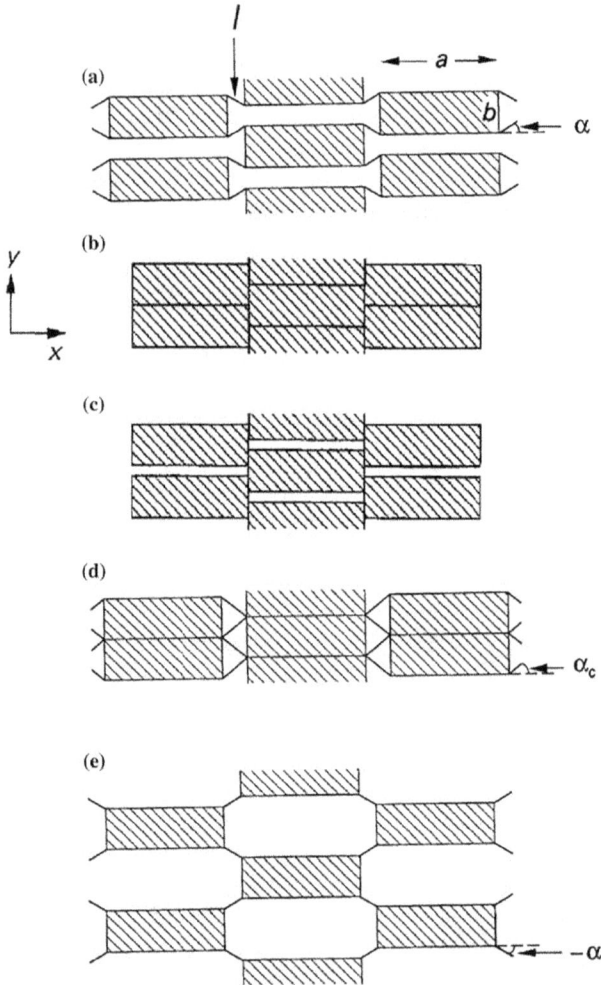

FIGURE 1.24 Visual representation of the nodule-fibril (NF) model, illustrating (a) common parameters within a partially extended network, and (b) a completely densified network with $l = b/2$, $\alpha_0 = 90°$, and (c) undeformed, partially open networks that occur when $l < b/2$ where nodules touching along the x-axis, (d) $l > b/2$ with nodules in contact along the y-axis, and (e) negative fibril angle.

Alderson and Evans' nodule-fibril microstructural model explains the deformation mechanisms involving fibril stretching, flexure, and hinging separately [3]. As shown in Figure 1.24, the length of the main axis of the idealized nodule is represented by the symbol "a," while the length of the minor axis is indicated by "b." The geometry of the fibril is determined by its length, denoted as "l," and its inclination to the x-axis at an angle of "α." When subjected to tensile stress in the x-direction, the fibrils undergo hinging, causing a decrease in the angle "α" until it reaches its minimum value of $\alpha = 0$ degrees [2].

Figures 1.25 provide schematic representations illustrating deformation caused by fibril hinging mode subjected to tension along the x-direction and compression along the y-direction, respectively. Alderson and Evans [3] conducted an analysis utilizing this model, which allowed for the determination of the Poisson's ratio and Young's moduli resulting from fibril hinging. These findings played a crucial

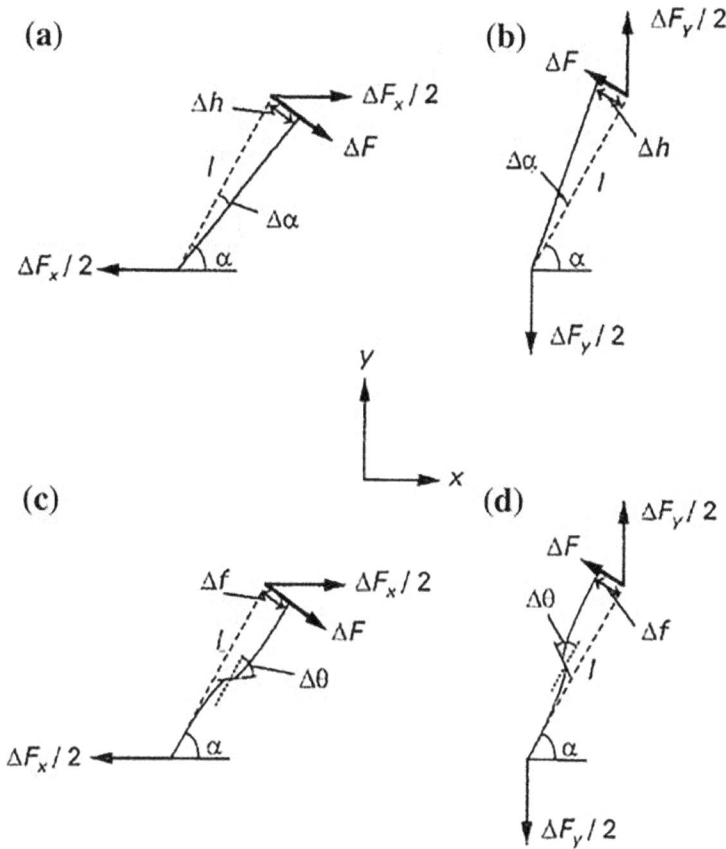

FIGURE 1.25 (a) Fibril hinging under tension of the NF network along the x-axis; (b) fibril hinging under compression of the NF network along the y-axis; (c) fibril flexure under tension of the NF network along the x-axis; and (d) fibril bending caused by NF network compression along the y-axis [2].

role in understanding the mechanical properties associated with this deformation mechanism.

$$v_{xy} = -\frac{\cos\alpha\,(a + l\cos\alpha)}{\sin\alpha\,(b - l\sin\alpha)} = \frac{1}{v_{yx}} \qquad (1.13)$$

$$E_x = \frac{K_h}{l^2 \sin^2\alpha}\left(\frac{a + l\cos\alpha}{b - l\sin\alpha}\right) \qquad (1.14)$$

$$E_y = \frac{K_h}{l^2 \cos^2\alpha}\left(\frac{b - l\sin\alpha}{a + l\cos\alpha}\right) \qquad (1.15)$$

As a result, the engineering Poisson's ratio and engineering Young's moduli were determined as follows:

$$v_{xy}^e = -\frac{\cos\alpha\,(a + l\cos\alpha_0)}{\sin\alpha\,(b - l\sin\alpha_0)} = \frac{1}{v_{yx}^e} \qquad (1.16)$$

$$E_x^e = \frac{K_h}{l^2 \sin^2\alpha}\left(\frac{a + l\cos\alpha_0}{b - l\sin\alpha_0}\right) \qquad (1.17)$$

$$E_y^e = \frac{K_h}{l^2 \cos^2\alpha}\left(\frac{b - l\sin\alpha_0}{a + l\cos\alpha_0}\right) \qquad (1.18)$$

The value of the hinge force coefficient is determined by the following equation:

$$l\Delta F = K_h \Delta\alpha \qquad (1.19)$$

for an angular change $\Delta\alpha$ in response to a variation in force ΔF while α_0 represents the initial angle. The analysis conducted by Alderson and Evans [3], with reference to Figure 1.25c, d, yielded the same Poisson's ratio expression obtained through fibril hinging, as illustrated in equations 1.20 and 1.21. They determined the Young's moduli resulting from fibril flexure as follows:

$$E_x = \frac{3K_f}{l^2 \sin^2\alpha}\left(\frac{a + l\cos\alpha}{b - l\sin\alpha}\right) \qquad (1.20)$$

$$E_y = \frac{3K_f}{l^2 \cos^2\alpha}\left(\frac{b - l\sin\alpha}{a + l\cos\alpha}\right) \qquad (1.21)$$

The engineering Young's moduli were determined as:

$$E_x^e = \frac{3K_f}{l^2 \sin^2\alpha}\left(\frac{a + l\cos\alpha_0}{b - l\sin\alpha_0}\right) \qquad (1.22)$$

$$E_y^e = \frac{3K_f}{l^2 \cos^2\alpha}\left(\frac{b - l\sin\alpha_0}{a + l\cos\alpha_0}\right) \qquad (1.23)$$

The coefficient of flexure force can be obtained from:

$$K_f \Delta\theta = \Delta M = \frac{l\Delta F}{2} \tag{1.24}$$

and $\Delta\theta$ represents the slight variation ($\Delta\theta = \tan\Delta\theta$) in the fibril's midpoint's slope with respect to its orientation, as shown in Figure 1.25c, d. The stretching mode of deformation for the fibril is depicted in Figure 1.26. Using this terminology, Alderson and Evans [3] calculated the values for Poisson's ratio and Young's moduli in the following manner:

$$v_{xy} = \frac{\sin\alpha\left(a + l\cos\alpha\right)}{\cos\alpha\left(b - l\sin\alpha\right)} = \frac{1}{v_{yx}} \tag{1.25}$$

$$E_x = \frac{K_s}{\cos^2\alpha}\left(\frac{a + l\cos\alpha}{b - l\sin\alpha}\right)$$
$$E_y = \frac{K_s}{\sin^2\alpha}\left(\frac{b - l\sin\alpha}{a + l\cos\alpha}\right) \tag{1.26}$$

Young's moduli and the corresponding engineering Poisson's ratio are defined as:

$$v_{xy}^e = \frac{\sin\alpha\left(a + l\cos\alpha_0\right)}{\cos\alpha\left(b - l\sin\alpha_0\right)} = \frac{1}{v_{yx}^e}$$

$$E_x^e = \frac{K_s}{\cos^2\alpha}\left(\frac{a + l\cos\alpha_0}{b - l\sin\alpha_0}\right) \tag{1.27}$$

$$E_y^e = \frac{K_s}{\sin^2\alpha}\left(\frac{b - l\sin\alpha_0}{a + l\cos\alpha_0}\right)$$

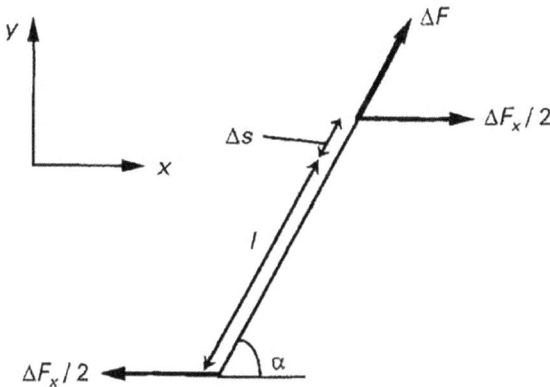

FIGURE 1.26   Fibril extension along the x-axis due to tension of the nodule-fibril network [3].

in which the stretching force coefficient is defined as:

$$K_s = \frac{\Delta F}{\Delta s} \qquad (1.28)$$

for an extension of Δs in response to ΔF.

Figure 1.27 depicts a comparative analysis of analytical models and experimental findings for polytetrafluoroethylene (PTFE). These visual representations offer insights into the relationship between theoretical predictions and actual observations of PTFE's behavior. Meanwhile, Figure 1.27c provides a similar examination for ultra-high molecular weight polyethylene (UHMWPE), enabling researchers to assess the accuracy of theoretical models in predicting UHMWPE's mechanical response. In their 1995 study, Alderson and Evans discovered that as the angle α tends towards 0 degrees, the dominance of fibril stretching becomes increasingly pronounced. Consequently, two important consequences arise: the transition from primarily hinging to stretching modes occurs at higher strains due to longer fibril length, and this transition is spread out over a range of strain values rather than a specific point [3].

Fibril flexure and hinging produce the exact Poisson's ratio values and Young's modulus trends, according to Alderson and Evans' [3] research, which takes into

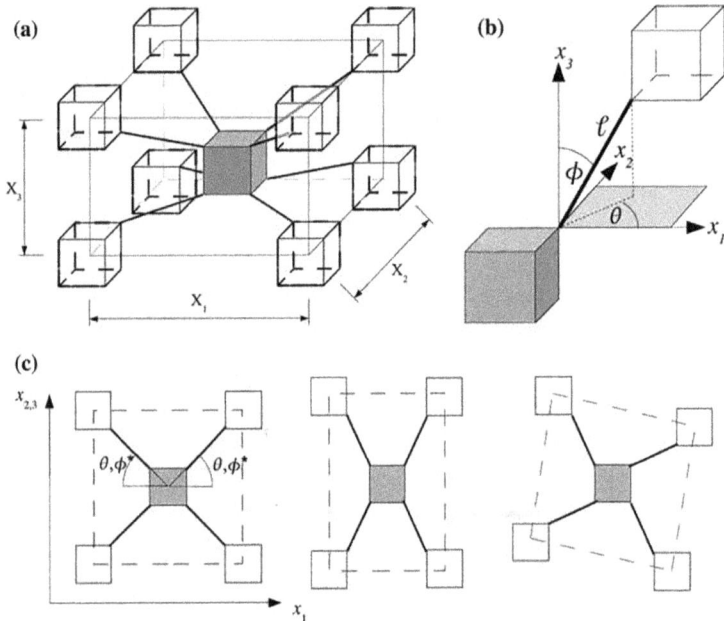

FIGURE 1.27 Schematic diagrams of a generalized 3D tethered nodule are presented, illustrating: (a) a central nodule linked to eight other nodules through corner fibrils; (b) geometric properties and characteristics; and (c) a projection of the nodule in either the x_1-x_2 axis or the x_1-x_3 axis, featuring the angle θ or the projected angle ϕ^* indicated on both the left and right sides.

account fibril stretching, flexure, and hinging individually. Alderson and Evans [3] employed deformation mechanisms involving both stretching and hinging to accomplish the desired features associated with loading along the x-direction:

$$v_{xy} = \frac{\left[1-l^2\left(K_s/K_h\right)\right]\sin\alpha\cos\alpha}{l^2\left(K_s/K_h\right)\sin^2\alpha+\cos^2\alpha}\left(\frac{a+l\cos\alpha}{b-l\sin\alpha}\right)$$

$$v_{xy}^e = \frac{\left[1-l^2\left(K_s/K_h\right)\right]\sin\alpha\cos\alpha}{l^2\left(K_s/K_h\right)\sin^2\alpha+\cos^2\alpha}\left(\frac{a+l_0\cos\alpha_0}{b-l_0\sin\alpha_0}\right)$$

$$E_x = \left(\frac{l^2\sin^2\alpha}{K_h}+\frac{\cos^2\alpha}{K_s}\right)^{-1}\frac{a+l\cos\alpha}{b-l\sin\alpha}$$

$$E_x^e = \left(\frac{l^2\sin^2\alpha}{K_h}+\frac{\cos^2\alpha}{K_s}\right)^{-1}\frac{a+l_0\cos\alpha_0}{b-l_0\sin\alpha_0}$$

(1.29)

and likewise in the y-direction:

$$v_{yx} = \frac{\left[1-l^2\left(K_s/K_h\right)\right]\sin\alpha\cos\alpha}{l^2\left(K_s/K_h\right)\cos^2\alpha+\sin^2\alpha}\left(\frac{b-l\sin\alpha}{a+l\cos\alpha}\right)$$

$$v_{yx}^e = \frac{\left[1-l^2\left(K_s/K_h\right)\right]\sin\alpha\cos\alpha}{l^2\left(K_s/K_h\right)\cos^2\alpha+\sin^2\alpha}\left(\frac{b-l_0\sin\alpha_0}{a+l_0\cos\alpha_0}\right)$$

(1.30)

Gaspar et al. [5] formulated a series of equations to describe the elasticity properties of a material based on the 3D tethered nodule and several idealized models: the idealized stretching model. These models were simplified versions of existing models. In Figure 1.27a and b, the connecting rods have equal lengths denoted as "l," and their projections on the x-axis are labeled as $X_1, X_2,$ and X_3. The angle between the rods and the X_3 direction is represented by ϕ, while the angle between the projection of the rods on the $x_1 - x_2$ plane and the x_1 direction is denoted as θ. The stiffness values, k_l, k_θ, and k_ϕ, are defined as follows, considering an incremental force of dF_l, dF_θ, and dF_ϕ, resulting in corresponding incremental changes of dl, dθ, and dϕ.

The depiction includes a reference cell on the left, along with two distinct symmetry cases in the middle and right, demonstrating the deformation of the marked angle. In the first case (middle), both the left and right angles change simultaneously. In the second case (right), the left and right angles undergo changes with opposite signs [4],

$$k_l = \frac{dF_l}{dl}$$

$$k_\theta = \frac{dM_\theta}{d\theta} = \frac{ldF_\theta}{d\theta}$$

$$k_\phi = \frac{dM_\phi}{d\phi} = \frac{ldF_\phi}{d\phi}$$

(1.31)

respectively. The following elastic moduli were derived by Gaspar et al. [5]:

$$E_1^l = \frac{2k_l}{\cos^2\theta\sin^2\phi}\left(\frac{X_1}{X_2 X_3}\right)$$

$$E_2^l = \frac{2k_l}{\sin^2\theta\sin^2\phi}\left(\frac{X_2}{X_1 X_3}\right)$$

$$E_3^l = \frac{2k_l}{\cos^2\phi}\left(\frac{X_3}{X_1 X_2}\right)$$

$$\nu_{31}^l = \left(\nu_{13}^l\right)^{-1} = -\cos\theta\tan\phi\left(\frac{X_3}{X_1}\right) \tag{1.32}$$

$$\nu_{32}^l = \left(\nu_{23}^l\right)^{-1} = -\sin\theta\tan\phi\left(\frac{X_3}{X_2}\right)$$

$$\nu_{12}^l = \left(\nu_{21}^l\right)^{-1} = -\tan\theta\left(\frac{X_1}{X_2}\right)$$

According to an idealized stretching model, these elastic moduli are determined as follows:

$$E_1^\phi = \frac{2k_\phi}{l^2\cos^2\theta\cos^2\phi}\left(\frac{X_1}{X_2 X_3}\right)$$

$$E_2^\phi = \frac{2k_\phi}{l^2\sin^2\theta\cos^2\phi}\left(\frac{X_2}{X_1 X_3}\right)$$

$$E_3^\phi = \frac{2k_\phi}{l^2\sin^2\phi}\left(\frac{X_3}{X_1 X_2}\right)$$

$$\nu_{31}^\phi = \left(\nu_{13}^\phi\right)^{-1} = \cos\theta\cos\phi\left(\frac{X_3}{X_1}\right) \tag{1.33}$$

$$\nu_{32}^\phi = \left(\nu_{23}^\phi\right)^{-1} = \sin\theta\cos\phi\left(\frac{X_3}{X_2}\right)$$

$$\nu_{12}^\phi = \left(\nu_{21}^\phi\right)^{-1} = -\tan\theta\left(\frac{X_1}{X_2}\right)$$

Using the idealized ϕ hinging model, the subsequent elastic moduli are as follows:

$$E_1^\theta = \frac{2k_\theta}{l^2\sin^2\theta\sin\phi}\left(\frac{X_1}{X_2 X_3}\right)$$

$$E_2^\theta = \frac{2k_\theta}{l^2\cos^2\theta\sin\phi}\left(\frac{X_2}{X_1 X_3}\right) \tag{1.34}$$

$$\nu_{12}^\theta = \left(\nu_{21}^\theta\right)^{-1} = \cot\theta\left(\frac{X_1}{X_2}\right)$$

based on the idealized θ hinging model (Figure 1.28).

The elastic moduli, taking into consideration all three types of deformation, are presented by Gaspar et al. [5] as follows:

$$\frac{1}{E_1} = \frac{X_2 X_3}{2X_1} \left(\frac{\cos^2\theta \sin^2\phi}{k_l} + \frac{l^2 \cos^2\theta \cos^2\phi}{k_\phi} + \frac{l^2 \sin^2\theta \sin\phi}{k_\theta} \right)$$

$$\frac{1}{E_2} = \frac{X_1 X_3}{2X_2} \left(\frac{\sin^2\theta \sin^2\phi}{k_l} + \frac{l^2 \sin^2\theta \cos^2\phi}{k_\phi} + \frac{l^2 \cos^2\theta \sin\phi}{k_\theta} \right) \qquad (1.35)$$

$$\frac{1}{E_3} = \frac{X_1 X_2}{2X_3} \left(\frac{\cos^2\phi}{k_l} + \frac{l^2 \sin^2\phi}{k_\phi} \right)$$

$$v_{12} = -\frac{X_1}{X_2} \frac{\dfrac{\sin\theta \sin\phi}{k_l} + \dfrac{l^2 \sin\theta \cos\phi \cot\phi}{k_\phi} - \dfrac{l^2 \sin\theta}{k_\theta}}{\dfrac{\cos\theta \sin\phi}{k_l} + \dfrac{l^2 \cos\theta \cos\phi \cot\phi}{k_\phi} + \dfrac{l^2 \sin\theta \tan\theta}{k_\theta}} \qquad (1.36)$$

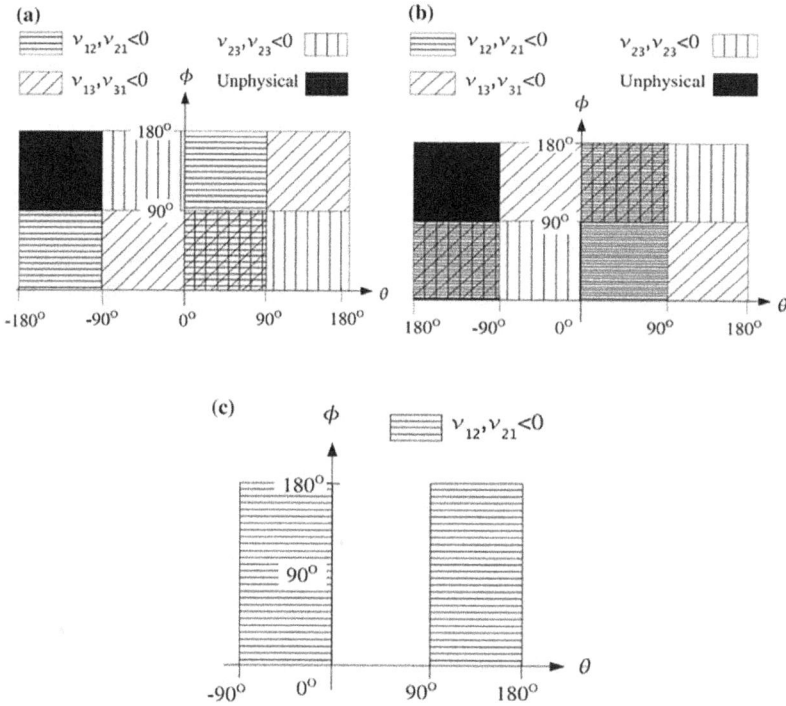

FIGURE 1.28   Phase diagrams: illustrating areas with negative Poisson's ratio: (a) specifically related to the stretching of fibrils, (b) only for ϕ -deformation, and (c) for θ-bending only.

$$v_{21} = -\frac{X_2}{X_1} \frac{\dfrac{\cos\theta\sin\phi}{k_l} + \dfrac{l^2\cos\theta\cot\phi\cos\phi}{k_\phi} - \dfrac{l^2\cos\theta}{k_\theta}}{\dfrac{\sin\theta\sin\phi}{k_l} + \dfrac{l^2\sin\theta\cos\phi\cot\phi}{k_\phi} + \dfrac{l^2\cos\theta\cot\theta}{k_\theta}} \tag{1.37}$$

$$v_{13} = -\frac{X_1}{X_3} \frac{\dfrac{\cos\phi}{k_l} - \dfrac{l^2\cos\phi}{k_\phi}}{\dfrac{\cos\theta\sin\phi}{k_l} + \dfrac{l^2\cos\theta\cos\phi\cot\phi}{k_\phi} + \dfrac{l^2\sin\theta\tan\theta}{k_\theta}} \tag{1.38}$$

$$v_{23} = -\frac{X_2}{X_3} \frac{\dfrac{\cos\phi}{k_l} - \dfrac{l^2\cos\phi}{k_\phi}}{\dfrac{\sin\theta\sin\phi}{k_l} + \dfrac{l^2\sin\theta\cos\phi\cot\phi}{k_\phi} + \dfrac{l^2\cos\theta\cot\theta}{k_\theta}} \tag{1.39}$$

$$v_{31} = -\frac{X_3}{X_1} \frac{\dfrac{\cos\theta\sin\phi}{k_l} - \dfrac{l^2\cos\theta\sin\phi}{k_\phi}}{\dfrac{\cos\phi}{k_l} + \dfrac{l^2\sin\phi\tan\phi}{k_\phi}}$$

$$v_{32} = -\frac{X_3}{X_2} \frac{\dfrac{\sin\theta\sin\phi}{k_l} - \dfrac{l^2\sin\theta\sin\phi}{k_\phi}}{\dfrac{\cos\phi}{k_l} + \dfrac{l^2\sin\phi\tan\phi}{k_\phi}} \tag{1.40}$$

Auxetic behavior can be obtained in these kinds of structures through the rotation of rigid polygons that are linked with each other through hinges. The rotating square, the rotating rectangle, the rotating parallelogram and rhombi, the rotating triangle, and the rotating tetrahedral are only some of the examples.

In their pioneering work on auxetic behavior, Grima and Evans [6] illustrated the concept by rotating interconnected squares, revealing a fascinating property: $v = -1$. By employing the geometric model shown in Figure 1.29a, the researchers successfully derived a stress-strain relationship in two dimensions, contributing to our understanding of this unique mechanical phenomenon.

$$\begin{Bmatrix} \varepsilon_1 \\ \varepsilon_2 \\ \gamma_{12} \end{Bmatrix} = \begin{bmatrix} s_{11} & s_{12} & s_{13} \\ s_{21} & s_{22} & s_{23} \\ s_{31} & s_{32} & s_{33} \end{bmatrix} \begin{Bmatrix} \sigma_1 \\ \sigma_2 \\ \tau_{12} \end{Bmatrix} \tag{1.41}$$

(a)

(b)

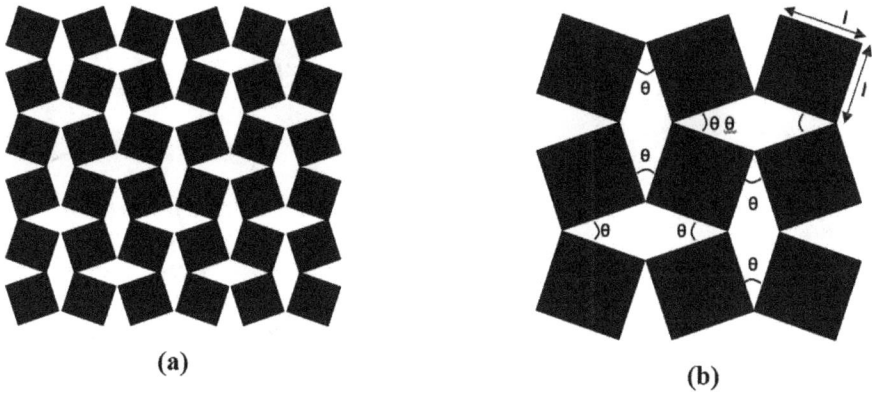

FIGURE 1.29 (a) An overview and (b) specific details of connected rotating squares and their auxetic properties.

Typically, the compliance matrix in equation (1.41) is represented as:

$$S = \begin{bmatrix} \dfrac{1}{E_1} & -\dfrac{v_{21}}{E_2} & \dfrac{\eta_{31}}{G_{12}} \\[2ex] -\dfrac{v_{12}}{E_1} & \dfrac{1}{E_2} & \dfrac{\eta_{32}}{G_{12}} \\[2ex] \dfrac{\eta_{13}}{E_1} & \dfrac{\eta_{23}}{E_2} & \dfrac{1}{G_{12}} \end{bmatrix} \tag{1.42}$$

where η_{ij} are the shear coupling coefficients.

In their study, Grima and Evans (2000) presented the compliance matrix for rigid hinge-connected rotating squares with side lengths of 1 unit, where the hinges possess a rotational stiffness constant K_h.

$$S = \frac{1}{E} \begin{bmatrix} 110 \\ 110 \\ 000 \end{bmatrix} \tag{1.43}$$

the effective Young's modulus is defined as:

$$E = \frac{8K_h}{zl^2 \left(1 - \sin\theta\right)} \tag{1.44}$$

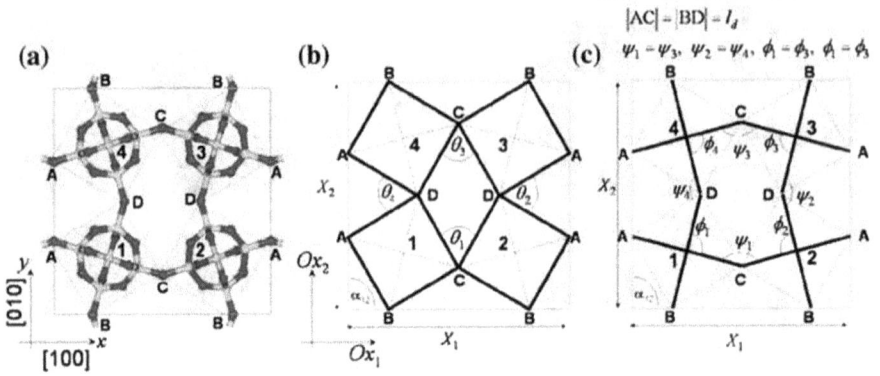

FIGURE 1.30 The variables being observed are defined in relation to the zeolite THO, as illustrated in the diagram. In the diagram, (a) represents the zeolite THO, (b) shows the original structure known as the "rotating squares," indicated by bold squares, and (c) displays the new structure where the squares are replaced by their diagonals, indicated by bold diagonals, $\psi_1 = \psi_3, \psi_2 = \psi_4, \phi_1 = \phi_3$, and $\phi_2 = \phi_4$.

where z is the squares' thickness and θ is the angle shown in Figure 1.30b. Building upon this initial research, Grima et al. [6] further explored the auxetic properties of semi-rigid squares that undergo rotational motion, as depicted in Figure 1.30.

For the on-axis elastic moduli, Grima et al. [6] developed the analytical model as follows:

$$v_{12} = v_{21} = -cot\left(\frac{\psi_1}{2}\right) tan\left(\frac{\psi_2}{2}\right)\left[1 + 4\left(\frac{k_\psi}{k_\phi}\right)\right]^{-1}$$

$$E_1 = E_2 = \frac{8k_\psi\left(k_\phi + 2k_\psi\right)}{l_d^2 z\left(k_\phi + 4k_\psi\right)} \frac{sin\left(\frac{\psi_2}{2}\right)}{sin\left(\frac{\psi_1}{2}\right)cos^2\left(\frac{\psi_2}{2}\right)} \qquad (1.45)$$

$$G_{12} = \frac{\frac{k_\phi}{zl_d^2}}{sin\left(\frac{\psi_1}{2}\right)sin\left(\frac{\psi_2}{2}\right)sin\phi_1}$$

where l_d is the square's diagonal length (i.e., AC or BD in Figure 1.30), k_ψ and k_ϕ are the rotational stiffness constants that restrict the angles ψ and ϕ from changing respectively, and z is the squares' thickness. When the initial angle is equal to $\phi = \pi/2$ and there is no strain (i.e., $\psi_1 = \psi_1 = \psi$), the elastic moduli equations reduce to

$$v_{12} = v_{21} = -\left[1 + 4\left(\frac{k_\psi}{k_\phi}\right)\right]^{-1}$$

$$E_1 = E_2 = \frac{8k_\psi\left(k_\phi + 2k_\psi\right)}{l_d^2 z\left(k_\phi + 4k_\psi\right)} sec^2\left(\frac{\psi}{2}\right)$$

$$G_{12} = \frac{k_\phi}{zl_d^2}\left[\sin^2\left(\frac{\psi}{2}\right)\right]^{-1}.$$

(1.46)

According to Grima et al. [6], the off-axis elastic moduli at an angle ζ around the third direction are

$$\frac{1}{E_1^\zeta} = \frac{m^4}{E_1} + \frac{n^4}{E_2} - m^2 n^2\left(2\frac{v_{12}}{E_1} - \frac{1}{G_{12}}\right)$$

$$v_{12}^\zeta = E_1^\zeta\left[\left(m^4 + n^4\right)\frac{v_{12}}{E_1} - m^2 n^2\left(\frac{1}{E_1} + \frac{1}{E_2} - \frac{1}{G_{12}}\right)\right]$$

$$\frac{1}{G_{12}^\zeta} = \frac{m^4 + n^4}{G_{12}} + 2m^2 n^2\left(\frac{2}{E_1} + \frac{2}{E_2} + \frac{4v_{12}}{E_1} + \frac{1}{G_{12}}\right)$$

(1.47)

where $m = \cos(\zeta)$ and $n = \sin(\zeta)$. Figure 1.31 illustrates how the parameters k_ψ and k_ϕ influence the off-axis Poisson's ratio v_{12}^ζ. On the other hand, Figure 1.31 presents a comparison between the analytical model and molecular modeling results regarding v_{12}^ζ.

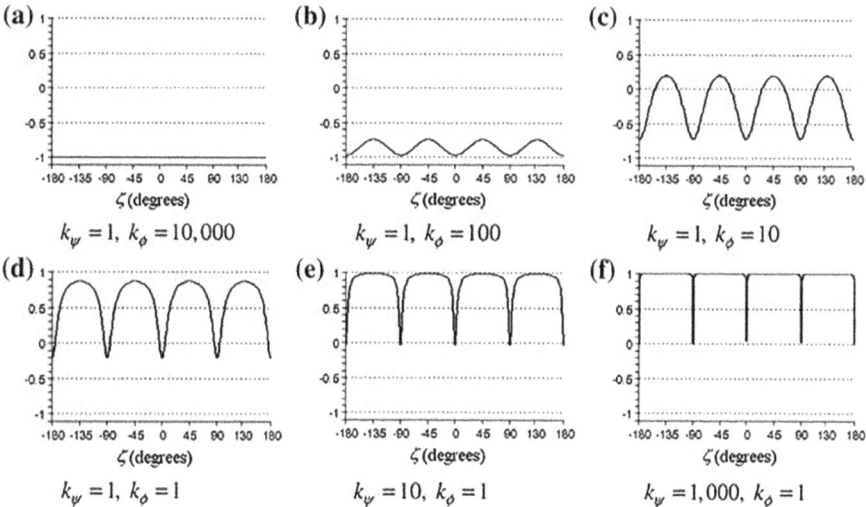

FIGURE 1.31 Off-axis plots for different combinations of K_ψ, K_ϕ, $\psi = 145°$.

Later, Grima et al. [6] developed an alternative mechanism of deformation in which the deformation occurs by varying the square side lengths. In this mode, the angles of the system remain unchanged, but the side lengths of the squares do vary, resulting in shapes that can be either rectangles or squares of different sizes. In their discussion of deformation, Grima et al. [6] considered the shapes formed by the rotating squares. Through their investigation, they determined that there are two possible orientations.

Grima and Evans [6] developed a rotating triangle model, alongside the rotating square model, which can be seen in Figure 1.32a, for the purpose of researching auxetic behavior. Grima and Evans [6] used the diagrams in Figure 1.32b and adopted rotational stiffness coefficient (K_h) and the length of the triangle side (l), in order to get the subsequent elastic properties for rotating rigid triangles:

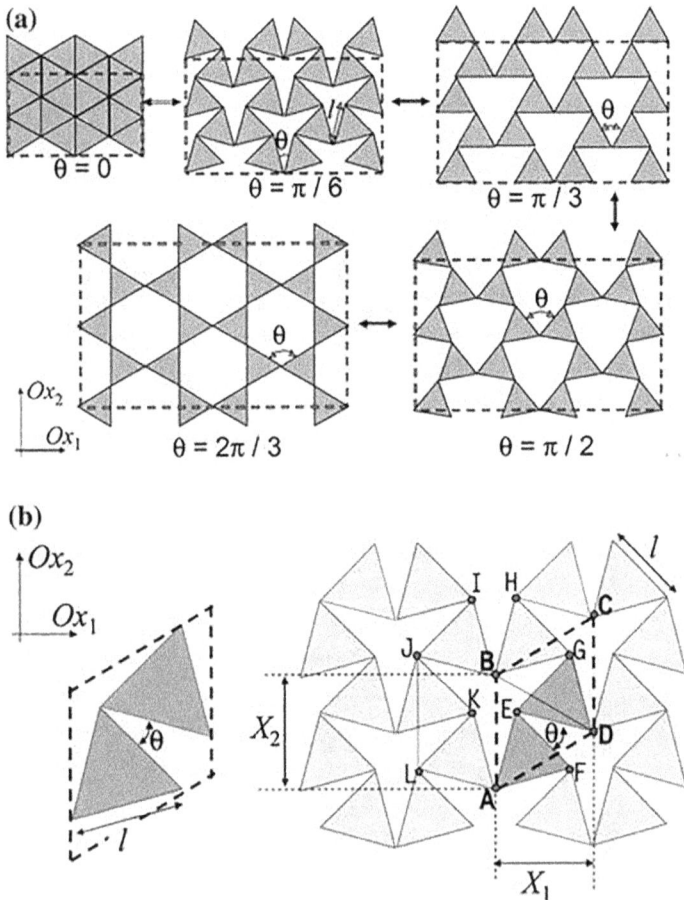

FIGURE 1.32   (a) Rotating triangles exhibiting auxetic characteristics, (b) characterization of a common repetitive unit and unit cell (left), and explanation of geometric parameters (right).

rotational stiffness coefficient (K_h) and the length of the triangle's side (l)

$$v_{12} = v_{21} = -1 \tag{1.48}$$

and

$$E_1 = E_2 = \frac{4\sqrt{3}K_h}{l^2}\left[1+\cos\left(\frac{\pi}{3}+\theta\right)\right]^{-1} \tag{1.49}$$

Grima et al. [6] demonstrated through simulations using FE models that sheets produced from widely accessible conventional non-crystalline materials with star- or triangle-shaped perforations can exhibit auxetic behavior, a characteristic that may be observed when the sheets are subjected to both tension and compression. The researchers then tried to explain this behavior in terms of analytical models based on "rotating rigid triangles," and it was demonstrated that one could adjust both the magnitude and sign of the Poisson's ratio by carefully choosing the shape and density of the perforations.

The discovery provides an efficient and cost-effective approach to create adaptable systems of any size, capable of being adjusted to exhibit specific Poisson's ratio values, either auxetic or non-auxetic, in order to fulfill specific practical needs. These systems can be customized to possess precise Poisson's ratio values, aligning with specific practical requirements. Grima et al. [6] introduced and examined a model based on rotating scalene rigid triangles in relation to one another. This model has demonstrated its ability to generate a wide range of Poisson's ratio values, influenced by the dimensions and angles of the triangles. Additionally, it can provide insights into the behavior of various materials such as auxetic foams and their surface density. Due to its versatile nature, this model can be widely applied.

This model demonstrates that, as compared to the comparable idealized rotating rigid triangles model, the degree of auxeticity tends to be less pronounced when other deformation mechanisms are allowed, leading to more accurate predictions of the Poisson's ratios.

Alderson and Evans [3] created a 3D version of the rotating squares, rectangles, and triangles in order to comprehend the auxetic properties of α-cristobalite through simultaneous rotation and dilatation. According to Figure 1.33a, the used unit cell is tetragonal and contains four regular tetrahedron of uniform size and edge length l. In the extended network, each corner is shared by two tetrahedra since the tetrahedra are connected at the corners. As shown in Figure 1.33b,c, each tetrahedron is tilted about its tilt axis by an angle δ, where $\delta = 0$ corresponds to the case where the top and bottom edges of each tetrahedron are perpendicular to the x_3 axis.

The rotating tetrahedral model (RTM) proposes that tetrahedra are rigid and can rotate collectively around the tilt axis while preserving network connectivity (Figure 1.34). An alternative formulation suggests that applying an external load leads to a change in δ [3]. This deformation mechanism builds upon previous concepts that involved simulating lattice parameter variations in silica structures during phase transitions or thermal expansion by utilizing the rotation of stiff SiO_4

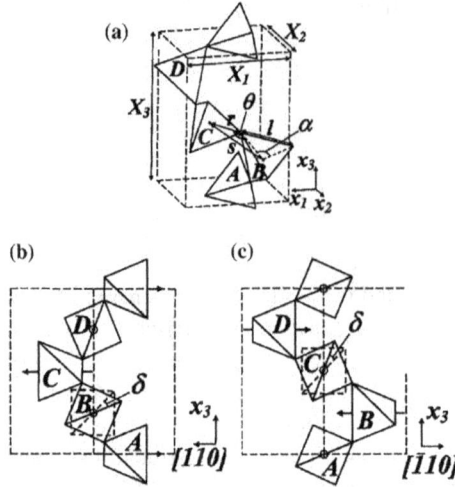

FIGURE 1.33 (a) The geometrical parameters and coordinate system of the tetrahedral framework unit cell. The unit cell consists of four tetrahedra labeled as A, B, C, and D. (b) In the b $x_3 - [-110]$ projection of the unit cell, the tetrahedral axes and the "untitled" tetrahedron (B) are depicted to determine the tilt angle δ. (c) The $x_3 - [-110]$ projection of the unit cell displays the tetrahedral axes and the "untilted" tetrahedron (C) for the purpose of defining the tilt angle δ.

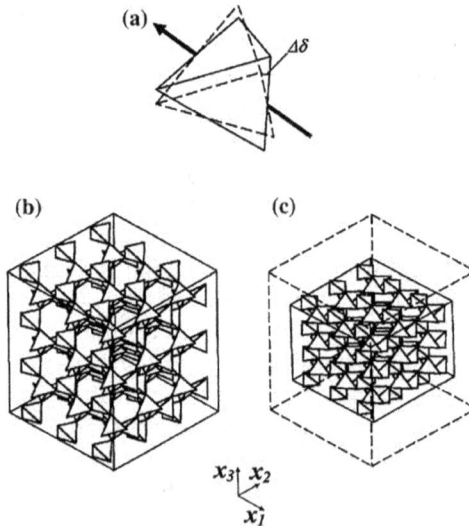

FIGURE 1.34 (a) Mechanism for rotating tetrahedra: this mechanism allows for the rotation of tetrahedra around the tilt axis, passing through the centers of two opposite edges. The size of the tetrahedron remains constant, while the orientation changes during rotation.
(b) Fully expanded 3x3x3 extended tetrahedral network: this refers to a 3x3x3 extended tetrahedral network that is fully expanded, indicated by a tilt angle of 0 degrees (δ = 0°).
(c) Fully densified 3x3x3 extended tetrahedral network: this describes a 3x3x3 extended tetrahedral network that is fully densified, indicated by a tilt angle of 45 degrees (δ = 45°).

tetrahedra. Such changes in the tetrahedral rotation can occur as silica undergoes thermal expansion.

When tetrahedra are subjected to an external force, it is assumed that they will maintain their initial orientation but undergo a change in size while preserving their tetrahedral shape. The concept of size change was first introduced by Alderson and Evans [3], now referred to as the second mode of deformation or the dilating tetra-hedral model (DTM) (refer to Figure 1.35 for additional details). The third mode of deformation is explained by the contemporaneous tetrahedral model (CTM), which combines both the RTM and the DTM simultaneously. Consequently, the Poisson's ratio v_{ij} was determined for each mode under the following conditions:

$$RTM : dl = 0$$
$$DTM : d\delta = 0 \tag{1.50}$$
$$CTM : l\frac{d\delta}{dl} = \kappa$$

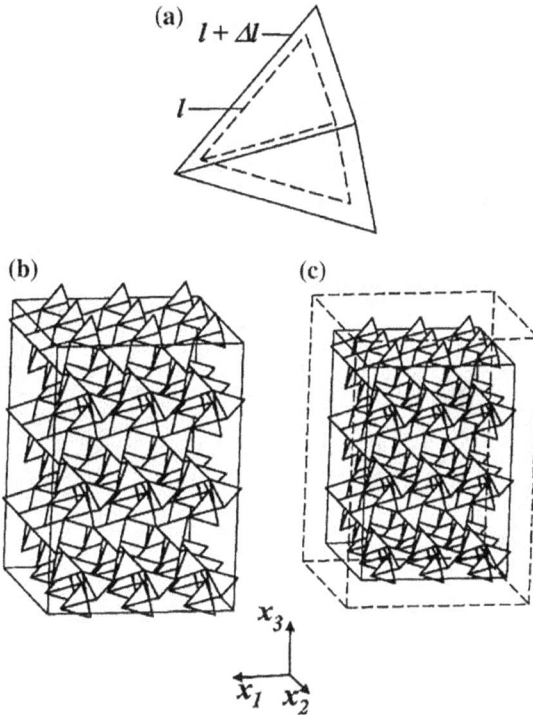

FIGURE 1.35   (a) Mechanism for tetrahedra dilatation deformation: this mechanism involves the dilatation of tetrahedra, where the size of the tetrahedron changes while the orientation remains constant during the dilatation process.

(b) Pre-dilatation 3x3x3 extended tetrahedral network: This refers to a 3x3x3 extended tetra-hedral network before the tetrahedra undergo dilatation.

(c) Post-contraction 3x3x3 extended tetrahedral network: This describes the 3x3x3 extended tetrahedral network after the tetrahedra have undergone contraction. The value of δ, which represents the dilatation, has the same value as described in (b).

where κ is a weight parameter that determines the relative strengths of the two simultaneous deformation mechanisms. The following results were discovered by Alderson and Evans [3]:

$$v_{12} = v_{21} = -1 \tag{1.51}$$

for DTM, RTM, and hence also for CTM modes of deformation, while

$$v_{31} = -1 \tag{1.52}$$

based on the DTM mode of deformation,

$$v_{31} = -\frac{\cos\delta}{1+\cos\delta} \tag{1.53}$$

based on the RTM mode of deformation, and

$$v_{31} = -\frac{\cos\delta}{1+\cos\delta}\left(\frac{1+\cos\delta - \kappa\sin\delta}{\cos\delta - \kappa\sin\delta}\right) \tag{1.54}$$

for the CTM.

CTM possesses a unique feature where, for a given value of δ, there is a distinct range of values κ that allows the attainment of positive values for v_{31}. This occurs despite the fact that the individual mechanisms in the model exhibit auxetic properties when operating independently, as shown in Figure 1.36.

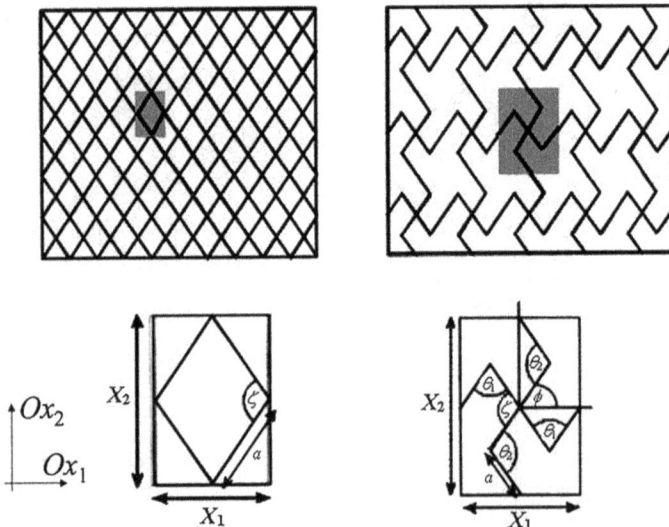

FIGURE 1.36   The unit cells selected for the intact version and the cut version, along with their geometrical properties, are shown in the idealized networks of the intact version and the cut version (top left and top right), respectively.

A modeling approach that involves a missing rib to simulate the elastic properties of auxetic foams was proposed. This missing rib model, also referred to as the cut version, exhibits auxetic behavior. It is based on the intact model, which represents the conventional version, and serves as the basis for the idealized microstructure of the missing rib model. A detailed depiction of this model can be seen in Figure 1.36 (top). By analyzing the corresponding reference figure (Figure 1.36, bottom left), the following elastic moduli to characterize the behavior of the missing rib model is proposed:

$$v_{21} = \frac{1}{v_{12}} = tan^2\left(\frac{\zeta}{2}\right) \tag{1.55}$$

$$E_1 = \frac{4k_\zeta}{a^2} cot\left(\frac{\zeta}{2}\right) csc^2\left(\frac{\zeta}{2}\right)$$

$$E_2 = \frac{4k_\zeta}{a^2} tan\left(\frac{\zeta}{2}\right) sec^2\left(\frac{\zeta}{2}\right) \tag{1.56}$$

for the intact model, where k_ζ is the spring constant that restraints the change to the angle ζ. The elastic moduli for the missing rib model were determined with respect to Figure 1.37 (bottom right). These moduli are obtained by considering the spring constant, denoted as k_θ, which restricts the alteration of the angle θ.

$$v_{21} = \frac{1}{v_{12}} = -tan(\phi)tan(\zeta - \phi)$$

$$E_1 = \frac{k_\theta}{4a^2} \frac{cot(\zeta - \phi)}{sin(\phi)sin(\zeta - \phi)}$$

$$E_2 = \frac{k_\theta}{4a^2} \frac{tan(\zeta - \phi)}{cos(\phi)cos(\zeta - \phi)} \tag{1.57}$$

Figure 1.37 presents a comparison of the Poisson's ratio function among various models: the hexagonal model for both conventional and auxetic foams, the intact model, and the intact and missing rib models, along with experimental results. The values of h for the hexagonal model are determined by fitting to the conventional and auxetic foams. Utilizing the best fit predictions, Figure 1.37 illustrates the relationship between true stress and true strain for both the honeycomb and rib models.

The chiral lattice model presents another approach to expain the auxetic behavior. This model consists of six ligaments that are tangentially attached to one another, connecting each rigid ring, as illustrated in Figure 1.37. By rotating the rigid rings counterclockwise, this model clearly demonstrates how tensile stress in a specific direction results in the elongation in the loading direction. This rotational motion of the rings causes the entire network to expand, resulting in the emergence of an in-plane auxetic feature. A 2D chiral honeycomb was analyzed both theoretically and experimentally, and the honeycomb exhibited a Poisson's ratio of $v = -1$ for in-plane deformation, as revealed in their study. In contrast to other known materials with

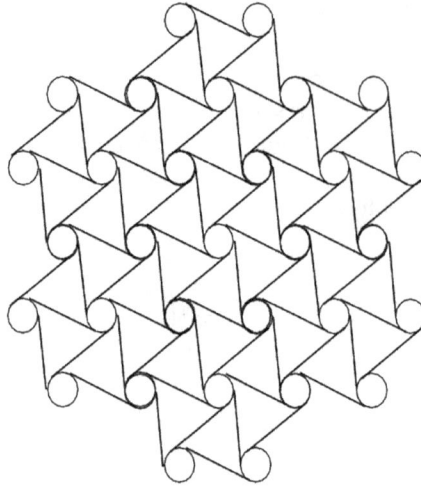

FIGURE 1.37   Chiral lattice auxetic behavior in hexagonal array.

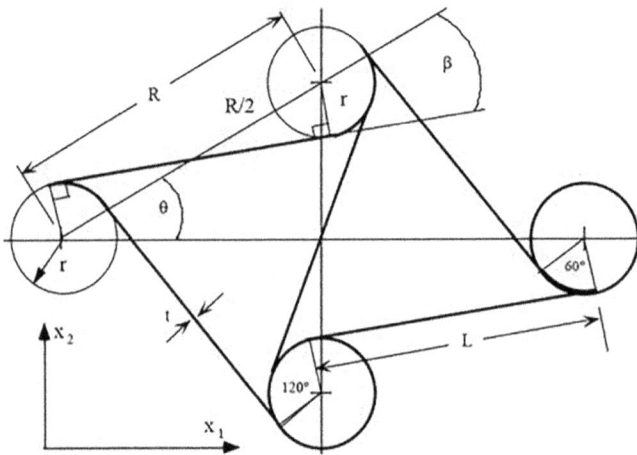

FIGURE 1.38   The geometry utilized for analyzing a chiral lattice.

negative Poisson's ratios, where a change in Poisson's ratio occurs in response to strain, this Poisson's ratio remains unaffected by a wide variety of strain and remains constant throughout.

In Figure 1.38, the ligament is illustrated in the form of a beam with dimensions of thickness t and length L, whilst the ring is denoted by its radius r. In previous studies it was determined the in-plane Young's modulus value for a base material that had a Young's modulus of E_s.

$$E = \sqrt{3}E_s \left(\frac{L}{r}\right)^2 \left(\frac{t}{L}\right)^3 \tag{1.58}$$

The honeycomb model studied by Gibson and Ashby (1988) exhibited a similar dependence on the term $(t / L)^3$.

To overcome the issue of indeterminacy with $v = -1$, an investigation was conducted on a chiral lattice using a micropolar-continuum model, which serves as an equivalent model. Spadoni and Ruzzene [7] formulated constitutive models for two distinct scenarios by utilizing the geometric terminology presented in Figure 1.40. In the first scenario, the nodes are expected to maintain their rigid shape, whereas in the second one, some degree of deformation is allowed. Although the constitutive relation for the deformable node scenario was established through numerical analysis using a finite element model, the constitutive relation for the rigid node case was found through analytical approaches. The constitutive relation for rigid nodes was discovered by Spadoni and Ruzzene [7] using a model that was quite similar to the micropolar-continuum model.

$$
\begin{Bmatrix} \sigma_{11} \\ \sigma_{22} \\ \sigma_{12} \\ \sigma_{21} \\ m_{13} \\ m_{23} \end{Bmatrix} = \begin{bmatrix} D_{11} & D_{12} & 0 & 0 & 0 & 0 \\ D_{21} & D_{22} & 0 & 0 & 0 & 0 \\ 0 & 0 & D_{33} & D_{34} & 0 & 0 \\ 0 & 0 & D_{43} & D_{44} & 0 & 0 \\ 0 & 0 & 0 & 0 & D_{55} & 0 \\ 0 & 0 & 0 & 0 & 0 & D_{66} \end{bmatrix} \begin{Bmatrix} \varepsilon_{11} \\ \varepsilon_{22} \\ \varepsilon_{12} \\ \varepsilon_{21} \\ \kappa_{13} \\ \kappa_{23} \end{Bmatrix} \tag{1.59}
$$

where

$$
D_{11} = D_{22} = \frac{\sqrt{3} E_s t}{4 L^3} \frac{\left(t^4 - L^4 \right) \cos^2 \beta + L^4 + 3 L^2 t^2}{\left(t^2 - L^2 \right) \cos^2 \beta + L^2}
$$

$$
D_{12} = D_{21} = -\frac{\sqrt{3} E_s t}{4 L^3} \frac{L^4 \tan^2 \beta + t^4 - L^2 t^2 \left(1 + \tan^2 \beta \right)}{L^2 \tan^2 \beta + t^2} \tag{1.60}
$$

$$
D_{33} = D_{44} = \frac{\sqrt{3} E_s t}{4 L^3} \frac{2 L^4 \tan^2 \beta + L^2 R^2 + t^2 \left(2 L^2 + R^2 \right)}{R^2}
$$

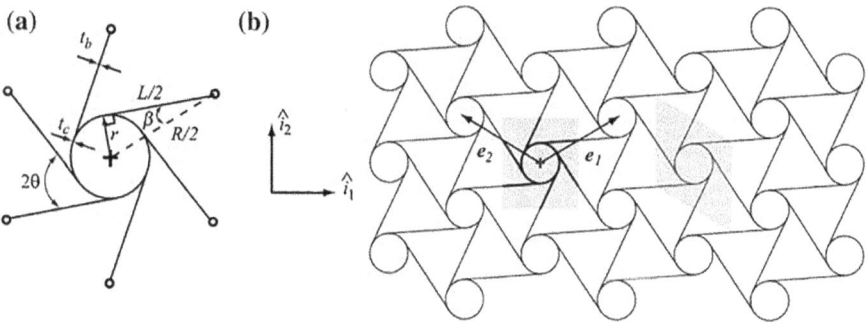

FIGURE 1.39 Geometry of a hexagonal chiral lattice: (a) unit cell and (b) unit volume with symmetry vectors.

(a)

(b)

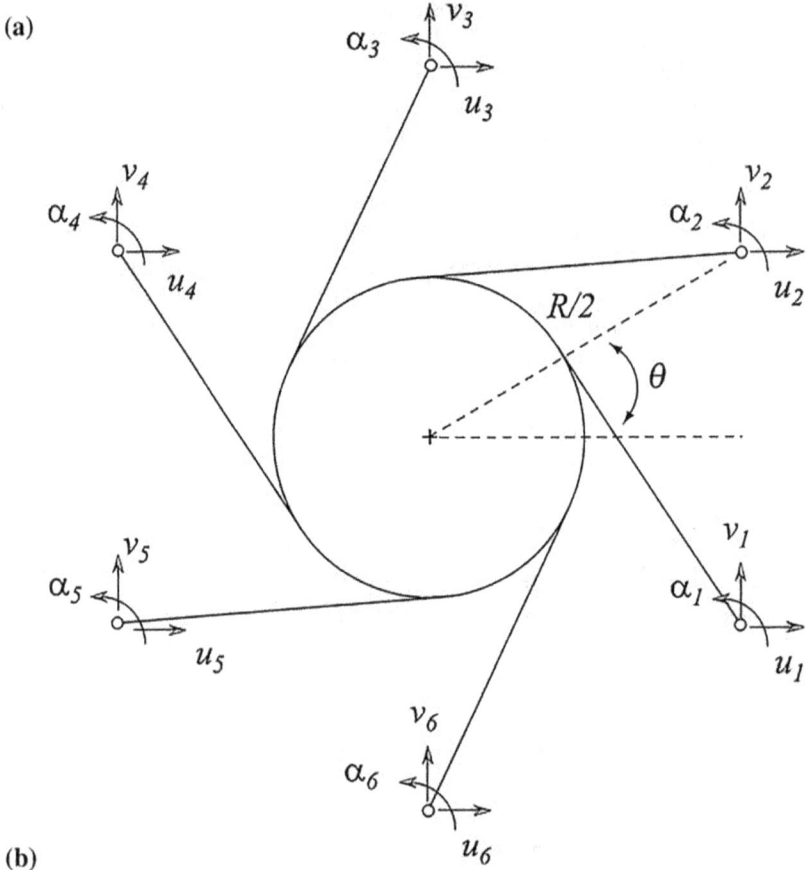

	ε_{11}	ε_{22}	ε_{12}	ε_{21}	κ_{13}	κ_{23}
u_1	$R\cos\theta/2$	0	$-R\sin\theta/2$	0	0	0
v_1	0	$-R\sin\theta/2$	0	$R\cos\theta/2$	0	0
α_1	$\hat{\alpha}_1$	$\hat{\alpha}_1$	$\hat{\alpha}_1+1/2$	$\hat{\alpha}_1-1/2$	$R\cos\theta/2$	$-R\sin\theta/2$
u_2	$R\cos\theta/2$	0	$R\sin\theta/2$	0	0	0
v_2	0	$R\sin\theta/2$	0	$R\cos\theta/2$	0	0
α_2	$\hat{\alpha}_2$	$\hat{\alpha}_2$	$\hat{\alpha}_2+1/2$	$\hat{\alpha}_2-1/2$	$R\cos\theta/2$	$R\sin\theta/2$
u_3	0	0	$R/2$	0	0	0
v_3	0	$R/2$	0	0	0	0
α_3	$\hat{\alpha}_3$	$\hat{\alpha}_3$	$\hat{\alpha}_3+1/2$	$\hat{\alpha}_3-1/2$	0	$R/2$

FIGURE 1.40 Configurations with deformable rings were studied using (a) an FE model of the unit cell with boundary degrees of freedom specified for each examined strain state and (b) imposed displacements corresponding to six independent strain states.

$$D_{34} = D_{43} = -\frac{\sqrt{3}E_s t}{4L^3} \frac{2L^4 \tan^2\beta - L^2 R^2 + t^2 \left(2L^2 - R^2\right)}{R^2}$$

$$D_{55} = D_{66} = \frac{\sqrt{3}E_s t}{4L^3} \left(L^4 \tan^2\beta + \frac{4}{3}L^2 t^2\right).$$

Spadoni and Ruzzene [7] formulated the engineering constants based on the stiffness coefficients $E = \left(D_{11}^2 - D_{12}^2\right)/D_{11}$ and $v = D_{12}/D_{12}$, using the approaches presented by Nakamura and Lakes (1995) as well as Yang and Huang (2001).

$$E = \frac{2\sqrt{3}\left[1+\left(\dfrac{t}{L}\right)^2\right]\left(\dfrac{t}{L}\right)^3}{\left(\dfrac{t}{L}\right)^4 \cos^2\beta + \sin^2\beta + 3\left(\dfrac{t}{L}\right)^2} E_s$$

$$v = \frac{4\left(\dfrac{t}{L}\right)^2}{\left(\dfrac{t}{L}\right)^4 \cos^2\beta + \sin^2\beta + 3\left(\dfrac{t}{L}\right)^2} - 1 \tag{1.61}$$

$$G = \frac{\sqrt{3}}{4}\frac{t}{L}\left[1+\left(\frac{t}{L}\right)^2\right]E_s.$$

Due to the challenges in deriving analytical formulas for deformable rings, Spadoni and Ruzzene [7] utilized a finite element model that incorporated prescribed displacements and rotations related to strain conditions, as illustrated in Figure 1.39. Figure 1.40 displays the Young's modulus, shear modulus, and Poisson's ratio using two different approaches: the analytical approach for a rigid node and the finite element approach for a deformable node, with the Young's and shear moduli represented in a normalized manner, i.e., $\overline{E}_m/(t/L)^3$ and $\overline{G}_m/(t/L)^3$, respectively in which

$$\begin{Bmatrix} \overline{E}_m \\ \overline{G}_m \end{Bmatrix} = \frac{1}{E_s}\begin{Bmatrix} E \\ G \end{Bmatrix} \tag{1.62}$$

Alderson [3] have also investigated the auxetic properties of chiral and anti-chiral lattices, specifically focusing on the chiral lattice as depicted in Figure 1.41. Analytical and finite element analyses were performed on Chen et al.'s anti-tetrachiral structural model [8]. The following moduli of elasticity were determined by Chen et al. [8] using the structure depicted in Figure 1.42:

$$v_{xy} = -\frac{L_x}{L_y} \tag{1.63}$$

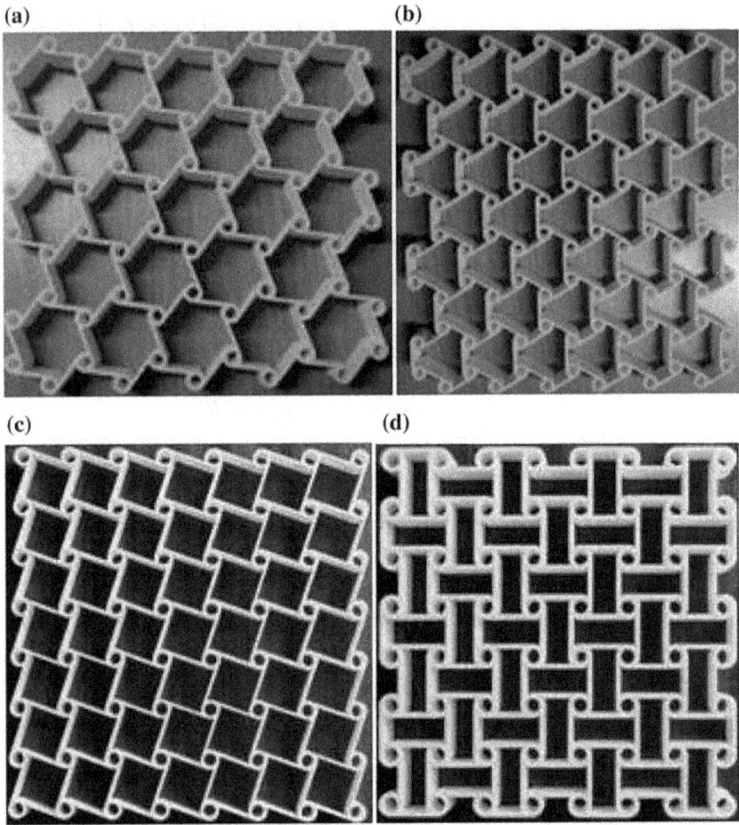

FIGURE 1.41 Four different types of rapid prototype chiral honeycombs: (a) trichiral, (b) anti-trichiral, (c) tetrachiral, and (d) anti-tetrachiral.

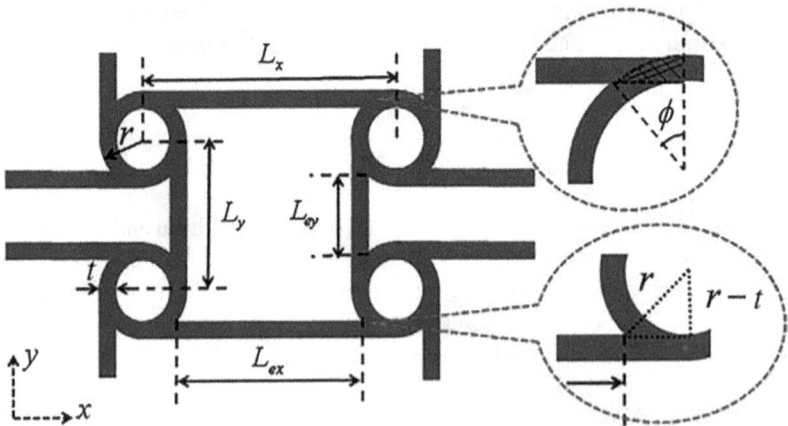

FIGURE 1.42 The geometric configuration of the anisotropic honeycomb unit cell, known as the anti-tetrachiral structure.

$$E_x = \frac{E_c \beta^3 \alpha_x}{12(1-\beta/2)^2 \alpha_y} \left(\frac{1}{\alpha_x - 2\sqrt{2\beta - \beta^2}} + \frac{1}{\alpha_y - 2\sqrt{2\beta - \beta^2}} \right)$$

$$E_y = \frac{E_c \beta^3 \alpha_y}{12(1-\beta/2)^2 \alpha_x} \left(\frac{1}{\alpha_x - 2\sqrt{2\beta - \beta^2}} + \frac{1}{\alpha_y - 2\sqrt{2\beta - \beta^2}} \right) \qquad (1.64)$$

$$E_z = \frac{\beta \left[\alpha_x + \alpha_y + \pi(2-\beta) \right] - 2 \left[\phi - (1-\beta)\sin\phi \right]}{\alpha_x \alpha_y} E_c$$

$$\phi = \cos^{-1}(1-\beta) \qquad (1.65)$$

and the dimensionless parameters are given by

$$\begin{Bmatrix} \alpha_x \\ \alpha_y \\ \beta \\ \gamma \end{Bmatrix} = \frac{1}{r} \begin{Bmatrix} L_x \\ L_y \\ t \\ b \end{Bmatrix} \qquad (1.66)$$

Here, E_c represents the core material's Young's modulus, while the cell depth is represented by b.

If the ligament lengths are the same ($\alpha_x = \alpha_y = \alpha$), it simplifies to $v_{xy} = -1$.

$$E_x = E_y = \frac{E_c \beta^3}{6(1-\beta/2)^2} \left(\frac{1}{\alpha - 2\sqrt{2\beta - \beta^2}} \right)$$

$$\qquad (1.67)$$

$$E_z = \frac{\beta[2\alpha + \pi(2-\beta)] - 2[\phi - (1-\beta)\sin\phi]}{\alpha^2} E_c$$

The FE model is shown in Figure 1.43a in order to compare it with the analytical model, and the experimental process is graphically summarized in Figure 1.43b, where Chen et al. [8] provide more information. The analytical and FE results of Chen et al. [8], Poisson's ratio are shown in Figure 1.43. These results, along with other findings presented in their work, demonstrate that by altering the longitudinal and transverse lengths of the ligaments, significant altercations of the in-plane negative Poisson's ratios can be anticipated. This offers valuable recommendations for developing and producing a novel core design for sandwich structures, applicable to various engineering applications [8].

The planar sheet model, which has recently garnered significant attention, can be conceptualized as a variant referred to as the crumpled sheets model. By employing meticulous optical microscopy and conducting Poisson's ratio measurements, it is successfully identified the occurrence of auxetic behavior, attributable to a microstructure characterized by crumpling throughout the thickness.

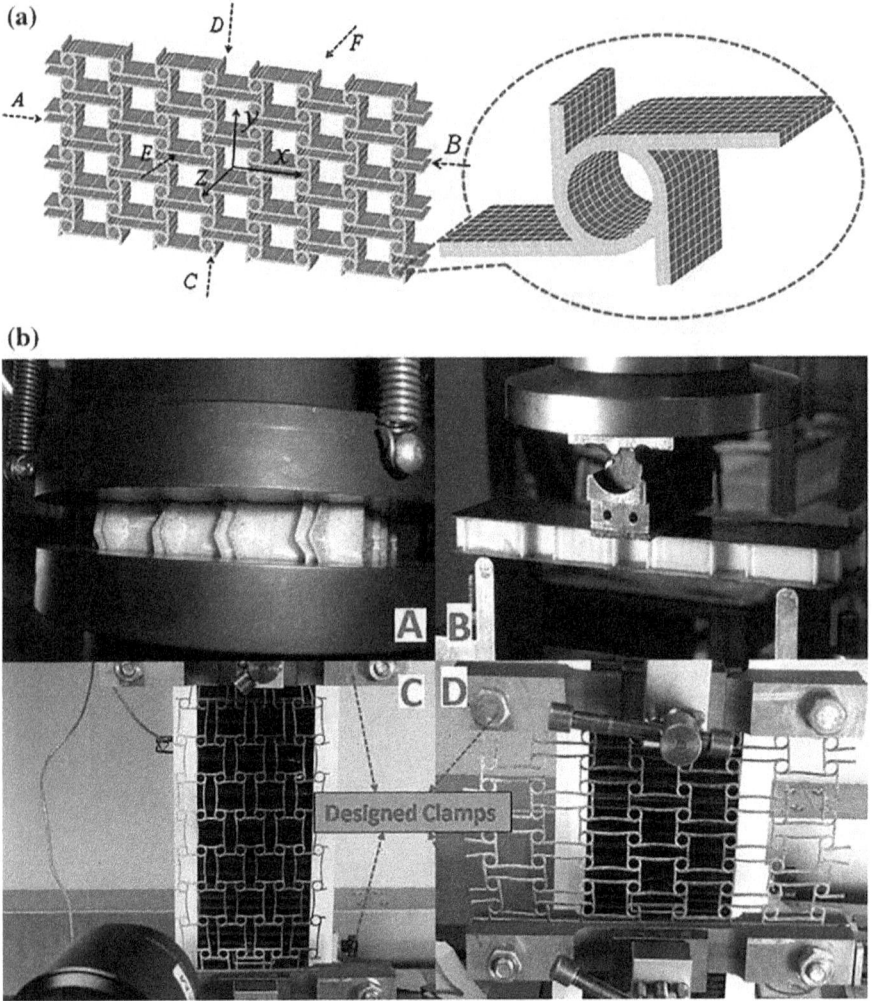

FIGURE 1.43 (a) A layout of a repeating unit cell in a finite element model. (b) The experimental setup for honeycomb structures, including: (A) compression tests performed in the flatwise direction; (B) three-point bending tests; (C) and (D) tensile tests.

Figure 1.44 is an illustration that shows the structural properties of the auxetic structure as well as its deformation. By making the following assumptions, the authors derived equations to estimate the structure's Poisson's ratio. These assumptions include: (1) the straight ligaments of the corrugated sheets remain straight during loading; (2) the elastic deformation of the structure is disregarded; (3) the thickness of the corrugated sheets (h) is negligible; and (4) the tubes are tightly attached to the corrugated sheets, ensuring no slippage occurs. The study concluded that the geometric features of the proposed structure influence its auxetic behavior.

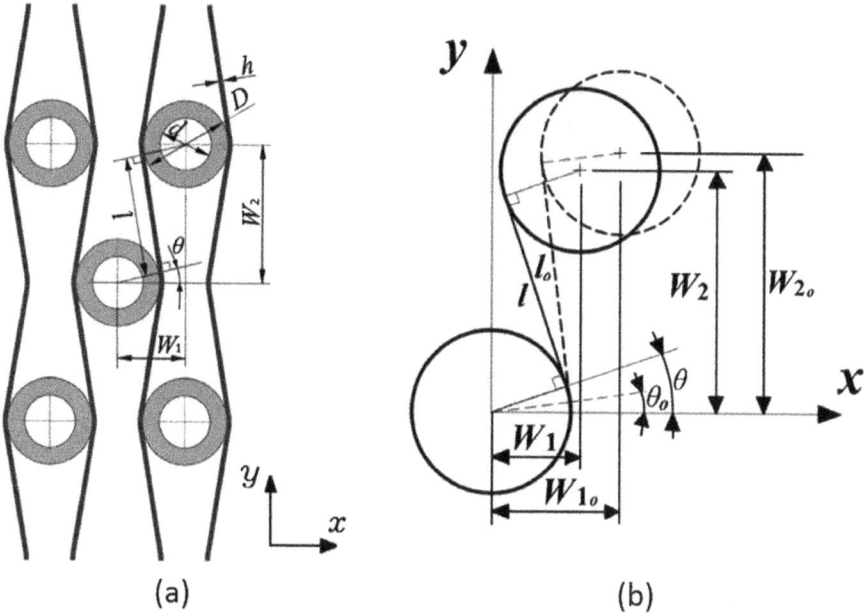

FIGURE 1.44　Auxetic structure with corrugated sheets.

It is possible to improve the auxetic behavior by reducing the crimped influence that the corrugated sheets have and by making the material more stable when it is undergoing the initial deformation.

$$v = -\frac{\varepsilon_x}{\varepsilon_y} = -a\frac{\sin\theta_0\left(b\cos\theta-1\right)-\sin\theta\left[\left(b\cos\theta_0-1\right)-\sin\theta_0\left(\theta-\theta_0\right)\right]}{\sin\theta_0\left(b\sin\theta-a\right)+\cos\theta\left[\left(b\cos\theta_0-1\right)-\sin\theta_0\left(\theta-\theta_0\right)\right]} \quad (1.68)$$

$$\theta_0 = \arcsin\frac{ab-\sqrt{1+a^2-b^2}}{1+a^2} \quad (1.69)$$

$$a = \frac{W_{20}}{W_{l_0}}, \ b = \frac{D}{W_{l_0}} \quad (1.70)$$

Numerous scholars have been drawn to the Japanese craft of kirigami in order to further their understanding of auxetic materials and structures. A hole with a parallelogram-shaped opening surrounded by two zigzag strips formed the unit cell of the designs.

The conclusion reached by the authors is that the dislocating zigzag strips found on the joining ridges of the Miura-ori serve to preserve and enhance its features. Extensive research conducted utilized both finite element methods and experiments to examine Miura-ori patterned sheets. They specifically focused on the deformation of these sheets under three different types of tests: out-of-plane compression,

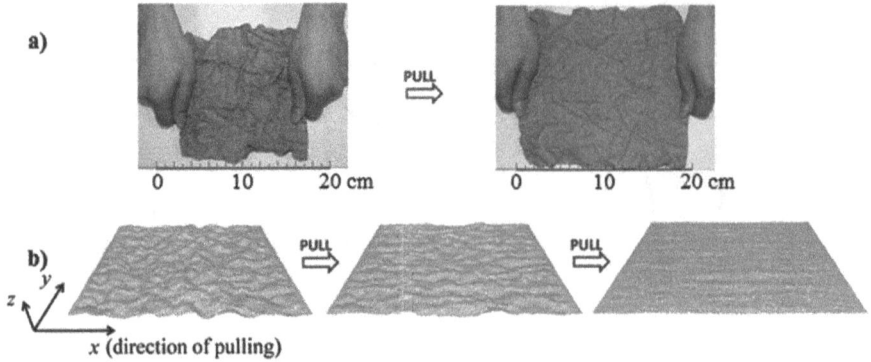

FIGURE 1.45 Examples of typical crumpled sheet conformations: (a) a crumpled piece of paper at various applied strain levels; (b) a graphene sheet with 3.0% defects at various applied strain levels.

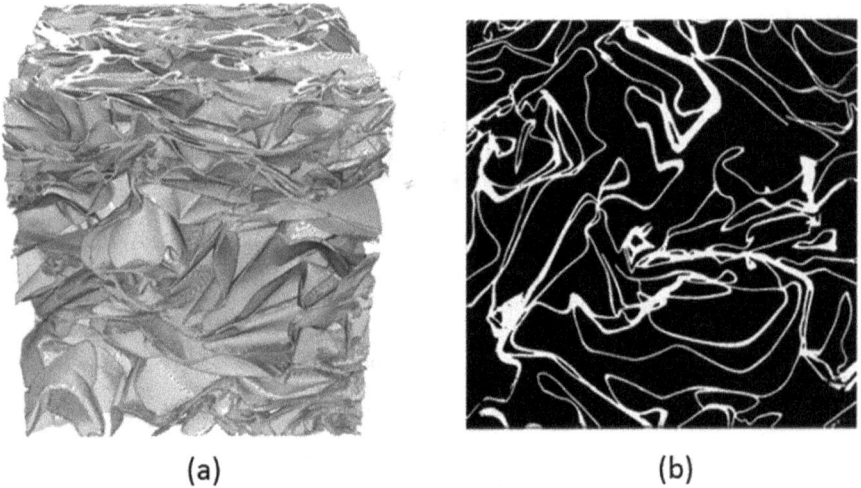

FIGURE 1.46 Images of a typical crumpled aluminum foil can be observed in the following micrographs: (a) a 3D representation and (b) a 2D segmented image.

three-point bending, and in-plane compression. While it is widely accepted that the auxetic properties of paper structures are influenced by the cellulose fiber network structure within the sheets, additional investigation is still required to determine the impact of various materials and processing variables on auxetic behavior.

These materials possess remarkable flexibility and minimal hysteresis, akin to conventional foams. However, unlike foams, these materials do not exhibit a plateau in their mechanical properties after reaching the yield stress.

In the past few years, there has been an evident increase in the interest surrounding the utilization of perforated sheet designs as innovative cellular materials and structures that possess auxetic properties. According to [6], conventional materials with

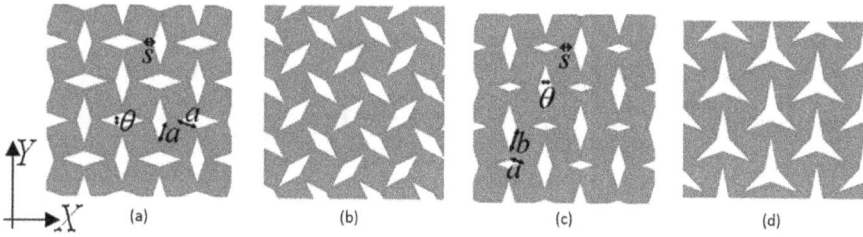

FIGURE 1.47 Some examples of perforated sheets with negative Poisson's ratio: (a) when the diamond-shaped inclusions are aligned parallel to the sides of the square, forming a 0-degree angle; (b) when the diamond-shaped inclusions are oriented at a 45-degree angle relative to the sides of the square; (c) when diamond-shaped inclusions of two distinct sizes are incorporated into the sheet; and (d) when a star-shaped inclusion is present within the sheet.

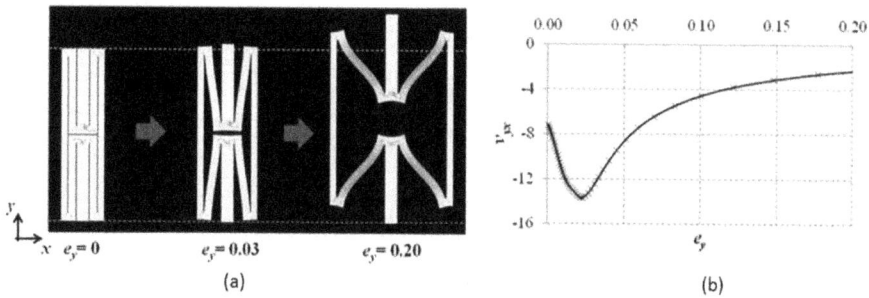

FIGURE 1.48 Illustration of (a) a typical perforated sheet with auxetic properties at different levels of strain in the y-direction; (b) comparison of Poisson's ratio with engineering strain for the corresponding sheet.

perforations in diamond or star shapes can display auxetic behavior when subjected to both tension and compression. An alternative perspective on the mechanism is to view it as an extension constructed using rigidly rotating components. Figures 1.48 and 1.49 provide examples of commonly employed perforated sheet patterns. This figure represents the function of Poisson's ratio, which can be found below:

$$v_{xy} = \left(v_{yx}\right)^{-1} = \frac{a^2\cos^2\left(\dfrac{\theta}{2}\right) - b^2\sin^2\left(\dfrac{\theta}{2}\right)}{a^2\sin^2\left(\dfrac{\theta}{2}\right) - b^2\cos^2\left(\dfrac{\theta}{2}\right)}, \qquad (1.71)$$

The lengths of the rectangles are represented by **a** and **b**, while the angle formed between them is indicated as θ.

Grima et al. [6] used molecular simulation to demonstrate the auxeticity of networked calix [5] arene polymers based on an "egg rack" structure (Figures 1.49 and 1.50).

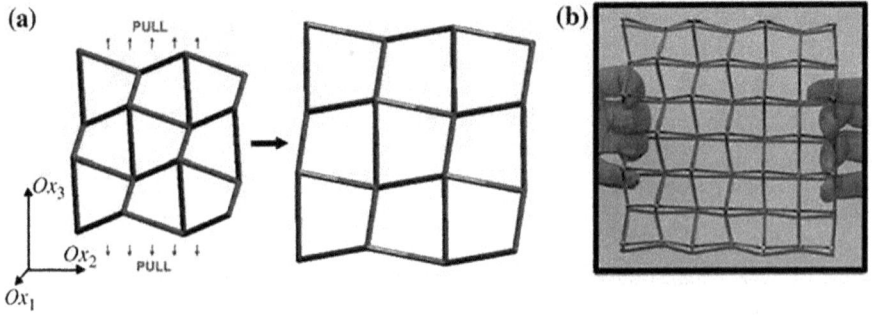

FIGURE 1.49 The model macrostructure is shown in schematic form in (a), and the commercially available "egg rack" structure, which possesses the identical shape, is shown in (b).

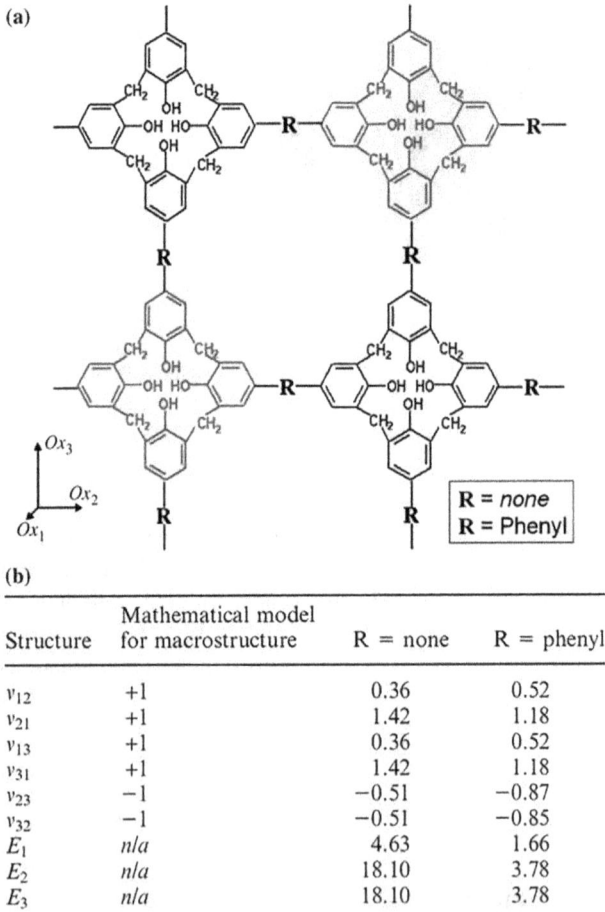

(a)

(b)

Structure	Mathematical model for macrostructure	R = none	R = phenyl
V_{12}	+1	0.36	0.52
V_{21}	+1	1.42	1.18
V_{13}	+1	0.36	0.52
V_{31}	+1	1.42	1.18
V_{23}	−1	−0.51	−0.87
V_{32}	−1	−0.51	−0.85
E_1	n/a	4.63	1.66
E_2	n/a	18.10	3.78
E_3	n/a	18.10	3.78

FIGURE 1.50 (a) Structural formulas of repeating units in "double calix" molecular structures and (b) simulation results for Poisson's ratios (V_{ij}) and Young's moduli (E_i) in single crystalline structures: a comparison with predicted values for the idealized folded macrostructure.

1.6   UNIT CELL AND THE ENTIRE MIURA-ORI PATTERN

Figure 1.51a illustrates a folded Miura-ori (n_1, n_2) with x_3 as the out-of-plane dimension and n_1 (=11) vertices in x_1 direction and n_2 (=11) vertices in x_2 direction. The planar state that corresponds to it is illustrated in Figure 1.1b. Several congruent rigid parallelogram faces (four of which are highlighted in grey in Figure 1.1b) are connected by edges that can be folded in a way that resembles mountain and valley creases to form the geometry of a Miura-ori. The unit cell of the Miura-ori, a

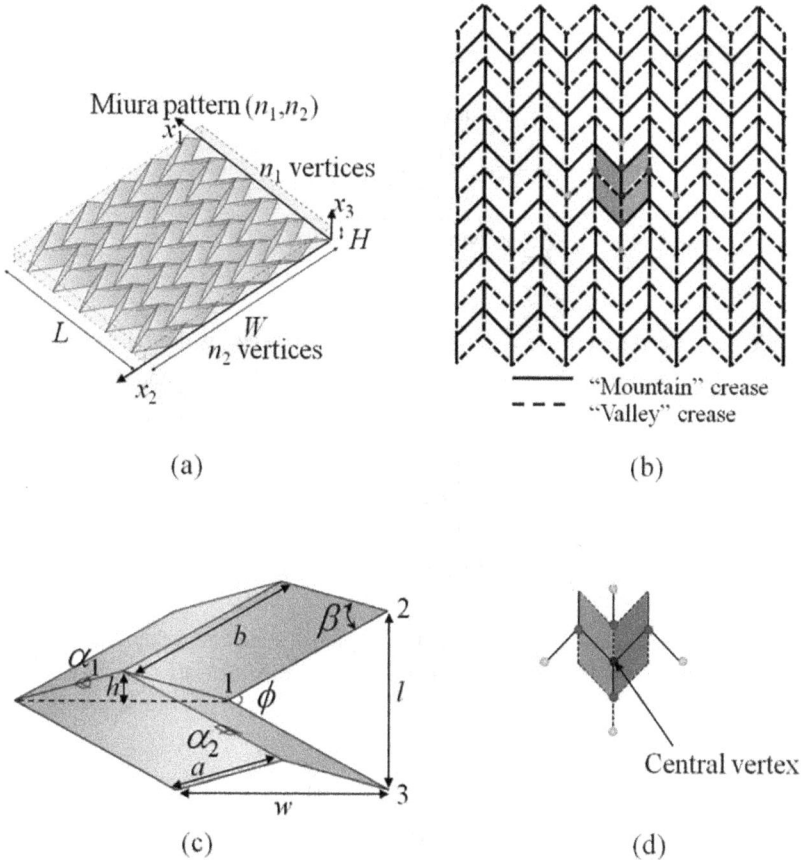

(a)

(b)

(c)

(d)

FIGURE 1.51 Visual representations of Miura-ori. (a) A folded Miura-ori is shown, characterized by n_1 vertices in the x_1 direction and n_2 vertices in the x_2 direction, with the x_3 direction perpendicular to the plane. In this specific example, the values are $n_1 = 11$, $n_2 = 11$, $\beta = 45$ degrees, and a ratio of a/b = 1/√2. (b) Displays the Miura-ori in its flat state, corresponding to the folded state in (a). The solid lines represent "mountain" creases on the top surface, while the dashed lines depict "valley" creases on the bottom surface. (c) Focuses on a unit cell of the Miura-ori, presenting two dihedral angles, α_1 and α_2. Each parallelogram within the unit cell has a short side length of **a**, a long side length of **b**, and an acute angle of β. The projected angle between the two ridges is denoted as ϕ. The dimensions of the unit cell are l, w, and h in the x_1, x_2, and x_3 directions, respectively. Lastly, (d) highlights a non-local element of the Miura-ori, specifically emphasizing the central vertex.

periodic structure, is depicted in Figure 1.1c. It consists of four congruent parallelo-grams that have acute angles of $\beta \in [0°,90°]$ and lengths **a**, **b**.

The rigid origami condition states that in a Miura-ori structure with n = 4 edges at a single vertex, there is only one degree of freedom when there are n-3 degrees of freedom overall. Therefore, by defining the characteristics of a parallelogram face using β, a, and b, we can effectively describe the folding of the Miura-ori unit cell using a single parameter $\phi \in [0^\wedge\circ, 2\beta]$. This parameter represents the angle of projection between two ridges. When the Miura-ori is in its planar state, ϕ is set to 2β, indicating a fully unfolded configuration. On the other hand, when the Miura-ori is fully collapsed, ϕ is set to $0^\wedge\circ$, signifying that the structure is folded completely.

The dimensions of the unit cell are:

$$l = 2b\sin(\phi/2),$$

$$w = 2a\frac{\cos\beta}{\cos(\phi/2)} \tag{1.72}$$

$$h = \frac{a\sqrt{\sin^2\beta - \sin^2(\phi/2)}}{\cos(\phi/2)}$$

It should be noticed that the length of the "tail" $b\cos(\phi/2)$ is not taken into consid-eration by the unit cell. Only two dihedral angles, $\alpha_1 \in [0°,180°]$ and $\alpha_2 \in [0°,180°]$, are needed to characterize this pattern's periodicity, and these angles are given by:

$$\alpha_1 = \cos^{-1}\left[1 - 2\frac{\sin^2(\phi/2)}{\sin^2\beta}\right],$$

$$\alpha_2 = \cos^{-1}\left[1 - 2\cot^2\beta\tan^2(\phi/2)\right] \tag{1.73}$$

The angles have a value of 180 degrees in the planar state and 0 degrees in the fully collapsed state. Upon the containment of the entire Miura-ori structure within a conceptual box with the dotted lines as its limits (Figure 1.1a), the Miura-ori's total dimension is then determined by:

$$L = (n_1 - 1)b\sin(\phi/2),$$

$$W = (n_2 - 1)a\frac{\cos\beta}{\cos(\phi/2)} + b\cos(\phi/2), \tag{1.74}$$

$$H = \frac{a\sqrt{\sin^2\beta - \sin^2(\phi/2)}}{\cos(\phi/2)}$$

Therefore, the size of the conceptual volume taken up by this Miura-ori can be expressed as:

$$V = L \times H \times W. \tag{1.75}$$

It is evident that the size of the Miura-ori in the x_2 direction (W) is not directly proportional to the size of the unit cell (w), despite its periodic nature. This is due to the presence of a "tail" with a length of $b\cos(\phi/2)$. As a result, it is inaccurate to use the unit cell, especially for smaller patterns, to investigate how the size of a full Miura-ori changes. Based on intuitive observations and theoretical investigations employing the unit cell, it is believed that the Miura-ori pattern exhibits a negative in-plane Poisson's ratio (as depicted in Figure 1.51c).

One way to precisely describe Poisson's ratio is by considering the size of an entire Miura-ori structure rather than just the unit cell. In particular, the in-plane Poisson's ratio v_{12} is defined as $v_{12} = -\dfrac{\varepsilon_{11}}{\varepsilon_{22}}\Big|_{\varepsilon_{22}\to 0}$, where $\varepsilon_{11} = \dfrac{dL}{L}$ and $\varepsilon_{22} = \dfrac{dW}{W}$ represent the infinitesimal strains in X and Y directions, respectively. By employing equation (1.2), we can calculate the in-plane Poisson's ratio, v_{12}

$$v_{12} = -\cot^2(\phi/2)\frac{(n_2-1)\eta\cos\beta + \cos^2(\phi/2)}{(n_2-1)\eta\cos\beta - \cos^2(\phi/2)}, \tag{1.76}$$

in which $\eta = a/b$. The reciprocal of v_{12}, denoted as v_{21}, represents another in-plane Poisson's ratio. Figure 1.52a visually depicts the contour of v_{12} concerning the angle ϕ and a combination parameter $(n_2-1)\eta\cos\beta$. It is evident that v_{12} can exhibit both negative and positive values, which is in contrast to the frequently observed negative in-plane Poisson's ratio.

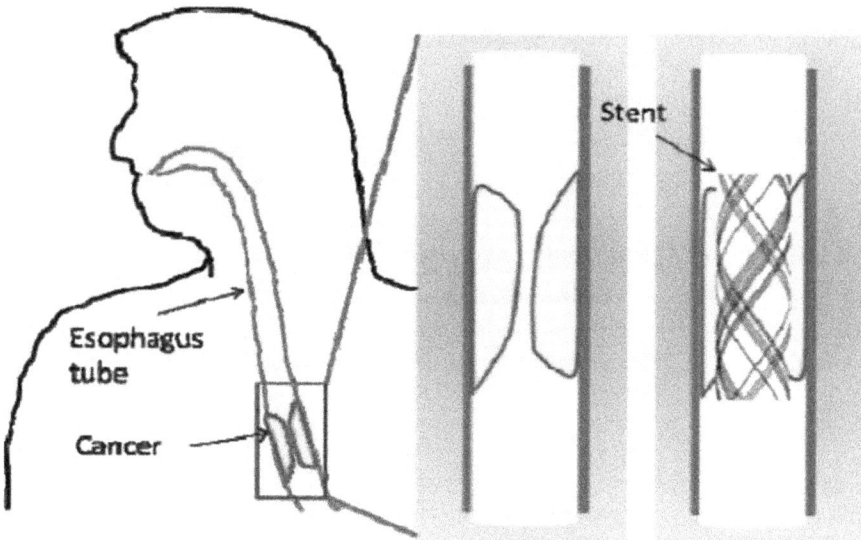

FIGURE 1.52 A constricted section of the esophagus can be held open using an auxetic stent.

1.7 APPLICATION OF METAMATERIALS AND AUXETIC MATERIALS

Auxetic materials and structures display peculiar behavior when subjected to deformation, rendering them highly desirable for various applications. Their extraordinary characteristics open up a wide range of potential uses for these intelligent materials. Currently, they are employed in diverse and innovative fields such as soft robotics, biomedicine, soft electronics, and acoustics. Auxetic materials are particularly advantageous for enhancing shock absorption, synclastic curvature, and shear performance. Consequently, they find significant value in industries such as aerospace, automotive, defense, and sports. A comprehensive overview of the manifold applications of auxetic components from various sources can be found in Table 1.2. This compilation offers insights into the extensive utilization of auxetic materials across numerous sectors, underscoring their versatility and utility [9].

1.7.1 MEDICAL APPLICATION

Auxetic materials have been recommended as crucial components in the functioning of foldable medical devices such as angioplasty stents, annuloplasty rings, and esophageal stents. Esophageal cancer can lead to the formation of tumors that block the passage of food, making it one of the top ten deadliest cancers worldwide. Given this context, the utilization of auxetics in the design and construction of such medical devices holds immense importance for improving patient outcomes and addressing the challenges posed by esophageal cancer. The auxetic stent could be used to lengthen patients' lives by expanding their esophagus and relieving their pain, as shown in Figure 1.53. Thin nanofiber membranes with an auxetic pattern were shown to have an almost ten-fold higher elongation capacity than that of control samples of non-auxetic nanofiber. As a result, it was thought that the auxetic designed thin nanofiber membrane may be used for a variety of biomedical purposes, such as tissue engineering.

TABLE 1.2
An Overview of Auxetic Materials' Applications

Field	Application and Rationale
Aerospace	Thermal insulation, nose-cones for aircraft, engine vanes, vibration absorbers
Automotive	Bumper, cushion, thermal protection, sounds and vibration absorber parts, fastener
Biomedical	Bandage, wound pressure pad, dental floss, artificial blood vessel, artificial skin drug-release unit, ligament anchors, surgical implants
Composite	Fiber reinforcement (because it reduces the cracking between fiber and matrix)
Military (Defense)	Helmet, bulletproof vest, knee pad, glove, protective gear (better impact property)
Sensors/actuators	Hydrophone, piezoelectric devices, various sensors
Textile (Industry)	Fibers, functional fabric, color-change straps or fabrics

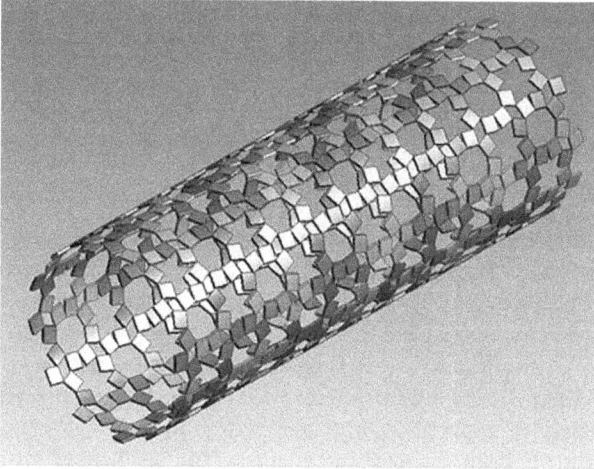

FIGURE 1.53   Hierarchical stent.

1.7.2   Application in Aerospace Engineering

The increased adoption of composite materials in aerospace engineering is being driven by the advantages they offer, such as high performance at a comparatively lower weight and cost. For example, since 1999, the tail of the American Boeing C-17 has been made from composites instead of aluminum, resulting in a substantial reduction in weight and the number of fasteners required. Nevertheless, conventional composites might not always be able to satisfy the strict criteria of aircraft engineering. This is why auxetic composites are preferred over traditional composites for aviation applications due to their benefits. One of the key qualities of auxetic composites is their higher shear modulus, which is essential for aircraft subjected to high levels of shear force during flight. Using materials with greater shear strength can improve aircraft performance. In the past, aircraft structures primarily consisted of heavy metals, but now, thanks to the development of auxetic technology, auxetic composites with a superior ratio of strength to weight can be used in the construction of aviation components, replacing the need for metals. Aircraft encounters with objects in the sky, such as birds, pose a significant risk during flight. Therefore, the aircraft body must have high impact resistance.

The engines attached to the wings of commercial aircraft generate excessive noise that can be transmitted to the fuselage via the stiffeners. This transmission of noise has a significant adverse effect on passenger comfort. To address this, a new type of auxetic composite was developed for the stiffener, consisting of an auxetic core, a damping layer, and a constraining layer. The auxetic core's surface should face the structural member, while the other surface should face away from it. A constraining layer can be placed on top of the second surface, with a damping layer in between. By utilizing these auxetic composites instead of traditional methods that involve using large layers of metallic material, the aircraft can save both energy and weight. Moreover, the increased formability of auxetic composites simplifies the process of fabricating intricate shapes and curved panels. This characteristic positions auxetic composites as a viable substitute for conventional composites in the production of aviation components in the upcoming years.

1.7.3 ADVANCED SMART SENSOR AND FILTER

Auxetic materials offer exciting possibilities for creating piezoceramic sensors, with the potential to greatly enhance the performance of piezoelectric actuators. The composites in figure are made of piezoelectric ceramic rods that are embedded in a passive polymer matrix. These composites facilitate the conversion of mechanical stress into electrical signals and, conversely, converting electrical signals back into mechanical stress. Many investigations have been conducted on Galfenol, which is also referred to as iron-gallium, an auxetic smart material possessing magneto-elastic properties. A material's ability to recover its original shape after being deformed plastically is known as shape memory. The mechanical characteristics of auxetic foams were found to be considerably influenced by the shape memory effect, and was introduced as an innovative concept for a smart cellular solid, suggesting a novel functional structure that combines the chiral honeycomb topology with shape memory alloys (SMA). Jacobs et al. created, produced, and tested a deployable SMA antenna that was inspired by an auxetic structure.

Smart filters can be designed by leveraging the advantageous permeability characteristics inherent in auxetic materials. Alderson et al. conducted basic studies that demonstrate the superior attributes of auxetic materials compared to conventional materials in applications such as filter defouling and the controlled regulation of pore sizes. By employing smart filters, it becomes feasible to effectively control passage pressure [3].

1.7.4 NAIL DESIGN

The authors also proposed the use of auxetic materials for nails. This proposition was based on the assumption that, when driven in, auxetic nails would become thinner, and when pulled out, they would grow thicker, as depicted in Figure 1.55. However, until recently, there had been a lack of research on auxetic nails. Recently, Ren et al., as illustrated in Figure 1.56, have pioneered the conception, creation, and

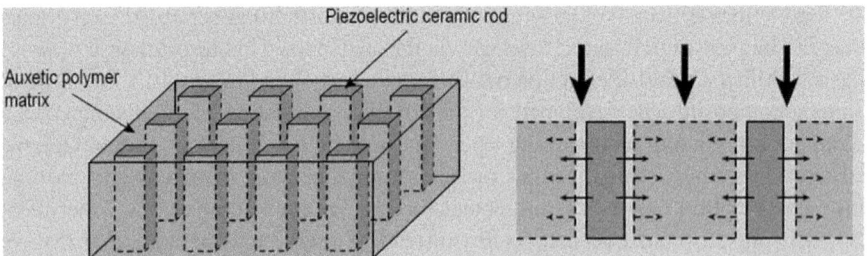

FIGURE 1.54   Sensor with auxetic and piezoelectric properties.

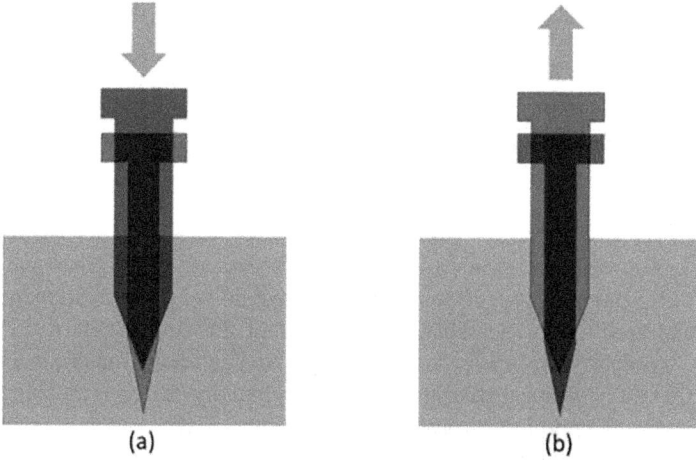

FIGURE 1.55   An example of auxeticity for auxetic nails is shown in (a) the push-in and (b) the pull-out. (The grey and red nails show the nail configurations before and after deformation, respectively.)

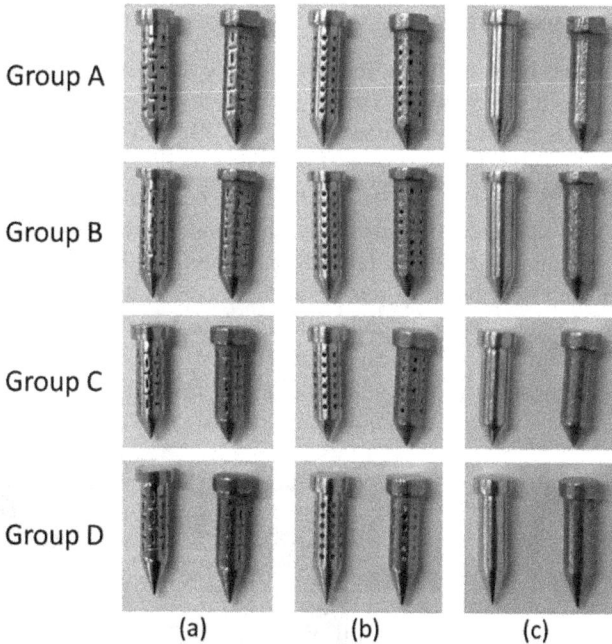

FIGURE 1.56   Twelve different nail types, divided into four groups, were 3D printed using brass and stainless steel materials: (a) auxetic nails (ANs); (b) nails with circular holes (CNs); and (c) solid nails (SNs). (The nails with the gold and silver colors were printed using brass and stainless steel, respectively.) (scaling bar: 10 mm)

experimental investigation of the first-ever auxetic nails. The research revealed that auxetic nails generally do not typically demonstrate a notable advantage compared to conventional nails in terms of push-in and pull-out capabilities. Challenges still remain in the development and design of metallic auxetic nails that can demonstrate improved performance in push-in and pull-out scenarios [10].

1.7.5 TEXTILE

Auxetic textile materials have gained significant traction due to their capacity to offer comfort, improved energy absorption, substantial volume alteration, resistance to wear, and excellent drapeability, leading to their growing popularity. There are generally two main methods for manufacturing auxetic textile materials. First, one can directly knit or weave textiles using fibers specifically designed for auxetic properties. The second approach involves weaving or knitting regular fibers in a way that can result in the production of auxetic textiles, which is known as "auxetic weaving." Recently, a newly identified 56-type 3D fabric with auxetic characteristics was recognized for exhibiting auxetic behavior in all directions within the fabric plane. Figure 1.58 shows several sports shoes that the Under Armour and Adidas companies have unveiled, as well as auxetic textiles that have just entered the market, like Gore-Tex and polytetrafluorethylene.

1.7.6 FABRICATION METHODS

The NPR effect has been studied at the macroscopic, mesoscopic, microscopic, and even molecular levels because the theory of elasticity is scale-independent. Many conceptual frameworks were created as a result. Only a small number of NPR material types, including skeleton materials, materials resembling foam, composite laminates, and a specific helical yarn, have been successfully manufactured. Since they were created using existing materials, they may typically be produced using common component materials and fabrication techniques. Numerous NPR materials have been proposed, but they remain at the theoretical phase due to the challenges involved in manufacturing their intricate microstructures, particularly when it comes to 3D disordered structures. As current manufacturing processes advance, additive manufacturing approaches may offer a promising way to create intricate

(a)   (b)

FIGURE 1.57   Athletes' shoes with (a) auxetic skin and (b) auxetic sole.

NPR structures. Indeed, some attempts have already been made at both the macro- and micro-scales, but the works are still restricted to periodic skeleton structures, which are typically two-phase composite systems. In reality, the most recent additive manufacturing processes may create a component using several materials, which suggests that it is now possible to create 3D disordered, graded, multi-phase meta- materials [10].

1.7.7 CHALLENGES OF AUXETIC MATERIALS

To begin with, the manufacturing expense of auxetic materials remains prohibi- tively high. The majority of 3D auxetic materials, which exhibit consistent auxetic behavior, are currently being developed through 3D printing technology. While 3D printing enables efficient fabrication and allows engineers to focus on design rather than manufacturing processes, there is a significant challenge in achieving mass pro- duction at a lower cost. This limitation greatly hinders the extensive use of auxetic materials.

Additionally, all examined auxetic materials and structures display a notable amount of porosity in their geometric configuration. This porosity adversely affects their mechanical strength when subjected to loads or impacts. In simple terms, when comparing auxetic materials to solid materials, auxetic materials initially trade off mechanical performance to obtain desirable characteristics, for instance, improved shear resistance, greater indentation resistance, and enhanced energy absorption. Consequently, in most cases, the achieved auxetic behavior is insufficient to com- pensate for the mechanical performance lost due to the porous microstructure of auxetic materials. This reduction in benefits, particularly for the purpose of energy absorption or protective devices, significantly undermines the practicality of auxetic behavior. However, this issue is seldom mentioned because auxetic materials lack a porous microstructure.

While the characteristics of auxetic materials are highly desirable, their practical implementation is not especially viable, particularly when taking into account cost and performance factors. Additionally, these advantageous traits of auxetic materi- als are not actually necessary in many applications. This helps explain why, despite the creation and manufacture of a certain quantity of auxetic materials, only a small fraction of them have been effectively utilized.

REFERENCES

1. Hou, X. & Silberschmidt, V. V., 2015. Metamaterials with negative poisson's ratio: A review of mechanical properties and deformation mechanisms. *Mechanics of Advanced Materials: Analysis of Properties and Performance*, 155–179.
2. Lim, T. C., 2015. *Auxetic Materials and Structures* (Vol. 11, pp. 383–387). Singapore: Springer.
3. Kolken, H. M. & Zadpoor, A. A., 2017. Auxetic mechanical metamaterials. *RSC Advances*, 7(9), pp. 5111–5129.
4. Zhai, Z., Wu, L. & Jiang, H., 2021. Mechanical metamaterials based on origami and kirigami. *Applied Physics Reviews*, 8(4).

5. Gaspar, N., Smith, C. W., Alderson, A., Grima, J. N. & Evans, K. E., 2011. A generalised three-dimensional tethered-nodule model for auxetic materials. *Journal of Materials Science*, 46, pp. 372–384.

6. Grima, J. N. & Evans, K. E., 2006. Auxetic behavior from rotating triangles. *Journal of Materials Science*, 41(10), pp. 3193–3196.

7. Spadoni, A. & Ruzzene, M., 2012. Elasto-static micropolar behavior of a chiral auxetic lattice. *Journal of the Mechanics and Physics of Solids*, 60(1), pp. 156–171.

8. Chen, Y. J., Scarpa, F., Liu, Y. J. & Leng, J. S., 2013. Elasticity of anti-tetrachiral anisotropic lattices. *International Journal of Solids and Structures*, 50(6), pp. 996–1004.

9. Lv, C., Krishnaraju, D., Konjevod, G., Yu, H. & Jiang, H., 2014. Origami based mechanical metamaterials. *Scientific Reports*, 4(1), p. 5979.

10. Ren, X., Das, R., Tran, P., Ngo, T. D. & Xie, Y. M., 2018. Auxetic metamaterials and structures: A review. *Smart Materials and Structures*, 27(2), p. 023001.

2 Static Analysis of Auxetic Structures

2.1 BACKGROUND

In contrast to conventional solid materials, there exists a distinct category of substances that possess a unique property of inducing a negative Poisson's ratio, specifically in one lateral direction. The utilization of auxetic nanocomposites offers the potential to augment the rigidity of structural components while concurrently diminishing their mass. The assessment of ultra-light structures can be simplified by employing metamaterials that demonstrate geometry-dependent properties. Given this justification, the present chapter is organized to investigate the nonlinear bending characteristics displayed by slender beams made of auxetic nanocomposites that have been enhanced with carbon nanotubes (CNTs). The potential to produce meta-nanomaterials that are exceptionally lightweight can be achieved through the combination of outstanding characteristics present in polymer nanocomposites (PNCs) and metamaterials. These materials exhibit enhanced sensitivity, rendering them suitable for utilization in wearable strain sensors. Hence, it is crucial to acquire reliable data pertaining to the performance of structural components that are manufactured using meta-nanomaterials before their design is undertaken. In addition to the objective of weight reduction, there are frequently circumstances that necessitate the development of structural components capable of withstanding substantial external forces.

To simultaneously achieve both objectives, it is necessary to employ auxetic nanocomposite structures as a viable approach. Given this particular concern, a thorough investigation is conducted to analyze the detrimental impacts of pores within the microstructure and the undulating arrangement of CNTs on the nonlinear bending properties of thick beams constructed from auxetic CNT-reinforced polymers. This study constitutes the initial investigation into these phenomena. The utilization of the Eshelby-Mori-Tanaka homogenization scheme is employed in order to analyze the impact of nanotube entanglement within the matrix on the manipulation of beam deflection. This chapter also examines the impact of varying porosity distributions on the mechanical behavior of the system. Moreover, this research integrates a computationally efficient homogenization method to consider the influence of waviness on the reinforcement mechanism in the auxetic PNC material. The problem at hand is characterized by the integration of the Euler-Bernoulli theory, which pertains to thin beams, and the higher-order shear deformation theory (HSDT). This integration involves the inclusion of a nonlinear definition of the strain tensor, as prescribed by von Kármán's theory. Following this, the well-known Navier's exact solution is employed to tackle the problem concerning thick beams with simple supports at both ends. Moreover, the present chapter delves into the investigation of the post-buckling characteristics exhibited by slender beams fabricated from auxetic CNTR nanocomposites. The determination of the effective modulus of meta-nanocomposites is achieved through the utilization of a bi-stage micromechanical homogenization technique.

The objective of this approach is to assess the impact of CNT agglomerates on the determination of modulus. Afterwards, the strain-displacement relations developed by von Kármán will be utilized in combination with the Euler-Bernoulli beam theory to calculate the nonlinear strain of the continuous system. The governing equation of the problem will be derived through the application of the principle of virtual work, taking into account its nonlinear nature. The subsequent procedure involves the utilization of the analytical method proposed by Galerkin to determine the nonlinear buckling load of auxetic nanocomposite beams with simply supported and clamped ends. Following this, a collection of representative case studies is presented with the aim of providing a point of reference. This chapter highlights the primary focal points concerning the improvement of the meta-nanocomposite beam's ability to withstand buckling-mode failure. This is achieved by meticulously choosing small auxeticity angles. Furthermore, empirical data suggests that the utilization of a wide auxetic lattice leads to a compromise in structural integrity, especially when exposed to smaller buckling loads. Furthermore, previous studies have shown that the buckling resistance of the auxetic nanocomposite beam can be affected by the presence of the agglomeration phenomenon.

The results of the current investigation illustrate the notable impact of the auxeticity angle on the extent of deflection. Smaller angles are correlated with lower deflections. Moreover, based on empirical evidence, it is not recommended to employ wide auxetic lattices for the purpose of minimizing the deflection of continua. When seeking to minimize the deflection of a system, it is highly recommended to utilize metamaterials with increased lattice thicknesses in a reversed configuration. The findings also indicate a notable augmentation in the flexural deformation of the auxetic PNC beam as the coefficient of porosity is raised. Furthermore, previous studies have provided evidence that the integration of cells possessing substantial wall thickness within the re-entrant lattice framework can proficiently govern the overall deflection of the system. In addition, disregarding the non-ideal morphology of the nanofillers would lead to a considerably larger deflection of the meta-nanomaterial beam in comparison to the deflection anticipated by ideal calculations. The findings derived from this investigation can be utilized as benchmarks for the advancement of advanced strain sensors employing polymer nanocomposites.

2.2   AUXETIC NANOCOMPOSITE BEAMS: STATIC ANALYSIS

2.2.1   Bending Characteristics of Auxetic Nanocomposite Beams

This chapter will employ a polymer nanocomposite reinforced with CNTs in an auxetic arrangement as the constituent material. The nanocomposite material employed in this study features a re-entrant lattice structure, with epoxy serving as the host matrix. The nanoparticles employed for enhancing the epoxy matrix consist of single-walled CNTs. To investigate the impact of bundles of CNTs on the stiffness of the nanocomposite, we make the assumption that the bonding between the polymer and CNTs is ideal. Hence, it is postulated that the likelihood of the fiber debonding phenomenon is negligible. In recent years, nanocomposites have emerged as highly favorable options for the development of advanced structural elements due to their exceptional properties [1]. These highly rigid nanomaterials have the capability to exhibit significant natural frequencies [2–4], withstand extreme compressive loads [5–12], and reduce deflection amplitudes [13–23] in continuous systems.

In this study, a two-step algorithm will be presented for determining the effective material properties of the meta-nanocomposite. Let us consider the CNTR polymer nanocomposite beam as having a re-entrant lattice shape. In order to enhance comprehension of the geometric characteristics of the lattice, it is advisable to refer to Figure 2.1. In the magnified depiction of this diagram, it is possible to readily ascertain the geometric characteristics of the auxetic unit cell. The angle θ is commonly

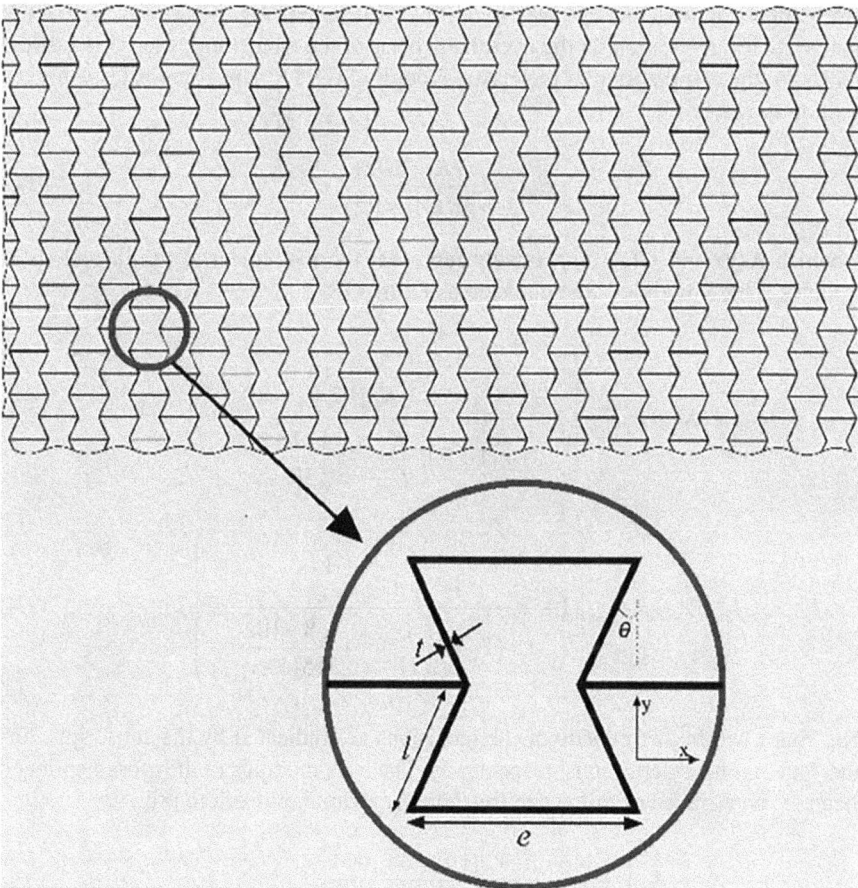

FIGURE 2.1 A schematic representation of the re-entrant lattice utilized in the study. The zoomed view also displays the geometrical parameters relevant to the determination of material properties.

referred to as the auxeticity angle. Furthermore, the vertical and inclined lengths of the cell rib are denoted as e and l, respectively. The cellular lattice exhibits a visual representation of the thickness t of its constituent cell walls.

To facilitate the process of homogenization, a retrogressive methodology will be employed. The effective stiffness of a polymer nanocomposite is denoted as E_{PNC}. To determine the effective stiffness of an auxetic lattice made from this material in the longitudinal direction, the following formula can be utilized [8]:

$$E_{eff} = E_{PNC} (t/l)^3 \frac{\cos\theta}{(e/l + \sin\theta)\sin^2\theta} \tag{2.1}$$

in which t, l, e, and θ are geometrical parameters governing the degree of auxeticity in the nanomaterial. In the paragraph before that, the definitions of these parameters were given. It is required to get the exact parameter for the CNTR nanocomposite in order to calculate the elasticity modulus using the relationship given. The Eshelby-Mori-Tanaka approach is one way to do this while capturing the negative impact of nanoparticle aggregation on the overall stiffness of the nanocomposite. This method allows for the computation of the elastic moduli of CNTR nanocomposites using the relation shown in [8]:

$$E_{PNC} = \frac{9K(z)G(z)}{3K(z)+G(z)} \tag{2.2}$$

in which $K(z)$ and $G(z)$ respectively represent the nanomaterial's bulk and shear modules. These modules are enhanced as follows [8]:

$$K(z) = K_{out} \left(1 + \frac{\mu\left(\frac{K_{in}}{K_{out}} - 1\right)}{1 + (1-\mu)\left(\frac{K_{in}}{K_{out}} - 1\right)\frac{1+v_{out}}{3(1-v_{out})}} \right) \tag{2.3}$$

$$G(z) = G_{out} \left(1 + \frac{\mu\left(\frac{G_{in}}{G_{out}} - 1\right)}{1 + (1-\mu)\left(\frac{G_{in}}{G_{out}} - 1\right)\frac{8-10v_{out}}{15(1-v_{out})}} \right) \tag{2.4}$$

The zones within and outside of the inclusions are indicated by the subscripts "in" and "out" in these definitions, respectively. The bulk modulus in the aforementioned locations is retrieved as follows in this homogenization procedure [8]:

$$\begin{aligned} K_{in}(z) &= K_m + \frac{V_r\eta(\delta_r - 3K_m\alpha_r)}{3(\mu - V_r\eta + V_r\eta\alpha_r)}, \\ K_{out}(z) &= K_m + \frac{V_r(1-\eta)(\delta_r - 3K_m\alpha_r)}{3(1-\mu - V_r(1-\eta)(\alpha_r - 1))} \end{aligned} \tag{2.5}$$

Additionally, [8] is used to calculate the shear modulus in the same zones:

$$G_{in}(z) = G_m + \frac{V_r \eta \left(\eta_r - 2G_m \beta_r \right)}{2 \left(\mu - V_r \eta + V_r \eta \beta_r \right)},$$

$$G_{out}(z) = G_m + \frac{V_r \left(1 - \eta \right) \left(\eta_r - 2G_m \beta_r \right)}{3 \left(1 - \mu - V_r \left(1 - \eta \right) + V_r \left(1 - \eta \right) \beta_r \right)} \tag{2.6}$$

where α_r, β_r, δ_r, and η_r are stiff components whose definitions are presented in Appendix 2A. Besides, η and μ are known as agglomeration parameters whose main purpose is to govern the nanoparticles' distribution in nanocomposites' composition. The term η is defined as the volume fraction of the CNTs bundled within the inclusions, whereas μ denotes the inclusions' volume fraction. It is worth regarding that the algebraic inequity $\mu \leq \eta$ is always valid. Plus, the Poisson's ratio corresponding with the outside region is computed with the following formula [8]:

$$V_{out} = \frac{3K_{out} - 2G_{out}}{6K_{out} + 2G_{out}} \tag{2.7}$$

In addition, the volume fraction of the reinforcing nanofillers (V_r) is achieved with the aid of the following identity [8]:

$$V_r(z) = \left[\frac{\rho_r}{w_r \rho_m} - \frac{\rho_r}{\rho_m} + 1 \right]^{-1} \left(\frac{z}{h} + \frac{1}{2} \right)^p \tag{2.8}$$

where ρ_m and ρ_r respectively denote the matrix and reinforcing elements corresponding mass densities. Also, the gradient index covering the geometry-dependency of the nanocomposite's stiffness is depicted with p. The nanoparticles' mass fraction (w_r) is calculated through [8]:

$$w_r = \frac{M_r}{M_r + M_m} \tag{2.9}$$

The motion of the structure's equation will be derived here. The structure is made of a beam with a rectangular cross-section, which is shown in Figure 2.2 with the symbols L, b, and h for length, width, and thickness.

In this section, we will investigate the mechanical properties of slender beams with high slenderness ratios. Hence, the impact of shear-induced deformation on the system's reaction to mechanical stimuli is not accounted for [13]. For this reason, the classical beam model is utilized in formulating the motion of the structure. In

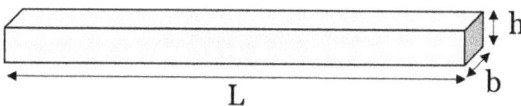

FIGURE 2.2 Schematic view of a beam element and its cross-section.

this theoretical framework, the displacement field of the beam is considered to be as stated in reference [14].

$$u_x(x,z) = u(x) - \left(z - z^*\right)\frac{\partial w(x)}{\partial x},$$

$$u_y(x,z) = 0, \tag{2.10}$$

$$u_z(x,z) = w(x)$$

The longitudinal movement and through-the-thickness deflection of the neutral axis of the continua are denoted as $u(x)$ and $w(x)$, respectively. In this analysis, the precise position of the neutral axis of the beam is being considered. In order to ascertain the location of this particular point, the subsequent definition is employed:

$$z^* = \frac{\int_{-h/2}^{h/2} zE_{eff}\,dz}{\int_{-h/2}^{h/2} E_{eff}\,dz} \tag{2.11}$$

In order to determine the nonlinear normal strain of the beam, it is necessary to utilize equation (2.10). By following this procedure, the nonlinear strain in the longitudinal direction can be obtained as stated in reference [14]:

$$\varepsilon_{xx} = \frac{\partial u}{\partial x} + \frac{1}{2}\left(\frac{\partial w}{\partial x}\right)^2 - \left(z - z^*\right)\frac{\partial^2 w}{\partial x^2} \tag{2.12}$$

The initial two elements of the aforementioned equation represent longitudinal elongation, while the last term quantifies the curvature of the beam. By employing the principle of virtual work, it is possible to derive the equations of motion. According to this principle, any random fluctuation in the total energy of the structure should be equal to zero [13].

$$\delta(U+V) = 0 \tag{2.13}$$

The beam's internal energy and the work done on the structure by external loading are denoted as U and V, respectively. The variation of strain energy can be obtained by applying the constitutive equation of linearly elastic solids and referring to equation (2.12) [14].

$$\delta U = \int_0^L \left\{ N\left[\frac{\partial \delta u}{\partial x} + \frac{\partial w}{\partial x}\frac{\partial \delta w}{\partial x}\right] - M\frac{\partial^2 \delta w}{\partial x^2} \right\} dx \tag{2.14}$$

where the stress resultants N and M are:

$$\begin{bmatrix} N \\ M \end{bmatrix} = \int_A \begin{bmatrix} 1 \\ z - z^* \end{bmatrix} \sigma_{xx}\,dA \tag{2.15}$$

Currently, it is necessary to formulate the expression for the variation in the work performed by the external loads. Let us consider a predetermined bending load q that is exerted on the upper surface of the beam. Hence, the expression for the fluctuation in the work performed by the load on the system can be represented as [13].

$$\delta V = \int_0^L q \delta w \, dx \qquad (2.16)$$

If equations (2.14) and (2.16) are substituted into equation (2.13) and the coefficients of u and w are assumed to be zero, the resulting Euler-Lagrange relations can be obtained.

$$\frac{\partial N}{\partial x} = 0 \qquad (2.17)$$

$$\frac{\partial}{\partial x}\left\{N\frac{\partial w}{\partial x}\right\} - \frac{\partial^2 M}{\partial x^2} + q = 0 \qquad (2.18)$$

The equations governing the motion of a beam were previously derived. Currently, there is a need to obtain suitable expressions for the normal force (N) and the bending moment (M) in order to establish the governing equations. The implementation of the relation for the auxetic nanocomposite beam in isotropic solids is in accordance with Hook's law [13].

$$\sigma_{xx} = E_{eff}\varepsilon_{xx} \qquad (2.19)$$

By performing the integration of the aforementioned definition across the cross-sectional area of the system, while also taking into account the established definitions of N and M, it becomes straightforward to express the following:

$$N = A_{xx}\left[\frac{\partial u}{\partial x} + \frac{1}{2}\left(\frac{\partial w}{\partial x}\right)^2\right] \qquad (2.20)$$

$$M = -D_{xx}\frac{\partial^2 w}{\partial x^2} \qquad (2.21)$$

In the above relations, A_{xx} and D_{xx} respectively stand for the rigidities of extensional and bending types. These terms are defined as:

$$\begin{bmatrix} A_{xx} \\ D_{xx} \end{bmatrix} = \int_A \begin{bmatrix} 1 \\ \left(z-z^*\right)^2 \end{bmatrix} E_{eff} \, dA \qquad (2.22)$$

By examining equation (2.17), it can be deduced that the value of N remains constant. Hence, it is possible to express equation (2.18) in an alternative form as follows:

$$N\frac{\partial^2 w}{\partial x^2} - \frac{\partial^2 M}{\partial x^2} + q = 0 \qquad (2.23)$$

By substituting equations (2.20) and (2.21) into equation (2.23), the following expression is obtained:

$$A_{xx}\left[\frac{\partial u}{\partial x}+\frac{1}{2}\left(\frac{\partial w}{\partial x}\right)^2\right]\frac{\partial^2 w}{\partial x^2}+D_{xx}\frac{\partial^4 w}{\partial x^4}+q=0 \qquad (2.24)$$

The aforementioned equation provides a description of the motion of the structure. Given the assumption that the constant term is of significant magnitude c, we can express this as follows:

$$\frac{\partial u}{\partial x}=\frac{c}{A_{xx}}-\frac{1}{2}\left(\frac{\partial w}{\partial x}\right)^2 \rightarrow u(x)=\frac{cx}{A_{xx}}-\frac{1}{2}\int_0^x\left(\frac{\partial w}{\partial x}\right)^2 dx+d \qquad (2.25)$$

The unknown variable, denoted as d, will be determined through the utilization of boundary constraints. The existing configuration is considered to be supported in a simply supported manner, with axially static supports. Therefore, it can be asserted:

$$u(0)=0 \rightarrow d=0 \qquad (2.26)$$

$$u(L)=0 \rightarrow c=\frac{A_{xx}}{2L}\int_0^L\left(\frac{\partial w}{\partial x}\right)^2 dx \qquad (2.27)$$

Therefore, it is permissible to assert:

$$\frac{\partial u}{\partial x}+\frac{1}{2}\left(\frac{\partial w}{\partial x}\right)^2=\frac{1}{2L}\int_0^L\left(\frac{\partial w}{\partial x}\right)^2 dx \qquad (2.28)$$

Upon substituting the aforementioned identity into equation (2.24), it is evident that:

$$\frac{A_{xx}}{2L}\int_0^L\left(\frac{\partial w}{\partial x}\right)^2 dx\frac{\partial^2 w}{\partial x^2}+D_{xx}\frac{\partial^4 w}{\partial x^4}+q=0 \qquad (2.29)$$

In this study, we aim to address the issue of nonlinear bending in auxetic nanocomposite beams. To date, numerous numerical and analytical techniques have been employed by scholars to investigate the static and dynamic behaviors of composite structures [15]. In this study, we utilize the Navier method, known for its efficiency, to address the nonlinear bending phenomenon observed in auxetic nanocomposite beams. The bending deflection of a fully simply supported beam will be expressed in a multi-term form, as stated by the method referenced in [15].

$$w(x)=\sum_{n=1}^{\infty}W_n\sin\lambda x, \qquad \lambda=\frac{n\pi}{L} \qquad (2.30a)$$

The axial mode number and deflection amplitude corresponding to the nth mode are denoted as n and W_n, respectively. Moreover, it is assumed that the bending load applied to the continuous system conforms to the expression presented in reference [15].

$$q(x) = \sum_{n=1}^{\infty} Q_n \sin \lambda x \qquad (2.30\text{b})$$

In accordance with the aforementioned definition, the loading amplitude can be attained by employing the subsequent formula:

$$Q_n = \frac{2}{L} \int_0^L q(x) \sin \lambda x \, dx \qquad (2.30\text{c})$$

This chapter will examine three loading mechanisms, namely uniform, sinusoidal, and concentrated. Hence, the loading amplitude for each of the aforementioned loadings can be derived by implementing the definition provided in reference [15].

$$Q_n = q_0 \ (n=1), \ \text{Sinusoidal Load} \ q_0$$

$$Q_n = \frac{4q_0}{n\pi} (n=1,3,5,\cdots), \ \text{Uniform Load} \ q_0 \qquad (2.30\text{d})$$

$$Q_n = \frac{2q_0}{L} \sin\left(\frac{n\pi}{2}\right)(n=1,2,3,\cdots), \ \text{Point Load} \ q_0 \ \text{at} \ x = \frac{L}{2}$$

After substituting equations (2.30a) and (2.30b) into equation (2.29), the resulting identity can be obtained:

$$-\frac{A_{xx}}{2L}\sum_{n=1}^{\infty} W_n^3 \lambda^4 \int_0^L \cos^2 \lambda x \, dx \sin \lambda x + D_{xx}\sum_{n=1}^{\infty} W_n \lambda^4 \sin \lambda x$$
$$+\sum_{n=1}^{\infty} Q_n \sin \lambda x = 0 \qquad (2.30\text{f})$$

Upon conducting mathematical simplifications, the aforementioned identity can be reformulated in the subsequent manner:

$$\sum_{n=1}^{\infty} K_n^L W_n + \sum_{n=1}^{\infty} K_n^{NL} W_n^3 = \sum_{n=1}^{\infty} Q_n \qquad (2.30\text{g})$$

The linear and nonlinear stiffnesses associated with any arbitrary mode number are denoted as K_n^L and K_n^{NL}, respectively. By algebraically manipulating equation (2.30g), one can obtain the deflection amplitude at any natural mode. Subsequently, the aggregation of the deflections of the natural modes of the system provides us with the comprehensive bending deflection of the auxetic nanocomposite beam.

This section aims to present a series of illustrative case studies that will be utilized to examine the nonlinear bending behavior of a thin beam composed of auxetic CNTR nanocomposites. It is important to note that in this particular section, the length-to-thickness ratio of the beam is assumed to be equal to 100, unless a different value is explicitly stated. The rationale behind this decision is to fulfill the initial assumption that the analyzed beam is thin. Furthermore, in the absence of any other specified information, it is assumed that the vertical-to-inclined length, thickness-to-inclined length, and angle of auxeticity of the cell rib in the lattice are denoted as $e/l = 4$, $t/l = 0.0138571$, and $\theta = -50$ respectively. In addition, the auxetic nanocomposite structure will experience bending loads with a specific amplitude $q_0 = 10$ kPa. In the subsequent discussion, the dimensionless expression for the deflection of the beam is observed in order to simplify the analysis. The dimensionless deflection in this study is postulated to be:

$$W = \frac{wE_m h^3}{q_0 L^4} \tag{2.30h}$$

The initial stage involves the verification of the existing modeling's validity within the context of Figure 2.3. The present figure depicts the relationship between the nondimensional, nonlinear deflection of polymer nanocomposite beams reinforced with graphene platelets (GPLs) and the dimensionless load. The present approach effectively reproduces the nonlinear deflection curve as depicted in the illustration, as reported in reference [17]. It is evident that the curves exhibit a high degree of proximity, with only minute variations observed in large dimensionless load values.

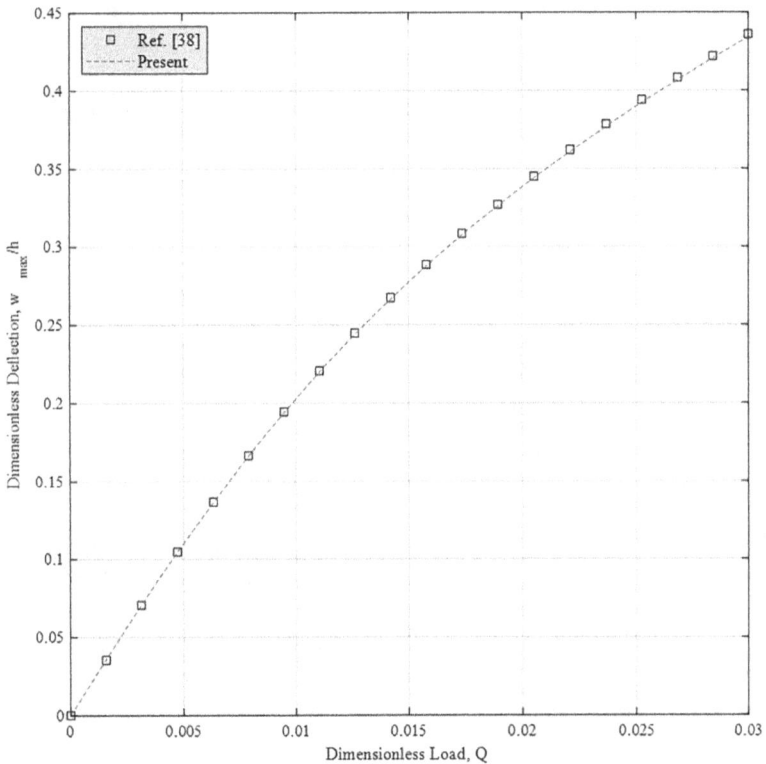

FIGURE 2.3 Comparison of the nonlinear bending response of polymer nanocomposite beams reinforced with uniformly distributed GPLs (f_i=0.5%, N=10, L/h=20).

The disparity arises due to the greater significance of shear deformation deflection in the overall deformation of the continuous system in such instances. In light of the model's foundation in the classical theory of thin beams, it is anticipated that the deflections it predicts will be smaller when compared to those obtained using the Timoshenko beam theory [17] under similar conditions. This phenomenon can be attributed to the more rigid and inflexible nature of classical theories in contrast to first-order or higher-order theories. However, it is important to note that this curve is plotted for beams of moderate thickness with $L/h = 20$. In future case studies, however, we will focus on the analysis of slender beams. Therefore, the information presented in this text can be deemed reliable.

The primary objective of showcasing Figure 2.4 is to observe and analyze the combined effects of the gradient index and angle of auxeticity on the Young's modulus

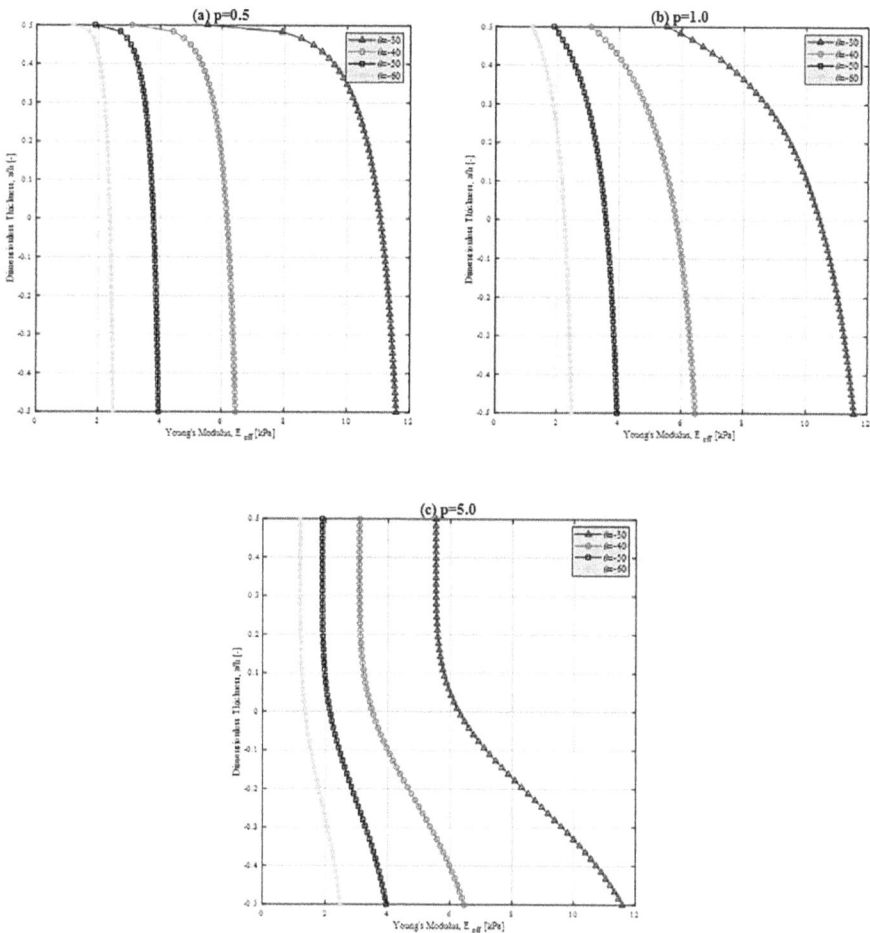

FIGURE 2.4 Variation of the Young's modulus of the auxetic CNTR nanocomposite material across the thickness of the structure for (a) p=0.5, (b) p=1.0, and (c) p=5.0 while different auxeticity angles are considered (η=1, μ=0.4, w_r=0.1).

of the constituent auxetic nanomaterial. Based on the data presented in the figure, it can be observed that the selection of a softer auxetic nanocomposite for theoretical calculations is contingent upon assigning a higher magnitude to the gradient index. This phenomenon can be observed across various thickness levels. Furthermore, this figure illustrates the detrimental effects of incorporating the auxeticity angle on the stiffness properties of the meta-nanocomposite. In conclusion, it can be stated that the softest behavior is observed when the highest values are assigned to both the gradient index and auxeticity angle.

Figure 2.5 is dedicated to examining the fluctuation in the deflection of the beam as the agglomeration parameter is altered, while also considering various loadings. It is evident that concentrated point loading induces the highest deflection in the meta-nanomaterial beam, while sinusoidal and uniform loadings produce lower deflections in descending order.

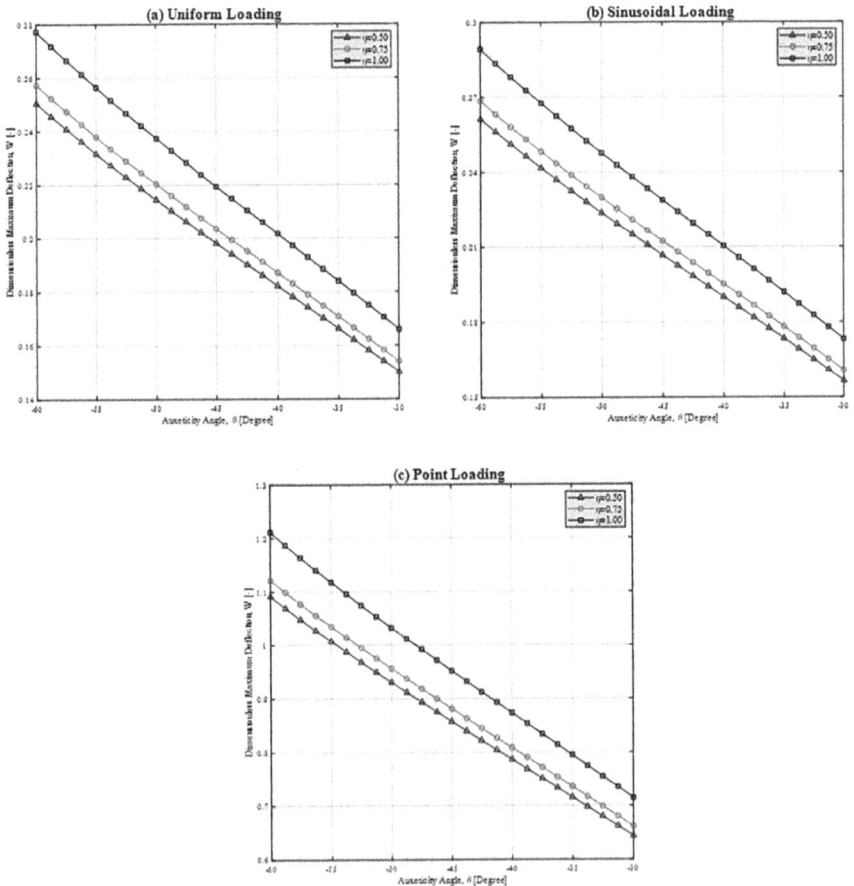

FIGURE 2.5   Influence of volume fraction of the CNTs inside the clusters on the variation of the maximum deflection of auxetic PNC beams subjected to (a) uniform, (b) sinusoidal, and (c) point loading ($\mu=0.4$, $w_r=0.1$, $p=5$).

Furthermore, it is evident that the magnitude of deflection will be amplified at any desired angle of auxeticity when the proportion of CNTs aggregated within the clusters is heightened. The underlying physical explanation for this observed phenomenon can be attributed to the detrimental effects of entanglement between CNTs on the reinforcing mechanism within polymer nanocomposites. Consequently, the detrimental effect on the comparable stiffness of the auxetic nanocomposite leads to an increase in the deflection of the continuous material due to the inverse relationship between deflection and stiffness in linearly elastic solids. Moreover, it can be demonstrated that auxetic CNTR nanocomposite beams exhibiting broader auxeticity angles will undergo more significant deflections. The primary factor contributing to this phenomenon can be readily comprehended by consulting Figure 2.4. This trend has emerged due to the diminishing impact of the auxeticity angle on the stiffness of meta-nanomaterials across various thicknesses.

In contrast, Figure 2.6 highlights the impact of the volume fraction of clusters α, which is another parameter of agglomeration, on the deflection response of auxetic nanocomposite beams subjected to uniform loading. Based on the data presented in the figure, it is evident that an increase in the agglomeration parameter results in

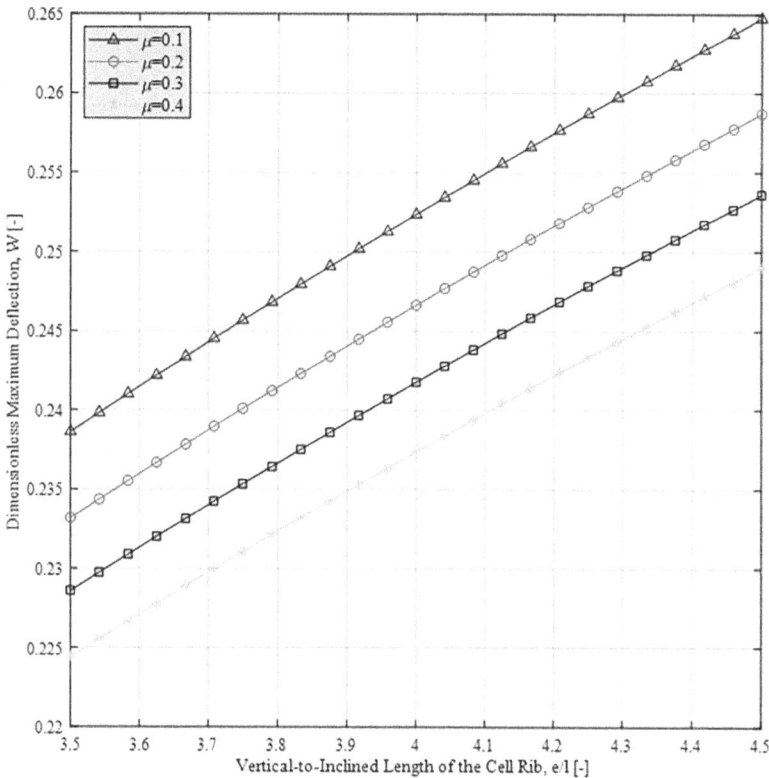

FIGURE 2.6 Influence of volume fraction of the clusters on the variation of the maximum deflection of auxetic PNC beams subjected to uniform loading ($\eta=1$, $w_r=0.1$, $p=5$).

a reduction in the deflection amplitude of the auxetic CNTR beam. The observed decline in the amplitude of the beam's deflection can be attributed to the beneficial impact of incorporating the ratio of inclusions on the stiffness of the meta-nanomaterial constituents. To clarify, by increasing the assigned values $\propto$, the preexisting clusters will experience an expansion in their spatial extent. Consequently, the auxetic nanocomposite will approach a state wherein a significant cluster is formed within its microstructure, thereby entangling all nanoparticles within it. While the agglomeration of CNTs typically has a detrimental effect on the stiffness of nanocomposites, its influence in this particular scenario will be mitigated. Hence, it is inherent that the deflection of the beam is diminished when considering the inverse correlation between deflection and stiffness. Additionally, this diagram illustrates the promotion of utilizing re-entrant lattices featuring wide cells to enhance the deflection amplitude of the beam. It is important to consider that the utilization of narrow meta-nanomaterials is advisable when the objective is to reduce the magnitude of flexural deformation in a structure.

As depicted in the preceding illustration, Figure 2.7 displays the fluctuation in deflection amplitude of a beam composed of meta-nanomaterial when subjected to

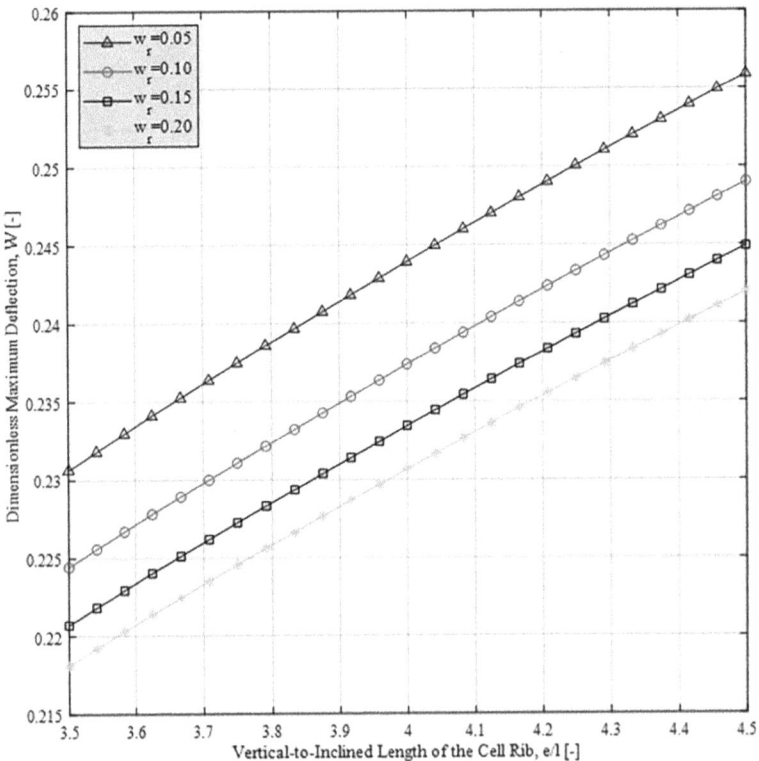

FIGURE 2.7 Influence of mass fraction of the CNTs on the variation of the maximum deflection of auxetic PNC beams subjected to uniform loading ($\eta=1$, $\mu=0.4$, $p=5$).

uniform loading. In the present diagram, however, the mass fraction of CNTs has been altered in order to investigate the influence of this particular parameter on the nonlinear bending deflection of the beam-type element. Based on the information presented in the image, it can be inferred that an increase in the proportion of reinforcing elements utilized in the composition of the CNTR meta-nanocomposite beam may result in a reduction in deflection amplitude. The primary factor contributing to this phenomenon is the exceptional rigidity exhibited by CNTs. In essence, the nanofillers can be described as highly rigid nanomaterials. Their unique characteristic enables them to enhance the overall stiffness of the auxetic nanomaterial when a small quantity of them is incorporated into the composition of the meta-nanomaterial. If one recalls the inverse relationship between deflection and stiffness, it appears reasonable to expect the observation of such a trend. Furthermore, it has been established that the magnitude of deflection increases when a larger unit cell is chosen for the design of the meta-nanomaterial component.

Figure 2.8 examines the combined effects of cell rib thickness and gradient index on the bending deflection of auxetic CNTR nanocomposite beams under uniform

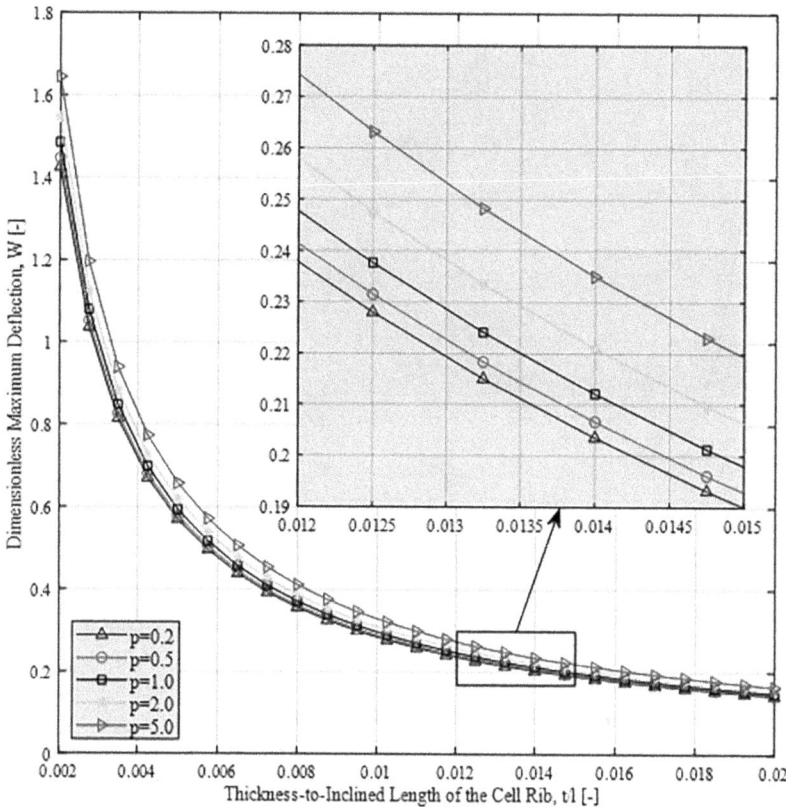

FIGURE 2.8 Influence of gradient index on the variation of the maximum deflection of auxetic PNC beams subjected to uniform loading ($\eta=1, \mu=0.4, w_r=0.1$).

traction. The figure illustrates that the inclusion of a gradient index leads to a notice-able augmentation in the nonlinear bending deflection of the meta-nanomaterial beam. The underlying justification for this matter pertains to the diminishing influ-ence of the gradient index on the rigidity of the auxetic nanocomposite, as depicted in Figure 2.4. Additionally, the presented image demonstrates the significant influ-ence of the wall thickness of the auxetic lattice composed of meta-nanomaterial on the deflection of the beam. It is observed that thicker lattices possess the capa-bility to effectively mitigate the deformation of the beam. According to the pro-vided diagram, it can be inferred that the designed auxetic nanocomposite structure exhibits enhanced resistance to bending stimuli across a range of deformation mag-nitudes. The observed phenomenon can be attributed to the strong positive relation-ship between the thickness of the walls in the auxetic lattice and the rigidity of the meta-nanomaterial. The designed element exhibits increased stiffness, resulting in a decrease in the flexural deformation of the beam.

As the concluding case study, Figure 2.9 is provided to examine the influence of the degree of carbon nanotubes' entanglement within the inclusions on the nonlinear

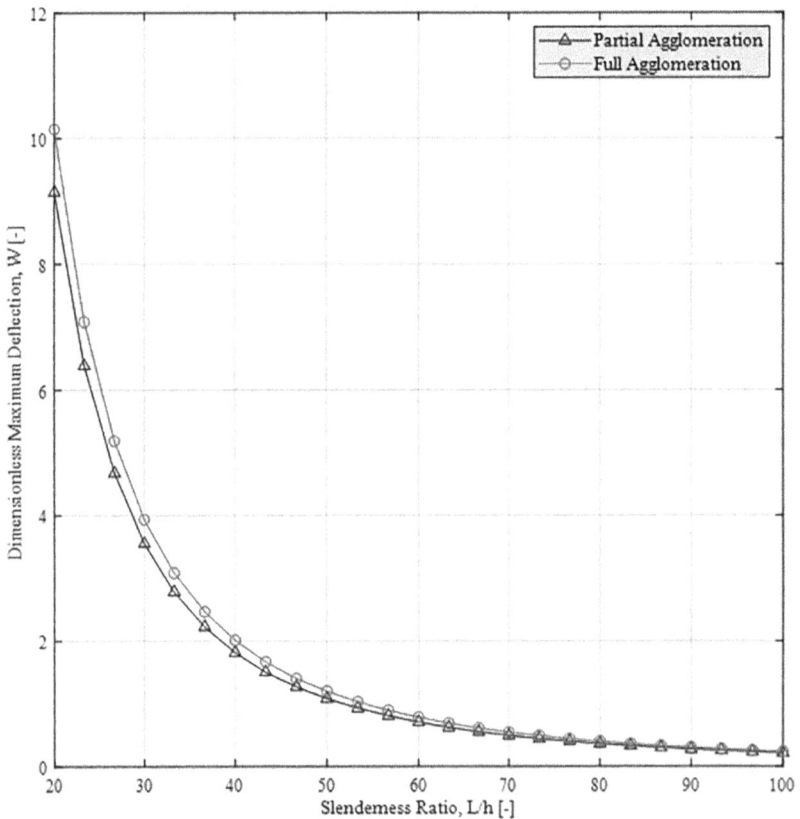

FIGURE 2.9 Influence of agglomeration pattern on the variation of the maximum deflec-tion of auxetic PNC beams subjected to uniform loading ($\mu=0.4$, $w_r=0.1$, $p=5$).

bending deflection of auxetic nanocomposite systems with varying slenderness ratios. Based on the diagram provided, it can be observed that in the event of complete aggregation, the beam undergoes larger deflections. The cause of this outcome can be attributed to the extremely low stiffness exhibited by the auxetic nanocomposite beam under these specific conditions. Moreover, when considering the definition of the dimensionless deflection as stated in equation (2.30h), it is evident that the inclusion of the slenderness ratio would result in reduced dimensionless deflections for the auxetic CNTR beam.

2.2.2 Porosity Effects

In the present study, a polymer nanocomposite reinforced with CNTs will be employed as the constituent material, featuring an auxetic arrangement. In order to achieve this objective, a three-step algorithm will be presented in this context. Initially, this study aims to elucidate the methodology for determining the material properties of meta-nanomaterial when provided with the specifications of a porous nanomaterial. Subsequently, we will proceed with the extraction of properties pertaining to porous CNTR materials, given that the equivalent properties of the nonporous counterpart are already established. In the concluding phase of this section, the Halpin-Tsai scheme will be employed to obtain the elastic modulus of a nonporous nanocomposite when employing curved nanotubes in the fabrication process. The CNTR polymer nanocomposite can be considered to possess a re-entrant lattice structure. In order to enhance comprehension of the geometric characteristics of the lattice, it is advisable to refer to Figure 2.1. In the magnified depiction of this diagram, it is possible to readily ascertain the geometric characteristics of the auxetic unit cell. The angle denoted as θ is commonly referred to as the auxeticity angle. Additionally, the vertical and inclined lengths of the cell rib are denoted as e and l, respectively. The thickness of the cell wall in this lattice is depicted using the symbol "a". To facilitate the process of homogenization, a retrogressive approach will be employed. In order to proceed, it is necessary to consider the Young's moduli and shear moduli of the nanomaterial as equivalent values denoted by E_{PNC} and G_{PNC}, respectively. The calculation of the effective modules, which indicate the stretch and shear properties of the auxetic nanocomposite formed in polymer nanocomposite auxetic lattices, can be performed as shown [16]:

$$E_{eff} = E_{PNC} \left(t/l\right)^3 \frac{\cos\theta}{\left(e/l + \sin\theta\right)\sin^2\theta} \tag{2.31}$$

$$G_{eff} = G_{PNC} \left(t/l\right) \frac{\cos\theta}{\left(e/l + \sin\theta\right)} \tag{2.32}$$

The geometrical parameters, denoted t, l, e, and θ, play a crucial role in determining the level of auxeticity exhibited by the nanomaterial. The present moment necessitates the extraction of the normal and shear moduli of CNTR polymer nanocomposites. This study aims to investigate the significant influence of voids on the microstructure of the CNTR material. To date, numerous endeavors have been undertaken to elucidate the influence of porosity on the equivalent properties of both homogeneous and heterogeneous materials. The implementation of the simple rule of mixture was observed in certain endeavors aimed at altering the modulus of porous materials [22]. These studies focused on analyzing the volume fraction allocated to the pores and examining the detrimental impact of porosity on the stiffness and density of the material. To address this issue, a simplified mixture's law was employed. Subsequently, it was discovered by researchers that the modulus of the porous material exhibits a correlation with the density of the sample in both its porous and nonporous states [23–24]. The proposed methodology demonstrates a higher degree of reliability in predicting a reduced modulus for the material, surpassing the limitations of the simplified rule of mixture. The distribution of pores in the material is not taken into account in any of the aforementioned models. In previous investigations, the saturated porous model was utilized [25–27]. In this study, we employ the latter approach to estimate the characteristics pertaining to the majority of nanomaterial when the distribution of large and small pores throughout its thickness is altered. In the context of porosity distribution, it is observed that larger pores tend to be situated in proximity to the neutral axis, while smaller pores are found in close proximity to the upper and lower edges of the structure. Porosity of type II, on the other hand, suggests that larger pores should be situated at a considerable distance from the neutral axis. Both of the aforementioned distributions exhibit symmetric patterns. In the ultimate distribution profile, it is postulated that pores are uniformly dispersed throughout the media. In accordance with the findings presented in reference [28], the stiffness and mass density of the porous CNTR nanocomposite can be determined by utilizing the following mathematical relationships:

$$
\begin{cases}
E_{PNC} = E_e \left[1 - e_1 \cos\left(\dfrac{\pi z}{h} \right) \right] \\
\rho_{PNC} = \rho_e \left[1 - e_{m1} \cos\left(\dfrac{\pi z}{h} \right) \right]
\end{cases}, \text{ Porosity Type-I} \tag{2.33}
$$

$$
\begin{cases}
E_{PNC} = E_e \left[1 - e_2 \left\{ 1 - \cos\left(\dfrac{\pi z}{h} \right) \right\} \right] \\
\rho_{PNC} = \rho_e \left[1 - e_{m2} \left\{ 1 - \cos\left(\dfrac{\pi z}{h} \right) \right\} \right]
\end{cases}, \text{ Porosity Type-II} \tag{2.34}
$$

$$
\begin{cases}
E_{PNC} = E_e e_3 \\
\rho_{PNC} = \rho_e e_{m3}
\end{cases}, \text{ Porosity Type-III} \tag{2.35}
$$

In the above relations, porosity coefficients corresponding with types I, II, and III of porosity distribution are respectively shown with e_1, e_2, and e_3. Also, e_{m1}, e_{m2}, and e_{m3} are the mass coefficients of type I, II, and III porous nanomaterials. These coefficients are responsible for modifying the mass density of the porous nanocomposite. It is worth regarding that E_e and ρ_e indicate on the moduli and density of the nonporous nanocomposite, respectively. It is necessary to mention that the Poisson's ratio remains unchangeable and is not affected by the existence of pores in the CNTR nanomaterial (i.e., $v_{PNC} = v_e$). With the aid of above relations and by considering the relationship between tangent moduli, Poisson's ratio, and shear moduli in linearly elastic solids of isotropic kind, the shear modulus of porous nanocomposite can be simply found by using $G_{PNC} = E_{PNC}/2(1 + v_{PNC})$.

The porosity coefficients for types I, II, and III of porosity distribution are denoted as e_1, e_2, and e_3. Also, e_{m1}, e_{m2}, and e_{m3} respectively, in the aforementioned relations. The mass coefficients of type I, II, and III porous nanomaterials are denoted by E_e and ρ_e, respectively. The coefficients in question are accountable for altering the mass density of the porous nanocomposite. It is imperative to note that the Poisson's ratio remains constant and is not influenced by the presence of pores in the CNTR nanomaterial. By utilizing the aforementioned relationships and taking into account the correlation between tangent moduli, Poisson's ratio, and shear moduli in linearly elastic isotropic solids, the shear modulus of a porous nanocomposite can be determined straightforwardly using the given equation using $G_{PNC} = E_{PNC}/2(1 + v_{PNC})$.

It is important to acknowledge that the only coefficient currently known to us is the porosity coefficient, while the determination of other coefficients relies on this information. The determination of these coefficients can be achieved by taking into account the principle that the overall mass of the porous nanomaterial remains unaffected by the type of porosity. Therefore, the identity presented below is documented in reference [16].

$$\int_0^{h/2} \sqrt{1 - e_1 \cos\left(\frac{\pi z}{h}\right)}\, dz = \int_0^{h/2} \sqrt{1 - e_2\left[1 - \cos\left(\frac{\pi z}{h}\right)\right]}\, dz = \int_0^{h/2} \sqrt{e_3}\, dz \qquad (2.36)$$

Furthermore, the mass coefficients associated with each type of porosity distribution can be linked to their respective porosity coefficients through the following identities [16].

$$1 - e_{m1} \cos\left(\frac{\pi z}{h}\right) = \sqrt{1 - e_1 \cos\left(\frac{\pi z}{h}\right)} \qquad (2.37)$$

$$1 - e_{m2}\left[1 - \cos\left(\frac{\pi z}{h}\right)\right] = \sqrt{1 - e_2\left[1 - \cos\left(\frac{\pi z}{h}\right)\right]} \qquad (2.38)$$

$$e_{m3} = \sqrt{e_3} \qquad (2.39)$$

In this section, we will proceed with the derivation of the characteristics pertaining to nonporous carbon nanotube reinforced nanocomposites, wherein the nanofillers exhibit a nonlinear morphology. The modeling framework employed in this investigation incorporates a micromechanical scheme to account for the curved shape of the CNTs. The current model being utilized is an updated iteration of the Halpin-Tsai algorithm. The proposed method incorporates a waviness coefficient to consider the curved arrangement of nanotubes when evaluating the modulus of the nanomaterial. Furthermore, the incorporation of randomly aligned wavy CNTs within the polymer matrix is also considered. The extensional moduli of the polymer nanocomposite are determined using the approach described in reference [16].

$$E_e = \frac{1 + C\eta V_r}{1 - \eta V_r} E_m \qquad (2.40)$$

where

$$\eta = \frac{C_w \left[\alpha E_{CNT}/E_m \right] - 1}{C_w \left[\alpha E_{CNT}/E_m \right] + C} \qquad (2.41)$$

In the aforementioned relationships, the control of CNT loading will be achieved through the manipulation of the volume fraction of CNTs. The variables E_m and E_{CNT} represent the elastic moduli of the polymer and CNT, respectively. Furthermore, this model accurately represents the geometric configuration of the reinforcing nanofillers by incorporating a coefficient C, which is defined as twice the length of the ideal CNTs divided by their diameter ($C = 2l_{CNT}/d_{CNT}$). Due to the significant disparity between the length of the reinforcing elements and the thickness of the beam, it is necessary to establish the orientation factor at a specific value on $\alpha = 1/6$. The influence of the curved morphology of CNTs on the Young's modulus of the nanocomposite will be monitored by means of the waviness coefficient C_w. The coefficient in question exhibits variability within the range of zero to one, with its maximum value corresponding to CNTs that possess an ideal straight shape. It is imperative to assert that despite the method's simplicity, it exhibits a remarkable level of accuracy when dealing with minimal quantities of nanofillers. The assertion mentioned above was substantiated in reference [29] through a comparative analysis of the changes in the modulus of elasticity resulting from the incorporation of CNTs into the matrix. The modeling results were compared to experimental measurements to validate the claim. The findings of this comparison are documented in Figure 2.10. Based on the present data, it is postulated that by considering the waviness coefficient to fall within a specified range $0.3 < C_w < 0.4$, the estimated modulus values will closely align with those derived from experimental observations. It is important to note that Figure 2.10 displays the waviness coefficient using the symbol K_w instead of C_w. In conclusion, it is noteworthy to mention that the determination of mass density and Poisson's ratio for the nonporous nanocomposite can be achieved by employing the modified Halpin-Tsai method, which can be expressed as follows:

$$v_e = v_{CNT} V_r + v_m \left(1 - V_r \right) \qquad (2.42)$$

$$\rho_e = \rho_{CNT} V_r + \rho_m \left(1 - V_r \right) \qquad (2.43)$$

Hence, the shear modulus can be calculated by utilizing the corresponding equation. $G_e = E_e/2(1 + v_e)$.

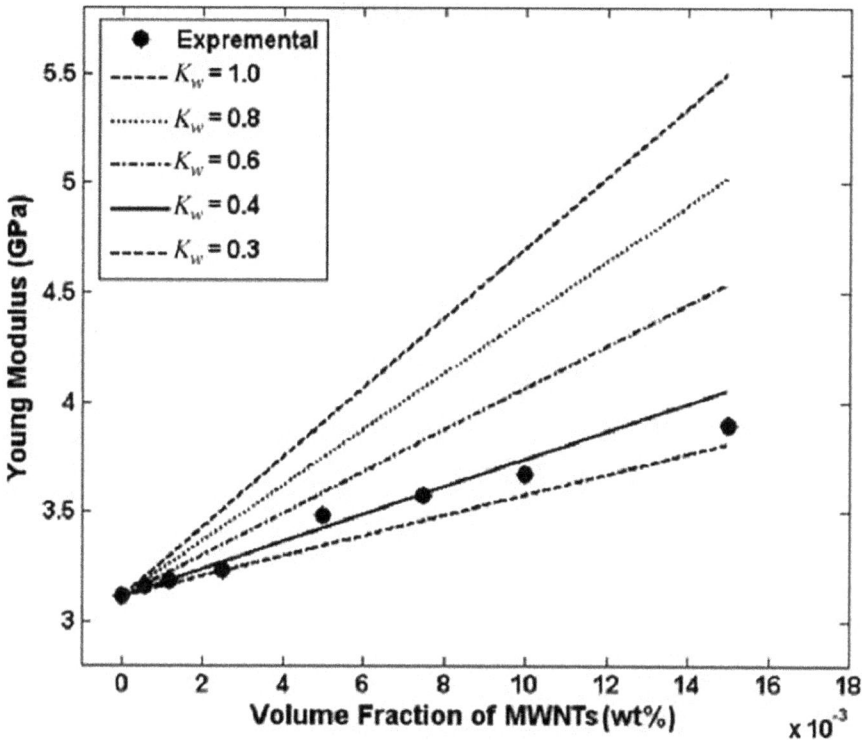

FIGURE 2.10   Variation of Young's modulus of CNTR polymers versus volume fraction of the CNTs for different waviness coefficients.

Figure 2.11 illustrates the effects of distribution and coefficient of porosity on the estimation of elasticity and shear modules in the context of auxetic CNTR nanocomposites. Based on the presented figure, it is evident that the normal and torsional stiffness of the auxetic nanomaterial will decrease when a larger magnitude is assigned to the porosity coefficient. According to the depicted illustration, the most unfavorable state is observed in auxetic nanocomposites characterized by the presence of type II pores distributed throughout the thickness. Within porous meta-nanomaterials, it can be observed that the material's stiffness reaches its minimum value at the upper and lower edges of the continua. The significant distance between these axes and the neutral axis contributes to their crucial role in determining the cross-sectional rigidities of the continua. Hence, it is anticipated that type II porous auxetic nanocomposites will exhibit more pronounced deflections compared to type III and type I porous meta-nanomaterials, in that order.

In this analysis, we will derive the mathematical relationships governing the motion of the structure. The structure under consideration is a beam with a rectangular cross-section. The dimensions of the beam are denoted as length (L), width (b), and thickness (h). This chapter aims to examine the mechanical characteristics of thick beams with a low slenderness ratio. The impact of shear-induced deflection

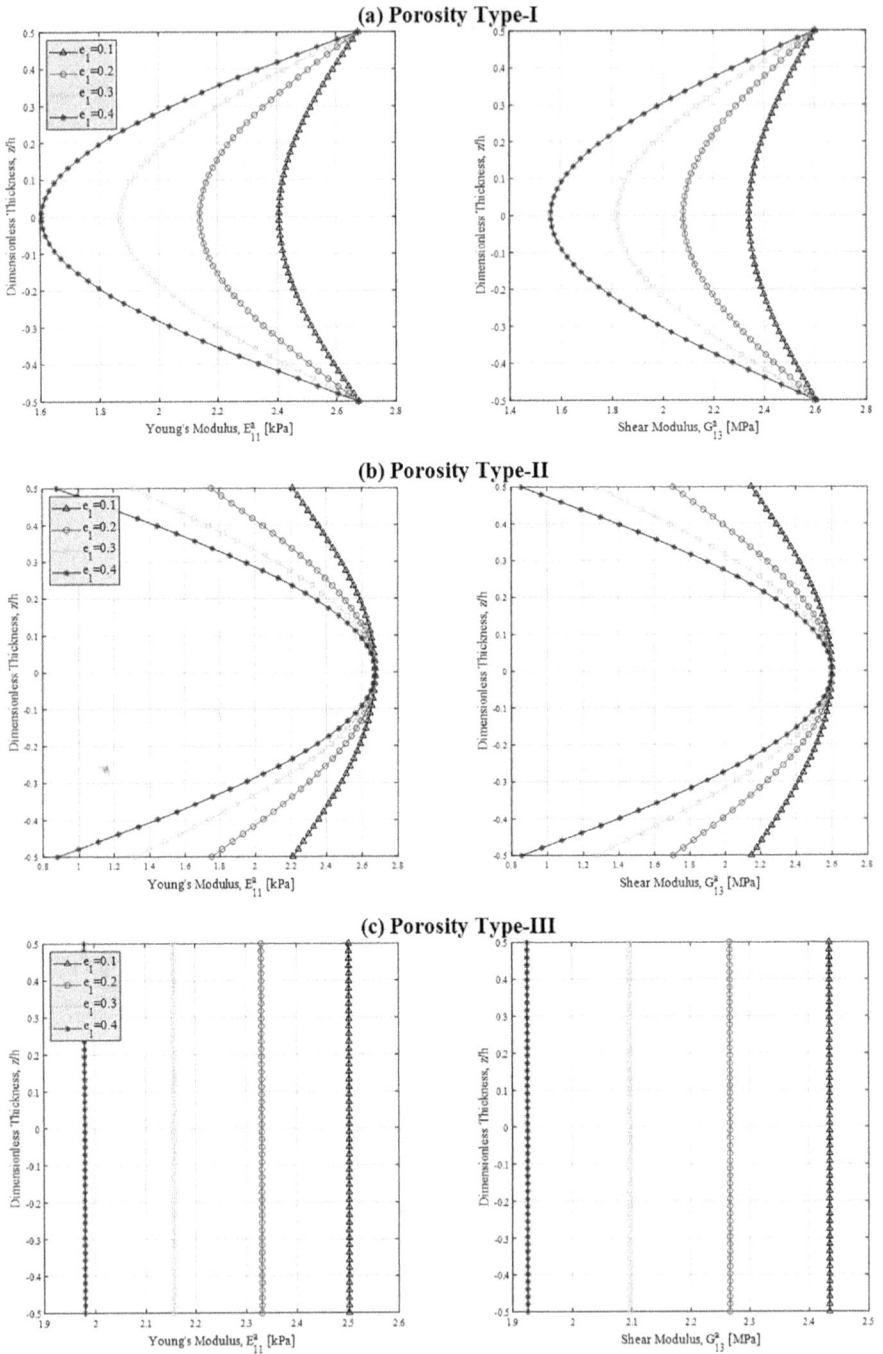

FIGURE 2.11 Effect of porosity coefficient e_1 on the variation of Young's modulus and shear modulus of the auxetic PNC materials with (a) type I, (b) type II, and (c) type III ($V_r=1\%$, $C_w=0.35$).

on the elastic properties of thin-type structures can be disregarded, as stated in reference [30]. Nevertheless, this assumption results in a decrease in precision when monitoring the response of a thick-type beam. Numerous high-speed data transmission (HSDT) techniques have been identified, with the Timoshenko's Shear Deformation Theory (TSDT) developed by Reddy being widely recognized as the most renowned [31]. The conventional form of TSDT for beams involves the utilization of three field variables, which correspond to the longitudinal deformation, bending deflection, and distortion of the cross-section. In this study, a modified version of the theory will be employed, which incorporates an axial displacement along with two distinct variables representing bending and shear deflections. The displacement field of the beam can be mathematically expressed as shown in equation (2.44) [28].

$$u_x(x,z) = u(x) - (z - \tilde{z})\frac{\partial w_b(x)}{\partial x} - (f(z) - \hat{z})\frac{\partial w_s(x)}{\partial x},$$

$$u_z(x,z) = w_b(x) + w_s(x)$$

(2.44)

The variable $u(x)$ represents the longitudinal deformation of the beam's neutral axis. On the other hand, $w_b(x)$ and $w_s(x)$ represent the deflections caused by bending and shear, respectively. Additionally, we possess:

$$\tilde{z} = \frac{\int_{-h/2}^{h/2} z E_{eff} dz}{\int_{-h/2}^{h/2} E_{eff} dz}, \qquad \hat{z} = \frac{\int_{-h/2}^{h/2} f(z) E_{eff} dz}{\int_{-h/2}^{h/2} E_{eff} dz}$$

(2.45)

The shape function $f(z) = z - 4z^3/3h^2$ responsible for accounting for the profiles of shear stress and strain across the flexural direction is described in reference [31]. The field introduced in equation (2.44) can now be utilized to extract identities that facilitate the enhancement of the structure's normal and shear strain, while also accounting for the effects of von Kármán type geometrical nonlinearity on the motion of the beam. The strain-displacement relations for the shear deformable beam can be expressed as follows:

$$\varepsilon_{xx} = \frac{\partial u}{\partial x} + \frac{1}{2}\left(\frac{\partial (w_b + w_s)}{\partial x}\right)^2 - (z - \tilde{z})\frac{\partial^2 w_b}{\partial x^2} - (f(z) - \hat{z})\frac{\partial^2 w_s}{\partial x^2},$$

$$\gamma_{xz} = 2\varepsilon_{xz} = g(z)\frac{\partial w_s}{\partial x}$$

(2.46)

In the present study, the principle of virtual work will be employed to derive the motion equations. According to this principle, it is necessary for the total energy of any given structure to be equal to zero when changes occur. In the context of mathematics, as denoted by reference [28],

$$\delta(U + V) = 0$$

(2.47)

where U and V represent the strain energy of the beam and the work done by external loads, respectively. By applying the constitutive relations of linearly elastic solids and referring to the definitions introduced in equation (2.46), the expression for δU can be readily obtained.

$$\delta U = \int_0^L \left\{ \begin{array}{l} N_{xx}\left[\dfrac{\partial \delta u}{\partial x} + \dfrac{\partial(w_b + w_s)}{\partial x}\dfrac{\partial \delta(w_b + w_s)}{\partial x}\right] \\ -M_{xx}^b\dfrac{\partial^2 \delta w_b}{\partial x^2} - M_{xx}^s\dfrac{\partial^2 \delta w_s}{\partial x^2} + Q_{xz}\dfrac{\partial \delta w_s}{\partial x} \end{array} \right\} dx \qquad (2.48)$$

The stress-resultants that have been previously discussed are mathematically represented as follows:

$$\begin{bmatrix} N_{xx} \\ M_{xx}^b \\ M_{xx}^s \end{bmatrix} = \int_A \begin{bmatrix} 1 \\ z - \bar{z} \\ f(z) - \tilde{z} \end{bmatrix} \sigma_{xx} dA, \qquad Q_{xz} = \int_A g(z)\sigma_{xz} dA \qquad (2.49)$$

Currently, the formulation of the work performed on the system is being undertaken. Let us consider a scenario where a uniform bending force is applied to the top edge of the beam. Thus, it can be observed that we have reference [17].

$$\delta V = \int_0^L q\delta(w_b + w_s)dx \qquad (2.50)$$

Upon substituting equations (2.48) and (2.50) into equation (2.47), the resultant motion relations are improved.

$$\frac{\partial N_{xx}}{\partial x} = 0 \qquad (2.51)$$

$$\frac{\partial}{\partial x}\left\{N_{xx}\frac{\partial(w_b + w_s)}{\partial x}\right\} - \frac{\partial^2 M_{xx}^b}{\partial x^2} + q = 0 \qquad (2.52)$$

$$\frac{\partial}{\partial x}\left\{N_{xx}\frac{\partial(w_b + w_s)}{\partial x}\right\} - \frac{\partial^2 M_{xx}^s}{\partial x^2} + \frac{\partial Q_{xz}}{\partial x} + q = 0 \qquad (2.53)$$

In the preceding sections, the equations governing the motion of the beam were derived. In this section, we will explore additional relationships that describe the inherent characteristics of beam-type structures. In accordance with Hook's law, the following one-dimensional identities pertain to the meta-nanocomposite system:

$$\sigma_{xx} = E_{eff}\varepsilon_{xx}, \qquad \sigma_{xz} = G_{eff}\gamma_{xz} \qquad (2.54)$$

By performing integration over the cross-sectional area of the structure and referring to equation (2.49), the following expressions are improved:

$$N_{xx} = A_{xx}\left[\frac{\partial u}{\partial x} + \frac{1}{2}\left(\frac{\partial(w_b + w_s)}{\partial x}\right)^2\right] \qquad (2.55)$$

$$M_{xx}^b = -D_{xx} \frac{\partial^2 w_b}{\partial x^2} - D_{xx}^s \frac{\partial^2 w_s}{\partial x^2} \tag{2.56}$$

$$M_{xx}^s = -D_{xx}^s \frac{\partial^2 w_b}{\partial x^2} - H_{xx}^s \frac{\partial^2 w_s}{\partial x^2} \tag{2.57}$$

$$Q_{xz} = A_{xz} \frac{\partial w_s}{\partial x} \tag{2.58}$$

Due to careful analysis of the location of the neutral axis in equation (2.45), the terms in the aforementioned relations with integrands having odd powers of z do not exist. It is simple to compute the cross-sectional rigidities of the beam-type element using the following definition:

$$\begin{bmatrix} A_{xx} \\ D_{xx} \\ D_{xx}^s \\ H_{xx}^s \end{bmatrix} = \int_A \begin{bmatrix} 1 \\ (z-\overline{z})^2 \\ (z-\overline{z})(f(z)-\overline{z}) \\ (f(z)-\overline{z})^2 \end{bmatrix} E_{eff} dA, \quad A_{xz} = \int_A g^2(z) G_{eff} dA \tag{2.59}$$

Now, a quick glance at equation (2.51) demonstrates this $N_{xx} = \text{cte}$. In order to construct equations (2.52) and (2.53) again, do the following:

$$N_{xx} \frac{\partial^2 (w_b + w_s)}{\partial x^2} - \frac{\partial^2 M_{xx}^b}{\partial x^2} + q = 0 \tag{2.60}$$

$$N_{xx} \frac{\partial^2 (w_b + w_s)}{\partial x^2} - \frac{\partial^2 M_{xx}^s}{\partial x^2} + \frac{\partial Q_{xz}}{\partial x} + q = 0 \tag{2.61}$$

The following relationships will be reached if the equations (2.55) to (2.58) are introduced into equations (2.61) and (2.62):

$$A_{xx} \left[\frac{\partial u}{\partial x} + \frac{1}{2} \left(\frac{\partial (w_b + w_s)}{\partial x} \right)^2 \right] \frac{\partial^2 (w_b + w_s)}{\partial x^2} + D_{xx} \frac{\partial^4 w_b}{\partial x^4} + D_{xx}^s \frac{\partial^4 w_s}{\partial x^4} + q = 0 \tag{2.62}$$

$$A_{xx} \left[\frac{\partial u}{\partial x} + \frac{1}{2} \left(\frac{\partial (w_b + w_s)}{\partial x} \right)^2 \right] \frac{\partial^2 (w_b + w_s)}{\partial x^2} + D_{xx}^s \frac{\partial^4 w_b}{\partial x^4} + H_{xx}^s \frac{\partial^4 w_s}{\partial x^4}$$
$$+ A_{xz} \frac{\partial^2 w_s}{\partial x^2} + q = 0 \tag{2.63}$$

The aforementioned relations define how the beam moves. Assuming that c is the constant magnitude of the normal load, it results in:

$$\frac{\partial u}{\partial x} = \frac{c}{A_{xx}} - \frac{1}{2} \left(\frac{\partial (w_b + w_s)}{\partial x} \right)^2 \rightarrow u(x) = \frac{cx}{A_{xx}} - \frac{1}{2} \int_0^x \left(\frac{\partial (w_b + w_s)}{\partial x} \right)^2 dx + d \tag{2.64}$$

It contains d an unknown that may be discovered using boundary conditions (BCs). The beam is thought to be just minimally supported, hence we have:

$$u(0) = 0 \rightarrow d = 0 \tag{2.65}$$

$$u(L) = 0 \rightarrow c = \frac{A_{xx}}{2L} \int_0^L \left(\frac{\partial(w_b + w_s)}{\partial x} \right)^2 dx \tag{2.66}$$

The findings above lead one to the conclusion that:

$$\frac{\partial u}{\partial x} + \frac{1}{2} \left(\frac{\partial(w_b + w_s)}{\partial x} \right)^2 = \frac{1}{2L} \int_0^L \left(\frac{\partial(w_b + w_s)}{\partial x} \right)^2 dx \tag{2.67}$$

In equations (2.62) and (2.63), the aforementioned identity may be substituted to show:

$$\frac{A_{xx}}{2L} \int_0^L \left(\frac{\partial(w_b + w_s)}{\partial x} \right)^2 dx \frac{\partial^2(w_b + w_s)}{\partial x^2} + D_{xx} \frac{\partial^4 w_b}{\partial x^4} + D_{xx}^s \frac{\partial^4 w_s}{\partial x^4} + q = 0 \tag{2.68}$$

$$\frac{A_{xx}}{2L} \int_0^L \left(\frac{\partial(w_b + w_s)}{\partial x} \right)^2 dx \frac{\partial^2(w_b + w_s)}{\partial x^2} + D_{xx}^s \frac{\partial^4 w_b}{\partial x^4} + H_{xx}^s \frac{\partial^4 w_s}{\partial x^4}$$
$$+ A_{xz} \frac{\partial^2 w_s}{\partial x^2} + q = 0 \tag{2.69}$$

It is tried to solve the nonlinear deformation issue of auxetic nanocomposite beams in the present portion of the chapter. Researchers have used a variety of numerical and analytical techniques up to this point to examine the time-dependent and time-independent responses of nonhomogeneous structures [23]. Here, the well-known Navier's method will be used to find a solution. This method assumes that a completely simply supported structure's bending and shear deflections have the shape of the bellow series [23]:

$$\begin{Bmatrix} w_b(x) \\ w_s(x) \end{Bmatrix} = \sum_{n=1}^{\infty} \begin{Bmatrix} W_{bn} \\ W_{sn} \end{Bmatrix} \sin \lambda x, \qquad \lambda = \frac{n\pi}{L} \tag{2.70}$$

where W_{bn} and W_{sn} are the bending and shear deflection amplitudes that correspond to the nth mode, respectively, and n is the axial mode number. Additionally, the inserted loading is predicated on the following expression:

$$q(x) = \sum_{n=1}^{\infty} Q_n \sin \lambda x \tag{2.71}$$

According to the definition above, the loading amplitude may be calculated using the relation below:

$$Q_n = \frac{2}{L} \int_0^L q(x) \sin \lambda x \, dx \tag{2.72}$$

The continua in this text are assumed to be under uniform loading, hence the amount of force applied is defined by:

$$Q_n = \frac{4q_0}{n\pi} \qquad (n = 1,3,5,\cdots) \tag{2.73}$$

The following linked nonlinear equations may be obtained after equations (2.70) and (2.71) are replaced in equations (2.68) and (2.69):

$$-\frac{A_{xx}}{2L}\sum_{n=1}^{\infty}\left(W_{bn}+W_{sn}\right)^3 \lambda^4 \int_0^L \cos^2 \lambda x\, dx \sin \lambda x + D_{xx}\sum_{n=1}^{\infty}W_{bn}\lambda^4 \sin \lambda x$$

$$+D_{xx}^s \sum_{n=1}^{\infty}W_{sn}\lambda^4 \sin \lambda x + \sum_{n=1}^{\infty}Q_n \sin \lambda x = 0 \tag{2.74a}$$

$$-\frac{A_{xx}}{2L}\sum_{n=1}^{\infty}\left(W_{bn}+W_{sn}\right)^3 \lambda^4 \int_0^L \cos^2 \lambda x\, dx \sin \lambda x + D_{xx}^s\sum_{n=1}^{\infty}W_{bn}\lambda^4 \sin \lambda x$$

$$+H_{xx}^s \sum_{n=1}^{\infty}W_{sn}\lambda^4 \sin \lambda x - A_{xz}\sum_{n=1}^{\infty}W_{sn}\lambda^2 \sin \lambda x + \sum_{n=1}^{\infty}Q_n \sin \lambda x = 0 \tag{2.74b}$$

The coupled differential equations mentioned above may be stated differently by using the following mathematical simplifications:

$$\sum_{n=1}^{\infty}K_{bbn}^L W_{bn} + \sum_{n=1}^{\infty}K_{bsn}^L W_{sn} + \sum_{n=1}^{\infty}K_n^{NL}\left(W_{bn}+W_{sn}\right)^3 = \sum_{n=1}^{\infty}Q_n \tag{2.75a}$$

$$\sum_{n=1}^{\infty}K_{sbn}^L W_{bn} + \sum_{n=1}^{\infty}K_{ssn}^L W_{sn} + \sum_{n=1}^{\infty}K_n^{NL}\left(W_{bn}+W_{sn}\right)^3 = \sum_{n=1}^{\infty}Q_n \tag{2.75b}$$

where the linear stiffnesses for any specified mode number of the first and second equations' bending and shear deflections, respectively, are K_{bbn}^L, K_{bsn}^L, K_{sbn}^L, and K_{ssn}^L. The nonlinear stiffness K_n^{NL} is also linked to any random mode number. The deflection amplitude at any natural mode may be obtained by concurrently solving equations (2.75a) and (2.75b) for W_{bn} and W_{sn}. The overall deflection of the auxetic nanocomposite beam is then obtained by adding the deflections of the system's inherent modes.

To examine the nonlinear bending response of a shear deformable beam made of auxetic CNTR nanocomposite, a collection of interesting case studies is provided. Unless otherwise stated, the slenderness ratio and thickness of the beam shall be maintained at $L/h = 10$ and $h = 2$ mm in this section. This shape was chosen mostly because the modeling that is being used is capable of tracking thick structures' nonlinear tendencies. Furthermore, if not otherwise stated, the vertical-to-inclined length, thickness-to-inclined length, and angle of auxeticity of the lattice's cell rib are each assumed to be $e/l = 4$, $t/l = 0.0138571$, and $\theta = -45$ respectively. For reference, Table 2.1 lists the input materials for resin and CNTs.

The diameter of the CNTs is taken into account $d_{CNT} = 1.356$ nm in Table 2.1. It is important to note that this figure is provided in accordance with the definition

TABLE 2.1

Material Properties of Polymer Matrix and CNTs [18]

Polymer's properties

$E_m = 2.1$ GPa, $v_m = 0.34$, $\rho_m = 1150$ kg/m^3

CNTs' properties

$E_{CNT} = 450$ GPa, $v_{CNT} = 0.2$, $\rho_{CNT} = 2237$ kg/m^3, $t_{CNT} = 0.34$ nm, $d_{CNT} = 1.356$ nm

---

$d_{CNT} = a\sqrt{m^2 + mn + n^2}/\pi$, also where graphene's lattice constant is $a = \sqrt{3}a_{C\text{-}C}$; where the length of the $a_{C\text{-}C}$ carbon-carbon bond in graphene is 0.142 nm. The elements of the chiral vector of the CNT are the variables m and n. The analysis will be made simpler in the sections that follow by using the dimensionless version of the beam's deflection and loading. The following parameters are considered to be dimensionless in this study:

$$W = \frac{w(L/2)}{h}, \qquad Q = \frac{q_0 L^2}{A_{xx}^m h} \tag{2.76}$$

where A_{xx}^m is the bare matrix's stretching stiffness (A_{xx}). The correctness of the newly incorporated modeling will be evaluated in Step 1 using Figure 2.12. Plotted against a dimensionless load in this picture is the change in the nondimensional, nonlinear deflection of nanocomposite beams consisting of reinforced polymer with graphene platelets (GPLs). This figure shows how current modeling may effectively recreate the nonlinear deflection curve described in reference [17]. The curves' close proximity and little variances in large amounts of dimensionless load may clearly be seen. This discrepancy results from the fact that, under these circumstances, the shear deformation's contribution to the overall deformation of the continuous system is greater. Given that the current model is based on the HSDT of thick beams, it predicts deflections that are more accurate than those found using FSDT [17] under these conditions. While reference [17] used the FE-assisted Ritz approach to extract the solution to the nonlinear issue, we applied an analytical precise solution in this work.

Figure 2.13's primary goal is to examine how the distribution of porosity affects the nonlinear load-deflection curve of beams made of porous meta-nanomaterials. This picture makes it abundantly evident that type II porosity distribution exhibits the most deflection for any given arbitrary dimensionless stress. This tendency is brought about by the fact that, when compared to other forms of porosity distributions, this type has a stronger destructive effect on the CNTR meta-nanomaterial's modulus. By looking at Figure 2.11, this physical explanation may be seen clearly.

Additionally, Figure 2.14 covers the effects of auxeticity angle and waviness problem on the deflection response of the non-ideal meta-nanomaterial beam. Type III porosity distribution is taken into account in this figure. The figure shows that increasing the nonlinear deflection of the auxetic nanomaterial structure results from raising the absolute value of the angle of auxeticity. This leads to the conclusion that

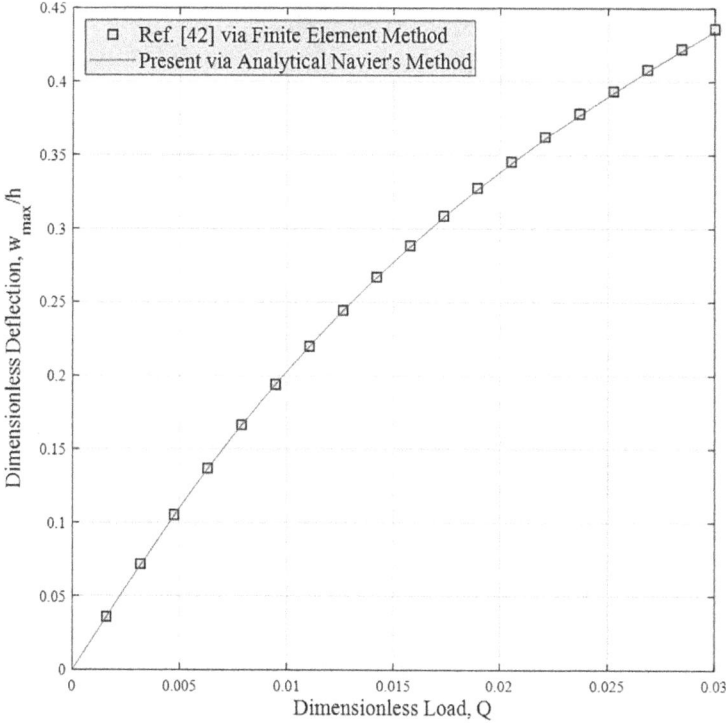

FIGURE 2.12 Comparison of the nonlinear bending response of polymer nanocomposite beams reinforced with uniformly distributed GPLs (f_i=0.5%, N=10, L/h=20).

changing the auxeticity angle may be used to control how the porous meta-nanomaterial beam deforms. This is the result of using metamaterials in structural instrument design, which may allow the designer extra degrees of freedom. However, this figure shows how the waviness problem is having an increasing effect on the nonlinear deflection of the porous auxetic nanocomposite structure. In actuality, this is because the stiffness increase mechanism in the polymer nanocomposite is negatively impacted by the curved shape of the nanoparticles. As a result, the stiffness will decrease, and since stiffness and deflection in linearly elastic substances have an inverse relationship, the maximum deflection of the structure will rise. In other words, when wavy CNTs are present in the microstructure of the auxetic CNTR nanomaterial, the structure acts in a softer way.

The nonlinear deflection of the beam is studied in Figure 2.15 by plotting the effect of the wall thickness of the auxetic lattice of the meta-nanomaterials made from ideal or non-ideal nanofillers. Similar to the previous figure, it is clear from this one that the presence of wavy nanotubes in a polymer nanocomposite forces it to withstand greater deflections since their waviness has a negative impact on the nanocomposite's equivalent modulus. Additionally, it is shown that altering the auxetic lattice's wall thickness allows one to adjust the meta-nanomaterial structure's deflection. In other words, increasing the thickness-to-inclined length ratio causes an

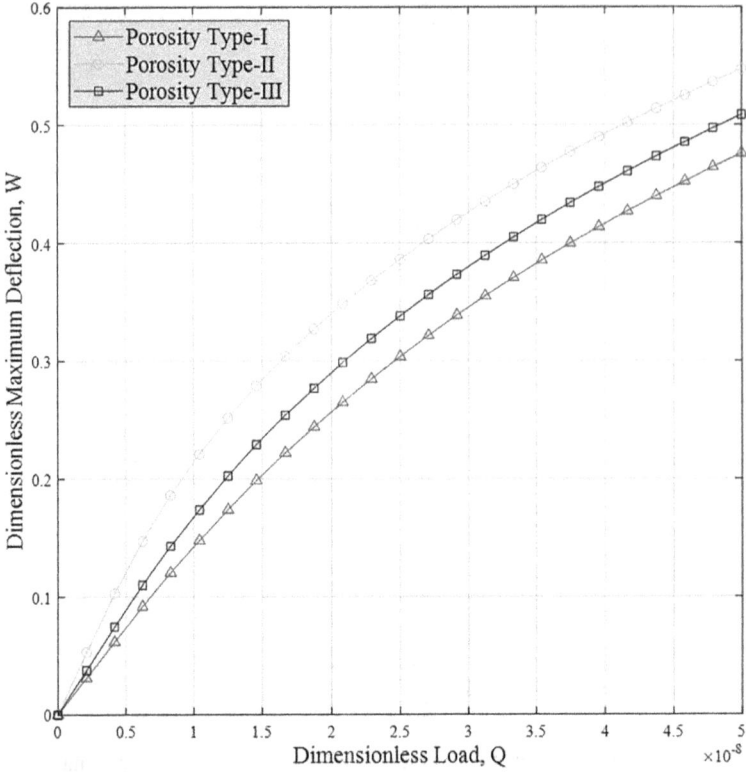

FIGURE 2.13 Influence of the porosity type on the deflection-load curve of auxetic porous PNC beams reinforced via non-straight CNTs (V_r=1%, e_1=0.5).

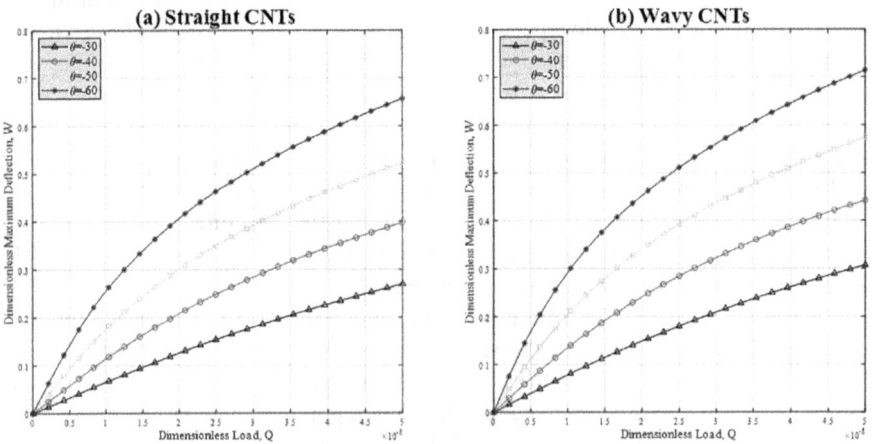

FIGURE 2.14 Effect of the auxeticity angle on the deflection-load curve of the auxetic porous PNC beams reinforced with (a) straight and (b) wavy CNTs (V_r=1%, e_1=0.5). In this figure, porosity type III is considered for all cases.

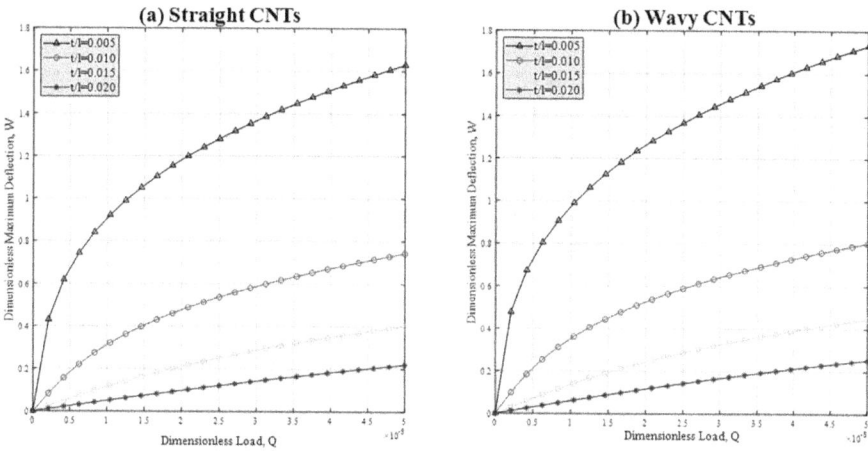

FIGURE 2.15 Effect of the thickness-to-inclined length of the cell rib on the deflection-load curve of the auxetic porous PNC beams reinforced with (a) straight and (b) wavy CNTs (V_r=1%, e_1=0.5). In this figure, porosity type III is considered for all cases.

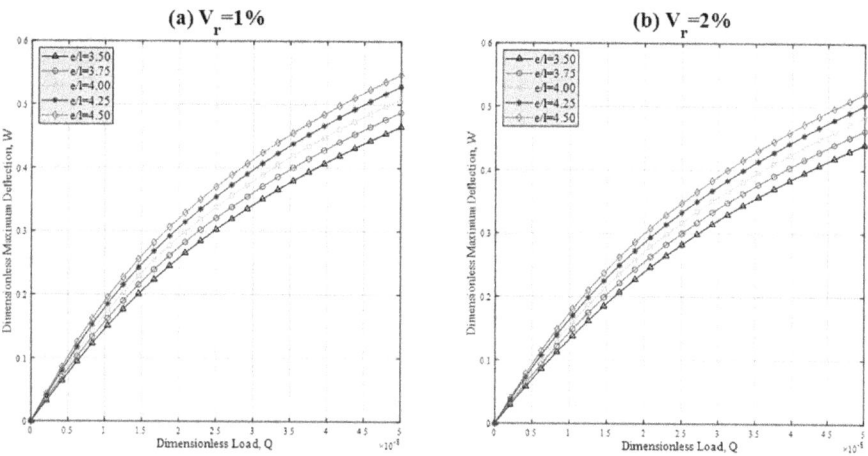

FIGURE 2.16 Effect of the vertical-to-inclined length of the cell rib on the deflection-load curve of the auxetic porous PNC beams reinforced with wavy CNTs at loadings of (a) V_r=1% and (b) V_r=2% (C_w=0.35, e_1=0.5). In this figure, porosity type III is considered for all cases.

improvement in the longitudinal and torsional stiffness of the beam, which reduces the nonlinear deformation of the meta-nanomaterial beam.

Additionally, Figure 2.16 considers the problem of analyzing the impact of cell rib's vertical-to-inclined length on the nonlinear deflection behaviors of non-ideal meta-nanomaterial beams. Different values are ascribed to e/l in this picture, and it is accepted that the inclusion of this dimensionless parameter results in an increase in the auxetic CNTR beam's deflection. By making use of equations (2.1) and (2.2), it is simple to deduce the reasoning for this tendency. The meta-nanomaterial's longitudinal and torsional stiffnesses are reduced as the width of the cell rib is added. As a result, the cross-sectional rigidities that contribute to the structure's stiffness will be reduced. As a result of the inverse connection between deflection and stiffness, it is only normal to see a rise in the beam's deflection. This figure also shows that if more nanoparticles are included in the composition of the nanocomposite, the nonlinear deflection of the porous meta-nanomaterial structure may be reduced. The ultra-stiffness of polymer nanocomposite materials makes this result achievable.

The analysis of the quantitative impact of the porosity coefficient on the modification of the deflection-load curve of porous meta-nanomaterial beams with different kinds of porosity distributions is the subject of the next case study. According to Figure 2.17, increasing the porosity coefficient will result in a greater nonlinear

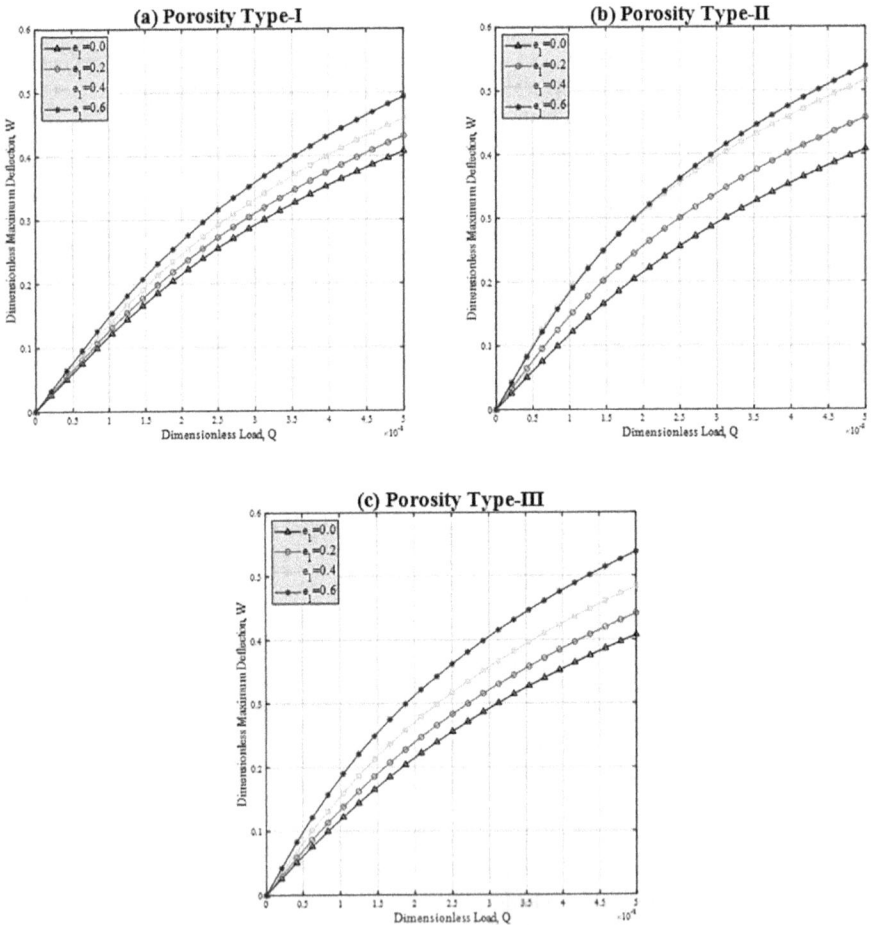

FIGURE 2.17 Effect of the porosity coefficient $e1$ on the deflection-load curve of the auxetic porous PNC beams reinforced with wavy CNTs while (a) type I, (b) type II, and (c) type III of porosity distribution is considered (C_w=0.35, V_r=1%).

deflection of the porous auxetic nanocomposite structure for all porosity types and at high loading amplitudes. The underlying logic of this problem is the detrimental effect that holes have on the equivalent stiffness of the meat-nanomaterial's micro-structure. Again, it can be seen that the highest deflection here coincides with porosity types II, III, and I, in that order.

2.2.3 POSTBUCKLING CHARACTERISTICS OF AUXETIC NANOCOMPOSITE BEAMS

According to Figure 2.2, auxetic CNTR nanocomposite beams with length L and a rectangular cross-section with width b and thickness h are taken into consideration in this work. In this study, it will be considered that the cross-sectional measurements of the beam are fixed at $h = 2$ mm. It is presumed that the beam-type structure is long enough to belong within the category of thin structures. Along with its structural characteristics, the nanocomposite material is created using a re-entrant lattice model, whose schematic representation is shown in Figure 2.1. By adjusting the angle, also known as the auxeticity angle, the level of auxeticity in the meta-nanocomposite may be customized. Now, using the method presented in section 2.2.1 and equations 2.1 to 2.29, it is simple to extract the problem's decoupled governing equation. As a result, we have:

$$\frac{A_{xx}}{2L}\int_0^L \left(\frac{\partial w}{\partial x}\right)^2 dx \frac{\partial^2 w}{\partial x^2} - P_b \frac{\partial^2 w}{\partial x^2} + D_{xx}\frac{\partial^4 w}{\partial x^4} = 0 \tag{2.77}$$

In this part, the nonlinear buckling load of the meta-nanocomposite beam for both S-S and C-C BCs will be determined using the Galerkin's analytical approach. Using this approach, the beam's bending deflection may be thought of as [14]:

$$w(x) = \sum_{m=1}^{\infty} W_m X_m(x) \tag{2.78}$$

In the formulation above, W_m denotes the amplitude of the beam's deformation and $X_m(x)$ is the eigen function, fulfilling the problem's primary BCs in the longitudinal direction. For S-S and C-C beams, the following functions are advised in accordance with the limits imposed by each of the considered set of BCs on the ends of the structure [14]:

$$\text{S-S Beam: } X_m(x) = \sin\left(\frac{m\pi x}{L}\right)$$
$$\text{C-C Beam: } X_m(x) = \sin^2\left(\frac{m\pi x}{L}\right) \tag{2.79}$$

The following identity may be written by replacing the bending deflection from equation (2.78) in the problem's governing equation, namely equation (2.77), and taking into account the orthogonality of the continuous system's natural modes:

$$K_L W_m + K_{NL} W_m^3 - P_b W_m = 0 \tag{2.80}$$

where

$$K_L = D_{xx} r_{40}, \qquad K_{NL} = \frac{A_{xx}}{2L} r_{11} r_{20} \qquad (2.81)$$

r_{ij}'s are BC-based integrals in the aforementioned definitions, and the method for calculating them is as follows:

$$r_{ij} = \int_0^L X_m^{(i)}(x) X_m^{(j)}(x) dx \qquad (2.82)$$

where $X_m^{(i)}(x)$ represents the eigen function's $X_m(x)$ ith derivative with respect to x. The nonlinear buckling load of the meta-nanocomposite beam may be readily determined by calculating equation (2.80) for P_b.

In this section of the text, numerical examples will be shown and analyzed to show how variations in the auxetic nanocomposite's characteristics might affect the nonlinear buckling load of the beam. In these experiments, the beam's slenderness ratio is fixed $L/h = 100$ in order to confirm the structure's thinness hypothesis. The thickness-to-inclined length, vertical-to-inclined length, and auxeticity angle of the re-entrant cell rib are assumed to be 0.01, 4, and −45 degrees, respectively, if other values are not specified. Table 2.1 provides information on the materials' characteristics that were utilized to create the epoxy matrix and CNTs. It should be noted that the following nondimensional form of the nonlinear buckling stress is monitored in future studies:

$$P_{cr} = \frac{P_b L^2}{E_m h^3} \qquad (2.83)$$

By comparing the postbuckling route of GPLR nanocomposite beams with those previously developed in reference [32], the correctness of the material delivered is first assessed. In this comparison, the GPLs' weight percentage is maintained at 0.3%, and it is assumed that they are spread equally throughout the polymeric matrix. Figure 2.18 makes it clear that the findings of our investigation are remarkably similar to those published in [32]. It must be noted that the slenderness ratio of the beam, which is $L/h = 10$, is what causes the minor discrepancies between our findings and those published in [32]. Because the structure in this instance is thick, shear distortion has a significant impact on the mechanical response. The presented statistics may be relied upon with confidence since we employ slenderness ratios as large as 100 in the case studies that follow.

The effects of gradient index p and auxeticity angle on the meta-nanocomposite's Young's modulus are emphasized in Figure 2.19. Based on this example, it is clear that an increase in the gradient index causes the Young's modulus of the auxetic nanocomposite in any arbitrary thickness to decrease. This pattern demonstrates that, in order to offer a safer structure that can withstand compressive excitations with a greater safety factor, large gradient indices should be taken into account in theoretical calculations. Additionally, it is well known that increasing the auxeticity

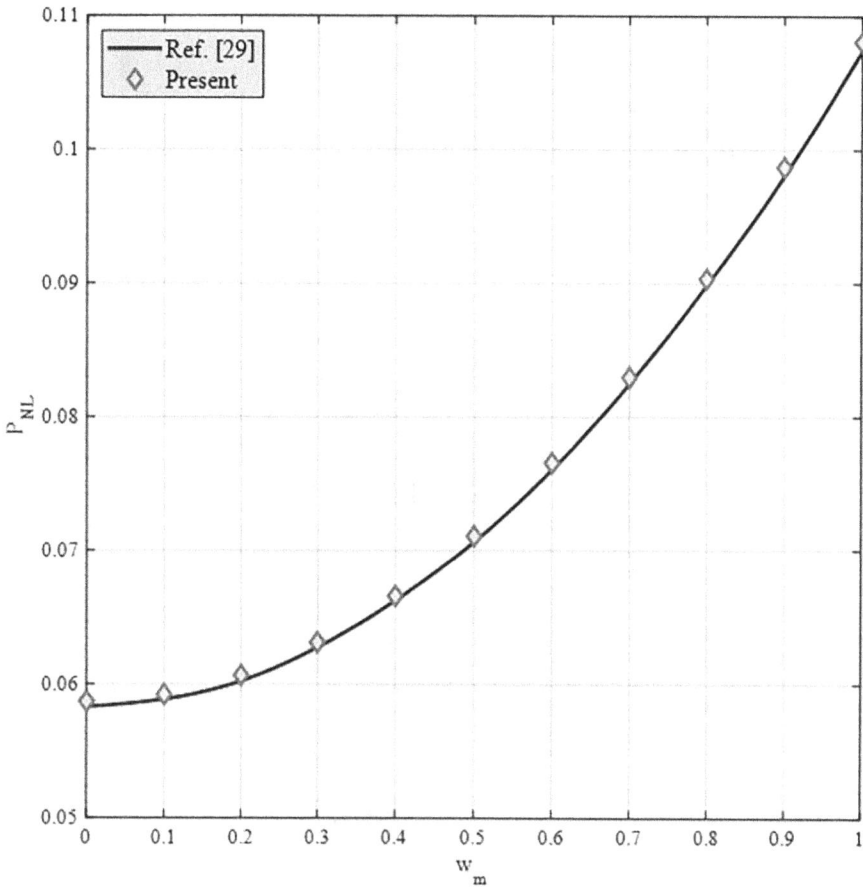

FIGURE 2.18 Comparison of the postbuckling paths of fully simply supported PNC beams reinforced via GPLs with those reported in [32] (W_{GPL}=0.3%, L/h=10).

angle leads in a reduction in the overall stiffness of the meta-nanomaterial. Designing CNTR auxetic lattices with large auxeticity angles is thus not advised [33].

The analysis of the effects of the auxeticity angle and agglomeration parameter on the nonlinear buckling load of meta-nanocomposite beams is shown in Figure 2.20 [34–50]. According to this graph, if a large value is given to the volume percentage of the CNTs within the clusters, the auxetic nanocomposite beam will break under weaker buckling stimulations. The adverse impact of CNT aggregation on the rigidity of the meta-nanocomposite is the underlying physical cause of this problem. In other words, when the volume proportion of CNTs within the clusters increases, the agglomeration phenomena will have an impact on the stiffness-enhancement process. Reversed, it can be demonstrated that the postbuckling endurance of the meta-nanocomposite beam will increase at any agglomeration parameter by selecting an auxetic lattice with in magnitude smaller. The use of reduced auxeticity angles has a favorable impact on the meta-nanocomposite's Young's modulus, which is the

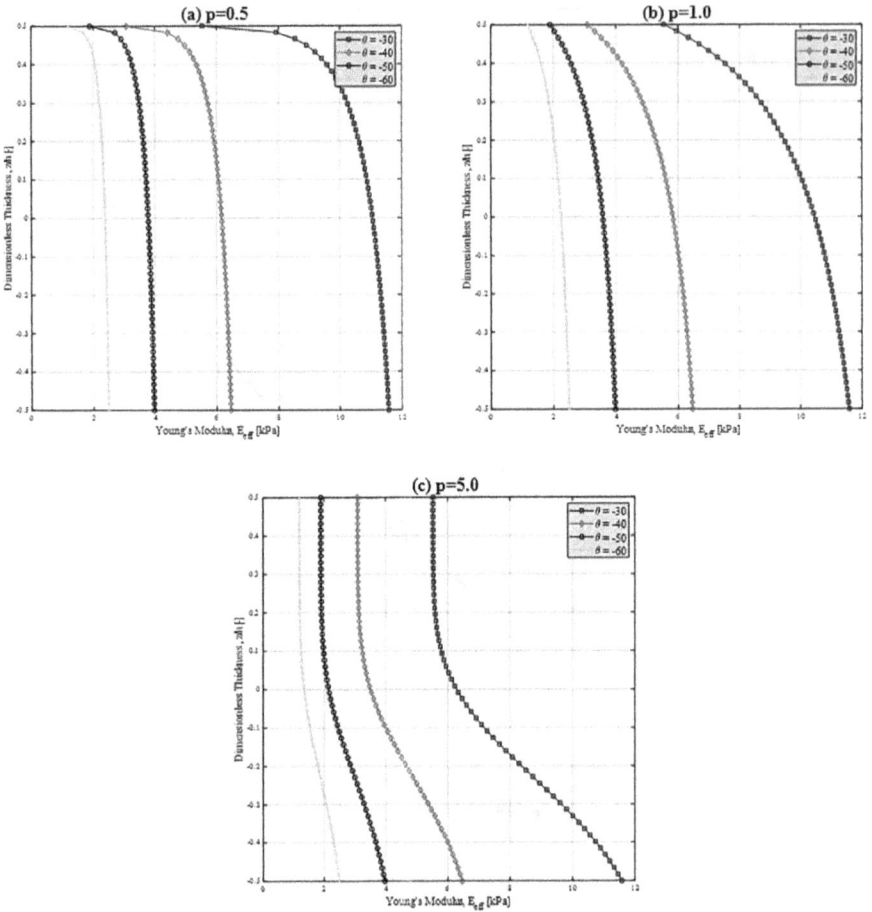

FIGURE 2.19 Through-the-thickness variation of the Young's modulus of the auxetic PNC for (a) p=0.5, (b) p=1.0, and (c) p=5.0 whenever different values are assigned to the auxeticity angle (η=1, μ=0.4, w_r=0.1).

primary cause of this trend. Figure 2.19 provides a quantitative illustration of this impact. In addition to these results, this picture demonstrates that clamped ends on beams make them more resistant to buckling-mode failure than are beams with simple supports. The enhanced structural stiffness of the clamped BC compared to the merely supported one is the cause of the C-C beams' superior resistance to compression. Notably, this tendency will hold true for Figures 2.21 to 2.24 as well. So, the issue's physical reason won't be brought up again.

Figure 2.21 is presented with a focus on how the other agglomeration parameter (μ) affects the meta-nanocomposite beam's static stability. Based on this picture, it can be seen that if large values are given to the volume fraction of the clusters, then more axial compression is needed for the beam to reach the postbuckling condition. This tendency may be explained by keeping in mind that the stiffness of

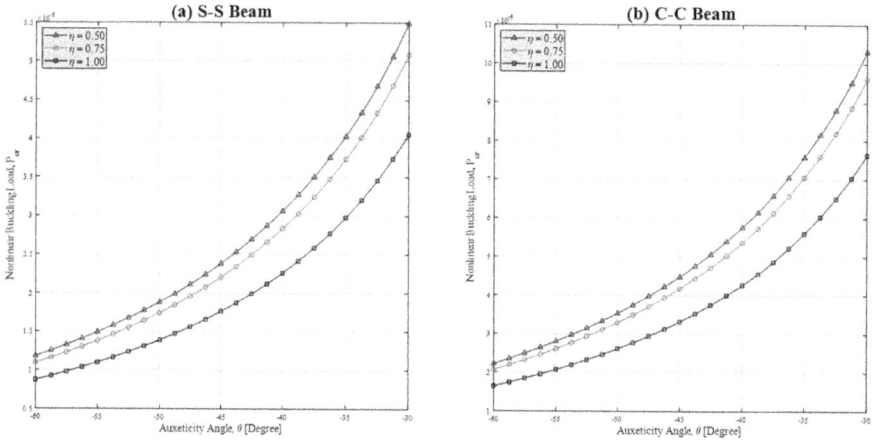

FIGURE 2.20   Variation of the nonlinear buckling load of (a) S-S and (b) C-C auxetic PNC beams versus auxeticity angle while different values are assigned to the volume fraction of the CNTs inside the clusters ($W=h$, $\mu=0.4$, $w_r=0.1$, $p=5$).

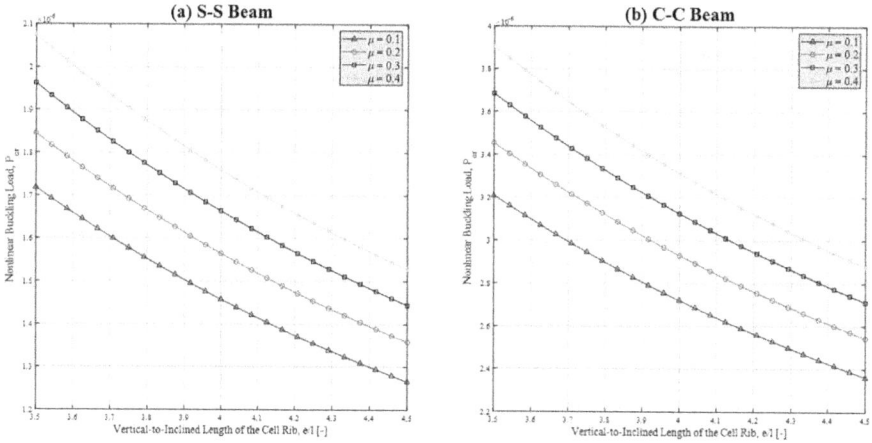

FIGURE 2.21   Variation of the nonlinear buckling load of (a) S-S and (b) C-C auxetic PNC beams versus vertical-to-inclined length of the cell rib while different values are assigned to the volume fraction of the clusters ($W=h$, $\eta=1$, $w_r=0.1$, $p=5$, $\theta=-45$).

the meta-nanocomposite would be less impacted by the aggregation process as the agglomeration parameter is increased. An increase in μ stiffens the auxetic nanocomposite. This figure also shows the effect of the re-entrant lattice's vertical-to-inclined length on the continuous system's postbuckling performance. It is obvious that the auxetic nanocomposite system will be less resistant to buckling-mode stimulations as the vertical-to-inclined length increases. This trend began with the effect of adding e/l softening on the meta-nanocomposite's equivalent Young's modulus. Therefore, it is expected to see a similar decline in the continua's nonlinear buckling load. Once again, it is shown that clamped beams are superior to other options in terms of axial compression resistance.

Similar to Figure 2.21, Figure 2.22 shows how the nonlinear buckling load of the meta-nanocomposite beams varies with the vertical-to-inclined length of the cell rib when the mass percent of the CNTs is changed. This graph illustrates the beneficial effect of the CNTs' mass fraction on the beam's postbuckling load. In other words, a stiffer meta-nanocomposite is produced by implementing a larger CNT content in the auxetic nanocomposite's composition. This pattern is easily justifiable if the direct correlation between buckling load and stiffness is kept in mind. Additionally, it is shown that adopting broader re-entrant cell ribs reduces the meta-nanocomposite beam's resistance to compressive excitations. You may find the physical cause of this problem by thinking back to the previous sentence. Additionally, it is clear that C-C beams have higher buckling resistance than S-S ones.

Additionally, Figure 2.23 covers how the thickness of the re-entrant cell rib and gradient index affect the nonlinear buckling stress of the auxetic nanocomposite beam. This example shows that by increasing the thickness-to-inclined length of the cell rib, the continuum will be more resistant to buckling-mode failure. The auxetic beam will be stiffer in this situation because a larger content of the meta-nanocomposite will be used. Therefore, given the clear correlation between buckling load and stiffness, it is not surprising to detect such a hardening tendency. Additionally, it is shown that by giving the gradient index a higher value, the nonlinear buckling load of the beam in any specified BC will be decreased. This tendency may be explained by keeping in mind the gradient index's detrimental impact on the meta-nanocomposite's effective modulus at any given thickness (see Figure 2.19).

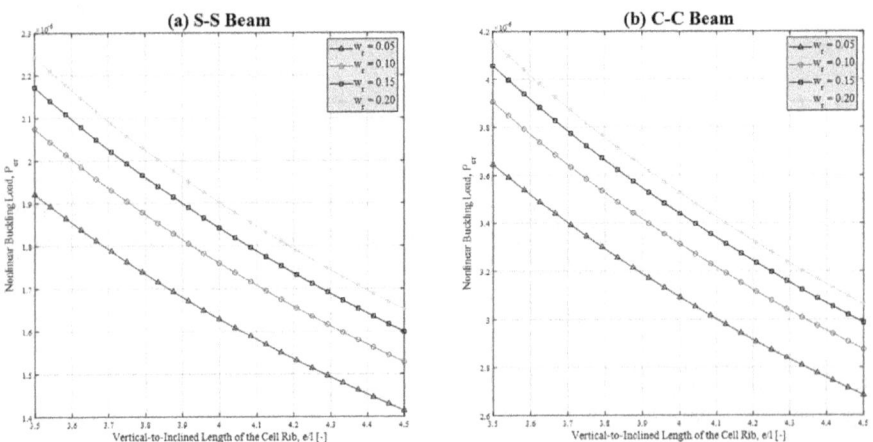

FIGURE 2.22 Variation of the nonlinear buckling load of (a) S-S and (b) C-C auxetic PNC beams versus vertical-to-inclined length of the cell rib while different values are assigned to the mass fraction of the CNTs ($W=h$, $\eta=1$, $\mu=0.4$, $p=5$, $\theta=-45$).

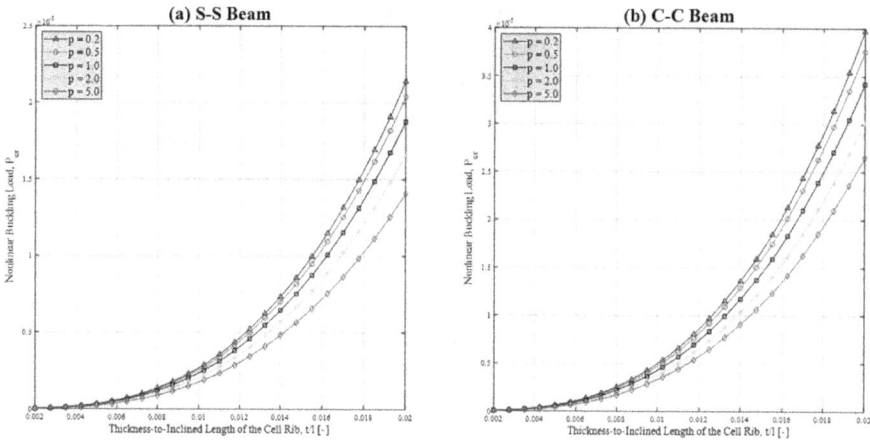

FIGURE 2.23 Variation of the nonlinear buckling load of (a) S-S and (b) C-C auxetic PNC beams versus thickness-to-inclined length of the cell rib while different values are assigned to the gradient index ($W=h$, $\eta=1$, $\mu=0.4$, $w_r=0.1$, $\theta=-45$).

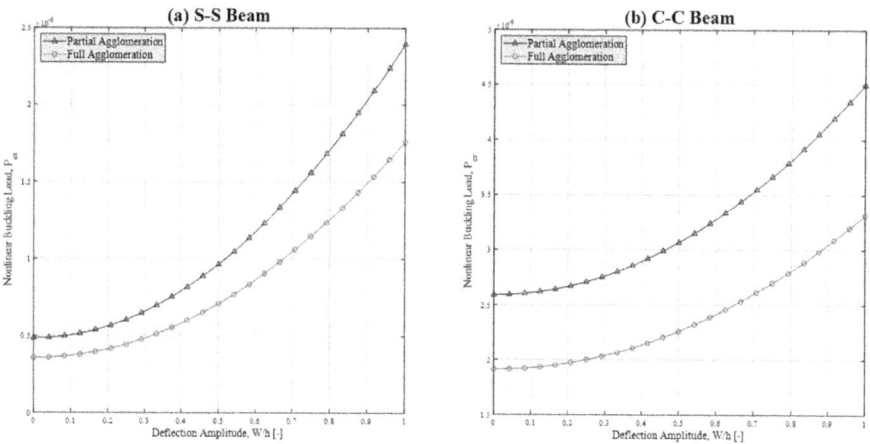

FIGURE 2.24 Postbuckling path of (a) S-S and (b) C-C auxetic PNC beams while different patterns of nanofillers' agglomeration are considered ($\mu=0.4$, $w_r=0.1$, $p=5$, $\theta=-45$, $t/l=0.01$, $e/l=4$).

The influence of the agglomeration pattern on the auxetic nanocomposite beams' postbuckling course is lastly shown in Figure 2.24. Two situations are taken into account in this figure. Partial agglomeration refers to the arrangement of certain CNTs inside clusters in the first scenario; complete agglomeration refers to the arrangement of all nanofillers. It is clear that once the whole agglomeration pattern is used instead of a partial one, the meta-nanocomposite beam's postbuckling path will be pushed lower. The rigidity of the meta-nanocomposite is negatively impacted by complete agglomeration more than partial agglomeration, which is the physical cause of this problem. In other words, the stiffness of the nanomaterial would be more adversely impacted if all of the CNTs are entangled in the clusters. The tendency that has been seen is therefore entirely reasonable. Another time, it is confirmed that using a clamped support rather than a simple one allows for the application of larger compressive excitations to the auxetic nanocomposite beam.

2.3 SUMMARY

This chapter's main goal was to investigate how the auxeticity and aggregation of CNTs affect the nonlinear bending and postbuckling responses of meta-nanomaterial beams. In this context, a two-level micromechanical method was developed to determine the effective Young's moduli of the auxetic CNTR nanomaterial hosted by a polymer. The nonlinear bending issue was put forward using the traditional beam theory. The issue was then resolved analytically to track changes in the beam's amplitude of deflection. The purpose of the second section of this chapter was to investigate the nonlinear deflection properties of porous auxetic CNTR polymer beams that were bent uniformly. We used a refined-type HSDT and paired it with von Kármán's theorem in the context of an energy-based algorithm to obtain the motion relations of the thick meta-nanomaterial structures since we wanted to deliver trustworthy data for thick beams. This chapter's last section examined how the nonlinear bending and buckling loads of thin meta-nanocomposite beams varied. In order to do this, a two-step micromechanical technique was used to determine the equivalent modulus of the auxetic CNTR nanocomposite with regard to the influence of the CNT aggregation on the modulus estimate. After that, the issue was formulated using the conventional theory of thin beams. The postbuckling load of the beam was reached after the issue was resolved using the analytical Galerkin approach. The highlights that may be noted are as follows, according to this thorough study [34–50]:

- The utilization of narrow cells to create a re-entrant lattice can offer advantages in improving the resilience of the meta-nanomaterial structure against bending excitations.
- If a thicker lattice is employed for the meta-nanomaterial, it is possible to decrease the bending response of the system.
- The nonlinear flexural deflection of the auxetic nanocomposite structure may experience an increase in the event that aggregation of the CNTs occurs within the clusters.
- It is advisable to utilize functionalized CNTs in the development of these sophisticated meta-nanomaterials in order to mitigate excessive deflections.
- The deformation of the designed meta-nanomaterial structure can be regulated through the utilization of thick-walled lattices composed of narrow nanocomposites.
- The findings of this study demonstrate that by altering the geometrical specifications of the cell rib in the auxetic network, it is possible to manipulate the deflection amplitude of the beam.

- Thick-walled lattices exhibit exceptional potential as suitable options for addressing deflection reduction scenarios.
- Wide cell ribs are necessary in order to achieve a flexible meta-nanomaterial beam.
- The influence of non-straight reinforcing nanofillers on the nonlinear deflection of porous auxetic CNTR beams was effectively demonstrated.
- It has been demonstrated that it is more advantageous to avoid the presence of large pores located at the upper and lower edges of the meta-nanomaterial structure. In the event that such an undesirable condition arises, the magnitude of the nonlinear deflection will surpass that which is anticipated by ideal theoretical calculations.
- If the presence of agglomerates of CNTs is disregarded in the static design, the compressive excitations required for buckling-mode failure are lower than the theoretical predictions.
- The postbuckling load of the meta-nanocomposite can be increased by employing cell ribs with greater thickness.
- It is advisable to employ smaller auxeticity angles in order to enhance the buckling load resistance of the auxetic nanocomposite structure.
- The utilization of an excessively broad auxetic lattice is not suitable for the design of a meta-nanocomposite structure intended to withstand compressive stimuli.
- The current study's key findings can be utilized in the development of wearable strain sensors using meta-nanomaterials, which exhibit high sensitivity due to the negative Poisson's ratio characteristic of these nanocomposites.

REFERENCES

1. Ebrahimi, F. & Dabbagh, A., 2020. *Mechanics of Nanocomposites: Homogenization and Analysis.* 1st ed. Boca Raton, FL: CRC Press.
2. Ebrahimi, F. & Dabbagh, A., 2022. Vibration analysis of multi-scale hybrid nanocomposite shells by considering nanofillers' aggregation. *Waves in Random and Complex Media*, 32(3), pp. 1060–1078.
3. Ebrahimi, F., Nopour, R. & Dabbagh, A., 2022. Effect of viscoelastic properties of polymer and wavy shape of the CNTs on the vibrational behaviors of CNT/glass fiber/polymer plates. *Engineering with Computers*, 38(5), pp. 4113–4126.
4. Ebrahimi, F., Nopour, R. & Dabbagh, A., 2022. Effects of polymer's viscoelastic properties and curved shape of the CNTs on the dynamic response of hybrid nanocomposite beams. *Waves in Random and Complex Media*, 2022, pp. 1–18.
5. Nopour, R., Ebrahimi, F., Dabbagh, A. & Aghdam, M. M., 2023. Nonlinear forced vibrations of three-phase nanocomposite shells considering matrix rheological behavior and nano-fiber waviness. *Engineering with Computers*, 39(1), pp. 557–574.
6. Ebrahimi, F., Nouraei, M., Dabbagh, A. & Rabczuk, T., 2019. Thermal buckling analysis of embedded graphene-oxide powder-reinforced nanocomposite plates. *Advances in Nano Research*, 7(5), pp. 293–310.
7. Ebrahimi, F., Dabbagh, A., Rastgoo, A. & Rabczuk, T., 2020. Agglomeration effects on static stability analysis of multi-scale hybrid nanocomposite plates. *Computers, Materials & Continua*, 63(1), pp. 41–64.

8. Dabbagh, A., Rastgoo, A. & Ebrahimi, F., 2021. Static stability analysis of agglomerated multi-scale hybrid nanocomposites via a refined theory. *Engineering with Computers*, 37(3), pp. 2225–2244.

9. Ebrahimi, F. & Dabbagh, A., 2021. An analytical solution for static stability of multi-scale hybrid nanocomposite plates. *Engineering with Computers*, 37(1), pp. 545–559.

10. Dabbagh, A., Rastgoo, A. & Ebrahimi, F., 2022. Post-buckling analysis of imperfect multi-scale hybrid nanocomposite beams rested on a nonlinear stiff substrate. *Engineering with Computers*, 38(1), pp. 301–314.

11. Ebrahimi, F., Dabbagh, A. & Rastgoo, A., 2023. Static stability analysis of multi-scale hybrid agglomerated nanocomposite shells. *Mechanics Based Design of Structures and Machines*, 51(1), pp. 501–517.

12. Ebrahimi, F., Dabbagh, A. & Taheri, M., 2021. Vibration analysis of porous metal foam plates rested on viscoelastic substrate. *Engineering with Computers*, 37(4), pp. 3727–3739.

13. Ebrahimi, F. & Dabbagh, A., 2022. *Mechanics of Multiscale Hybrid Nanocomposites*. 1st ed. Cambridge, MA: Elsevier.

14. Dabbagh, A. & Ebrahimi, F., 2021. Postbuckling analysis of meta-nanocomposite beams by considering the CNTs' agglomeration. *The European Physical Journal Plus*, 136(11), p. 1168.

15. Thai, H.-T. & Vo, T. P., 2012. Bending and free vibration of functionally graded beams using various higher-order shear deformation beam theories. *International Journal of Mechanical Sciences*, 62(1), pp. 57–66.

16. Ebrahimi, F. & Dabbagh, A., 2023. Porosity effects on static performance of Carbon Nanotube-Reinforced Meta-Nanocomposite structures. *Micromachines*, 14(7), p. 1402.

17. Feng, C., Kitipornchai, S. & Yang, J., 2017. Nonlinear bending of polymer nanocomposite beams reinforced with non-uniformly distributed graphene platelets (GPLs). *Composites Part B: Engineering*, 110, pp. 132–140.

18. Dabbagh, A., Rastgoo, A. & Ebrahimi, F., 2019. Finite element vibration analysis of multi-scale hybrid nanocomposite beams via a refined beam theory. *Thin-Walled Structures*, 140, pp. 304–317.

19. Ebrahimi, F. & Dabbagh, A., 2019. A novel porosity-based homogenization scheme for propagation of waves in axially-excited FG nanobeams. *Advances in Nano Research*, 7(6), pp. 379–390.

20. Ebrahimi, F. & Dabbagh, A., 2019. *Wave Propagation Analysis of Smart Nanostructures*. Boca Raton, FL: CRC Press.

21. Ebrahimi, F., Dabbagh, A. & Rastgoo, A., 2019. Vibration analysis of porous metal foam shells rested on an elastic substrate. *The Journal of Strain Analysis for Engineering Design*, 54(3), pp. 199–208.

22. Ebrahimi, F., Jafari, A. & Barati, M. R., 2017. Vibration analysis of magneto-electro-elastic heterogeneous porous material plates resting on elastic foundations. *Thin-Walled Structures*, 119, pp. 33–46.

23. Ebrahimi, F., Nouraei, M., Dabbagh, A., et al., 2019b. Thermal buckling analysis of embedded graphene-oxide powder-reinforced nanocomposite plates. *Advances in Nano Research*, 7(5), pp. 293–310.

24. Taheri, M. & Ebrahimi, F., 2020. Buckling analysis of CFRP plates: A porosity-dependent study considering the GPLs-reinforced interphase between fiber and matrix. *The European Physical Journal Plus*, 135, pp. 1–19.

25. Ebrahimi, F., Jafari, A. & Mahesh, V., 2019. Assessment of porosity influence on dynamic characteristics of smart heterogeneous magneto-electro-elastic plates. *Structural Engineering and Mechanics, An Int'l Journal*, 72(1), pp. 113–129.

26. Ebrahimi, F., Dabbagh, A., Rabczuk, T. & Tornabene, F., 2019. Analysis of propagation characteristics of elastic waves in heterogeneous nanobeams employing a new two-step porosity-dependent homogenization scheme. *Advances in Nano Research*, 7(2), p. 135.

27. Ebrahimi, F. & Dabbagh, A., 2019. Wave dispersion characteristics of heterogeneous nanoscale beams via a novel porosity-based homogenization scheme. *The European Physical Journal Plus*, 134, pp. 1–8.

28. Ebrahimi, F., Seyfi, A. & Dabbagh, A., 2019. A novel porosity-dependent homogenization procedure for wave dispersion in nonlocal strain gradient inhomogeneous nanobeams. *The European Physical Journal Plus*, 134, pp. 1–11.

29. Arasteh, R., Omidi, M., Rousta, A. H. A., et al., 2011. A study on effect of waviness on mechanical properties of multi-walled carbon nanotube/epoxy composites using modified Halpin—Tsai theory. *Journal of Macromolecular Science*, Part B 50(12), pp. 2464–2480.

30. Ebrahimi, F., Dabbagh, A. & Rabczuk, T., 2021. On wave dispersion characteristics of magnetostrictive sandwich nanoplates in thermal environments. *European Journal of Mechanics—A/Solids*, 85, p. 104130. https://doi.org/10.1016/j.euromechsol.2020.104130

31. Reddy, J. N., 1984. A simple higher-order theory for laminated composite plates. *Journal of Applied Mechanics*, 51(4), pp. 745–752.

32. Yang, J., Wu, H. & Kitipornchai, S., 2017. Buckling and postbuckling of functionally graded multilayer graphene platelet-reinforced composite beams. *Composite Structures*, 161, pp. 111–118.

33. Ebrahimi, F. & Dabbagh, A., 2022. *Mechanics of Multiscale Hybrid Nanocomposites*. Duxford: Elsevier. ISBN: 9780128196144

34. Ebrahimi, F., Barati, M. R. & Haghi, P., 2016. Thermal effects on wave propagation characteristics of rotating strain gradient temperature-dependent functionally graded nanoscale beams. *Journal of Thermal Stresses*, pp. 1–13.

35. Ebrahimi, F. & Dabbagh, A., 2016. On flexural wave propagation responses of smart FG magneto-electro-elastic nanoplates via nonlocal strain gradient theory. *Composite Structures* 162(2017): 281–293.

36. Ebrahimi, F. & Salari, E., 2015. Size-dependent thermo-electrical buckling analysis of functionally graded piezoelectric nanobeams. *Smart Materials and Structures*, 24(12), p. 125007.

37. Ebrahimi, F. & Barati, M. R., 2016. A unified formulation for dynamic analysis of nonlocal heterogeneous nanobeams in hygro-thermal environment. *Applied Physics A*, 122(9), p. 792.

38. Ebrahimi, F. & Barati, M. R., 2016. A nonlocal higher-order refined magneto-electro-viscoelastic beam model for dynamic analysis of smart nanostructures. *International Journal of Engineering Science*, 107, pp. 183–196.

39. Ebrahimi, F. & Barati, M. R., 2017. A nonlocal strain gradient refined beam model for buckling analysis of size-dependent shear-deformable curved FG nanobeams. *Composite Structures*, 159, pp. 174–182.

40. Ebrahimi, F. & Barati, M. R., 2016. Buckling analysis of nonlocal third-order shear deformable functionally graded piezoelectric nanobeams embedded in elastic medium. *Journal of the Brazilian Society of Mechanical Sciences and Engineering*, pp. 1–16.

41. Ebrahimi, F. & Barati, M. R., 2016. Magnetic field effects on buckling behavior of smart size-dependent graded nanoscale beams. *The European Physical Journal Plus*, 131(7), pp. 1–14.

42. Ebrahimi, F. & Barati, M. R., 2016. Buckling analysis of smart size-dependent higher order magneto-electro-thermo-elastic functionally graded nanosize beams. *Journal of Mechanics*, pp. 1–11.

43. Ebrahimi, F. & Barati, M. R., 2017. A nonlocal strain gradient refined beam model for buckling analysis of size-dependent shear-deformable curved FG nanobeams. *Composite Structures*, 159, pp. 174–182.

44. Ebrahimi, F., Barati, M. R. & Haghi, P., 2016. Thermal effects on wave propagation characteristics of rotating strain gradient temperature-dependent functionally graded nanoscale beams. *Journal of Thermal Stresses*, pp. 1–13.

45. Ebrahimi, F. & Dabbagh, A., 2017. Nonlocal strain gradient based wave dispersion behavior of smart rotating magneto-electro-elastic nanoplates. *Materials Research Express*, 4(2), p. 025003.

46. Ebrahimi, F. & Barati, M. R., 2016. Wave propagation analysis of quasi-3D FG nanobeams in thermal environment based on nonlocal strain gradient theory. *Applied Physics A*, 122(9), p. 843.

47. Ebrahimi, F., Barati, M. R. & Dabbagh, A., 2016. A nonlocal strain gradient theory for wave propagation analysis in temperature-dependent inhomogeneous nanoplates. *International Journal of Engineering Science*, 107, pp. 169–182.

48. Ebrahimi, F. & Barati, M. R., 2016. Flexural wave propagation analysis of embedded S-FGM nanobeams under longitudinal magnetic field based on nonlocal strain gradient theory. *Arabian Journal for Science and Engineering*, pp. 1–12.

49. Ebrahimi, F., Ghazali, M. & Dabbagh, A., 2022. Hygro-thermo-viscoelastic wave propagation analysis of FGM nanoshells via nonlocal strain gradient fractional time—space theory. *Waves in Random and Complex Media*, pp. 1–20.

50. Ebrahimi, F., Barati, M. R. & Dabbagh, A., 2016. Wave dispersion characteristics of axially loaded magneto-electro-elastic nanobeams. *Applied Physics A*, 122(11), p. 949.

APPENDICES

APPENDIX 2A

The stiff components used in Eqs. (2.5) and (2.6) can be calculated from following relation [14]:

$$\alpha_r = \frac{3(K_m + G_m) + k_r + l_r}{3(G_m + k_r)} \tag{A.1}$$

$$\beta_r = \frac{1}{5}\left(\frac{4G_m + 2k_r + l_r}{3(G_m + k_r)} + \frac{4G_m}{G_m + p_r} + \frac{2\left(G_m(3K_m + G_m) + G_m(3K_m + 7G_m)\right)}{G_m(3K_m + G_m) + m_r(3K_m + 7G_m)} \right) \tag{A.2}$$

$$\delta_r = \frac{1}{3}\left(n_r + 2l_r + \frac{(2k_r + l_r)(3K_m + G_m - l_r)}{G_m + k_r} \right) \tag{A.3}$$

$$\eta_r = \frac{1}{5}\left(\begin{array}{c} \dfrac{2}{3}(n_r - l_r) + \dfrac{8G_m p_r}{G_m + p_r} + \dfrac{(2k_r - l_r)(2G_m + l_r)}{3(G_m + k_r)} \\ + \dfrac{8m_r G_m(3K_m + 4G_m)}{3K_m(m_r + G_m) + G_m(7m_r + G_m)} \end{array} \right) \tag{A.4}$$

In these definitions, subscripts "*m*" and "*r*" respectively correspond with polymer (matrix) and CNT. The coefficients k_r, l_r, m_r, n_r, and p_r are the Hill's elastic constants of the CNTs.

3 Vibration and Dynamic Characteristics of Auxetic Structures

3.1 BACKGROUND

The primary aim of this chapter is to establish guidelines for controlling the vibrational and dynamic properties of auxetic structures, specifically those incorporating an auxetic or graphene-origami auxetic core. The initial segment of this chapter focuses on the examination of the torsional vibration characteristics exhibited by non-circular nanorods composed of auxetic honeycomb material. Hamilton's principle is employed in order to derive the kinematic relation of rods at the nanoscale. The analytical method is employed to solve the nonlocal derived governing equations of non-circular auxetic nanorods under various boundary conditions (B.C.). This study presents an analysis of the vibrational behavior of a novel sandwiched plate, incorporating a metamaterial core and electro-magnetically graded composite layers, using the Mindlin plate theory. The governing equations and associated boundary conditions for sandwich plates are derived by incorporating geometric nonlinearity using von Karman's theory, Maxwell's law, and Hamilton's principle. Subsequently, the governing equations of a sandwich plate are discretized utilizing the generalized differential quadrature method (GDQM). In this section, the present study employs the higher-order Reddy plate theory to examine the mathematically nonlinear oscillation of a unique piezoelectrically sandwich plate. This plate is composed of a center made of FG graphene origami auxetic (FG-GOA) material and layers of smart porous multi-scale hybrid (GPL/CF/PVDF) nanocomposite (SPHNC). Furthermore, the nonlinear frequency ratio and nonlinear dynamic characteristics of the hybrid nanocomposite plate are assessed with respect to the electrical voltage, the weight fraction of graphene platelets (GPL), and the volume fraction of carbon fibers.

This chapter focuses on the investigation of the time-dependent viscoelastic deflection characteristics of laminated composite structures. These structures are composed of an auxetic core, which is surrounded by layers of piezoelectric and gold materials at the bottom and top, respectively. Moreover, the primary objective of the present investigation is to regulate the vibrational characteristics of an auxetic plate that has been coated with a magnetostrictive substance. The findings suggest that the incorporation of an auxetic core into a magnetostrictive plate leads to an augmentation in the dimensionless natural frequency. The results obtained from this study can potentially contribute to advancements in the field of sensor and

DOI: 10.1201/9781003289296-3

actuator development, as well as in the area of vibration cancellation techniques. Additionally, the obtained results could serve as a foundational basis for subsequent research endeavors.

This chapter focuses on examining the nonlinear vibration characteristics of the FG-GOA sandwich doubly curved shell, which is integrated with multi-scale smart hybrid nanocomposite layers. The Homotopy Perturbation Method (HPM) is employed for this investigation. The present study proposes the utilization of multi-scale hybrid nanocomposite layers comprising graphene platelet (GPL), carbon fiber (CF), and polyvinylidene fluoride polymer (PVDF), employing the Halpin-Tsai model. The governing equation for the nonlinear vibration analysis of smart FG-GOA sandwich doubly curved shells is derived by applying the Hamilton principle and Maxwell rule, utilizing the third-order shear deformation theory (TSDT) and the von-Karman nonlinearity factor. Consequently, this chapter presents a comprehensive examination of various characteristics, such as the weight fraction of the graphene platelet, the diverse volume fraction of CF, the thicknesses of the smart layers, and the FG-GOA core. The nonlinear frequency ratio in each graph is significantly influenced by these factors. The findings derived from this investigation have the potential to make significant contributions to the progress of sensor and actuator technology, as well as the application of strategies aimed at mitigating vibrations.

3.2   AUXETIC RODS: VIBRATION ANALYSIS

The initial segment of this chapter delves into the examination of the torsional vibration characteristics exhibited by non-circular nanorods composed of auxetic honeycomb material. The analysis will focus on the triangular and elliptical cross-sections. Furthermore, it is noteworthy that auxetic honeycomb materials exhibit a unique characteristic known as a negative Poisson's ratio, which can have significant implications for the design and analysis of various structures. To account for the small size or scale effect, the constitutive relation of nanorods is adjusted using Eringen's nonlocal elasticity theory (ENET). Hamilton's principle is employed in order to derive the kinematic relation of rods at the nanoscale. The analytical method is employed to solve the nonlocal derived governing equations of non-circular auxetic nanorods under various boundary conditions. To ensure the accuracy of the obtained results, they are compared to previous investigations, and a strong correlation can be observed. This study investigates the effects of different parameters, including Clamped-Clamped (C-C) and Clamped-Free (C-F) boundary conditions, nonlocal parameter, inclined angles, and geometrical ratio. The findings are presented and visualized in a series of figures, which provide detailed insights into the subject matter.

In contemporary times, there is a growing imperative to design and fabricate materials that possess the capability to withstand and endure high levels of shock loading. Researchers and engineers employ a novel class of materials known as auxetic materials to address this issue. One notable characteristic of such materials is

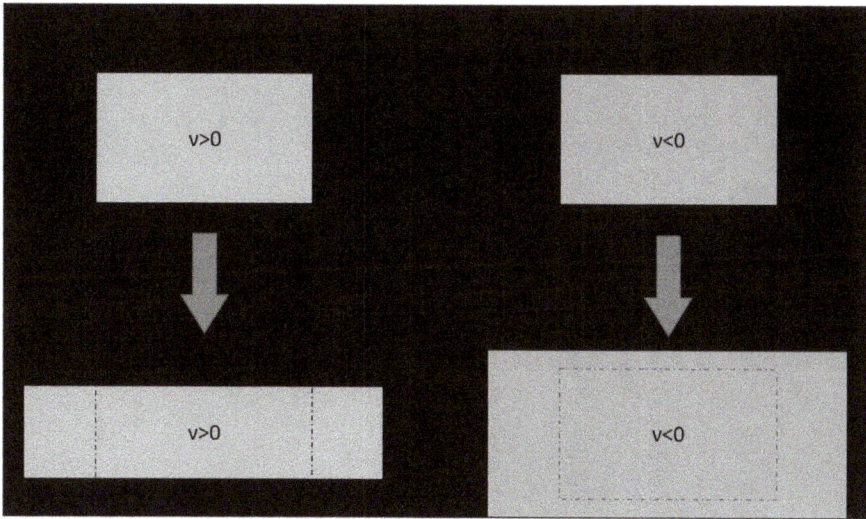

FIGURE 3.1   Comparison between behavior of conventional and auxetic materials against tensile.

their negative Poisson's ratio, which is a crucial parameter in the field of mechanical engineering for the design and analysis of structures with diverse applications. Figure 3.1 is presented to enhance the comprehension of readers regarding the disparities between the behavior of conventional materials and auxetic materials when subjected to tensile forces [1].

A range of mechanical analyses have been conducted to investigate the mechanical behavior of auxetic structures, including vibration, buckling, and wave propagation [1–16]. The utilization of nanostructures in engineering is driving novel designs, replacing conventional macroscale structures due to the unique characteristics exhibited by nanoscale elements. According to the findings presented, classical continuum theories demonstrate limitations in accurately assessing the behaviors of nanostructures. Therefore, in order to address these challenges, scale-dependent theories were formulated to investigate the static and dynamic behaviors of nanoscale continuum systems. One of the continuum theories that is dependent on scale is the theory of nonlocal elasticity, which was introduced by Eringen [17]. In this theoretical framework, it is imperative to consider the stress state at any given point within a continuum as a multi-unknown function of the strains exhibited by all neighboring points, including the strain of the point in question. Numerous researchers have employed this theory to examine diverse issues pertaining to the mechanical behavior of nanostructures. Furthermore, numerous inquiries have been conducted utilizing alternative scale-dependent theories, including nonlocal strain gradient theory (NSGT) and modified couple stress theory (MCST) [18–97]. Ebrahimi and Haghi [98] conducted an analysis on

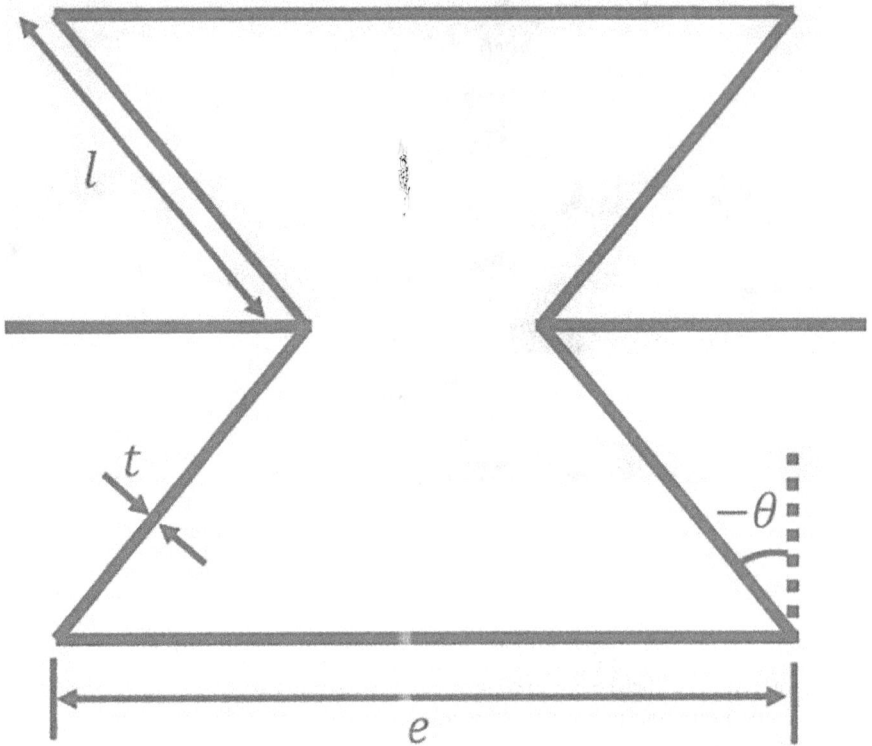

FIGURE 3.2 The cellular structure of the auxetic honeycomb core.

the dispersion of waves in rotating FG nanosize beams within the thermal environment, utilizing the framework of NSGT. In their study, Ebrahimi et al. [99] utilized
Eringen's nonlocal elasticity theory (ENET) to investigate the behavior of propa-
gated waves in a magneto-electro-elastic (MEE) nanotube. In addition, Ebrahimi
et al. conducted an analysis on the wave propagation characteristics of imperfect
nanoscale beams and plates made of functionally graded materials (FG). This analysis involved the implementation of the Nonlinear Spectral Galerkin Technique
(NSGT) [100, 101]. The torsional vibration analysis of non-circular rectangular
cross-section nanorods under various boundary conditions was investigated by
Khosravi et al. [102]. In this particular section, a reformulated auxetic honeycomb
structure has been employed as the constitutive material for the nanorod. The utilization of auxetic material represents a significant advantage in this study. Figure
3.2 illustrates multiple parameters of the auxetic unit. In the depicted diagram,
the variable θ denotes the angle of inclination. The extensive examination of this
significant parameter is presented in detail within the subsequent section on outcomes. The dimensions of the auxetic cell, specifically the inclined and horizontal
edges, are denoted as l and e, respectively [1].

The calculation of the effective mechanical properties of auxetic honeycomb materials can be achieved according to reference [39].

$$
\left\{
\begin{aligned}
E_1^a &= E\left(\frac{t}{l}\right)^3 \frac{\cos\theta}{\left(\dfrac{e}{l}+\sin\theta\right)\sin^2\theta} \\[2mm]
v_{12}^a &= \frac{\cos^2\theta}{\left(\dfrac{e}{l}+\sin\theta\right)\sin\theta} \\[2mm]
\rho^a &= \rho \frac{\dfrac{t}{l}\left(\dfrac{e}{l}+2\right)}{2\cos\theta\left(\dfrac{e}{l}+\sin\theta\right)} \\[2mm]
G_{12}^a &= E\left(\frac{t}{l}\right)^3 \frac{\left(\dfrac{e}{l}+\sin\theta\right)}{\left(\dfrac{e}{l}\right)^2\left(1+2\dfrac{e}{l}\right)\cos\theta}
\end{aligned}
\right.
\tag{3.1}
$$

In the context of these relationships, the material properties of Young's modulus, Poisson's ratio, density, and shear modulus are denoted by the symbols $E, v, \rho,$ and G, respectively. The notation a is used to represent the properties of auxetic materials. In this study, the variables $\frac{t}{l}$ and $\frac{e}{l}$ have been maintained at constant values, specifically 4 and 0.0138571, respectively. When fixed values are employed, Poisson's ratio can exhibit a negative value. In this model, the displacement field of an auxetic rod in the x-direction for any desired point can be defined as:

$$
u_1(x,t) = \psi(y,z)\frac{\partial\theta}{\partial x}
\tag{3.2a}
$$

$$
u_2(x,t) = -z\theta(x,t)
\tag{3.2b}
$$

$$
u_3(x,t) = y\theta(x,t)
\tag{3.2c}
$$

In these formulations, the variables u_1, u_2, and u_3 represent the displacements along the x, y, and z directions, respectively. Additionally, the symbols θ and ψ (y, z) are used to denote the Twix angle of the nanorods and the warping function, respectively. This chapter explores the phenomenon of torsional vibration in triangular and elliptical nanorods, with each nanorod exhibiting its own unique warping function. The warping functions can be expressed in the following form [1]:

$$
\psi_t(y,z) = \frac{1}{6\eta}y\left(3z^2 - y^2\right)
\tag{3.3}
$$

$$\psi_e(y,z) = y \times z\left(\frac{b^2 - a^2}{b^2 + a^2}\right) \tag{3.4}$$

The expressions for warping functions of triangular and elliptical nanorods are denoted as $\psi_t(y,z)$ and $\psi_e(y,z)$, respectively. Additionally, the variable η is defined as $\eta = \dfrac{\sqrt{3}\,\zeta}{6}$. If the center of gravity of the triangle is located at the point (0,0), then the vertices of the equilateral triangle can be represented as $\left(-\sqrt{3}\eta, -\eta\right), \left(\sqrt{3}\eta, \eta\right), \left(0, 2\eta\right)$. To enhance readers' comprehension, Figure 3.3 presents a schematic representation of non-circular nanorods.

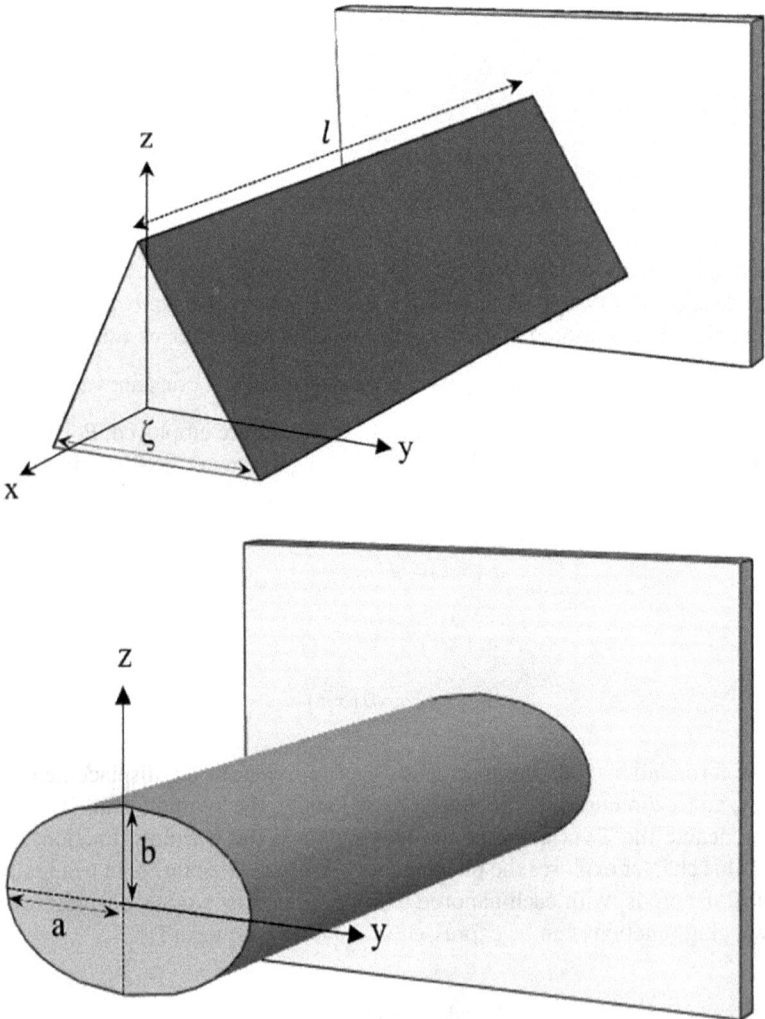

FIGURE 3.3   Geometrical representation of (a) triangular and (b) elliptical nanorods.

The non-zero elements of the strain field can be expressed as follows:

$$\varepsilon_{xy} = \frac{\partial u_1}{\partial y} + \frac{\partial u_2}{\partial x} = \left(\frac{\partial \psi}{\partial y} - z\right)\frac{\partial \theta}{\partial x} \tag{3.5}$$

$$\varepsilon_{xz} = \frac{\partial u_1}{\partial z} + \frac{\partial u_3}{\partial x} = \left(\frac{\partial \psi}{\partial z} + y\right)\frac{\partial \theta}{\partial x} \tag{3.6}$$

The governing equation for torsional vibration is derived by applying Hamilton's principle within the time interval from 0 to t.

$$\int_0^t \left[\delta U - \left(\delta T + \delta W_{nc}\right)\right] dt = 0 \tag{3.7}$$

In Hamilton's equation, the symbols U, T, and W_{nc} represent the quantities of strain energy, kinetic energy, and work performed by non-conservative external forces, respectively. The derivation of the variation of strain energy can be obtained from equation (3.7).

$$\delta \int_0^t U dt = \int_0^t \int_v \sigma_{ij} \delta \varepsilon_{ij} dV = \int_v \left\{\sigma_{xy} \delta \varepsilon_{xy} + \sigma_{xz} \delta \varepsilon_{xz}\right\} dV \tag{3.8}$$

The introduction of equations (3.5) to (3.6) into equation (3.8) results in:

$$\delta \int_0^t U dt = \int_0^t \int_v \sigma_{ij} \delta \varepsilon_{ij} dV = \int_0^t \int_0^t \left\{\left(M^T \frac{\partial \delta \theta}{\partial x}\right) dx\right.$$
$$\left. + \int_0^t \int_a \left(\sigma_{xy} \frac{\partial \delta \psi(y,z)}{\partial y} + \sigma_{xz} \frac{\partial \delta \psi(y,z)}{\partial z}\right)\frac{\partial \theta}{\partial x} dA dx\right\} dt \tag{3.9}$$

The expression for the moment of torsional torque can be represented as follows, in cases where the cross-section is not circular.

$$M^T = \int_A \left(\sigma_{xy}\left(\frac{\partial \psi(y,z)}{\partial y} - z\right) + \sigma_{xz}\left(\frac{\partial \psi(y,z)}{\partial z} + y\right)\right) dA \tag{3.10}$$

Equation (3.9) may be rewritten as follows:

$$\delta \int_0^t U dt = \int_0^t \int_v \sigma_{ij} \delta \varepsilon_{ij} dV = \int_0^t \left(M^T\Big|_0^l - \int_0^l \frac{\partial M^T}{\partial x}\delta\theta dx\right) dt \tag{3.11}$$

For further Hamiltonian components, the nanorod's kinetic energy is provided by:

$$T = \frac{1}{2}\int_v \rho \left\{\psi^2(y,z)\left(\frac{\partial^2 \theta}{\partial x \partial t}\right)^2 + \left(-z\frac{\partial \theta}{\partial t}\right)^2 + \left(y\frac{\partial \theta}{\partial t}\right)^2\right\} dV = I_a + I_b \tag{3.12}$$

where ρ, I_b, and I_a stand for density, polar inertia, and axial inertia, respectively, in the equation above, may be expressed as follows:

$$I_a = \frac{1}{2}\int_0^l\int_A \rho\psi^2(y,z)\left(\frac{\partial^2\theta}{\partial x\partial t}\right)^2 dAdx \qquad (3.13)$$

$$I_b = \frac{1}{2}\int_0^l \rho I_p\left(\frac{\partial\theta}{\partial t}\right)^2 dx = \frac{1}{2}\int_0^l I_o\left(\frac{\partial\theta}{\partial t}\right)^2 dx \qquad (3.14)$$

where the variables I_o and I_p, which represent the mass and polar moment of inertias, respectively, in equation (3.14), may be written as follows:

$$I_p = \int_A \left(y^2 + z^2\right)dA \qquad (3.15)$$

$$I_o = \rho\int_A \left(y^2 + z^2\right)dA = \rho I_p \qquad (3.16)$$

The following provides the initial variation of the mass and polar moments of inertia:

$$\delta\int_0^t I_a = \int_0^t\int_0^l \rho I_\psi \frac{\partial^4\theta}{\partial x^2\partial t^2}\delta\theta dxdt - \int_0^t \rho I_\psi \frac{\partial^3\theta}{\partial x\partial t^2}\delta\theta\Big|_0^l\, dt \qquad (3.17)$$

$$\delta\int_0^t I_b = \int_0^l I_o\left(\frac{\partial\theta}{\partial t}\delta\theta\Big|_0^l - \int_0^l \frac{\partial^2\theta}{\partial t^2} dx\right) \qquad (3.18)$$

where

$$I_\psi = \int_A \psi^2(y,z)dA \qquad (3.19)$$

Since there are no external non-conservative factors at play in this inquiry, it is acceptable to disregard any variance in the work produced by these forces. Hamilton's principle may be used to determine the equation of motion for auxetic non-circular nanorods by adding equations (3.11) and (3.12) into equation (3.7). The result is as follows:

$$\frac{\partial M^T}{\partial x} = I_o\frac{\partial^2\theta}{\partial t^2} - \rho I_\psi\frac{\partial^4\theta}{\partial x^2\partial t^2} \qquad (3.20)$$

The following provides the relation's boundary condition:

$$\left(M^T + \rho I_\psi\frac{\partial^3\theta}{\partial x\partial t^2}\right)\Big|_0^l = 0 \text{ or } \delta\theta\big|_0^l = 0 \qquad (3.21)$$

Based on Eringen's nonlocal micropolar elasticity [2], the following is a statement of the constitutive nonlocal relations of nanostructures:

$$\sigma_{xy} - \mu\frac{\partial^2\sigma_{xy}}{\partial x^2} = G\frac{\partial\theta}{\partial x}\left(\frac{\partial\psi(y,z)}{\partial y} - z\right) \qquad (3.22)$$

$$\sigma_{xz} - \mu \frac{\partial^2 \sigma_{xz}}{\partial x^2} = G \frac{\partial \theta}{\partial x} \left(\frac{\partial \psi(y,z)}{\partial y} + y \right) \tag{3.23}$$

Here, μ stands for the nonlocal parameter, which is equivalent to $(e_0 a)^2$, where a is the material constant and is the internal characteristic length. The dispersion curves of plane waves and atomic lattice dynamics are matched to experimentally calculate or approximate these characteristics. Thus, the following formula may be used to compute the nonlocal torsional vibration:

$$M^T - \mu \frac{\partial^2 M^T}{\partial x^2} = G \left(\int_A \left(\frac{\partial \psi(y,z)}{\partial y} - z \right)^2 + \left(\frac{\partial \psi(y,z)}{\partial z} + y \right)^2 dA \right) \frac{\partial \theta}{\partial x} \tag{3.22}$$

The nonlocal governing equation of the auxetic nanorod as reaction to torsional vibration may be stated as follows by coupling equation (3.20) with equation (3.22) and simplifying:

$$G \left(\int_A \left(\frac{\partial \psi(y,z)}{\partial y} - z \right)^2 + \left(\frac{\partial \psi(y,z)}{\partial z} + y \right)^2 dA \right) \frac{\partial^2 \theta}{\partial x^2} + \rho I_\psi \frac{\partial^4 \theta}{\partial x^2 \partial t^2}$$

$$- I_o \frac{\partial^2 \theta}{\partial t^2} + \mu \left(I_o \frac{\partial^4 \theta}{\partial t^2 \partial x^2} - \rho I_\psi \frac{\partial^6 \theta}{\partial x^4 \partial t^2} \right) = 0 \tag{3.23}$$

To extract the numerical results of the computed formulation, a generic solution has been offered in this section:

$$\theta(x,t) = \sum_{m=1}^\infty \Theta_m(x) e^{i\Omega_m t} \tag{3.24}$$

This formula expresses the nth mode shape as Θ_m, which may be expressed in the following form for two distinct boundary conditions:

$$\Theta_m = \Gamma_m \sin(\alpha x) \tag{3.25}$$

Using the two distinct boundary conditions, C-C and C-F, as examples, the parameter α may be written as follows:

$$\alpha = \begin{cases} \dfrac{m\pi}{l} & C-C \\ \dfrac{(2m-1)\pi}{2l} & C-F \end{cases} \tag{3.26}$$

In order to determine the natural frequency of nanorods with varying cross-sections, one has to use the appropriate warping function. It is possible to describe the natural frequency of the auxetic nanorod as follows by integrating equation. (3.25) into equation (3.23).

$$\Omega_m = \sqrt{\frac{G\left[\int_A \left(\frac{\partial \psi(y,z)}{\partial y} - z\right)^2 + \left(\frac{\partial \psi(y,z)}{\partial z} + y\right)^2 dA\right]\alpha^2}{\rho I_\psi \alpha^2 + I_0 + \mu\left(I_0\alpha^2 + \rho I_\psi \alpha^4\right)}} \tag{3.27}$$

The following formula is used to get the natural frequency's non-dimensional form:

$$\bar{\Omega}_m = \Omega_m \times l \times \sqrt{\frac{\rho}{G}} \tag{3.28}$$

This section examines, using various diagrams, the frequency of the auxetic triangular and elliptical nanorods that are subject to the C-C and C-F boundary constraints. The following charts illustrate the research and plotting of the effects of various factors on variations in natural frequency. Additionally, the findings of a comparison between the current model's non-dimensional natural frequencies and those from similar research will be shown. In Figure 3.4, the impact of various triangle edges on the change of the nonlocal parameter is examined for the (a) C-C and (b) C-F boundary conditions. As nonlocal parameters increase to greater levels, the auxetic nanorod's inherent frequencies drop. This reduction provides evidence for the classical theories' inaccuracy when compared to nonlocal formulization. Moreover, the reduction in natural frequency caused by an increase in the size of triangle edges is identical to that caused by a nonlocal parameter. The mechanism is the same when the diagrams are compared, but the rod that is subject to the C-C boundary condition has larger amounts of natural frequency. In fact, triangular edges and smaller nonlocal factors are required in order to achieve higher natural frequency values.

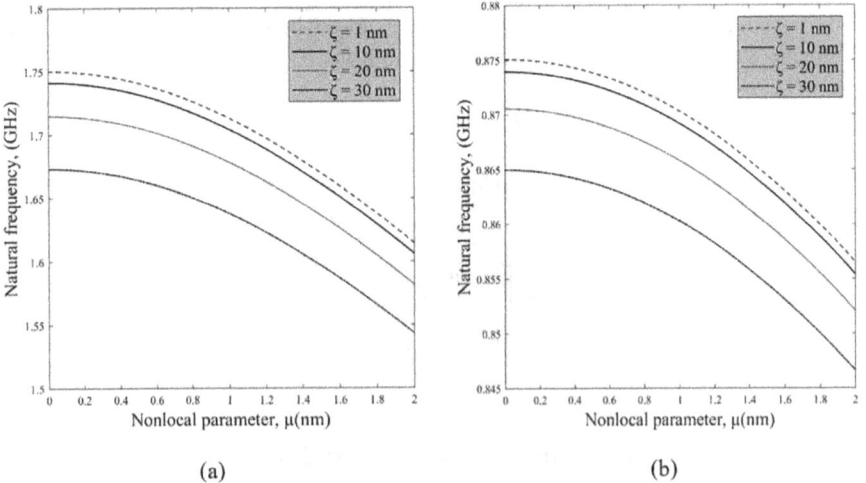

FIGURE 3.4 Variations in the basic natural frequency as a function of the nonlocal parameter for various triangle edge values under the following boundary conditions: (a) C-C, (b) C-F.

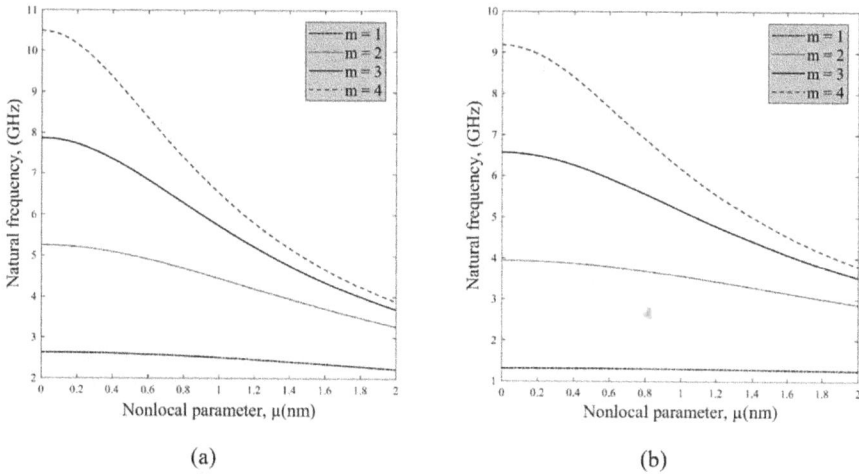

FIGURE 3.5 The alterations in the fundamental natural frequency are examined in relation to the nonlocal parameter, considering four distinct values of m under two boundary conditions: (a) C-C and (b) C-F.

The impact of the triangular nanorod's m parameter natural frequencies is examined in Figure 3.5, subject to different nonlocal parameter values and (a) C-C and (b) C-F boundary conditions. Increases in nonlocal factors cause the natural frequency to decrease. Furthermore, the observed rising influence of m is present. It is evident that when m grows, the impact of the nonlocal parameter on natural frequency may be mitigated. Although the natural frequency somewhat dropped when the boundary condition of the nanorod was changed from C-C to C-F, the overall behavior of the figure was maintained.

Figure 3.6 shows the effect of various inclination angles and nonlocal factors on the auxetic triangular nanorod's frequency fluctuation under varied boundary conditions. A decrease in natural frequency may be attributed to the use of larger inclined angle values in both boundary conditions. It is easy to observe how the nonlocal parameter has a decreasing influence on natural frequency in addition to earlier displays. This result demonstrates how nonlocality has a significant impact on nanostructures' mechanical reaction. It is evident that the natural frequencies of a triangular nanorod with a C-C boundary condition are greater than those with a C-F boundary condition.

The impact of various factors on the natural frequency fluctuation of an elliptical nanorod is shown in the following. Plots demonstrating the effects of geometrical ratio and various m on the elliptical nanorod's natural frequency can be found in Figures 3.7 and 3.8. Growing the geometric ratio has a favorable influence on natural frequencies at low geometric ratios up to a value of 1, when the elliptical nanorod is transformed into a circular nanorod. This phenomenon reaches its greatest extent at the geometric ratio of 1. Analogously to nonlocal theory, it is evident that at this moment, the natural frequency is exaggerated in traditional theory (i.e., $\mu=0$). This finding indicates that good estimation of plausible nonlocalities in the nanorod is

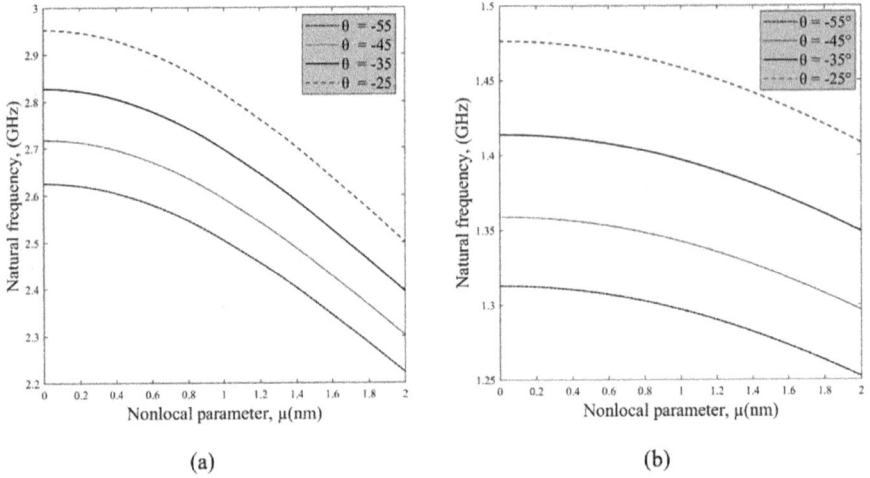

FIGURE 3.6 Fundamental natural frequency variations as a function of nonlocal parameter in an auxetic triangular nanorod with two distinct inclination angles θ, according to two boundary conditions: (a) C-C and (b) C-F.

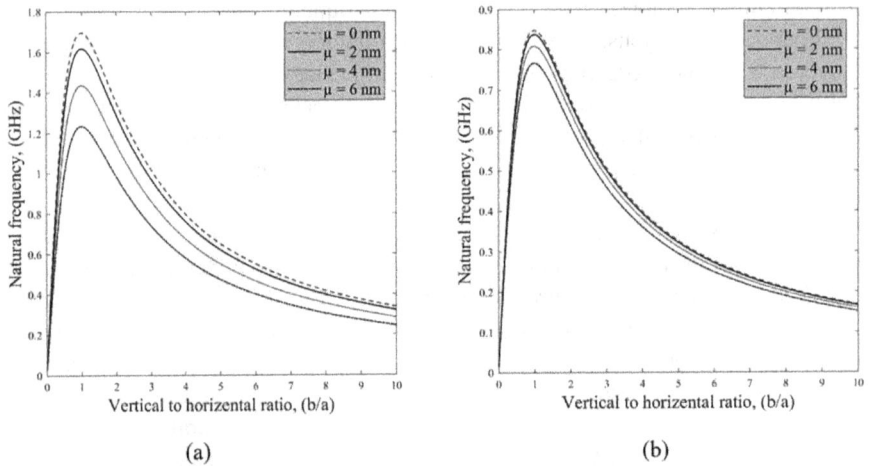

FIGURE 3.7 The alterations in the fundamental natural frequency are examined in relation to the geometrical ratio b/a, considering various nonlocal parameters μ for both (a) C-C and (b) C-F elliptical nanorods.

necessary for precise and physically stable outcomes. With varied values of the nonlocal factors, the second portion of the figure shows the decremental influence of the geometric ratio on natural frequency, in contrast to the first section. In comparison to a horizontal ellipse (i.e., a>b), the decrease in the moment of inertia is the cause of the drop in natural frequency. It is evident from Figure 3.8 that the natural

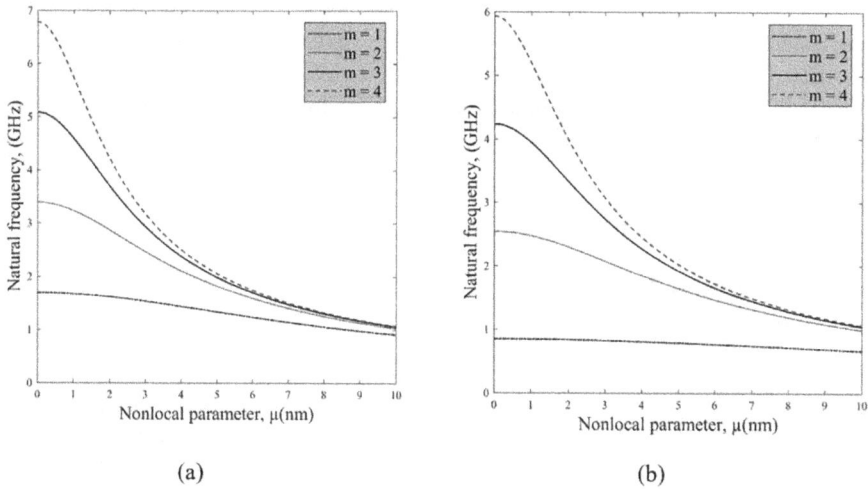

(a)   (b)

FIGURE 3.8 Variations in the fundamental natural frequency with respect to the nonlocal parameter μ and varying *m* for the (a) C-C and (b) C-F elliptical nanorods, respectively.

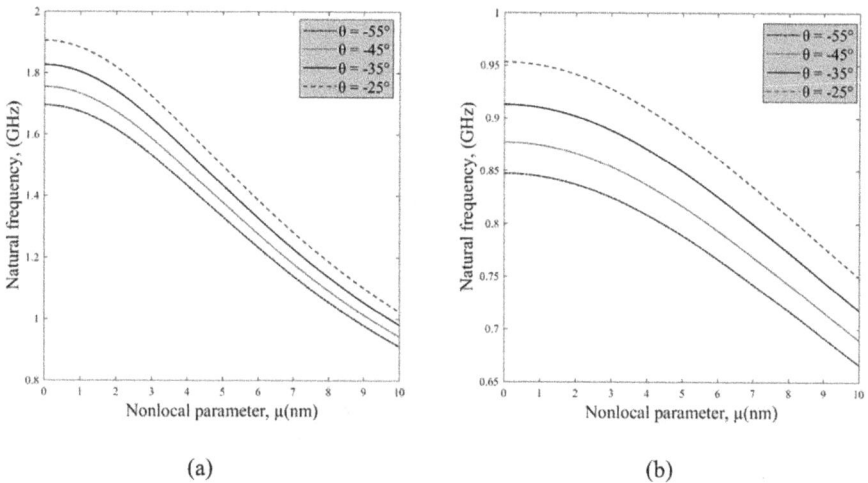

(a)   (b)

FIGURE 3.9 Variations in the fundamental natural frequency with four different inclined angles θ for the (a) C-C and (b) C-F elliptical nanorods, respectively, as a function of the nonlocal parameter μ.

frequency is inversely proportional to the nonlocal parameter and proportionate to the *m* parameter. It is evident from Figures 3.7 and 3.8 that the elliptical nanorod with C-C edges has a higher natural frequency than the one with C-F edges.

Figure 3.9 shows the effect of various inclination angles on natural frequency as a function of nonlocal parameter. Auxetic structure's inclined inclination significantly affects how the elliptical nanorod's natural frequency varies. Angle of inclination has a negligible effect on the auxetic nanorod's natural frequency fluctuation. Put another way, natural frequency decreases with increasing slanted angle. Moreover, the use of the C-C boundary condition may result in higher natural frequency values than the C-F boundary condition.

3.3 AUXETIC PLATES: VIBRATION ANALYSIS

3.3.1 VIBRATION ANALYSIS OF SMART SANDWICHED AUXETIC PLATE WITH MEE FACES

Figure 3.10 depicts a sandwich plate that is actuated using a piezo-electromagnetic mechanism. The upper and lower layers of the sandwich plate are supported by magneto-electro-elastic graded composite layers. In order to determine the effective properties of the smart layers, the micro-magneto-electro-mechanical model is introduced in the following manner:

$$V_p = \left(\frac{z}{h}\right)^{\eta}$$
$$V_m = 1 - V_p; \tag{3.29}$$

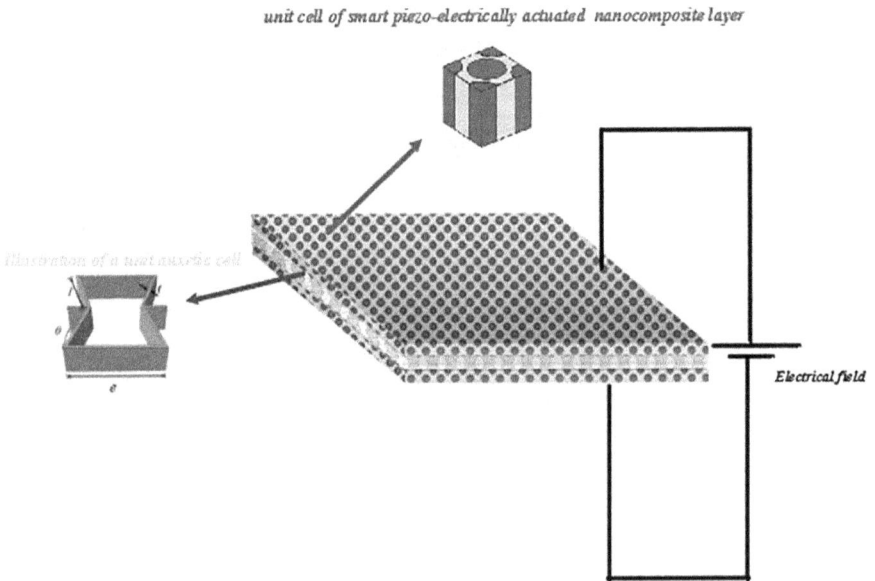

unit cell of smart piezo-electrically actuated nanocomposite layer

Electrical field

FIGURE 3.10 The diagram illustrating the configuration of the sandwich plate, which is actuated by piezo-magnetic elements and incorporates a metamaterial core.

$$C_{11} = C_{22} = k + m$$

$$C_{33} = C_{33}^m + V_p \left[C_{33}^p - C_{33}^m - \frac{V_m \left(C_{13}^m - C_{33}^p \right)^2}{V_m k_p + V_p k_m + m_m} \right]$$

$$C_{13} = C_{23} = C_{13}^m + \frac{V_p \left(C_{13}^p - C_{13}^m \right) \left(k_m + m_m \right)}{V_m k_p + V_p k_m + m_m}$$

$$C_{12} = k - m$$

$$C_{44} = C_{55} = C_{44}^m + j \left[\begin{array}{l} \left(e_{15}^p - e_{15}^m \right) (fe - gh) \\ + \left(C_{44}^p - C_{44}^m \right) (ih - fd) \\ + \left(q_{15}^p - q_{15}^m \right) (gd - ie) \end{array} \right]$$

$$C_{66}^m = m$$

(3.30)

$$k = \frac{k_m k_p + V_p k_p m_m + V_m k_m m_m}{V_m k_p + V_p k_m + m_m}$$

$$m = \frac{m_m \left(k_m m_p + V_p k_m m_p + V_m k_m m_m + 2 m_k m_m \right)}{V_m k_m m_p + k_m m_m + V_p k_m m_m + 2 V_m m_p m_m + 2 V_p m_m^2}$$

(3.31)

$$e_{13} = e_{23} = e_{13}^m + \frac{V_p \left(e_{13}^p - e_{13}^m \right) \left(k_m + m_m \right)}{V_m k_p + V_p k_m + m_m}$$

$$e_{33} = e_{33}^m + V_p \left[e_{33}^p - e_{33}^m - \frac{V_m \left(C_{13}^m - C_{13}^p \right) \left(e_{31}^p - e_{31}^m \right)}{V_m k_p + V_p k_m + m_m} \right]$$

$$e_{15} = e_{24} = e_{15}^m + j \left[\begin{array}{l} \left(e_{15}^p - e_{15}^m \right) (ga - fc) \\ + \left(C_{44}^p - C_{44}^m \right) (fb - ia) \\ + \left(q_{15}^p - q_{15}^m \right) (ic - gb) \end{array} \right]$$

(3.32)

$$q_{31} = q_{32} = q_{31}^m + \frac{V_p \left(q_{31}^p - q_{31}^m \right) \left(k_m + m_m \right)}{V_m k_p + V_p k_m + m_m}$$

$$q_{33} = q_{33}^m + V_p \left[q_{33}^p - q_{33}^m - \frac{V_m \left(C_{13}^m - C_{13}^p \right) \left(q_{31}^p - q_{31}^m \right)}{V_m k_p + V_p k_m + m_m} \right]$$

$$q_{15} = q_{24} = q_{15}^m + j \left[\begin{array}{l} \left(e_{15}^p - e_{15}^m \right) (ch - ae) \\ + \left(C_{44}^p - C_{44}^m \right) (ad - bh) \\ + \left(q_{15}^p - q_{15}^m \right) (be - cd) \end{array} \right]$$

(3.33)

$$s_{11} = s_{22} = s_{11}^m + j \begin{bmatrix} \left(s_{11}^p - s_{11}^m\right)(ga - fc) \\ +\left(e_{15}^p - e_{15}^m\right)(ia - fb) \\ +\left(r_{11}^p - r_{11}^m\right)(ic - gd) \end{bmatrix}$$

$$s_{33} = s_{33}^m + V_p \left[s_{33}^p - s_{33}^m - \frac{V_m \left(e_{31}^p - e_{31}^m\right)^2}{V_m k_p + V_p k_m + m_m} \right] \qquad (3.34)$$

$$s_{11} = s_{22} = s_{11}^m + j \begin{bmatrix} \left(s_{11}^p - s_{11}^m\right)(ga - fc) \\ +\left(e_{15}^p - e_{15}^m\right)(ia - fb) \\ +\left(r_{11}^p - r_{11}^m\right)(ic - gd) \end{bmatrix}$$

$$s_{33} = s_{33}^m + V_p \left[s_{33}^p - s_{33}^m - \frac{V_m \left(e_{31}^p - e_{31}^m\right)^2}{V_m k_p + V_p k_m + m_m} \right] \qquad (3.35)$$

$$r_{11} = r_{22} = r_{11}^m + j \begin{bmatrix} \left(d_{11}^p - d_{11}^m\right)(ch - ae) \\ +\left(q_{15}^p - q_{15}^m\right)(bh - ad) \\ +\left(r_{11}^p - r_{11}^m\right)(be - cd) \end{bmatrix}$$

$$r_{33} = r_{33}^m + V_p \left[r_{33}^p - r_{33}^m - \frac{V_m \left(r_{31}^p - r_{31}^m\right)^2}{V_m k_p + V_p k_m + m_m} \right] \qquad (3.36)$$

$$d_{11} = d_{22} = d_{11}^m + j \begin{bmatrix} \left(s_{11}^p - s_{11}^m\right)(ch - ae) \\ +\left(e_{15}^p - e_{15}^m\right)(bh - ad) \\ +\left(d_{11}^p - d_{11}^m\right)(be - cd) \end{bmatrix}$$

$$d_{33} = d_{33}^m + V_p \left[d_{33}^p - d_{33}^m - \frac{V_m \left(e_{31}^p - e_{31}^m\right)\left(q_{31}^p - q_{31}^m\right)}{V_m k_p + V_p k_m + m_m} \right] \qquad (3.37)$$

$$s_{11} = s_{22} = s_{11}^m + j \begin{bmatrix} \left(s_{11}^p - s_{11}^m\right)(ga - fc) \\ +\left(e_{15}^p - e_{15}^m\right)(ia - fb) \\ +\left(r_{11}^p - r_{11}^m\right)(ic - gd) \end{bmatrix}$$

$$s_{33} = s_{33}^m + V_p \left[s_{33}^p - s_{33}^m - \frac{V_m \left(e_{31}^p - e_{31}^m\right)^2}{V_m k_p + V_p k_m + m_m} \right]$$

$$r_{11} = r_{22} = r_{11}^m + j \begin{bmatrix} \left(d_{11}^p - d_{11}^m\right)(ch - ae) \\ + \left(q_{15}^p - q_{15}^m\right)(bh - ad) \\ + \left(r_{11}^p - r_{11}^m\right)(be - cd) \end{bmatrix}$$

$$r_{33} = r_{33}^m + V_p \left[r_{33}^p - r_{33}^m - \frac{V_m \left(r_{31}^p - r_{31}^m\right)^2}{V_m k_p + V_p k_m + m_m} \right]$$

$$d_{11} = d_{22} = d_{11}^m + j \begin{bmatrix} \left(s_{11}^p - s_{11}^m\right)(ch - ae) \\ + \left(e_{15}^p - e_{15}^m\right)(bh - ad) \\ + \left(d_{11}^p - d_{11}^m\right)(be - cd) \end{bmatrix}$$

$$d_{33} = d_{33}^m + V_p \left[d_{33}^p - d_{33}^m - \frac{V_m \left(e_{31}^p - e_{31}^m\right)\left(q_{31}^p - q_{31}^m\right)}{V_m k_p + V_p k_m + m_m} \right] \qquad (3.38)$$

$$j = 2V_p \frac{V_{44}^m s_{11}^m r_{11}^m + r_{11}^m \left(e_{15}^m\right)^2 + s_{11}^m \left(q_{15}^m\right)^2 - 2d_{11}^m e_{15}^m q_{15}^m - C_{44}^m \left(d_{11}^m\right)^2}{h(ic - gb) + e(fb - ia) + d(ga - fc)}$$

$$a = V_m \begin{bmatrix} \left(q_{15}^p - q_{15}^m\right)\left(s_{11}^m r_{11}^m - \left(d_{11}^m\right)^2\right) \\ -\left(d_{11}^p - d_{11}^m\right)\left(e_{15}^m r_{11}^m - d_{11}^m q_{15}^m\right) \\ -\left(r_{11}^p - r_{11}^m\right)\left(q_{15}^m s_{11}^m - d_{11}^m e_{15}^m\right) \end{bmatrix}$$

$$b = V_m \begin{bmatrix} \left(e_{15}^p - e_{15}^m\right)\left(s_{11}^m r_{11}^m - \left(d_{11}^m\right)^2\right) \\ -\left(d_{11}^p - d_{11}^m\right)\left(q_{15}^m s_{11}^m - d_{11}^m e_{15}^m\right) \\ -\left(s_{11}^p - s_{11}^m\right)\left(e_{15}^m r_{11}^m - d_{11}^m q_{15}^m\right) \end{bmatrix}$$

$$c = V_m \begin{bmatrix} \left(C_{44}^p - C_{44}^m\right)\left(s_{11}^m r_{11}^m - \left(q_{15}^m\right)^2\right) \\ +\left(e_{15}^p - e_{15}^m\right)\left(e_{15}^m r_{11}^m - d_{11}^m q_{15}^m\right) \\ -\left(q_{15}^p - q_{15}^m\right)\left(q_{15}^m s_{11}^m - d_{11}^m e_{15}^m\right) \end{bmatrix} + \frac{j}{2}\left[h(ic - gb) + e(fb - ia) + d(ga - fc)\right]$$

$$d = V_m \begin{bmatrix} -\left(d_{11}^p - d_{11}^m\right)\left(d_{11}^m C_{44}^m + e_{15}^m q_{15}^m\right) \\ +\left(e_{15}^p - e_{15}^m\right)\left(e_{15}^m r_{11}^m - d_{11}^m q_{15}^m\right) \\ -\left(s_{11}^p - s_{11}^m\right)\left(\left(q_{15}^m\right)^2 - C_{44}^m r_{15}^m\right) \end{bmatrix}$$

$$+ \frac{j}{2}\left[h(ic - gb) + e(fb - ia) + d(ga - fc)\right] \qquad (3.39)$$

$$e = V_m \begin{bmatrix} \left(q_{15}^p - q_{15}^m\right)\left(d_{11}^m C_{44}^m + q_{15}^m e_{15}^m\right) \\ +\left(C_{44}^p - C_{44}^m\right)\left(e_{15}^m r_{11}^m - d_{11}^m q_{15}^m\right) \\ -\left(e_{15}^p - e_{15}^m\right)\left(\left(q_{15}^m\right)^2 + C_{44}^m r_{11}^m\right) \end{bmatrix} \quad f = V_m \begin{bmatrix} -\left(d_{11}^p - d_{11}^m\right)\left(d_{11}^m C_{44}^m + e_{15}^m q_{15}^m\right) \\ +\left(q_{15}^p - q_{15}^m\right)\left(q_{15}^m s_{11}^m - d_{11}^m e_{15}^m\right) \\ +\left(r_{11}^p - r_{11}^m\right)\left(\left(e_{15}^m\right)^2 + C_{44}^m s_{11}^m\right) \end{bmatrix}$$

$$+\frac{j}{2}\left[h\left(ic - gb\right) + e\left(fb - ia\right) + d\left(ga - fc\right)\right]$$

$$g = V_m \begin{bmatrix} \left(e_{15}^p - e_{15}^m\right)\left(d_{11}^m C_{44}^m + q_{15}^m e_{15}^m\right) \\ +\left(C_{44}^p - C_{44}^m\right)\left(q_{15}^m s_{11}^m - d_{11}^m e_{15}^m\right) \\ -\left(q_{15}^p - q_{15}^m\right)\left(\left(e_{15}^m\right)^2 + C_{44}^m s_{11}^m\right) \end{bmatrix} \quad h = V_m \begin{bmatrix} \left(q_{15}^p - q_{15}^m\right)\left(e_{15}^m r_{11}^m + d_{11}^m q_{15}^m\right) \\ -\left(r_{44}^p - r_{44}^m\right)\left(d_{11}^m C_{44}^m - q_{15}^m e_{15}^m\right) \\ +\left(d_{11}^p - d_{11}^m\right)\left(\left(q_{15}^m\right)^2 + C_{44}^m r_{11}^m\right) \end{bmatrix}$$

$$i = V_m \begin{bmatrix} \left(e_{15}^p - e_{15}^m\right)\left(q_{15}^m s_{11}^m - d_{11}^m e_{15}^m\right) \\ -\left(s_{11}^p - s_{11}^m\right)\left(d_{11}^m C_{44}^m - q_{15}^m e_{15}^m\right) \\ +\left(d_{11}^p - d_{11}^m\right)\left(\left(e_{15}^m\right)^2 + C_{44}^m s_{11}^m\right) \end{bmatrix} \tag{3.40}$$

$$\rho = V_p \rho^p + V_m \rho^m \tag{3.41}$$

The variable η represents the volume percentage of the smart filler present in the matrix. Furthermore, the present mechanical properties of the metamaterial can be acquired in the following manner:

$$E_{11}^a = E_{Al}\left(t/l\right)^3 \frac{\cos\theta}{\left(e/l + \sin\theta\right)\sin^2\theta}; \quad E_{22}^a = E_{Al}\left(t/l\right)^3 \frac{\left(e/l + \sin\theta\right)}{\cos^3\theta} \tag{3.42a}$$

$$G_{12}^a = E_{Al}\left(t/l\right)^3 \frac{\left(e/l + \sin\theta\right)}{\left(e/l\right)^2\left(1 + 2e/l\right)\cos\theta}; v_{12}^a = \frac{\cos^2\theta}{\left(e/l + \sin\theta\right)\sin\theta} \tag{3.42b}$$

$$\rho_a = \rho_{Al}\frac{\left(t/l\right)\left(e/l + 2\right)}{2\cos\theta\left(e/l + \sin\theta\right)} \tag{3.42c}$$

In addition, the equations incorporate the inclusion of geometrical parameters such as the angle (θ), the length of the vertical cell to length ratio e/l, and the thickness-to-length ratio t/l. The sandwiched plate is governed by a magneto-electro-elastic constitutive model, which can be described by the following equations [33]:

$$\sigma_{ij} = \left[C_{ijkl}\varepsilon_{kl} - e_{mij}E_m - q_{nij}H_n\right] \tag{3–43a}$$

$$D_i = \left[e_{ikl}\varepsilon_{kl} + s_{im}E_m + d_{in}H_n\right] \tag{3–43b}$$

$$B_i = \left[q_{ikl}\varepsilon_{kl} + d_{im}E_m + r_{in}H_n\right] \tag{3–43c}$$

The symbols C_{ijkl}, $\sigma_{ij} - \varepsilon_{kl}$ are used to represent the elements of the elastic and stress-strain tensors, respectively. Moreover, the variables D_i and E_m denote the constituents of the electric displacement and electric field, correspondingly. Moreover, the symbols B_i and H_m represent the constituents of the magnetic displacement and magnetic field, correspondingly. The dielectric constants, piezoelectric constants, piezomagnetic constants, magnetoelectric coefficients, and magnetic constants are denoted by s_{im}, e_{ikl} q_{ikl}, d_{im}, and r_{in} respectively. In a similar vein, the stress-strain relationship of the intelligent sandwich plate can be expressed as:

$$
\begin{Bmatrix} \sigma_{xx} \\ \sigma_{yy} \\ \sigma_{yz} \\ \sigma_{zx} \\ \sigma_{xy} \end{Bmatrix} = \begin{pmatrix} c_{11} & c_{12} & 0 & 0 & 0 \\ c_{12} & c_{11} & 0 & 0 & 0 \\ 0 & 0 & c_{55} & 0 & 0 \\ 0 & 0 & 0 & c_{55} & 0 \\ 0 & 0 & 0 & 0 & c_{66} \end{pmatrix} \begin{Bmatrix} \varepsilon_{xx} \\ \varepsilon_{yy} \\ \varepsilon_{yz} \\ \varepsilon_{zx} \\ \varepsilon_{xy} \end{Bmatrix} - \begin{pmatrix} 0 & 0 & e_{31} \\ 0 & 0 & e_{31} \\ 0 & e_{15} & 0 \\ e_{15} & 0 & 0 \\ 0 & 0 & 0 \end{pmatrix} \begin{Bmatrix} E_x \\ E_y \\ E_z \end{Bmatrix}
$$
$$
- \begin{pmatrix} 0 & 0 & q_{31} \\ 0 & 0 & q_{31} \\ 0 & q_{15} & 0 \\ q_{15} & 0 & 0 \\ 0 & 0 & 0 \end{pmatrix} \begin{Bmatrix} H_x \\ H_y \\ H_z \end{Bmatrix}
$$

$$
\begin{Bmatrix} D_x \\ D_y \\ D_z \end{Bmatrix} = \begin{pmatrix} 0 & 0 & 0 & 0 & e_{15} & 0 \\ 0 & 0 & 0 & e_{15} & 0 & 0 \\ e_{31} & e_{31} & 0 & 0 & 0 & 0 \end{pmatrix} \begin{Bmatrix} \varepsilon_{xx} \\ \varepsilon_{yy} \\ \varepsilon_{zz} \\ \varepsilon_{yz} \\ \varepsilon_{zx} \\ \varepsilon_{xy} \end{Bmatrix} + \begin{pmatrix} s_{11} & 0 & 0 \\ 0 & s_{11} & 0 \\ 0 & 0 & s_{33} \end{pmatrix} \begin{Bmatrix} E_x \\ E_y \\ E_z \end{Bmatrix}
$$
$$
+ \begin{pmatrix} d_{11} & 0 & 0 \\ 0 & d_{11} & 0 \\ 0 & 0 & d_{33} \end{pmatrix} \begin{Bmatrix} H_x \\ H_y \\ H_z \end{Bmatrix}
$$

$$
\begin{Bmatrix} B_x \\ B_y \\ B_z \end{Bmatrix} = \begin{pmatrix} 0 & 0 & 0 & 0 & q_{15} & 0 \\ 0 & 0 & 0 & q_{15} & 0 & 0 \\ q_{31} & q_{31} & 0 & 0 & 0 & 0 \end{pmatrix} \begin{Bmatrix} \varepsilon_{xx} \\ \varepsilon_{yy} \\ \varepsilon_{zz} \\ \varepsilon_{yz} \\ \varepsilon_{zx} \\ \varepsilon_{xy} \end{Bmatrix} + \begin{pmatrix} d_{11} & 0 & 0 \\ 0 & d_{11} & 0 \\ 0 & 0 & d_{33} \end{pmatrix} \begin{Bmatrix} E_x \\ E_y \\ E_z \end{Bmatrix}
$$
$$
+ \begin{pmatrix} r_{11} & 0 & 0 \\ 0 & r_{11} & 0 \\ 0 & 0 & r_{33} \end{pmatrix} \begin{Bmatrix} H_x \\ H_y \\ H_z \end{Bmatrix} \tag{3.44}
$$

According to the Mindlin plate theory, the displacement field of the smart sandwich plate can be derived as indicated in references [34,35].

$$U(x,y,z,t) = u_0(x,y,t) - z\psi_x(x,y,t) \tag{3.45a}$$

$$V(x,y,z,t) = v_0(x,y,t) - z\psi_y(x,y,t) \tag{3.45b}$$

$$W(x,y,z,t) = w(x,y,t) \tag{3.45c}$$

The longitudinal displacements, transverse displacements, and lateral deflection of the smart sandwich plate are denoted by u_0, v_0, and w respectively. Furthermore, the rotations of the transverse normal along the x and y axes are indicated ψ_x and ψ_y respectively.

Moreover, in order to comply with Maxwell's law, the values of Φ and Ψ can be determined using the following equations:

$$\Phi(x,y,z,t) = -\cos(\zeta z)\phi(x,z,t) + \frac{2z}{h}V_0 \tag{3.46a}$$

$$\Psi(x,y,z,t) = -\cos(\zeta z)\psi(x,z,t) + \frac{2z}{h}\Omega_0 \tag{3.46b}$$

in which $\zeta = \dfrac{\pi}{h}$, and the external electric voltage and magnetic potential are represented by the symbols V_0 and Ω_0, respectively. Furthermore, the strains of the sandwich plate can be expressed as:

$$\begin{Bmatrix} \varepsilon_{xx} \\ \varepsilon_{yy} \\ \varepsilon_{xy} \end{Bmatrix} = \begin{Bmatrix} \varepsilon^0_{xx} \\ \varepsilon^0_{yy} \\ \varepsilon^0_{xy} \end{Bmatrix} + z \begin{Bmatrix} k^b_{xx} \\ k^b_{yy} \\ k^b_{xy} \end{Bmatrix}, \qquad \begin{Bmatrix} \gamma_{xz} \\ \gamma_{yz} \end{Bmatrix} = \begin{Bmatrix} \gamma^0_{xz} \\ \gamma^0_{yz} \end{Bmatrix} \tag{3.47}$$

The expression for the nonlinear strain relations of the piezo-electro-magnetically actuated system, incorporating the von Karman nonlinearity, can be formulated as follows:

$$\begin{Bmatrix} \varepsilon^0_{xx} \\ \varepsilon^0_{yy} \\ \varepsilon^0_{xy} \end{Bmatrix} = \begin{Bmatrix} \dfrac{\partial u_0}{\partial x} + \dfrac{1}{2}\left(\dfrac{\partial w}{\partial x}\right)^2 \\ \dfrac{\partial v_0}{\partial x} + \dfrac{1}{2}\left(\dfrac{\partial w}{\partial y}\right)^2 \\ \dfrac{\partial u_0}{\partial y} + \dfrac{\partial v_0}{\partial x} + \dfrac{\partial w}{\partial x}\dfrac{\partial w}{\partial y} \end{Bmatrix}, \begin{Bmatrix} k^b_{xx} \\ k^b_{yy} \\ k^b_{xy} \end{Bmatrix} = \begin{Bmatrix} -\dfrac{\partial^2 w}{\partial x^2} \\ -\dfrac{\partial^2 w}{\partial y^2} \\ -\dfrac{\partial^2 w}{\partial x \partial y} \end{Bmatrix}, \begin{Bmatrix} k^s_{xx} \\ k^s_{yy} \\ k^s_{xy} \end{Bmatrix} = \begin{Bmatrix} \dfrac{\partial \psi_x}{\partial x} \\ \dfrac{\partial \psi_y}{\partial y} \\ \dfrac{\partial \psi_x}{\partial y} + \dfrac{\partial \psi_y}{\partial x} \end{Bmatrix},$$

$$\begin{Bmatrix} \gamma^0_{xz} \\ \gamma^0_{yz} \end{Bmatrix} = \begin{Bmatrix} \dfrac{\partial w}{\partial x} + \psi_x \\ \dfrac{\partial w}{\partial y} + \psi_y \end{Bmatrix} \tag{3.48}$$

Furthermore, the magneto-electrical fields of the sandwich plate can be attained by:

$$E_x = -\frac{\partial \Phi}{\partial x} = \cos(\zeta z)\frac{\partial \phi}{\partial x} \tag{3.49a}$$

$$E_y = -\frac{\partial \Phi}{\partial y} = \cos(\zeta z)\frac{\partial \phi}{\partial y} \tag{3.49b}$$

$$E_z = -\frac{\partial \Phi}{\partial z} = -\zeta \sin(\zeta z)\phi - \frac{2V_E}{h} \tag{3.49c}$$

$$H_x = -\frac{\partial \Psi}{\partial x} = \cos(\zeta z)\frac{\partial \psi}{\partial x} \tag{3.49d}$$

$$H_y = -\frac{\partial \Psi}{\partial y} = \cos(\zeta z)\frac{\partial \psi}{\partial y} \tag{3.49e}$$

$$H_z = -\frac{\partial \Psi}{\partial z} = -\zeta \sin(\zeta z)\psi - \frac{2\Omega_H}{h} \tag{3.49f}$$

The expression for the piezo-electro-magnetically actuated plate using Hamilton's principle is given by:

$$\int_0^t \delta(\Pi_S - \Pi_K + \Pi_W)\,dt = 0 \tag{3.50}$$

The variables Π_S and Π_K represent the strain and kinetic energy, respectively. Furthermore, the external labor performed by the intelligent sandwich plate is represented by Π_W. Additionally, the attainment of strain energy is possible.

$$\Pi_S = \int_V (\sigma_{ij}\varepsilon_{ij})\,dV = \int_V \begin{pmatrix} \sigma_{xx}\varepsilon_{xx} + \sigma_{yy}\varepsilon_{yy} + \sigma_{xy}\varepsilon_{xy} + \sigma_{xz}\varepsilon_{xz} + \sigma_{yz}\varepsilon_{yz} \\ -D_x E_x - D_y E_y - D_z E_z \\ -B_x H_x - B_y H_y - B_z H_z \end{pmatrix} dV \tag{3.51}$$

By substituting equation (3.45) into equation (3.49), the expression for the strain energy can be obtained.

$$\Pi_S = \int_A \left(\begin{matrix} N_{xx}\varepsilon_{xx}^0 + M_{xx}\varepsilon_{xx}^1 + N_{yy}\varepsilon_{yy}^0 + M_{yy}\varepsilon_{yy}^1 \\ +2N_{xy}\varepsilon_{xy}^0 + 2M_{xy}\varepsilon_{xy}^1 + Q_{xz}\gamma_{xz}^0 + Q_{yz}\gamma_{yz}^0 \end{matrix} \right) dA$$

$$+ \int_{-h/2}^{h/2}\int_A \left(-D_x \cos(\zeta z)\delta(\frac{\partial \phi}{\partial x}) - D_y \cos(\zeta z)\delta(\frac{\partial \phi}{\partial y}) + D_z \sin(\zeta z)\delta\phi \right) dA\,dz$$

$$+ \int_{-h/2}^{h/2}\int_A \left(-B_x \cos(\zeta z)\delta(\frac{\partial \psi}{\partial x}) - B_y \cos(\zeta z)\delta(\frac{\partial \psi}{\partial y}) + B_z \sin(\zeta z)\delta\psi \right) dA\,dz \tag{3.52}$$

The forces and moments of the smart plate are derived from the following equations:

$$\left(N_{xx}, N_{yy}, N_{xy}\right) = \int_{-h/2}^{h/2} \left\{\sigma_{xx}, \sigma_{yy}, \sigma_{xy}\right\} dz$$

$$\left(M_{xx}, M_{yy}, M_{xy}\right) = \int_{-h/2}^{h/2} \left\{\sigma_{xx}, \sigma_{yy}, \sigma_{xy}\right\} z dz$$

$$\left(Q_{xz}, Q_{yz}\right) = \int_{-h/2}^{h/2} \left\{\sigma_{xz}, \sigma_{yz}\right\} dz \qquad (3.53)$$

Furthermore, the external work can be expressed in the following manner:

$$\delta\Pi_W = \frac{1}{2}\int_A \left(\begin{array}{c} N_x^E\left(\dfrac{\partial w}{\partial x}\right)^2 + N_y^E\left(\dfrac{\partial w}{\partial y}\right)^2 \\[3mm] +N_x^H\left(\dfrac{\partial w}{\partial x}\right)^2 + N_y^H\left(\dfrac{\partial w}{\partial y}\right)^2 \end{array} \right) dA \qquad (3.54)$$

where N_x^E, N_y^E, N_x^H, and N_y^H represent the external electrical and magnetic loads in the x and y-axes. Furthermore, the calculation of kinetic energy is derived as follows:

$$\Pi_K = \frac{1}{2}\int_A \left[I_0\left(\frac{\partial u_0}{\partial t}\right)^2 + I_0\left(\frac{\partial v_0}{\partial t}\right)^2 + I_0\left(\frac{\partial w}{\partial t}\right)^2 + I_2\left(\frac{\partial \psi_x}{\partial t}\right)^2 + I_2\left(\frac{\partial \psi_y}{\partial t}\right)^2 \right] dA \qquad (3.55)$$

The mass inertias are obtained as:

$$(I_0, I_1, I_2) = \int_{-h/2}^{h/2} (1, z, z^2)\rho(z)dz \qquad (3.56)$$

By substituting equations (3.50), (3.53), and (3.54) with equation (3.48), the mathematical representation of the dynamic behavior of the piezo-electro-magnetically actuated plate can be expressed as:

$$\frac{\partial N_{xx}}{\partial x} + \frac{\partial N_{xy}}{\partial y} = I_0\frac{\partial^2 u_0}{\partial t^2} \qquad (3.57a)$$

$$\frac{\partial N_{xy}}{\partial x} + \frac{\partial N_{yy}}{\partial y} = I_0\frac{\partial^2 v_0}{\partial t^2} \qquad (3.57b)$$

$$\frac{\partial}{\partial x}\left(N_{xx}\frac{\partial w}{\partial x} + N_{xy}\frac{\partial w}{\partial y}\right) + \frac{\partial}{\partial y}\left(N_{xy}\frac{\partial w}{\partial x} + N_{yy}\frac{\partial w}{\partial y}\right)$$
$$+\frac{\partial Q_{xz}}{\partial x} + \frac{\partial Q_{yz}}{\partial y} - \left(N^E + N^H\right)\nabla^2 w = I_0\frac{\partial^2 w}{\partial t^2} \qquad (3.57c)$$

$$\frac{\partial M_{xx}}{\partial x} + \frac{\partial M_{xy}}{\partial y} - Q_{xz} = I_0 \frac{\partial^2 \psi_x}{\partial t^2} \qquad (3.57d)$$

$$\frac{\partial M_{xy}}{\partial x} + \frac{\partial M_{yy}}{\partial y} - Q_{yz} = I_0 \frac{\partial^2 \psi_y}{\partial t^2} \qquad (3.57e)$$

$$\int_A \left(\cos(\zeta z) \frac{\partial D_x}{\partial x} + \cos(\zeta z) \frac{\partial D_y}{\partial y} + \zeta \sin(\zeta z) D_z \right) dA = 0 \qquad (3.57f)$$

$$\int_A \left(\cos(\zeta z) \frac{\partial B_x}{\partial x} + \cos(\zeta z) \frac{\partial B_y}{\partial y} + \zeta \sin(\zeta z) B_z \right) dA = 0 \qquad (3.57g)$$

The force and moment obtained are determined by the constitutive model in the following manner:

$$\begin{Bmatrix} N_{xx} \\ N_{yy} \\ N_{xy} \end{Bmatrix} = \begin{pmatrix} A_{11} & A_{12} & 0 \\ A_{21} & A_{22} & 0 \\ 0 & 0 & A_{66} \end{pmatrix} \begin{Bmatrix} \frac{\partial u_0}{\partial x} + \frac{1}{2} \left(\frac{\partial w}{\partial x} \right)^2 \\ \frac{\partial v_0}{\partial x} + \frac{1}{2} \left(\frac{\partial w}{\partial y} \right)^2 \\ \frac{\partial u_0}{\partial y} + \frac{\partial v_0}{\partial x} + \frac{\partial w}{\partial x} \frac{\partial w}{\partial y} \end{Bmatrix} + \begin{pmatrix} A_{31} \\ A_{31} \\ 0 \end{pmatrix} \phi - \begin{pmatrix} N_x^E \\ N_y^E \\ 0 \end{pmatrix} \qquad (3.58)$$

$$\begin{Bmatrix} M_{xx} \\ M_{yy} \\ M_{xy} \end{Bmatrix} = \begin{pmatrix} D_{11} & D_{12} & 0 \\ D_{21} & D_{22} & 0 \\ 0 & 0 & D_{66} \end{pmatrix} \begin{Bmatrix} \frac{\partial \psi_x}{\partial x} \\ \frac{\partial \psi_y}{\partial y} \\ \frac{\partial \psi_x}{\partial y} + \frac{\partial \psi_y}{\partial x} \end{Bmatrix} + \begin{pmatrix} E_{31} \\ E_{31} \\ 0 \end{pmatrix} \phi + \begin{pmatrix} Q_{31} \\ Q_{31} \\ 0 \end{pmatrix} \psi - \begin{pmatrix} M_x^E \\ M_y^E \\ 0 \end{pmatrix} \qquad (3.59)$$

$$\begin{Bmatrix} Q_{xz} \\ Q_{yz} \end{Bmatrix} = \begin{pmatrix} A_{44} & 0 \\ 0 & A_{55} \end{pmatrix} \begin{Bmatrix} \frac{\partial w}{\partial x} + \psi_x \\ \frac{\partial w}{\partial y} + \psi_y \end{Bmatrix} - E_{15} \begin{Bmatrix} \frac{\partial \phi}{\partial x} \\ \frac{\partial \phi}{\partial y} \end{Bmatrix} - Q_{15} \begin{Bmatrix} \frac{\partial \psi}{\partial x} \\ \frac{\partial \psi}{\partial y} \end{Bmatrix} \qquad (3.60)$$

$$\int_A \left(\begin{Bmatrix} D_x \\ D_y \end{Bmatrix} \cos(\zeta z) \right) dA = E_{15} \begin{Bmatrix} \frac{\partial w}{\partial x} + \psi_x \\ \frac{\partial w}{\partial y} + \psi_y \end{Bmatrix} + X_{11} \begin{Bmatrix} \frac{\partial \phi}{\partial x} \\ \frac{\partial \phi}{\partial y} \end{Bmatrix} + Y_{11} \begin{Bmatrix} \frac{\partial \psi}{\partial x} \\ \frac{\partial \psi}{\partial y} \end{Bmatrix} \qquad (3.61)$$

$$\int_A \left(D_z \, \zeta \sin(\zeta z) \right) dA = \left(E_{31} \left(\frac{\partial \psi_x}{\partial x} + \frac{\partial \psi_y}{\partial y} \right) - X_{33} \phi - Y_{33} \psi \right) \tag{3.62}$$

$$\int_A \left(\left\{ \begin{array}{c} B_x \\ B_y \end{array} \right\} \cos(\zeta z) \right) dA = \left(Q_{15} \left\{ \begin{array}{c} \dfrac{\partial w}{\partial x} + \psi_x \\ \dfrac{\partial w}{\partial y} + \psi_y \end{array} \right\} + Y_{11} \left\{ \begin{array}{c} \dfrac{\partial \phi}{\partial x} \\ \dfrac{\partial \phi}{\partial y} \end{array} \right\} + T_{11} \left\{ \begin{array}{c} \dfrac{\partial \psi}{\partial x} \\ \dfrac{\partial \psi}{\partial y} \end{array} \right\} \right) \tag{3.63}$$

$$\int_A \left(B_z \zeta \sin(\zeta z) \right) dA = \left(Q_{31} \left(\frac{\partial \psi_x}{\partial x} + \frac{\partial \psi_y}{\partial y} \right) - Y_{33} \phi - T_{33} \psi \right) \tag{3.64}$$

where

$$\begin{pmatrix} A_{11} & D_{11} \\ A_{12} & D_{12} \\ A_{44} & D_{44} \end{pmatrix} = \left(1, z^2 \right) \left\{ 2 \int_{-h/2}^{-h/2+h_{SC}} \left\{ \begin{array}{c} c_{11}^{SC} \\ c_{12}^{SC} \\ c_{66}^{SC} \end{array} \right\} dz + \int_{-h_{Auxetic}/2}^{h_{Auxetic}/2} \left\{ \begin{array}{c} c_{11}^{Auxetic} \\ c_{12}^{Auxetic} \\ c_{66}^{Auxetic} \end{array} \right\} dz \right\} \tag{3.65}$$

$$\{E_{31}, Q_{31}\} = 2 \int_{-h/2}^{-h/2+h_{SC}} \{e_{31}, q_{31}\} \zeta z \sin(\zeta z) dz$$

$$\{E_{15}, Q_{15}\} = 2 \int_{-h/2}^{-h/2+h_{SC}} \{e_{15}, q_{15}\} \cos(\zeta z) dz$$

$$\{X_{11}, Y_{11}, T_{11}\} = 2 \int_{-h/2}^{-h/2+h_{SC}} \{s_{11}, d_{11}, r_{11}\} \cos^2(\zeta z) dz$$

$$\{X_{33}, Y_{33}, T_{33}\} = 2 \int_{-h/2}^{-h/2+h_{SC}} \{s_{33}, d_{33}, r_{33}\} \zeta^2 \sin^2(\zeta z) dz \tag{3.66}$$

In addition, the forces and moments resulting from external electric voltage and magnetic potential, which are derived from external electric-magnetic work, can be expressed as:

$$N^E = N_x^E = N_y^E = -2 \int_{-h/2}^{-h/2+h_{SC}} \left(e_{31} \frac{2V_0}{h_{SC}} \right) dz$$

$$N^H = N_x^H = N_y^H = -2 \int_{-h/2}^{-h/2+h_{SC}} \left(q_{31} \frac{2\Omega_0}{h_{SC}} \right) dz \tag{3.67}$$

$$M^E = M_x^E = M_y^E = -2 \int_{-h/2}^{-h/2+h_{SC}} \left(e_{31} \frac{2V_0}{h_{SC}} \right) z dz$$

$$M^H = M_x^H = M_y^H = -2 \int_{-h/2}^{-h/2+h_{SC}} \left(q_{31} \frac{2\Omega_0}{h_{SC}} \right) z dz \tag{3.68}$$

The nonlinear governing equations of the piezo-electro-magnetically actuated plate can be expressed in the following forms by substituting equations (3.58)–(3.64) into equations (3.57a)–(3.57g).

$$
\left(
\begin{aligned}
& A_{11}\left(\frac{\partial^2 u_0}{\partial x^2} + \frac{\partial w}{\partial x}\frac{\partial^2 w}{\partial x^2}\right) + \left(A_{12} + A_{66}\right)\left(\frac{\partial^2 v_0}{\partial x \partial y} + \frac{\partial w}{\partial y}\frac{\partial^2 w}{\partial x \partial y}\right) \\
& \quad + A_{66}\left(\frac{\partial^2 u_0}{\partial y^2} + \frac{\partial w}{\partial x}\frac{\partial^2 w}{\partial y^2}\right)
\end{aligned}
\right) = I_0 \frac{\partial^2 u_0}{\partial t^2} \tag{3.69a}
$$

$$
\left(
\begin{aligned}
& A_{22}\left(\frac{\partial^2 v_0}{\partial y^2} + \frac{\partial w}{\partial y}\frac{\partial^2 w}{\partial y^2}\right) + \left(A_{12} + A_{66}\right)\left(\frac{\partial^2 u_0}{\partial x \partial y} + \frac{\partial w}{\partial x}\frac{\partial^2 w}{\partial x \partial y}\right) \\
& \quad + A_{66}\left(\frac{\partial^2 v_0}{\partial x^2} + \frac{\partial w}{\partial y}\frac{\partial^2 w}{\partial x^2}\right)
\end{aligned}
\right) = I_0 \frac{\partial^2 v_0}{\partial t^2} \tag{3.69b}
$$

$$
\left(
\begin{aligned}
& k_s A_{44}\left(\frac{\partial^2 w}{\partial x^2} + \frac{\partial^2 w}{\partial y^2} + \frac{\partial \psi_x}{\partial x} + \frac{\partial \psi_y}{\partial y}\right) \\
& - k_s E_{15}\left(\frac{\partial^2 \phi}{\partial x^2} + \frac{\partial^2 \phi}{\partial y^2}\right) - k_s Q_{15}\left(\frac{\partial^2 \psi}{\partial x^2} + \frac{\partial^2 \psi}{\partial y^2}\right) \\
& + \left(N^E + N^H\right)\left(\frac{\partial^2 w}{\partial x^2} + \frac{\partial^2 w}{\partial y^2}\right) + Z_1 + Z_2
\end{aligned}
\right) = I_0 \frac{\partial^2 w}{\partial t^2} \tag{3.69c}
$$

$$
\left(
\begin{aligned}
& D_{11}\frac{\partial^2 \psi_x}{\partial x^2} + D_{12}\frac{\partial^2 \psi_x}{\partial x \partial y} + D_{66}\left(\frac{\partial^2 \psi_x}{\partial y^2} + \frac{\partial^2 \psi_y}{\partial x \partial y}\right) \\
& + \left(E_{31} + k_s E_{15}\right)\frac{\partial \phi}{\partial x} + \left(Q_{31} + k_s Q_{15}\right)\frac{\partial \psi}{\partial x} - k_s A_{44}\left(\frac{\partial w}{\partial x} + \psi_x\right)
\end{aligned}
\right) = I_2 \frac{\partial^2 \psi_x}{\partial t^2} \tag{3.69d}
$$

$$
\left(
\begin{aligned}
& D_{11}\frac{\partial^2 \psi_y}{\partial y^2} + D_{12}\frac{\partial^2 \psi_x}{\partial x \partial y} + D_{66}\left(\frac{\partial^2 \psi_y}{\partial x^2} + \frac{\partial^2 \psi_x}{\partial x \partial y}\right) \\
& + \left(E_{31} + k_s E_{15}\right)\frac{\partial \phi}{\partial y} + \left(Q_{31} + k_s Q_{15}\right)\frac{\partial \psi}{\partial y} - k_s A_{44}\left(\frac{\partial w}{\partial y} + \psi_y\right)
\end{aligned}
\right) = I_2 \frac{\partial^2 \psi_y}{\partial t^2} \tag{3.69e}
$$

$$
\left(
\begin{aligned}
& E_{15}\left(\frac{\partial^2 w}{\partial x^2} + \frac{\partial^2 w}{\partial y^2} + \frac{\partial \psi_x}{\partial x} + \frac{\partial \psi_y}{\partial y}\right) \\
& + E_{31}\left(\frac{\partial \psi_x}{\partial x} + \frac{\partial \psi_y}{\partial y}\right) + X_{11}\left(\frac{\partial^2 \phi}{\partial x^2} + \frac{\partial^2 \phi}{\partial y^2}\right) - X_{33}\phi \\
& + Y_{11}\left(\frac{\partial^2 \psi}{\partial x^2} + \frac{\partial^2 \psi}{\partial y^2}\right) - Y_{33}\psi
\end{aligned}
\right) = 0 \tag{3.69f}
$$

$$\left(\begin{array}{c} Q_{15}\left(\dfrac{\partial^2 w}{\partial x^2} + \dfrac{\partial^2 w}{\partial y^2} + \dfrac{\partial \psi_x}{\partial x} + \dfrac{\partial \psi_y}{\partial y} \right) \\[4mm] +Q_{31}\left(\dfrac{\partial \psi_x}{\partial x} + \dfrac{\partial \psi_y}{\partial y} \right) + Y_{11}\left(\dfrac{\partial^2 \phi}{\partial x^2} + \dfrac{\partial^2 \phi}{\partial y^2} \right) - Y_{33}\phi \\[4mm] +T_{11}\left(\dfrac{\partial^2 \psi}{\partial x^2} + \dfrac{\partial^2 \psi}{\partial y^2} \right) - T_{33}\psi \end{array} \right) = 0 \qquad (3.69\text{g})$$

$$Z_1 = \left[A_{11}\left(\frac{\partial u_0}{\partial x} + \frac{1}{2}\left(\frac{\partial w}{\partial x}\right)^2 \right) + A_{12}\left(\frac{\partial v_0}{\partial y} + \frac{1}{2}\left(\frac{\partial w}{\partial y}\right)^2 \right) \right]\left(\frac{\partial^2 w}{\partial x^2} \right)$$

$$+ \left[A_{11}\left(\frac{\partial^2 u_0}{\partial x^2} + \frac{\partial w}{\partial x}\frac{\partial^2 w}{\partial x^2} \right) + A_{12}\left(\frac{\partial^2 v_0}{\partial x \partial y} + \frac{\partial w}{\partial y}\frac{\partial^2 w}{\partial x \partial y} \right) \right]\left(\frac{\partial w}{\partial x} \right)$$

$$+ \left[A_{12}\left(\frac{\partial u_0}{\partial x} + \frac{1}{2}\left(\frac{\partial w}{\partial x}\right)^2 \right) + A_{11}\left(\frac{\partial v_0}{\partial y} + \frac{1}{2}\left(\frac{\partial w}{\partial y}\right)^2 \right) \right]\left(\frac{\partial^2 w}{\partial y^2} \right)$$

$$+ \left[A_{12}\left(\frac{\partial^2 u_0}{\partial x \partial y} + \frac{\partial w}{\partial x}\frac{\partial^2 w}{\partial x \partial y} \right) + A_{11}\left(\frac{\partial^2 v_0}{\partial y^2} + \frac{\partial w}{\partial y}\frac{\partial^2 w}{\partial y^2} \right) \right]\left(\frac{\partial w}{\partial y} \right) \qquad (3.69\text{h})$$

$$Z_2 = 2A_{66}\left(\frac{\partial u_0}{\partial y} + \frac{\partial v_0}{\partial x} + \frac{\partial w}{\partial x}\frac{\partial w}{\partial y} \right)\frac{\partial^2 w}{\partial x \partial y}$$

$$+ A_{66}\left(\frac{\partial^2 u_0}{\partial x \partial y} + \frac{\partial^2 v_0}{\partial x^2} + \frac{\partial w}{\partial y}\frac{\partial^2 w}{\partial x^2} + \frac{\partial w}{\partial x}\frac{\partial^2 w}{\partial x \partial y} \right)\frac{\partial w}{\partial y}$$

$$+ A_{66}\left(\frac{\partial^2 u_0}{\partial y^2} + \frac{\partial^2 v_0}{\partial x \partial y} + \frac{\partial w}{\partial y}\frac{\partial^2 w}{\partial x \partial y} + \frac{\partial w}{\partial x}\frac{\partial^2 w}{\partial y^2} \right)\frac{\partial w}{\partial x} \qquad (3.69\text{i})$$

In this context, it is assumed that the piezoelectric and magnetic potentials are negligible at the boundaries of the sandwich plates, $\phi = \psi = 0$. Consequently, the equations describing the behavior of piezo-electro-magnetically actuated plates with varying boundary conditions can be expressed as stated in references [34,35].

(a) piezo-electro-magnetically actuated plate via clamped edges (CCCC):

$$u = v = w = \psi_x = \psi_y = \phi = \psi = 0 \qquad at \quad x = 0, a$$
$$u = v = w = \psi_x = \psi_y = \phi = \psi = 0 \qquad at \quad y = 0, b \qquad (3.70\text{a})$$

(b) piezo-electro-magnetically actuated plate via simply-supported edges (SSSS):

$$u = v = w = M_{xx} = \psi_y = \phi = \psi = 0 \qquad at \quad x = 0, a$$
$$u = v = w = \psi_x = M_{yy} = \phi = \psi = 0 \qquad at \quad y = 0, b \qquad (3.70\text{b})$$

One popular technique for examining plate and shell vibration is the GDQM. The displacement field's m-th derivative is provided by:

$$\frac{\partial^m f(x)}{\partial x^m} = \sum_{k=1}^{n} C_{ik}^{(m)} f(x) \tag{3.71}$$

Furthermore, the partial derivative of the two variable functions f (x,y) on the x and y domains, or Kronecker product ($\otimes$), would be expressed as follows:

$$\frac{\partial^{m+n} f(x,y)}{\partial x^m \partial y^n} = \left(C_y^{(n)} \otimes C_x^{(m)} \right) f(x,y) \tag{3.72}$$

where m and n, respectively, represent the grid points in the x and y axes. Using the cosine form grid distribution method of Chebyshev-Gauss-Lobatto, the GDQ grid points may be expressed as follows:

$$x_i = \frac{a}{2}\left(1-\cos\left(\frac{(i-1)}{(m-1)}\pi\right)\right) \quad i=1,2,3,...,m$$

$$y_i = \frac{b}{2}\left(1-\cos\left(\frac{(j-1)}{(n-1)}\pi\right)\right) \quad j=1,2,3,...,n \tag{3.73}$$

The governing equations would be stated as follows if derivatives of the discretized forms were used:

$$\left[\mathbf{K}_L + \mathbf{K}_{NL}\right]d + \left[\mathbf{M}\right]\ddot{d} = 0 \tag{3.74}$$

where $\ddot{d}$ denotes the second time-based differentiation. Additionally, the mass matrix is represented by $\mathbf{M}$, and the linear and nonlinear stiffness matrices are given by **KL** and **KNL**, respectively. Each of the matrices listed above has a size $7N_x N_y \times 7N_x N_y$. If we were to replace $d^* = d.e^{i\omega}$ in equation (3.72), the eigenvalue system would look like this:

$$\left(\left[\mathbf{K}_L + \mathbf{K}_{NL}\right] - \omega^2 \left[\mathbf{M}\right]\right)\{d\} = 0 \tag{3.75}$$

and d is obtained as:

$$\{d\} = \left\{u, v, w, \psi_x, \psi_y, \phi, \psi\right\}^T \tag{3.76}$$

The primary method of detecting the eigenvalue and eigenvector (w^*_{max}) from equation (3.67) is to ignore the **KNL** (nonlinear stiffness matrix). Subsequently, the linear eigenvectors (**K$_L$**) approximate the nonlinear stiffness matrix (KNL), and the new eigenvalue and eigenvector are identified from the nonlinear eigenvalue system (3.75). Until the difference between the eigenvalues determined from two consecutive iterations is smaller than 1e-4, this procedure is repeated.

$$\left\|\left[\mathbf{K}_L + \mathbf{K}_{NL}\right] - \omega^2 \left[\mathbf{M}\right]\right\| = 0 \tag{3.77}$$

The nonlinear vibration analysis of the piezo-electro-magnetically actuated plate with varying boundary conditions, including CCCC and SSSS, is detected by the figures and tables provided in this section. The sandwich plate with piezo-electro-magnetically actuation is composed of an auxetic metamaterial core and intelligent layers. Additionally, the following are typical assumptions about the sandwich plate's geometrical and magneto-electro-mechanical parameters.

The auxetic metamaterial cell's parameters are $\theta = -55$, $e/l = 4$, and $t/l = 0.0138571$. Auxetic metamaterial core material is also assumed to be aluminum with: $E_{Al} = 70$ GPa $\rho_{Al} = 2705$ kg$/$m^3, and $v_{Al} = 0.3$. Additionally, Table 3.1 reports the characteristics of the magneto-electro-mechanical layers (reinforcement and matrix) [22].

Figure 3.11 shows the impact of a smart thickness (h_{SC}) on the nonlinear frequency ratio of a rectangular sandwich plate that is magneto-electro-elastic. As shown in this figure, increasing the $w$max from 0.1 to 0.4 reduces the nonlinear frequency ratio (ω_{nl}/ω_l). In both CCCC and SSSS B.C., it is evident that increasing the h_{SC} from 0.01*b to 0.015*b reduces the nonlinear frequency ratio. The value of (ω_{nl}/ω_l) in SSSS B.C. is often shown to be greater than in CCCC B.C.

Figure 3.12 illustrates how geometrical parameters (θ) affect a magneto-electro-elastic rectangular sandwich plate's frequency ratio. As this figure shows, increasing the amplitude $w$max from 0.1 to 0.5 reduces ($\theta = -55$, $\theta = -75$) and increases ($\theta =-15$, $\theta = -35$) (ω_{nl}/ω_l). It is shown that CCCC B.C. has a lower nonlinear frequency ratio than SSSS B.C.

Figure 3.13 shows how the power index (η) affects an MEE rectangular plate's nonlinear frequency ratio. This shows that increasing the amplitude $w$max from 0.1 to 0.6 reduces (ω_{nl}/ω_l). It is evident that in both CCCC and SSSS B.C., the nonlinear frequency ratio increases as the η increases from 2 to 6. Notably, it is shown that CCCC has a lower nonlinear frequency ratio than SSSS B.C.

TABLE 3.1
Material Properties $BaTiO_3$ and $CoFe_2O_4$

Properties	$BaTiO_3$	$CoFe_2O_4$
$c_{11} = c_{22}$ (GPa)	166	286
c_{33}	162	269.5
$c_{13} = c_{23}$	78	170.5
c_{12}	77	173
c_{55}	43	45.3
c_{66}	44.5	56.5
$e_{31}(Cm^{-2})$	−4.4	0
e_{33}	18.6	0
e_{15}	11.6	0
s_{11} (10^{-9} C^2 m^{-2} N^{-1})	11.2	0.08
s_{33}	12.6	0.093
ρ (kgm^{-3})	5800	5300

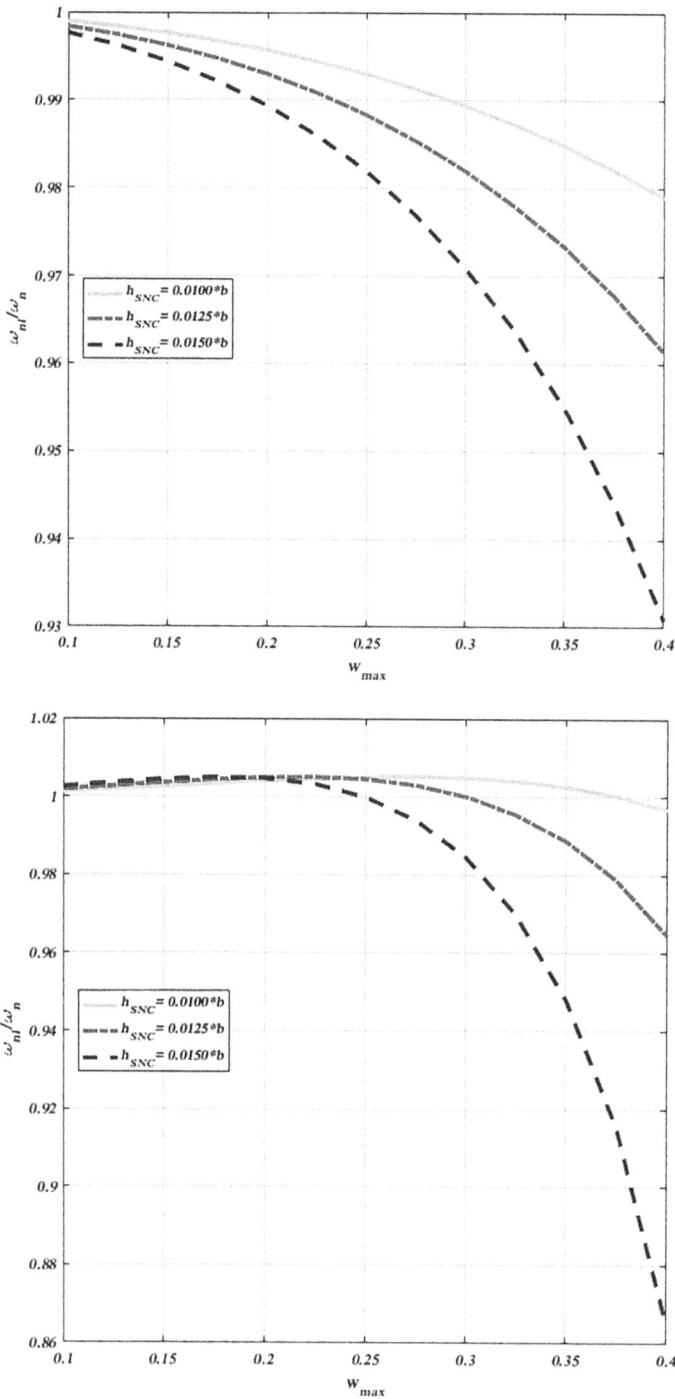

FIGURE 3.11 Impact of piezo-electro-magnetically actuated thickness (hSC) variation on the sandwich plate's nonlinear frequency ratio under various B.C. conditions, including (a) CCCC and (b) SSS.

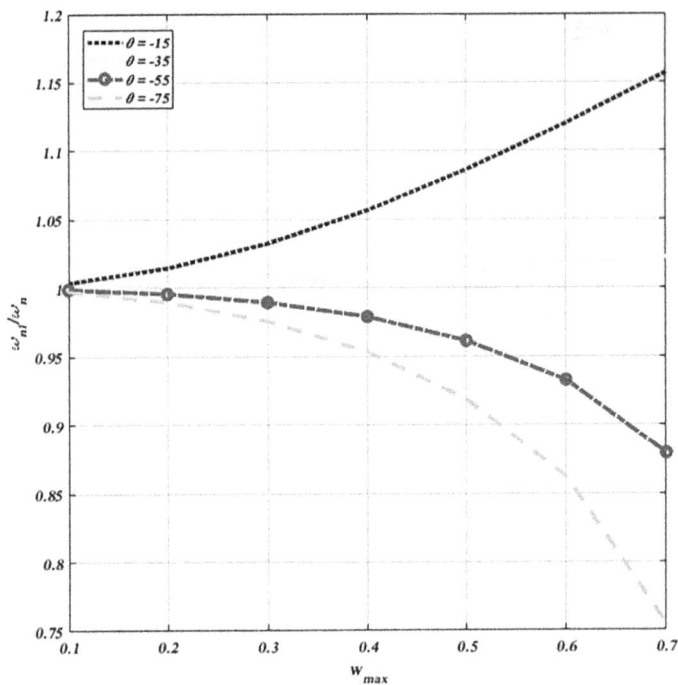

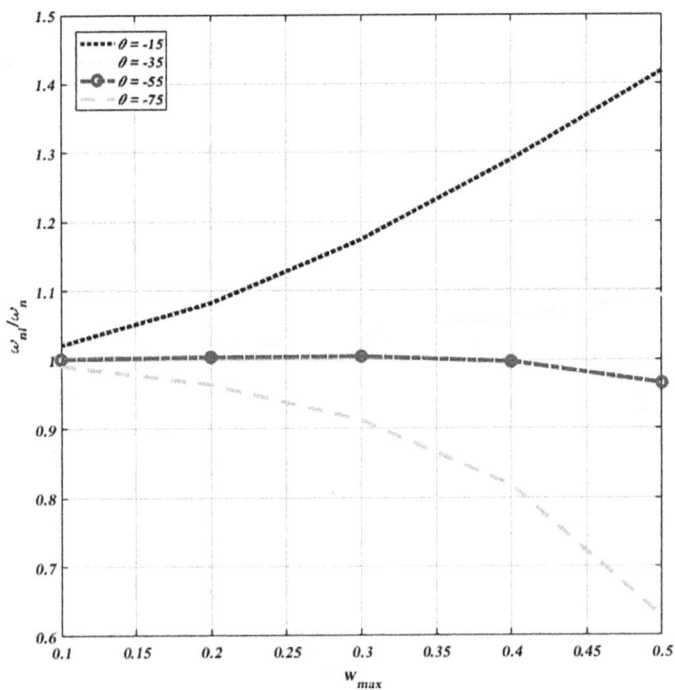

FIGURE 3.12 Influence of (θ) change on the nonlinear frequency ratio of the sandwich plate under several B.C. (a) CCCC and (b) SSSS.

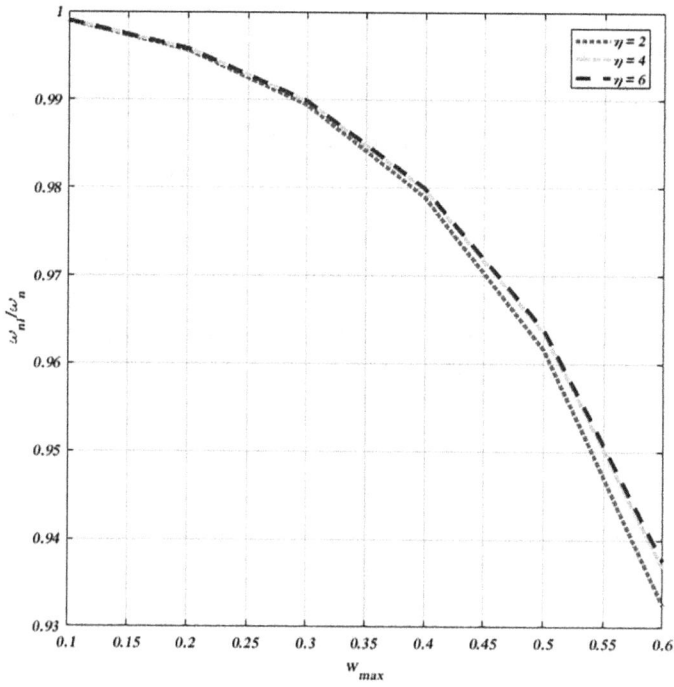

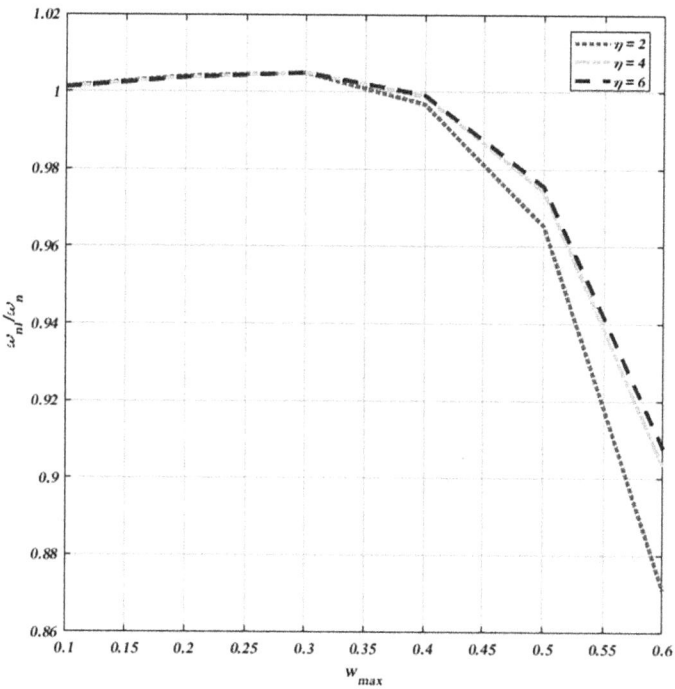

FIGURE 3.13   Influence of FG power index (η) change on the (ω_{nl}/ω_l) of the smart plate under several B.C. (a) CCCC and (b) SSSS.

Figure 3.14 shows how geometrical parameters (*e/l*) affect an MEE sandwich plate's nonlinear frequency ratio. This shows that increasing the amplitude *w*max from 0.1 to 1 for CCCC B.C. reduces (ω_{nl}/ω_l). Furthermore, the nonlinear frequency ratio for the SSSS rose initially before declining. In both CCCC and SSSS B.C., it is evident that an increase in (*e/l*) from 3.8 to 4.4 results in a nonlinear frequency ratio. The nonlinear frequency ratio in CCCC may often be found to be lower than in SSSS B.C.

Figure 3.15 shows the impact of (b) on the nonlinear frequency ratio of a rectangular MEE sandwich plate. As shown in this figure, increasing the amplitude *w*max from 0.1 to 0.6 reduces the (ω_{nl}/ω_l). Clearly, in both CCCC and SSSS B.C., the nonlinear frequency ratio decreases as (b) grows from 2 to 4. It goes without saying that the nonlinear frequency ratio in CCCC is less than in SSSS B.C.

Figure 3.16 shows the influence of auxetic core thicknesses and geometric parameters (θ) on the nonlinear frequency ratio of the MEE rectangular plate. This figure

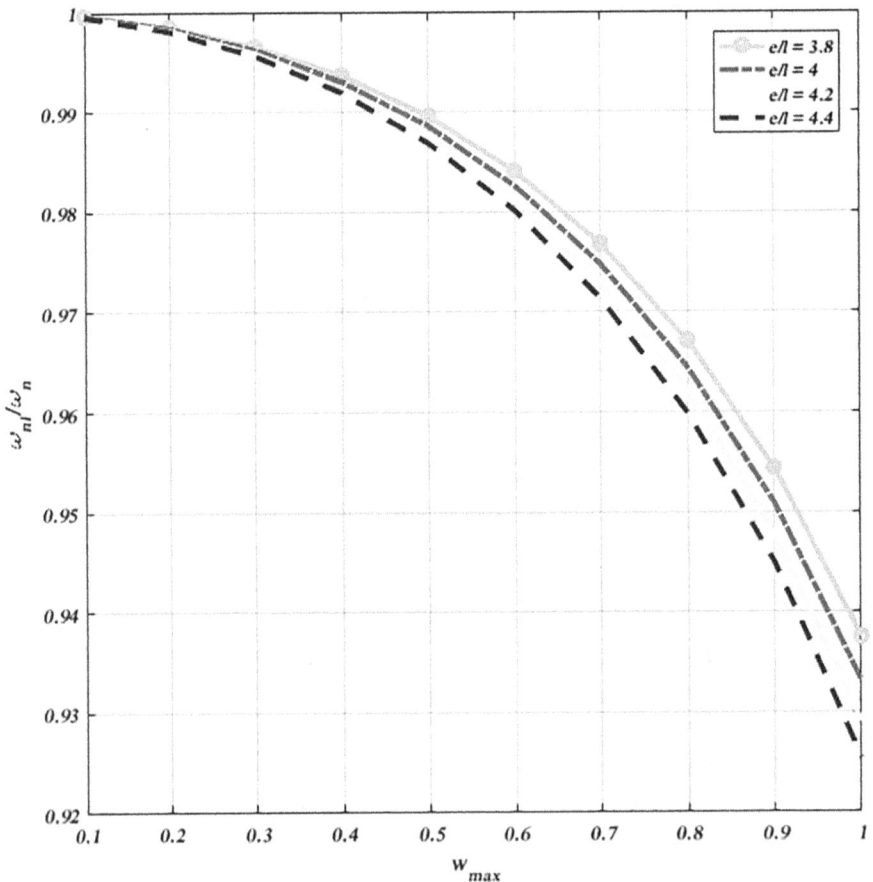

FIGURE 3.14 Influence of (*e/l*) change on the nonlinear frequency ratio of the sandwich plate under several B.C. (a) CCCC and (b) SSSS.

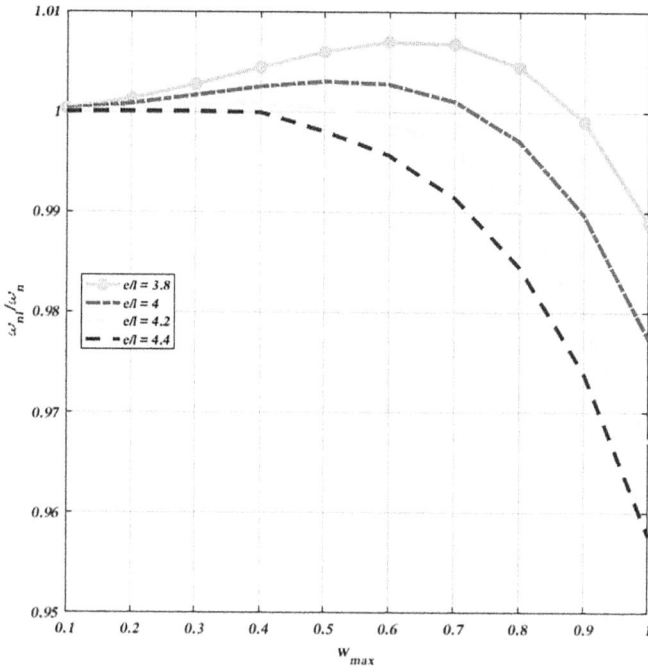

FIGURE 3.14 (Continued)

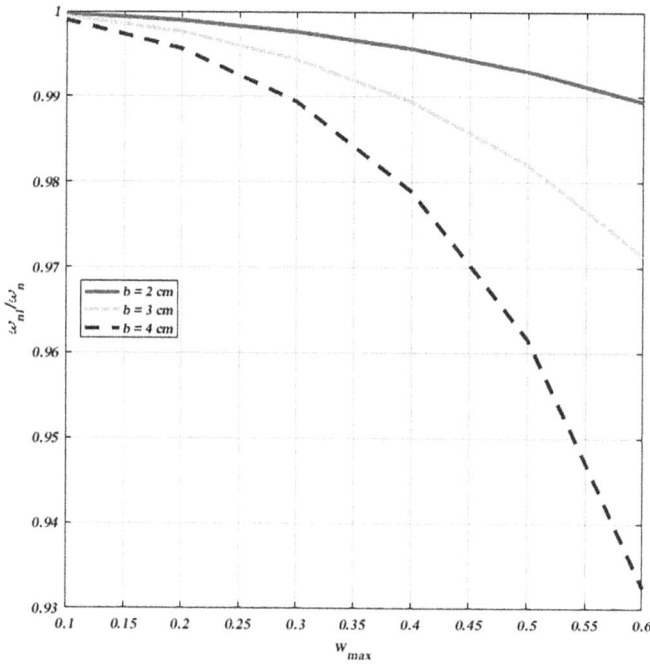

FIGURE 3.15 Influence of length (*b*) change on the nonlinear frequency ratio of the sandwich plate under several B.C. (a) CCCC and (b) SSSS.

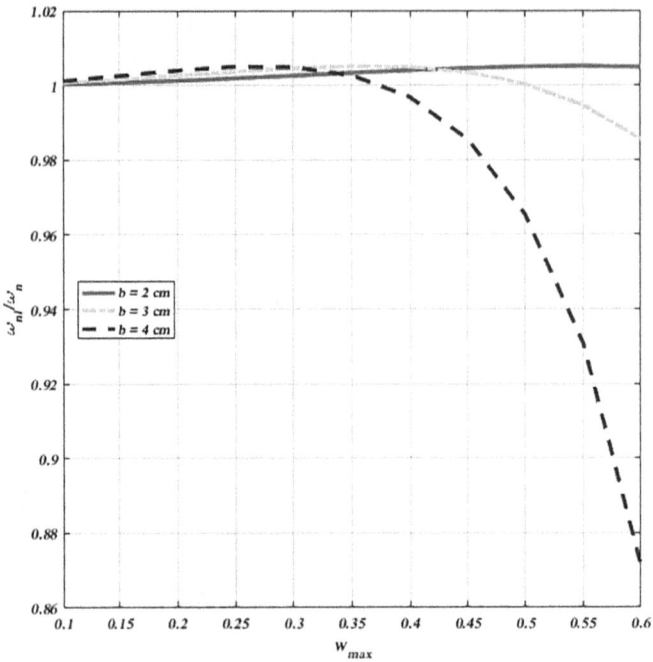

FIGURE 3.15   (Continued)

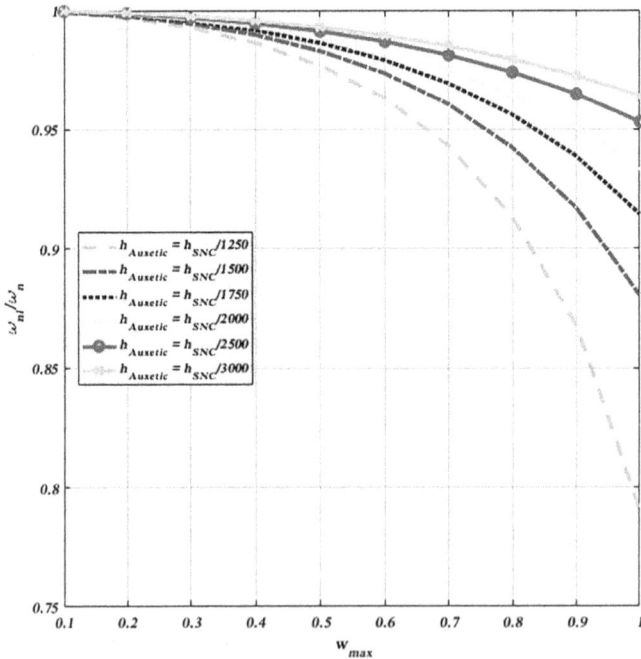

FIGURE 3.16   Influence of various thicknesses of auxetic core on the nonlinear frequency ratio of the sandwich plate under (θ) and several B.C. (a) CCCC- $\theta=-55$, (b) SSSS- $\theta=-55$, (c) CCCC- $\theta=-15$, and (d) SSSS- $\theta=-15$.

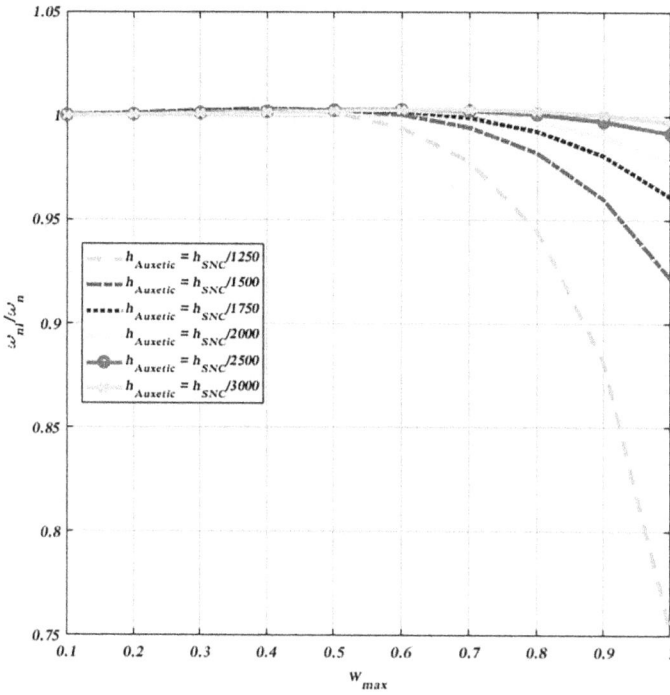

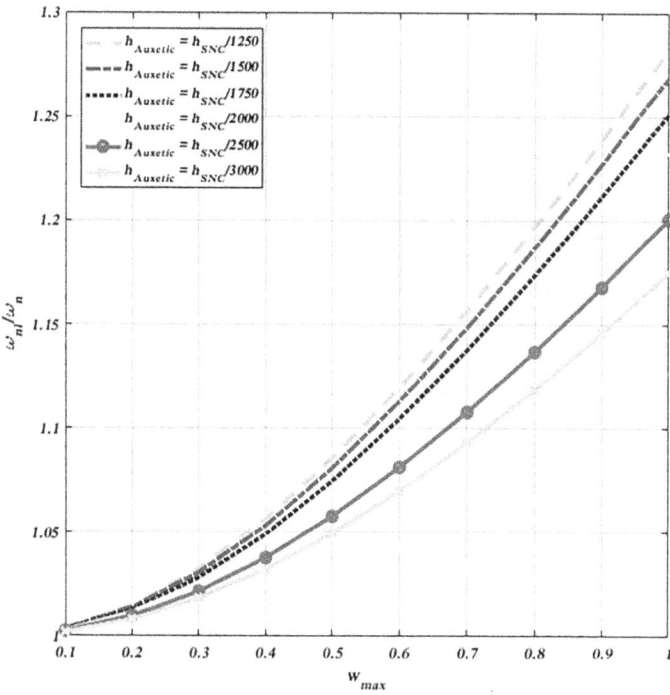

FIGURE 3.16   (Continued)

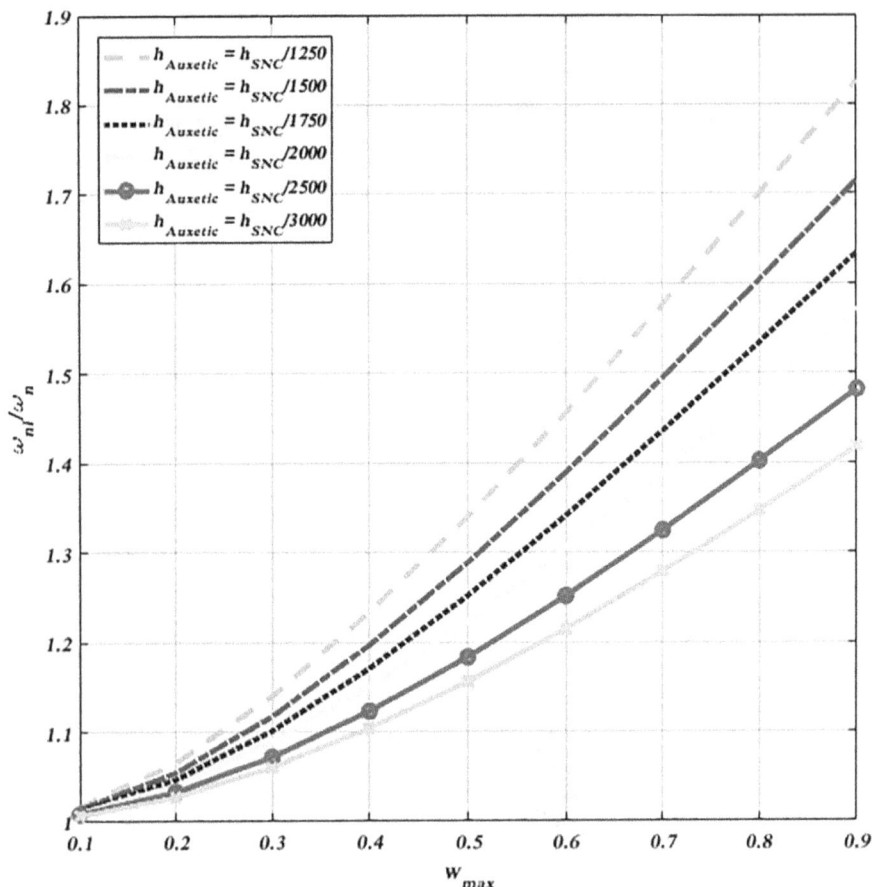

FIGURE 3.16 (Continued)

illustrates how increasing the amplitude $w$max from 0.1 to 1 reduces (ω_{nl}/ω_l) for $(\theta = -55)$. Additionally, (ω_{nl}/ω_l) is growing as a result of the auxetic core's thickness reduction. Additionally, for $(\theta = -15)$, (ω_{nl}/ω_l) is raised by increasing the $w$max from 0.1 to 1. Moreover, nonlinear frequency ratio increases as the auxetic core's thickness increases. It is clearly shown that CCCC has a lower nonlinear frequency ratio than SSSS B.C. Figure 3.17 illustrates how the nonlinear frequency ratio of an MEE sandwich plate is affected by the external voltage (V_0). As shown in this figure, increasing $w$max from 0.1 to 0.6 causes (ω_{nl}/ω_l) to decrease. In both CCCC and SSSS B.C., it is evident that an increase in (V_0) from -5 to 5 results in a higher nonlinear frequency ratio. It goes without saying that the nonlinear frequency ratio in CCCC is less than in SSSS B.C.

Figure 3.18 clearly illustrates how the nonlinear frequency ratio of an MEE rectangular plate is affected by the magnetic potential (Ω_0). As shown in this figure, increasing the amplitude $w$max from 0.1 to 0.6 results in a decrease in (ω_{nl}/ω_l). In both CCCC and SSSS B.C., it is evident that the nonlinear frequency ratio is larger

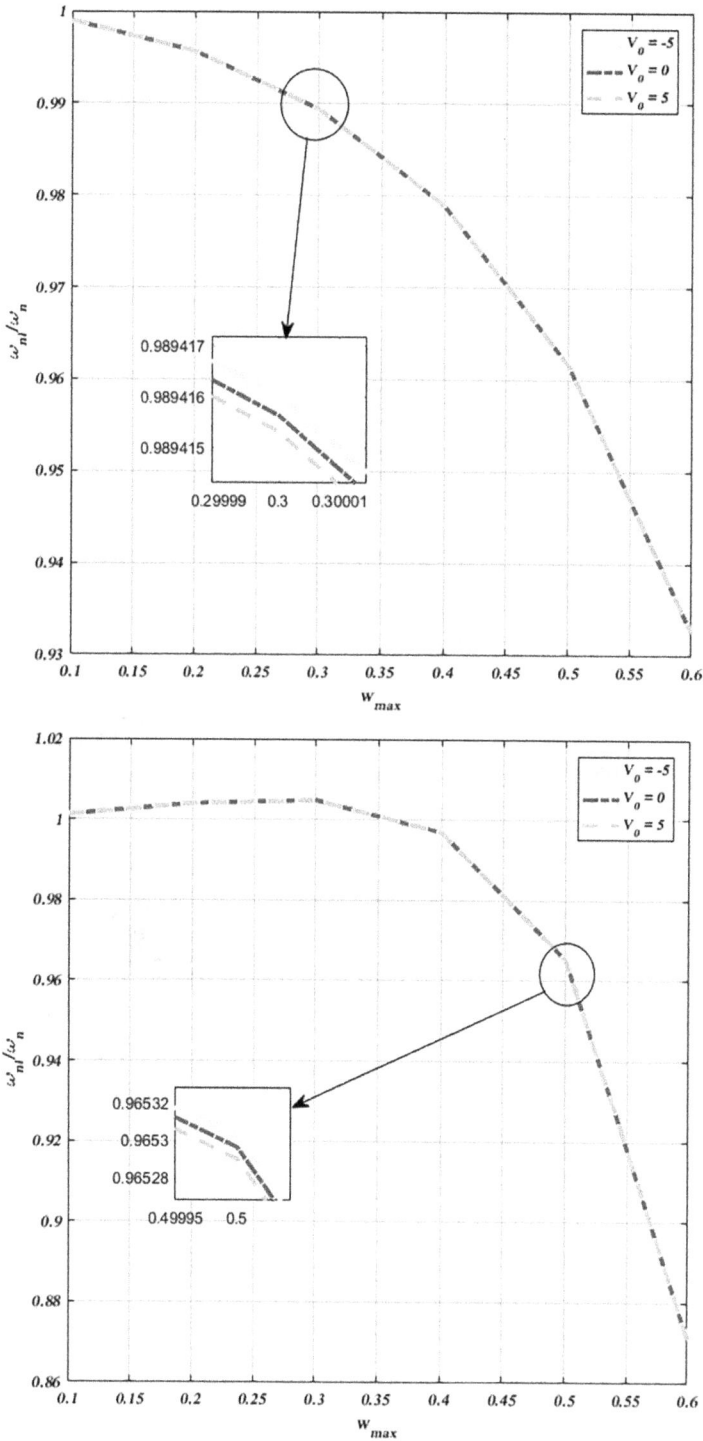

FIGURE 3.17 Influence of various external voltage (V_0) on the nonlinear frequency ratio of the sandwich plate under several B.C. (a) CCCC, (b) SSSS.

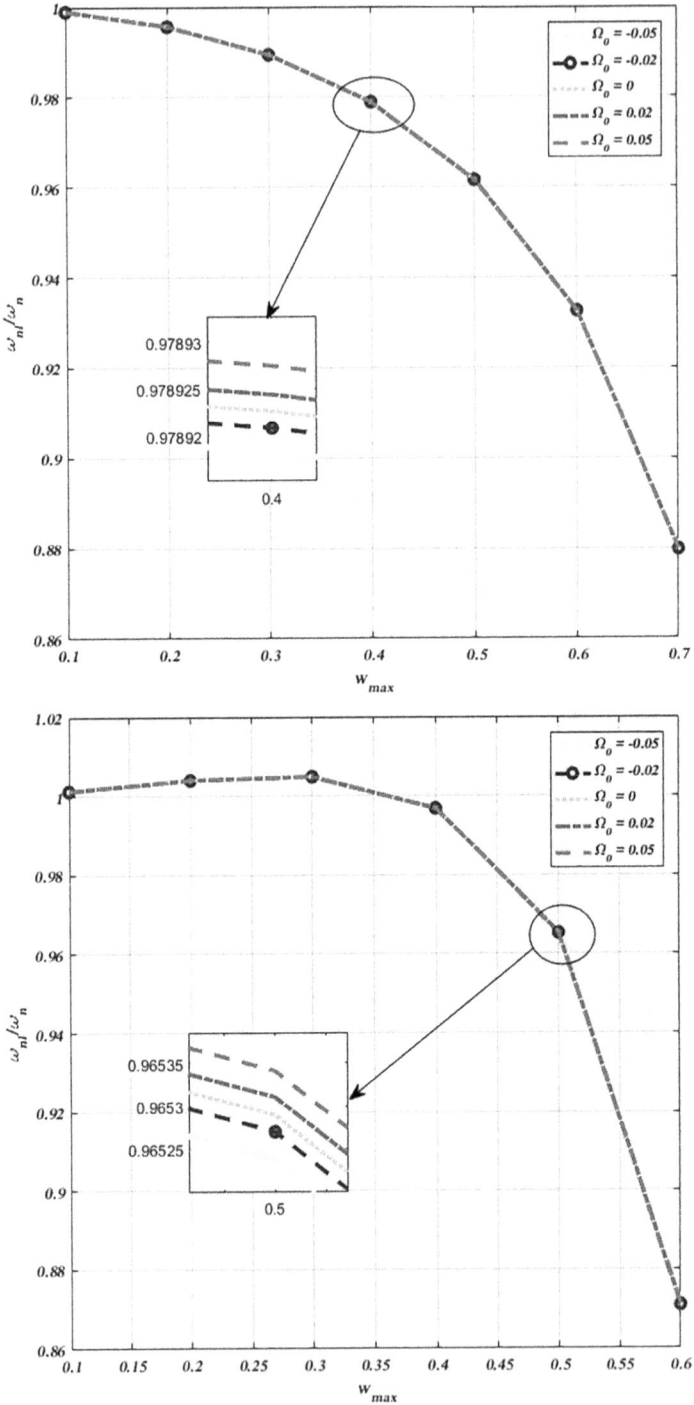

FIGURE 3.18 Impact of different magnetic potentials (Ω_0) on the sandwich plate's nonlinear frequency ratio under multiple B.C. (a) CCCC, (b) SSSS.

than it was before, while the magnetic potential (Ω_0) is increasing from -0.05 to 0.05. It can be seen that CCCC has a lower nonlinear frequency ratio than SSSS B.C.

3.3.2 VIBRATION ANALYSIS OF GRAPHENE ORIGAMI-ENABLED SANDWICHED AUXETIC PLATES WITH PIEZOELECTRIC FACE SHEETS

This section presents the nonlinear frequency deviation of a sandwiched auxetic plate enabled by graphene origami using higher-order Reddy plate theory, as seen in Figure 3.19. This clever multi-scale hybrid nanocomposite/metamaterial is composed of an FG-GOA core and permeable layers (GPL/CF/PVDF). The computed effective properties of the GPL are presented, based on the HT model and the ROM. Based on the hypotheses presented in this section, the GPL uniform pattern appears as follows [34]:

$$W_{GPL} = 100 W_C W_{GPL}^t \qquad (3.78)$$

where W_C and W_{GPL}^t indicate, respectively, the special characteristic of the graphene nanoplatelets and their weight %. One way to determine the volume fraction of graphene nanoplatelets is as follows:

$$V_{GPL} = \frac{W_{GPL}}{W_{GPL} + \left(\dfrac{\rho_{GPL}}{\rho_M}\right)(1 - W_{GPL})} \qquad (3.79)$$

ρ_{GPL} and ρ_M stand for the proportional concentrations of graphene- and PVDF-based nanoplatelets, respectively. We can compute the effective elasticity of the smart graphene platelet nanocomposite (SNC) as follows using the Halpin-Tsai model:

$$E_{SNCM} = \left(\frac{3}{8} E_L + \frac{5}{8} E_T\right) \qquad (3.80)$$

where E_L and E_T refer for the SNC's longitudinal and transverse moduli, respectively [104–109].

$$\{E_L, E_T\} = \left\{\frac{1 + \xi_L^{GPL} \eta_L^{GPL} V_{GPL}}{1 - \eta_L^{GPL} V_{GPL}}, \frac{1 + \xi_W^{GPL} \eta_W^{GPL} V_{GPL}}{1 - \eta_W^{GPL} V_{GPL}}\right\} E_M$$

$$\{\eta_L^{GPL}, \eta_W^{GPL}\} = \left\{\frac{\left(\dfrac{E_{GPL}}{E_M}\right) - 1}{\left(\dfrac{E_{GPL}}{E_M}\right) + \xi_L^{GPL}}, \frac{\left(\dfrac{E_{GPL}}{E_M}\right) - 1}{\left(\dfrac{E_{GPL}}{E_M}\right) + \xi_W^{GPL}}\right\}$$

$$\{\xi_L^{GPL}, \xi_W^{GPL}\} = \left\{\frac{2 l_{GPL}}{t_{GPL}}, \frac{2 w_{GPL}}{t_{GPL}}\right\} \qquad (3.81)$$

where the characters l_{GPL}, w_{GPL}, and t_{GPL} accordingly represent the graphene's length, breadth, and height. The rule of mixing (ROM) may also be used to compute the

Poisson's ratio, density, dielectric, and piezoelectric constants of the SNC layer. The following values may be found [104–109]:

$$V_{GPL} + V_M = 1 \tag{3.82a}$$

$$\rho_{SNC} = \left(V_M \rho^M + V_{GPL} \rho^{GPL} \right) \tag{3.82b}$$

$$v_{SNC} = V_M v^M + V_{GPL} v^{GPL} \tag{3.82c}$$

$$G_{SNC} = \left(\frac{E_{SNC}}{2\left(1 + v_{SNC}\right)} \right) \tag{3.82d}$$

$$e_{ij}^{SNC} = V_M e_{ij}^M + V_{GPL} e_{ij}^{GPL} \tag{3.82e}$$

$$s_{ij}^{SNC} = V_M s_{ij}^M + V_{GPL} s_{ij}^{GPL} \tag{3.82f}$$

The smart hybrid nanocomposite (SHNC) is strengthened in some areas by the employment of GPL and carbon fiber (CF). Instead, the mixed nanocomposite layers exhibit the following advantageous properties [104–109]:

$$V_{SNC} + V_F = 1 \tag{3.83a}$$

$$E_{SHNC} = \left(V_{SNC} E^{SNC} + V_F E^F \right) \tag{3.83b}$$

$$\rho_{SHNC} = \left(V_{SNC} \rho^{SNC} + V_F \rho^F \right) \tag{3.83c}$$

$$v_{SHNC} = \left(V_{SNC} v^{SNC} + V_F v^F \right) \tag{3.83d}$$

$$G_{SHNC} = \left(\frac{E_{SHNC}}{2\left(1 + v_{SHNC}\right)} \right) \tag{3.83e}$$

$$e_{ij}^{SHNC} = \left(V_{SNCM} e_{ij}^{SNC} + V_F e_{ij}^F \right) \tag{3.83f}$$

$$s_{ij}^{SHNC} = \left(V_{SNCM} s_{ij}^{SNC} + V_F s_{ij}^F \right) \tag{3.83g}$$

Then, using SHNC and F to represent the smart hybrid nanocomposite and fibers, respectively, we may approximate the properties of a smart porous hybrid nanocomposite as follows [104–109]:

$$E_{SPHNC} = E_{SHNC} \left(1 - e_0 \lambda \right) \tag{3.84a}$$

$$G_{SPHNC} = G_{SHNC} \left(1 - e_0 \lambda \right) \tag{3.84b}$$

$$\rho_{SPHNC} = \rho_{SHNC} \left(\sqrt{1 - e_0 \lambda} \right) \tag{3.84c}$$

where the smart porous hybrid nanocomposite was denoted by SPHNC. Just as e_0 represents permeability, so e_m represents mass density.

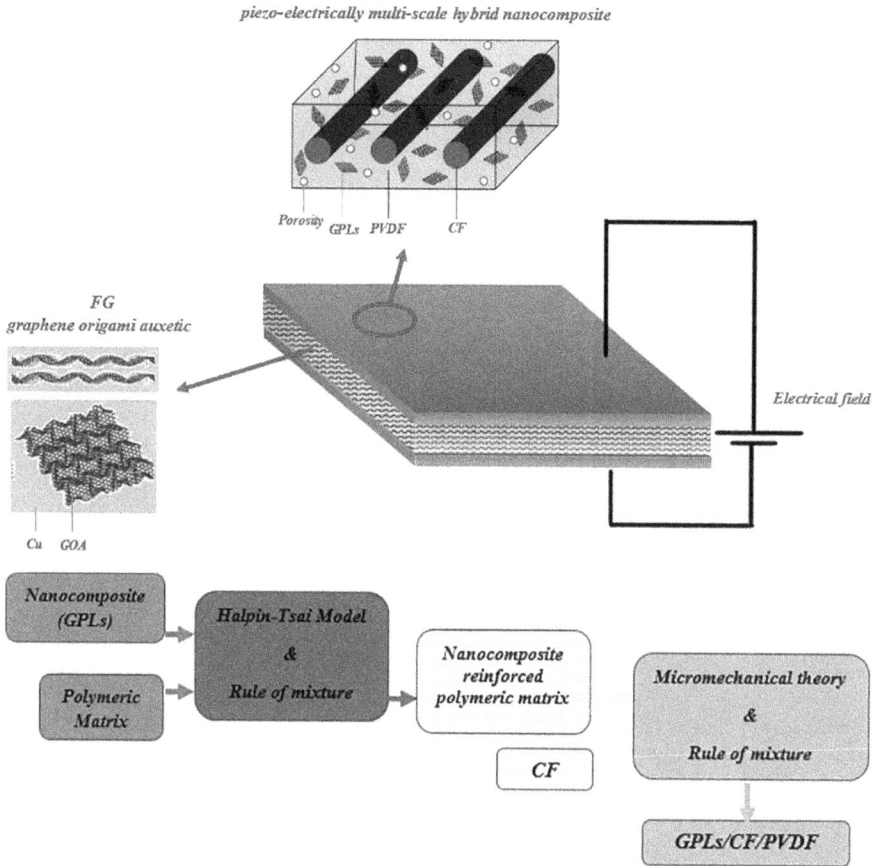

Porosity *GPLs* *PVDF*   *CF*

FG
graphene origami auxetic

Electrical field

Cu   *GOA*

Nanocomposite
(GPLs)   Halpin-Tsai Model

&   Nanocomposite
reinforced   Micromechanical theory
Polymeric   Rule of mixture   polymeric matrix
Matrix   &

CF   Rule of mixture

GPLs/CF/PVDF

FIGURE 3.19   (a) The schematic of the smart hybrid nanocomposite sandwich plate via FG graphene origami auxetic core. (b) Mixing process of GPLs/CF/PVDF piezoelectrically multi-scale hybrid nanocomposite.

$$e_m = 1 - \sqrt{1 - e_0} \tag{3.85}$$

Furthermore, the following conclusive forms may be provided for:

$$\lambda = \frac{1}{e_0} - \frac{1}{e_0} \left\{ \frac{2}{\pi} \sqrt{1 - e_0} - \frac{2}{\pi} + 1 \right\}^2 \tag{3.86}$$

Moreover, references [104–109] would provide the following properties of graphene origami auxetic (GOA):

$$E^{Auxetic} = E_{Cu} \frac{1 + \xi \eta V_{Gr}}{1 - \eta V_{Gr}} f_E \left(H_{Gr}, V_{Gr}, T \right) \tag{3.87a}$$

$$v^{Auxetic} = \left(V_{Gr} v_{Gr} + V_{Cu} v_{Cu} \right) f_v \left(H_{Gr}, V_{Gr}, T \right) \tag{3.87b}$$

$$\alpha^{Auxetic} = \left(V_{Gr}\alpha_{Gr} + V_{Cu}\alpha_{Cu}\right)f_{\alpha}\left(V_{Gr},T\right) \tag{3.87c}$$

$$\rho^{Auxetic} = \left(V_{Gr}\rho_{Gr} + V_{Cu}\rho_{Cu}\right)f_{\rho}\left(V_{Gr},T\right) \tag{3.87d}$$

$$f_E\left(H_{Gr},V_{Gr},T\right) = \begin{pmatrix} 1.11 - 1.22V_{Gr} - 0.134\left(\dfrac{T}{T_0}\right) \\ +0.559V_{Gr}\left(\dfrac{T}{T_0}\right) - 5.5H_{Gr}V_{Gr} \\ +38H_{Gr}V_{Gr}^2 - 20.6H_{Gr}^2V_{Gr}^2 \end{pmatrix}$$

$$f_v\left(H_{Gr},V_{Gr},T\right) = \begin{pmatrix} 1.01 - 1.43V_{Gr} + 0.165\left(\dfrac{T}{T_0}\right) \\ -16.8H_{Gr}V_{Gr} - 1.1H_{Gr}V_{Gr}\left(\dfrac{T}{T_0}\right) \\ +16H_{Gr}^2V_{Gr}^2 \end{pmatrix}$$

$$f_{\alpha}\left(V_{Gr},T\right) = \begin{pmatrix} 0.794 - 16.8V_{Gr}^2 - 0.0279\left(\dfrac{T}{T_0}\right)^2 \\ +0.182\left(\dfrac{T}{T_0}\right)\left(V_{Gr}+1\right) \end{pmatrix}$$

$$f_{\rho}\left(V_{Gr},T\right) = \left(1.01 - 2.01V_{Gr}^2 - 0.0131\left(\dfrac{T}{T_0}\right)\right) \tag{3.87e}$$

Both W_{Gr} and Cu, which indicate the weight percentage of graphene origami and Cu, respectively as $V_{Gr} = \dfrac{\rho_{Cu}W_{Gr}}{\rho_{Cu}W_{Gr} + \rho_{Gr}\left(1 - W_{Gr}\right)}$, $V_{Cu} = 1 - V_{Gr}$, and $f_{E,v,\alpha,\rho}\left(H_{Gr},V_{Gr},T\right)$, may be determined by molecular dynamics (MD), rendering them as valuable markers of material and dimension constants $\eta = \dfrac{\dfrac{E_{Gr}}{E_{Cu}} - 1}{\dfrac{E_{Gr}}{E_{Cu}} + \xi}, \xi = 2\left(\dfrac{l_{Gr}}{t_{Gr}}\right)$.

Furthermore, given the following unique FG distributions, the resultant form may be used to find properties of the FG graphene origami auxetic (FG-GOA):

$$V_{Gr}\left(z\right) = V_{Gr}\left|\frac{4z}{h_{FG-GOA}}\right|^n \tag{3.88}$$

The electro-elastic constitutive model of the sandwiched plate yields the following formulas [104–109]:

$$\sigma_{ij} = \left[C_{ijkl}\varepsilon_{kl} - e_{mij}E_m \right] \tag{3.89a}$$

$$D_i = \left[e_{ikl}\varepsilon_{kl} + s_{im}E_m \right] \tag{3.89b}$$

where the elements of the stress-strain matrix are represented by the symbols C_{ijkl}, $\sigma_{ij}-\varepsilon_{kl}$. Additionally, D_i and E_m are assigned to the electric potential and motion components, respectively. The dielectric and piezoelectric coefficients are denoted by s_{im} and e_{ikl}, respectively. Consequently, the stress-strain equation for the smart plate may be expressed as follows:

$$
\begin{Bmatrix} \sigma_{xx} \\ \sigma_{yy} \\ \sigma_{yz} \\ \sigma_{zx} \\ \sigma_{xy} \end{Bmatrix} =
\begin{pmatrix} C_{11} & C_{12} & 0 & 0 & 0 \\ C_{12} & C_{11} & 0 & 0 & 0 \\ 0 & 0 & C_{55} & 0 & 0 \\ 0 & 0 & 0 & C_{55} & 0 \\ 0 & 0 & 0 & 0 & C_{66} \end{pmatrix}
\begin{Bmatrix} \varepsilon_{xx} \\ \varepsilon_{yy} \\ \varepsilon_{yz} \\ \varepsilon_{zx} \\ \varepsilon_{xy} \end{Bmatrix} -
\begin{pmatrix} 0 & 0 & e_{31} \\ 0 & 0 & e_{31} \\ 0 & e_{15} & 0 \\ e_{15} & 0 & 0 \\ 0 & 0 & 0 \end{pmatrix}
\begin{Bmatrix} E_x \\ E_y \\ E_z \end{Bmatrix}
$$

$$\tag{3.90}$$

$$
\begin{Bmatrix} D_x \\ D_y \\ D_z \end{Bmatrix} =
\begin{pmatrix} 0 & 0 & 0 & 0 & e_{15} & 0 \\ 0 & 0 & 0 & e_{15} & 0 & 0 \\ e_{31} & e_{31} & 0 & 0 & 0 & 0 \end{pmatrix}
\begin{Bmatrix} \varepsilon_{xx} \\ \varepsilon_{yy} \\ \varepsilon_{zz} \\ \varepsilon_{yz} \\ \varepsilon_{zx} \\ \varepsilon_{xy} \end{Bmatrix} +
\begin{pmatrix} s_{11} & 0 & 0 \\ 0 & s_{11} & 0 \\ 0 & 0 & s_{33} \end{pmatrix}
\begin{Bmatrix} E_x \\ E_y \\ E_z \end{Bmatrix}
$$

We get the following formula for the displacement field of the smart sandwich plate using higher-order Reddy plate theory:

$$U(x,y,z,t) = u(x,y,t) - z\frac{\partial w(x,y,t)}{\partial x} + \Xi(z)\left(\psi_x(x,y,t) + \frac{\partial w(x,y,t)}{\partial x} \right) \tag{3.91a}$$

$$V(x,y,z,t) = v(x,y,t) - z\frac{\partial w(x,y,t)}{\partial x} + \Xi(z)\left(\psi_y(x,y,t) + \frac{\partial w(x,y,t)}{\partial x} \right) \tag{3.91b}$$

$$W(x,y,z,t) = w(x,y,t) \tag{3.91c}$$

$$\Xi(z) = \left(z - \frac{4z^3}{3h^2} \right) \tag{3.91d}$$

where u_0 denotes the sandwich plate's horizontal deviation, v_0 indicates its transverse displacement, and v_0 represents its longitudinal displacement, all as previously defined. Furthermore, transverse normal rotations along axes x and y were represented by $a\psi_x$ and ψ_y. It is possible to forecast the circumstances under which Maxwell's law will be met using the following formulas:

$$\Phi(x,y,z,t) = -\cos(\zeta z)\phi(x,z,t) + \frac{2z}{h}V_0 \qquad (3.92)$$

where V_0 represents the electrical energy derived from the external surroundings. Additionally, the strains on sandwich plates may be achieved as follows:

$$\begin{Bmatrix} \varepsilon_{xx} \\ \varepsilon_{yy} \\ \varepsilon_{xy} \end{Bmatrix} = \begin{Bmatrix} \varepsilon_{xx}^0 \\ \varepsilon_{yy}^0 \\ \varepsilon_{xy}^0 \end{Bmatrix} + z \begin{Bmatrix} k_{xx}^b \\ k_{yy}^b \\ k_{xy}^b \end{Bmatrix}, \qquad \begin{Bmatrix} \gamma_{xz} \\ \gamma_{yz} \end{Bmatrix} = \begin{Bmatrix} \gamma_{xz}^0 \\ \gamma_{yz}^0 \end{Bmatrix} \qquad (3.93a)$$

$$\Xi_1 = \Xi(z) - z \qquad \Xi_2 = \Xi(z) \qquad \Xi_3 = \frac{d\Xi(z)}{dz} \qquad (3.93b)$$

The von Karman strain of the piezoelectrically activated plate can be mathematically described by the following equation:

$$\begin{Bmatrix} \varepsilon_{xx}^0 \\ \varepsilon_{yy}^0 \\ \varepsilon_{xy}^0 \end{Bmatrix} = \begin{Bmatrix} \dfrac{\partial u}{\partial x} + \dfrac{1}{2}\left(\dfrac{\partial w}{\partial x}\right)^2 \\ \dfrac{\partial v}{\partial x} + \dfrac{1}{2}\left(\dfrac{\partial w}{\partial y}\right)^2 \\ \dfrac{\partial u}{\partial y} + \dfrac{\partial v}{\partial x} + \dfrac{\partial w}{\partial x}\dfrac{\partial w}{\partial y} \end{Bmatrix}, \quad \begin{Bmatrix} k_{xx}^b \\ k_{yy}^b \\ k_{xy}^b \end{Bmatrix} = \begin{Bmatrix} -\dfrac{\partial^2 w}{\partial x^2} \\ -\dfrac{\partial^2 w}{\partial y^2} \\ -\dfrac{\partial^2 w}{\partial x \partial y} \end{Bmatrix}, \quad \begin{Bmatrix} k_{xx}^s \\ k_{yy}^s \\ k_{xy}^s \end{Bmatrix} = \begin{Bmatrix} \dfrac{\partial \psi_x}{\partial x} \\ \dfrac{\partial \psi_y}{\partial y} \\ \dfrac{\partial \psi_x}{\partial y} + \dfrac{\partial \psi_y}{\partial x} \end{Bmatrix},$$

$$\begin{Bmatrix} \gamma_{xz}^0 \\ \gamma_{yz}^0 \end{Bmatrix} = \begin{Bmatrix} \dfrac{\partial w}{\partial x} + \psi_x \\ \dfrac{\partial w}{\partial y} + \psi_y \end{Bmatrix} \qquad (3.94)$$

Furthermore, the manipulation of electrical fields in the sandwich plate can be achieved through the implementation of the following methods:

$$E_x = -\frac{\partial \Phi}{\partial x} = \cos(\zeta z)\frac{\partial \phi}{\partial x} \qquad (3.95a)$$

$$E_y = -\frac{\partial \Phi}{\partial y} = \cos(\zeta z)\frac{\partial \phi}{\partial y} \qquad (3.95b)$$

$$E_z = -\frac{\partial \Phi}{\partial z} = -\zeta \sin(\zeta z)\phi - \frac{2V_E}{h} \qquad (3.95c)$$

When Hamilton's principle is applied to a plate that is activated by piezoelectricity, the aforementioned shape is employed to represent and explain the underlying concept.

$$\int_0^t \delta(\Pi_S - \Pi_K + \Pi_W)dt = 0 \qquad (3.96)$$

The strain and kinetic energy are denoted by Π_S and Π_K, respectively. Furthermore, the sign Π_W symbolizes the work performed by the exterior of the intelligent plate. The strain energy can indeed be calculated by:

$$\Pi_S = \int_V (\sigma_{ij}\varepsilon_{ij})\,dV = \int_V \left(\begin{array}{c} \sigma_{xx}\varepsilon_{xx} + \sigma_{yy}\varepsilon_{yy} + \sigma_{xy}\varepsilon_{xy} + \sigma_{xz}\varepsilon_{xz} + \sigma_{yz}\varepsilon_{yz} \\ -D_x E_x - D_y E_y - D_z E_z \end{array} \right) dV \quad (3.97)$$

If equations (3.91)–(3.93) were substituted into equation (3.95), the strain energy would be determined as follows:

$$\Pi_S = \int_A \left(\begin{array}{c} N_{xx}\varepsilon_{xx}^0 + M_{xx}\varepsilon_{xx}^1 + P_{xx}\varepsilon_{xx}^2 \\ +N_{yy}\varepsilon_{yy}^0 + M_{yy}\varepsilon_{yy}^1 + P_{yy}\varepsilon_{xx}^2 \\ +2N_{xy}\varepsilon_{xy}^0 + 2M_{xy}\varepsilon_{xy}^1 + 2P_{xy}\varepsilon_{xy}^2 \\ +Q_{xz}\gamma_{xz}^0 + Q_{yz}\gamma_{yz}^0 \end{array} \right) dA$$

$$+ \int_{-h/2}^{h/2} \int_A \left(-D_x \cos(\zeta z)\delta(\frac{\partial\phi}{\partial x}) - D_y \cos(\zeta z)\delta(\frac{\partial\phi}{\partial y}) + D_z \sin(\zeta z)\delta\phi \right) dA\,dz \quad (3.98)$$

By substituting equations (3.91)–(3.93) into equation (3.95), the calculation of the strain energy can be determined as follows:

$$\left(N_{xx}, N_{yy}, N_{xy} \right) = \int_{-h/2}^{h/2} \left\{ \sigma_{xx}, \sigma_{yy}, \sigma_{xy} \right\} dz$$

$$\left(P_{xx}, P_{yy}, P_{xy} \right) = \int_{-h/2}^{h/2} \left\{ \sigma_{xx}, \sigma_{yy}, \sigma_{xy} \right\} \Xi_1 dz$$

$$\left(M_{xx}, M_{yy}, M_{xy} \right) = \int_{-h/2}^{h/2} \left\{ \sigma_{xx}, \sigma_{yy}, \sigma_{xy} \right\} \Xi_2 dz$$

$$\left(Q_{xz}, Q_{yz} \right) = \int_{-h/2}^{h/2} \left\{ \sigma_{xz}, \sigma_{yz} \right\} \Xi_3 dz \quad (3.99)$$

In addition, the work done outside the organization is described as:

$$\delta\Pi_W = \frac{1}{2}\int_A \left[N_x^E \left(\frac{\partial w}{\partial x}\right)^2 + N_y^E \left(\frac{\partial w}{\partial y}\right)^2 \right] dA \quad (3.100)$$

The terms N_x^E and N_y^E denote the electrical activity occurring beyond the specified region in the x- and y-directions, respectively. Moreover, the acquisition of kinetic energy may be achieved by the following methods:

$$\Pi_K = \frac{1}{2}\int_A \left(\begin{array}{c} I_0\left(\frac{\partial u}{\partial t}\right)^2 + I_0\left(\frac{\partial v}{\partial t}\right)^2 + I_0\left(\frac{\partial w}{\partial t}\right)^2 + I_3\left(\frac{\partial\psi_x}{\partial t}\right)^2 + I_3\left(\frac{\partial\psi_y}{\partial t}\right)^2 \\ +I_1\left[\left(\frac{\partial^2 w}{\partial t\partial x}\right)^2 + \left(\frac{\partial^2 w}{\partial t\partial y}\right)^2\right] + 2I_2\left[\frac{\partial\psi_x}{\partial t}\frac{\partial^2 w}{\partial t\partial x} + \frac{\partial\psi_y}{\partial t}\frac{\partial^2 w}{\partial t\partial y}\right] \end{array} \right) dA \quad (3.101)$$

The mass inertias can be calculated as follows:

$$(I_0, I_1, I_2, I_3) = \int_{-h/2}^{h/2} (1, \Xi_1^2, \Xi_1\Xi_2, \Xi_2^2)\rho(z)dz \tag{3.102}$$

To get the equations of motion for the sandwich plate, it is necessary to use equation (3.96) in place of equations (3.95), (3.98), and (3.99).

$$\frac{\partial N_{xx}}{\partial x} + \frac{\partial N_{xy}}{\partial y} = I_0 \frac{\partial^2 u}{\partial t^2} \tag{3.103a}$$

$$\frac{\partial N_{xy}}{\partial x} + \frac{\partial N_{yy}}{\partial y} = I_0 \frac{\partial^2 v}{\partial t^2} \tag{3.103b}$$

$$\frac{\partial}{\partial x}\left(N_{xx}\frac{\partial w}{\partial x} + N_{xy}\frac{\partial w}{\partial y}\right) + \frac{\partial}{\partial y}\left(N_{xy}\frac{\partial w}{\partial x} + N_{yy}\frac{\partial w}{\partial y}\right)$$
$$+ \frac{\partial Q_{xz}}{\partial x} + \frac{\partial Q_{yz}}{\partial y} - \frac{\partial^2 P_{xx}}{\partial x^2} - \frac{\partial^2 P_{xy}}{\partial x \partial y} - \frac{\partial^2 P_{yy}}{\partial y^2} - N^E \nabla^2 w$$
$$= I_0 \frac{\partial^2 w}{\partial t^2} - I_1\left(\frac{\partial^4 w}{\partial t^2 \partial x^2} + \frac{\partial^4 w}{\partial t^2 \partial y^2}\right) + I_2\left(\frac{\partial^3 \psi_x}{\partial t^2 \partial x} + \frac{\partial^3 \psi_y}{\partial t^2 \partial y}\right) \tag{3.103c}$$

$$\frac{\partial M_{xx}}{\partial x} + \frac{\partial M_{xy}}{\partial y} - Q_{xz} = I_3\frac{\partial^2 \psi_x}{\partial t^2} + I_2\frac{\partial^3 w}{\partial t^2 \partial x} \tag{3.103d}$$

$$\frac{\partial M_{xy}}{\partial x} + \frac{\partial M_{yy}}{\partial y} - Q_{yz} = I_3\frac{\partial^2 \psi_y}{\partial t^2} + I_2\frac{\partial^3 w}{\partial t^2 \partial y} \tag{3.103e}$$

$$\int_A \left(\cos(\zeta z)\frac{\partial D_x}{\partial x} + \cos(\zeta z)\frac{\partial D_y}{\partial y} + \zeta \sin(\zeta z)D_z\right)dA = 0 \tag{3.103f}$$

Based on the constitutive equations, the forces and moments derived are as follows:

$$\begin{Bmatrix} N_{xx} \\ N_{yy} \\ N_{xy} \end{Bmatrix} = \begin{pmatrix} A_{11} & A_{12} & 0 \\ A_{21} & A_{22} & 0 \\ 0 & 0 & A_{66} \end{pmatrix} \begin{Bmatrix} \frac{\partial u}{\partial x} + \frac{1}{2}\left(\frac{\partial w}{\partial x}\right)^2 \\ \frac{\partial v}{\partial x} + \frac{1}{2}\left(\frac{\partial w}{\partial y}\right)^2 \\ \frac{\partial u}{\partial y} + \frac{\partial v}{\partial x} + \frac{\partial w}{\partial x}\frac{\partial w}{\partial y} \end{Bmatrix} + \begin{pmatrix} A_{31} \\ A_{31} \\ 0 \end{pmatrix}\phi - \begin{pmatrix} N_x^E \\ N_y^E \\ 0 \end{pmatrix} \tag{3.104a}$$

$$
\left\{\begin{array}{c} M_{xx} \\ M_{yy} \\ M_{xy} \end{array}\right\} = \left(\begin{pmatrix} C_{11} & C_{12} & 0 \\ C_{21} & C_{22} & 0 \\ 0 & 0 & C_{66} \end{pmatrix} \left\{\begin{array}{c} \dfrac{\partial^2 w}{\partial x^2} \\ \dfrac{\partial^2 w}{\partial y^2} \\ \dfrac{\partial^2 w}{\partial x \partial y} \end{array}\right\} + \right.
$$
$$
\left. \begin{pmatrix} D_{11} & D_{12} & 0 \\ D_{21} & D_{22} & 0 \\ 0 & 0 & D_{66} \end{pmatrix} \left\{\begin{array}{c} \dfrac{\partial \psi_x}{\partial x} \\ \dfrac{\partial \psi_y}{\partial y} \\ \dfrac{\partial \psi_x}{\partial y} + \dfrac{\partial \psi_y}{\partial x} \end{array}\right\} + \begin{pmatrix} E_{31} \\ E_{31} \\ 0 \end{pmatrix} \phi - \begin{pmatrix} M_x^E \\ M_y^E \\ 0 \end{pmatrix} \right) \tag{3.104b}
$$

$$
\left\{\begin{array}{c} P_{xx} \\ P_{yy} \\ P_{xy} \end{array}\right\} = \left(\begin{pmatrix} D_{11} & D_{12} & 0 \\ D_{21} & D_{22} & 0 \\ 0 & 0 & D_{66} \end{pmatrix} \left\{\begin{array}{c} \dfrac{\partial^2 w}{\partial x^2} \\ \dfrac{\partial^2 w}{\partial y^2} \\ \dfrac{\partial^2 w}{\partial x \partial y} \end{array}\right\} + \right.
$$
$$
\left. \begin{pmatrix} H_{11} & H_{12} & 0 \\ H_{21} & H_{22} & 0 \\ 0 & 0 & H_{66} \end{pmatrix} \left\{\begin{array}{c} \dfrac{\partial \psi_x}{\partial x} \\ \dfrac{\partial \psi_y}{\partial y} \\ \dfrac{\partial \psi_x}{\partial y} + \dfrac{\partial \psi_y}{\partial x} \end{array}\right\} + \begin{pmatrix} G_{31} \\ G_{31} \\ 0 \end{pmatrix} \phi - \begin{pmatrix} M_x^E \\ M_y^E \\ 0 \end{pmatrix} \right) \tag{3.104c}
$$

$$
\left\{\begin{array}{c} Q_{xz} \\ Q_{yz} \end{array}\right\} = \left(\begin{pmatrix} A_{44} & 0 \\ 0 & A_{55} \end{pmatrix} \left\{\begin{array}{c} \dfrac{\partial w}{\partial x} + \psi_x \\ \dfrac{\partial w}{\partial y} + \psi_y \end{array}\right\} - E_{15} \left\{\begin{array}{c} \dfrac{\partial \phi}{\partial x} \\ \dfrac{\partial \phi}{\partial y} \end{array}\right\} \right) \tag{3.104d}
$$

$$
\int_A \left(\left\{\begin{array}{c} D_x \\ D_y \end{array}\right\} \cos(\zeta z) \right) dA = \left(E_{15} \left\{\begin{array}{c} \dfrac{\partial w}{\partial x} + \psi_x \\ \dfrac{\partial w}{\partial y} + \psi_y \end{array}\right\} + X_{11} \left\{\begin{array}{c} \dfrac{\partial \phi}{\partial x} \\ \dfrac{\partial \phi}{\partial y} \end{array}\right\} \right) \tag{3.104e}
$$

$$\int_A \left(D_z \,\zeta\, \sin(\zeta\, z)\right) dA = \left(\begin{array}{c} E_{31}\left(\dfrac{\partial^2 w}{\partial x^2} + \dfrac{\partial^2 w}{\partial y^2}\right) \\[4mm] G_{31}\left(\dfrac{\partial \psi_x}{\partial x} + \dfrac{\partial \psi_y}{\partial y}\right) - X_{33}\phi \end{array}\right) \tag{3.104f}$$

where

$$\begin{bmatrix} A_{11} & D_{11} & C_{11} & H_{11} \\ A_{12} & D_{12} & C_{12} & H_{12} \\ A_{66} & D_{66} & C_{66} & H_{66} \\ A_{44} & D_{44} & C_{44} & H_{44} \\ A_{55} & D_{55} & C_{55} & H_{55} \end{bmatrix} = \left(1, \Xi_1 \Xi_2, \Xi_1^2, \Xi_2^2\right)\left(2 \int_{-h/2}^{-h/2+h_{SPHNC}} \begin{Bmatrix} c_{11}^{SHNC} \\ c_{12}^{SHNC} \\ c_{66}^{SHNC} \\ c_{44}^{SHNC} \\ c_{55}^{SHNC} \end{Bmatrix} dz\right.$$

$$\left. + \int_{-h_{FG-GOA}/2}^{h_{FG-GOA}/2} \begin{Bmatrix} c_{11}^{FG-GOA} \\ c_{12}^{FG-GOA} \\ c_{66}^{FG-GOA} \\ c_{44}^{FG-GOA} \\ c_{55}^{FG-GOA} \end{Bmatrix} dz\right) \tag{3.105}$$

$$E_{31} = 2 \int_{-h/2}^{-h/2+h_{SPHNC}} e_{31}\zeta\, \sin(\zeta\, z)\Xi_1 dz$$

$$G_{31} = 2 \int_{-h/2}^{-h/2+h_{SPHNC}} e_{31}\zeta\, \sin(\zeta\, z)\Xi_2 dz$$

$$E_{15} = 2 \int_{-h/2}^{-h/2+h_{SPHNC}} e_{15}\, \cos(\zeta\, z)\Xi_3 dz$$

$$\{X_{11}, Y_{11}\} = 2 \int_{-h/2}^{-h/2+h_{SPHNC}} \{s_{11}, d_{11}\} \cos^2(\zeta\, z)\Xi_3 dz$$

$$\{X_{33}, Y_{33}\} = 2 \int_{-h/2}^{-h/2+h_{SPHNC}} \{s_{33}, d_{33}\}\zeta^2 \sin^2(\zeta\, z) dz \tag{3.106}$$

Furthermore, the externally supplied electric voltage induces forces and moments that are determined by the external electric activity. These forces and moments may be mathematically described as follows:

$$N^E = N_x^E = N_y^E = -2 \int_{-h/2}^{-h/2+h_{SPHNC}} \left(e_{31}\frac{2V_0}{h_{SPHNC}}\right) dz \tag{3.107a}$$

$$M^E = M_x^E = M_y^E = -2 \int_{-h/2}^{-h/2+h_{SHNC}} \left(e_{31} \frac{2V_0}{h_{SPHNC}} \right) z dz \tag{3.107b}$$

The nonlinear governing equations of the piezoelectrically driven plate may be obtained by substituting formulas (3.104a)–(3.105) into equations (3.103a)–(3.103b).

$$\left[A_{11} \left(\frac{\partial^2 u}{\partial x^2} + \frac{\partial w}{\partial x}\frac{\partial^2 w}{\partial x^2} \right) + (A_{12}+A_{66}) \left(\frac{\partial^2 v}{\partial x \partial y} + \frac{\partial w}{\partial y}\frac{\partial^2 w}{\partial x \partial y} \right) + A_{66} \left(\frac{\partial^2 u}{\partial y^2} + \frac{\partial w}{\partial x}\frac{\partial^2 w}{\partial y^2} \right) \right] = I_0 \frac{\partial^2 u_0}{\partial t^2} \tag{3.108a}$$

$$\left[A_{22} \left(\frac{\partial^2 v}{\partial y^2} + \frac{\partial w}{\partial y}\frac{\partial^2 w}{\partial y^2} \right) + (A_{12}+A_{66}) \left(\frac{\partial^2 u}{\partial x \partial y} + \frac{\partial w}{\partial x}\frac{\partial^2 w}{\partial x \partial y} \right) + A_{66} \left(\frac{\partial^2 v}{\partial x^2} + \frac{\partial w}{\partial y}\frac{\partial^2 w}{\partial x^2} \right) \right] = I_0 \frac{\partial^2 v}{\partial t^2} \tag{3.108b}$$

$$\left[k_s A_{44} \left(\frac{\partial^2 w}{\partial x^2} + \frac{\partial^2 w}{\partial y^2} + \frac{\partial \psi_x}{\partial x} + \frac{\partial \psi_y}{\partial y} \right) - (k_s E_{15}+E_{31}) \left(\frac{\partial^2 \phi}{\partial x^2} + \frac{\partial^2 \phi}{\partial y^2} \right) - C_{11}\frac{\partial^4 w}{\partial x^4} - 2(C_{12}-2C_{11})\frac{\partial^4 w}{\partial x^2 \partial y^2} - C_{22}\frac{\partial^4 w}{\partial y^4} - D_{11}\frac{\partial^3 \psi_x}{\partial x^3} - (D_{12}+2D_{66}) \left(\frac{\partial^3 \psi_x}{\partial x \partial x^2} + \frac{\partial^3 \psi_y}{\partial x^2 \partial y} \right) - D_{22}\frac{\partial^3 \psi_y}{\partial y^3} + N^E \left(\frac{\partial^2 w}{\partial x^2} + \frac{\partial^2 w}{\partial y^2} \right) + Z_1 + Z_2 \right] = \left[I_0\frac{\partial^2 w}{\partial t^2} - I_1 \left(\frac{\partial^4 w}{\partial t^2 \partial x^2} + \frac{\partial^4 w}{\partial t^2 \partial y^2} \right) + I_2 \left(\frac{\partial^3 \psi_x}{\partial t^2 \partial x} + \frac{\partial^3 \psi_x}{\partial t^2 \partial y} \right) \right] \tag{3.108c}$$

$$\left[(G_{31}+k_s E_{15})\frac{\partial \phi}{\partial x} + (L_{31}+k_s Q_{15})\frac{\partial \psi}{\partial x} - k_s A_{44} \left(\frac{\partial w}{\partial x}+\psi_x \right) + D_{11}\frac{\partial^3 w}{\partial x^3} + (D_{12}+2D_{66})\frac{\partial^3 w}{\partial x \partial y^2} + H_{11}\frac{\partial^3 \psi_x}{\partial x^3} + (H_{12}+H_{66})\frac{\partial^2 \psi_y}{\partial x \partial y} + H_{66}\frac{\partial^3 \psi_x}{\partial y^3} \right] = \left(I_3\frac{\partial^2 \psi_x}{\partial t^2} + I_2\frac{\partial^3 w}{\partial t^2 \partial x} \right) \tag{3.108d}$$

$$\left(\begin{array}{l} +\left(G_{31}+k_s E_{15}\right)\dfrac{\partial \phi}{\partial x}+\left(L_{31}+k_s Q_{15}\right)\dfrac{\partial \psi}{\partial x} \\[2mm] -k_s A_{44}\left(\dfrac{\partial w}{\partial y}+\psi_y\right)+D_{22}\dfrac{\partial^3 w}{\partial y^3} \\[2mm] +\left(D_{12}+2D_{66}\right)\dfrac{\partial^3 w}{\partial x^2 \partial y}+H_{66}\dfrac{\partial^2 \psi_y}{\partial x^2} \\[2mm] +\left(H_{12}+H_{66}\right)\dfrac{\partial^2 \psi_x}{\partial x \partial y}+H_{22}\dfrac{\partial^2 \psi_y}{\partial y^2} \end{array}\right)=\left(I_3 \dfrac{\partial^2 \psi_y}{\partial t^2}+I_2 \dfrac{\partial^3 w}{\partial t^2 \partial y}\right) \qquad (3.108\mathrm{e})$$

$$\left(\begin{array}{l} E_{15}\left(\dfrac{\partial^2 w}{\partial x^2}+\dfrac{\partial^2 w}{\partial y^2}+\dfrac{\partial \psi_x}{\partial x}+\dfrac{\partial \psi_y}{\partial y}\right) \\[2mm] +G_{31}\left(\dfrac{\partial \psi_x}{\partial x}+\dfrac{\partial \psi_y}{\partial y}\right)+X_{11}\left(\dfrac{\partial^2 \phi}{\partial x^2}+\dfrac{\partial^2 \phi}{\partial y^2}\right) \\[2mm] +E_{31}\left(\dfrac{\partial^2 w}{\partial x^2}+\dfrac{\partial^2 w}{\partial y^2}\right)-X_{33}\phi \end{array}\right)=0 \qquad (3.108\mathrm{f})$$

$$Z_1 = \left[A_{11}\left(\dfrac{\partial u}{\partial x}+\dfrac{1}{2}\left(\dfrac{\partial w}{\partial x}\right)^2\right)+A_{12}\left(\dfrac{\partial v}{\partial y}+\dfrac{1}{2}\left(\dfrac{\partial w}{\partial y}\right)^2\right)\right]\left(\dfrac{\partial^2 w}{\partial x^2}\right)$$

$$+\left[A_{11}\left(\dfrac{\partial^2 u}{\partial x^2}+\dfrac{\partial w}{\partial x}\dfrac{\partial^2 w}{\partial x^2}\right)+A_{12}\left(\dfrac{\partial^2 v}{\partial x \partial y}+\dfrac{\partial w}{\partial y}\dfrac{\partial^2 w}{\partial x \partial y}\right)\right]\left(\dfrac{\partial w}{\partial x}\right)$$

$$+\left[A_{12}\left(\dfrac{\partial u}{\partial x}+\dfrac{1}{2}\left(\dfrac{\partial w}{\partial x}\right)^2\right)+A_{11}\left(\dfrac{\partial v}{\partial y}+\dfrac{1}{2}\left(\dfrac{\partial w}{\partial y}\right)^2\right)\right]\left(\dfrac{\partial^2 w}{\partial y^2}\right)$$

$$+\left[A_{12}\left(\dfrac{\partial^2 u}{\partial x \partial y}+\dfrac{\partial w}{\partial x}\dfrac{\partial^2 w}{\partial x \partial y}\right)+A_{11}\left(\dfrac{\partial^2 v}{\partial y^2}+\dfrac{\partial w}{\partial y}\dfrac{\partial^2 w}{\partial y^2}\right)\right]\left(\dfrac{\partial w}{\partial y}\right) \qquad (3.108\mathrm{g})$$

$$Z_2 = 2A_{66}\left(\dfrac{\partial u}{\partial y}+\dfrac{\partial v}{\partial x}+\dfrac{\partial w}{\partial x}\dfrac{\partial w}{\partial y}\right)\dfrac{\partial^2 w}{\partial x \partial y}$$

$$+A_{66}\left(\dfrac{\partial^2 u}{\partial x \partial y}+\dfrac{\partial^2 v}{\partial x^2}+\dfrac{\partial w}{\partial y}\dfrac{\partial^2 w}{\partial x^2}+\dfrac{\partial w}{\partial x}\dfrac{\partial^2 w}{\partial x \partial y}\right)\dfrac{\partial w}{\partial y}$$

$$+A_{66}\left(\dfrac{\partial^2 u}{\partial y^2}+\dfrac{\partial^2 v}{\partial x \partial y}+\dfrac{\partial w}{\partial y}\dfrac{\partial^2 w}{\partial x \partial y}+\dfrac{\partial w}{\partial x}\dfrac{\partial^2 w}{\partial y^2}\right)\dfrac{\partial w}{\partial x} \qquad (3.108\mathrm{h})$$

In this particular scenario, it is assumed that the piezoelectric effect is negligible at the boundaries of the plates. As a result, the terms of the plate with variable boundary conditions may be expressed in the following manner:

(c) FG-GOA-SPHNC plate with clamped edges (CCCC):

$$u = v = w = \frac{\partial w}{\partial x} = \psi_x = \psi_y = \phi = 0 \qquad at \qquad x = 0, a$$

$$u = v = w = \frac{\partial w}{\partial y} = \psi_x = \psi_y = \phi = 0 \qquad at \qquad y = 0, b \qquad (3.109a)$$

(d) FG-GOA-SPHNC plate with simply-supported edges (SSSS):

$$u = v = w = M_{xx} = P_{xx} = \psi_y = \phi = 0 \qquad at \qquad x = 0, a$$

$$u = v = w = \psi_x = M_{yy} = P_{yy} = \phi = 0 \qquad at \qquad y = 0, b \qquad (3.109b)$$

The evaluation of plate and panel resonance is often conducted via the use of the Generalized Differential Quadrature Method (GDQM), a widely utilized approach in this field. The mth derivative of the displacement field may be obtained using the following method:

$$\frac{\partial^m f(x)}{\partial x^m} = \sum_{k=1}^{n} C_{ik}^{(m)} f(x) \qquad (3.110)$$

Moreover, the Kronecker product ($\otimes$), which is sometimes referred to as the partial derivative of two input components, $f(x,y)$, may be represented as follows in the x and y domains:

$$\frac{\partial^{m+n} f(x,y)}{\partial x^m \partial y^n} = \left(C_y^{(n)} \otimes C_x^{(m)} \right) f(x,y) \qquad (3.111)$$

The variables m and n represent the locations on the x and y axes of the coordinate system, respectively. The GDQ nodes may be represented using the Chebyshev-Gauss-Lobatto via cosine format grid distribution model in the following manner:

$$x_i = \frac{a}{2}(1 - \cos(\frac{(i-1)}{(m-1)}\pi)) \qquad i = 1, 2, 3, ..., m$$

$$y_i = \frac{b}{2}(1 - \cos(\frac{(j-1)}{(n-1)}\pi)) \qquad j = 1, 2, 3, ..., n \qquad (3.112)$$

When using the derivatives of discretized forms inside the system of equations, the resulting expression may take the following form:

$$[K_L + K_{NL}]d + [M]\ddot{d} = 0 \qquad (3.113)$$

The symbol $\ddot{d}$ represents the second derivative with respect to time. In addition, it is worth noting that the nonlinear stiffness matrix is represented as $\mathbf{K}_{NL}$, whereas the linear stiffness matrix is marked as $\mathbf{K}_L$. Similarly, the mass matrix is indicated by M. Each of the aforementioned matrices is in a certain dimension $6N_xN_y \times 6N_xN_y$.

By substituting the expression $d*=d.e^{i\omega}$ into equation (3.113), the eigen platform may be represented as:

$$\left(\left[\mathbf{K}_L + \mathbf{K}_{NL}\right] - \omega^2\left[\mathbf{M}\right]\right)\{d\} = 0 \tag{3.114}$$

and d is gotten as

$$\{d\} = \{u, v, w, \psi_x, \psi_y, \phi\}^T \tag{3.115}$$

The eigenvalue and eigenvector are determined from the equations mostly by disregarding the $\mathbf{K}_{NL}$, which represents the nonlinear stiffening matrix. A novel eigenvalue/eigenvector may be determined from the nonlinear eigenvalue system after an estimation of the nonlinear stiffening matrix ($\mathbf{K}_{NL}$), even using linear eigenvectors ($\mathbf{K}_L$). Following each iteration, the discrepancy between the eigenvalues obtained in the previous iteration and the current iteration is examined to verify that it satisfies the criterion of being less than 1e-4.

$$\left|\left[\mathbf{K}_L + \mathbf{K}_{NL}\right] - \omega^2\left[\mathbf{M}\right]\right| = 0 \tag{3.116}$$

In the next part, readers will encounter many charts and tables that facilitate the identification of nonlinear dynamic analysis in a smart plate configuration with numerous boundary conditions. These boundary conditions include CCCC and SSSS. Table 3.2 provides comprehensive information about the geometric properties and electromechanical characteristics of both the FG-GOA metamaterial and the SPHNC sheets.

The influence of the height (h_{SPHNC}) of the smart porous hybrid nanocomposite layer on the frequency ratio and dimensionless frequency of the plate is seen in Figure 3.20. The augmentation of the dimensionless amplitude from 0.1 to 1 results in the

TABLE 3.2

Material Properties SPHNC and FG-GOA

SPHNC

Properties	Polyvinylidene fluoride	Properties	Graphene
c_{11} (GPa)	238.24	E(GPa)	985
c_{22}	23.6	e_{31} (Cm^{-2})	-0.221
c_{33}	10.64	e_{33}	0.221
c_{13}	2.19	e_{15}	-0.221
c_{23}	1.92	s_{33} ((10^{-2} C^2 m^{-2} N^{-1}))	1.106
c_{12}	3.98	ρ(kgm^{-3})	800

(Continued)

TABLE 3.2 (*Continued*)
Material Properties SPHNC and FG-GOA

SPHNC

Properties	Polyvinylidene fluoride	Properties	Graphene
c_{55}	4.4	v	0.186
c_{66}	6.43	$l(nm)$	3
e_{31} (Cm^{-2})	−0.13	$w(nm)$	1.8
e_{33}	−0.276	$h(nm)$	0.7
e_{15}	−0.135		
s_{11} $(10^{-2}\ C^2\ m^{-2}\ N^{-1})$	110.67	Properties	Carbon
s_{33}	106.07	$E(GPa)$	23.1
$\rho(kgm^{-3})$	1780	$\rho(kgm^{-3})$	1750
		v	0.2

GOA

Properties	Gr	Properties	Cu
$E(GPa)$	929.57	$E(GPa)$	65.79
$\rho(kgm^{-3})$	0.220	$\rho(kgm^{-3})$	8800
v	1800	v	0.387
$\alpha\ (10^{-6}\ K^{-2})$	−3.98	$\alpha\ (10^{-6}\ K^{-1})$	16.51
$l(nm)$	8.376		
$t(nm)$	0.34		

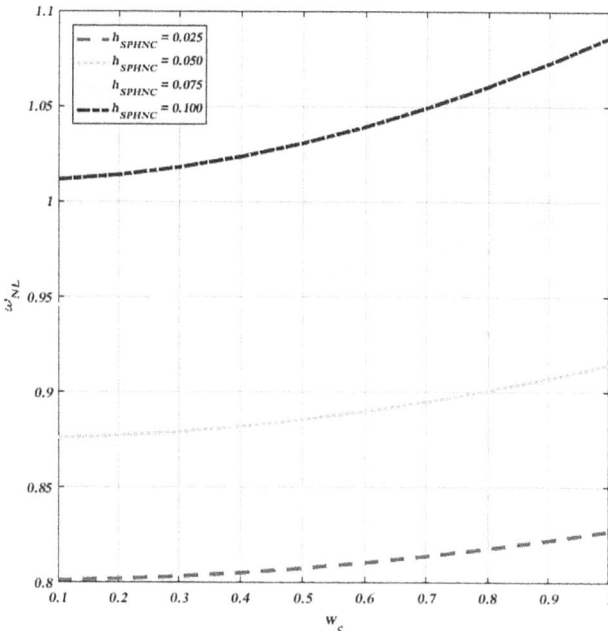

FIGURE 3.20 Influence of smart porous hybrid nanocomposite layers (h_{SPHNC}) change on the (ω_{nl}/ω_l) and dimensionless ω_{nl} of the plate in different B.C. (a) CCCC and (b) SSSS.

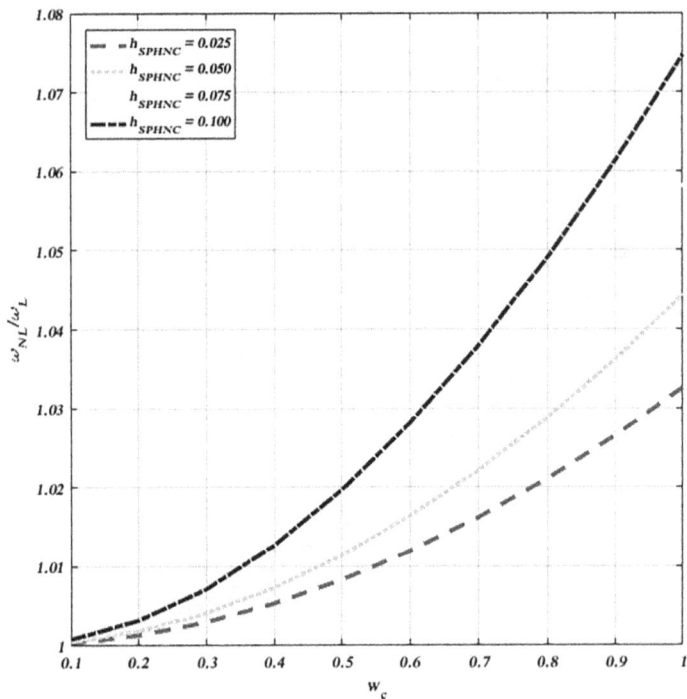

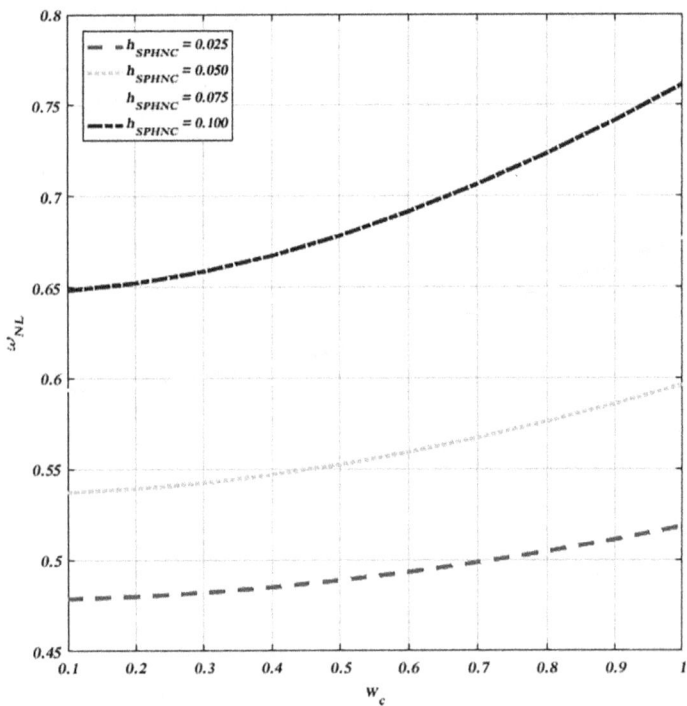

FIGURE 3.20 (Continued)

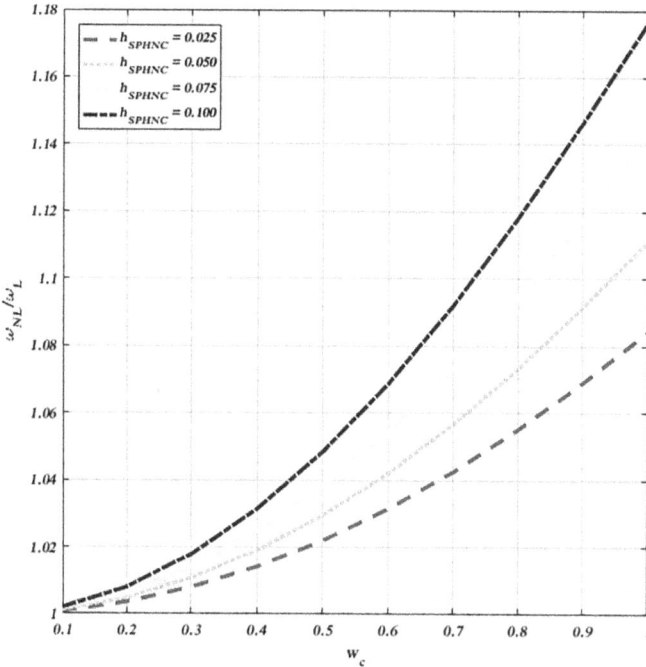

FIGURE 3.20 (Continued)

amplification of both the plate's frequency ratio and the frequency. When the h_{SPHNC} is increased from 0.025 to 0.1 in both CCCC and SSSS boundary conditions, there is a noticeable increase in both the frequency ratio and the frequency. As anticipated, the ratio of the natural frequency (w_{nl}) to the fundamental frequency (ω_l) is higher when subject to SSSS boundary conditions compared to CCCC boundary conditions.

The link between the thickness of the GOA core and the frequency ratio and frequency values of the smart plate is shown in Figure 3.21. As seen in this figure, an increase in the magnitude wmax from 0.1 to 1 results in the expansion of both the frequency ratio and the frequency. Furthermore, by the augmentation of the core thickness of the GOA, both the frequency ratio and the frequency exhibit a greater magnitude compared to their previous values. In contrast to SSSS B.C., CCCC has a comparatively lower frequency ratio value.

The graphical representation in Figure 3.22 illustrates the relationship between the frequency ratio and the frequency of the smart plate, with respect to the varying percentage weight composition of graphene nanoplatelets. The presented figure illustrates that augmenting the amplitude from 0.1 to 1 leads to greater values for both the frequency ratio and the frequency. It is evident that both CCCC and SSSS B.C. exhibit an escalation in frequency ratio when the value is elevated from 1% to 3%. The frequency ratio in CCCC may be determined to be lower than that of SSSS B.C., which is a noteworthy discovery.

The impact of the angle of the GOA folding degree on the frequency ratio and the nonlinear frequency of the smart hybrid plate is shown in Figure 3.23. Based on

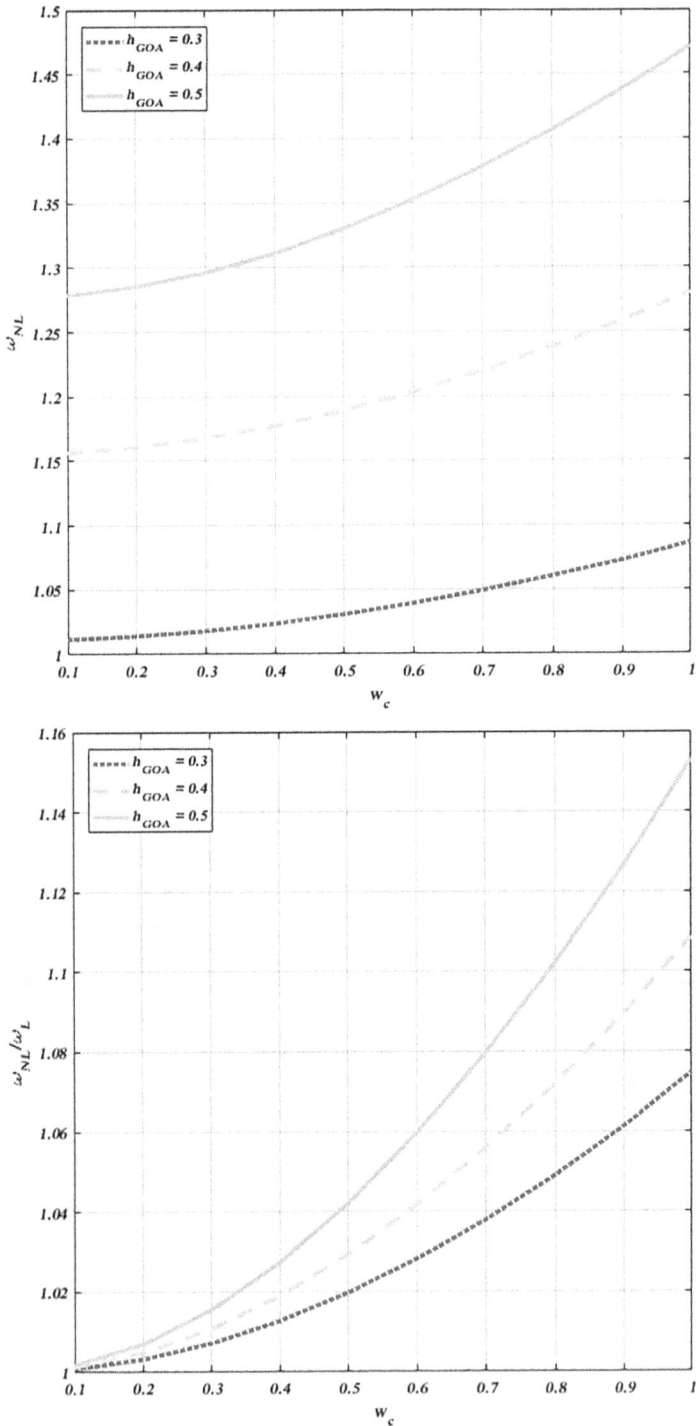

FIGURE 3.21 Influence of various thicknesses of GOA core on the nonlinear (ω_{nl}/ω_l) and dimensionless ω_{nl} of the plate in different B.C. (a) CCCC, (b) SSSS.

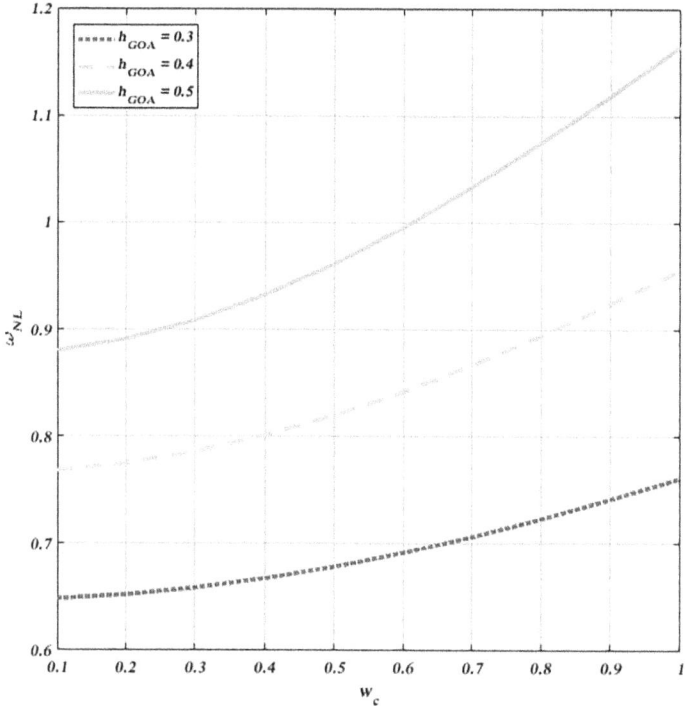

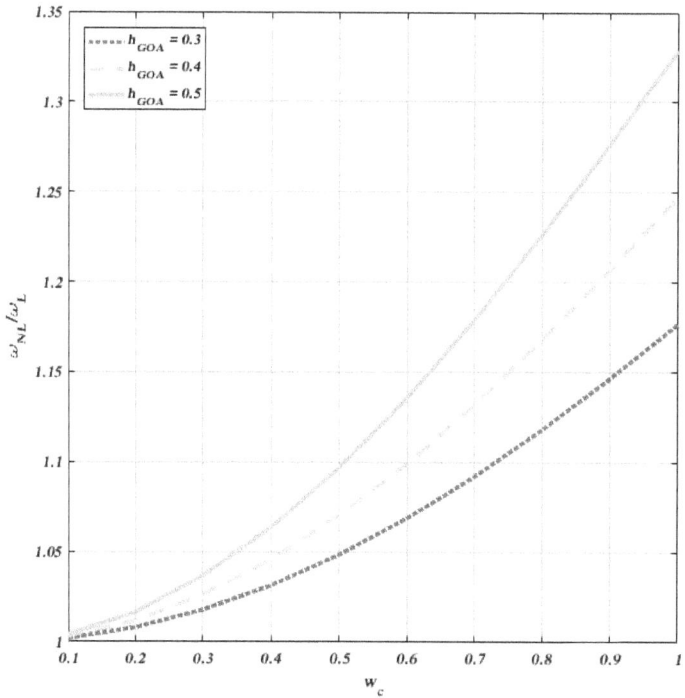

FIGURE 3.21 (Continued)

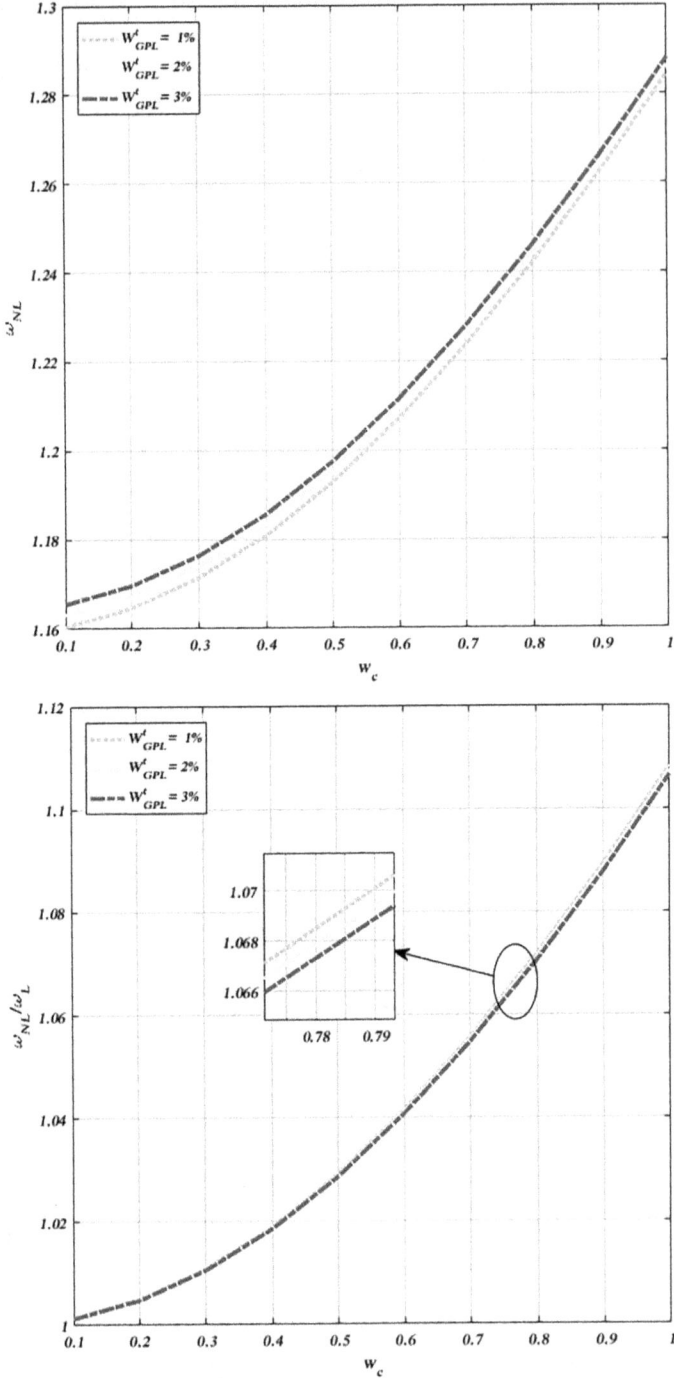

FIGURE 3.22 Influence of weight portion of graphene nanoplatelets on dimensionless w_{nl} of the plate in different B.C. (a) CCCC and (b) SSSS.

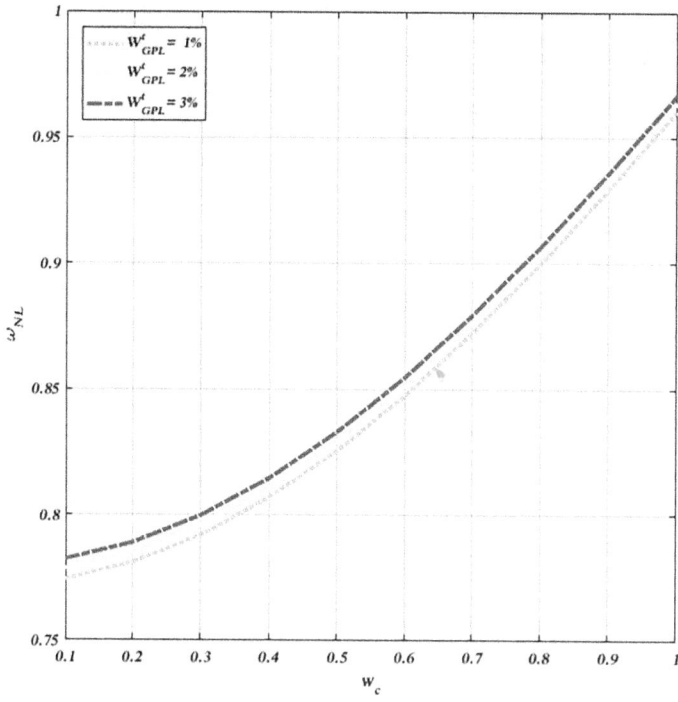

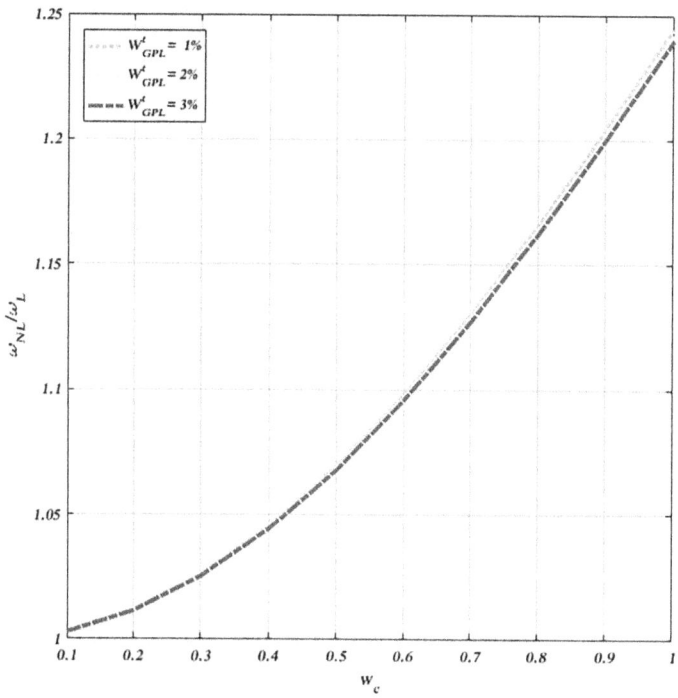

FIGURE 3.22 (Continued)

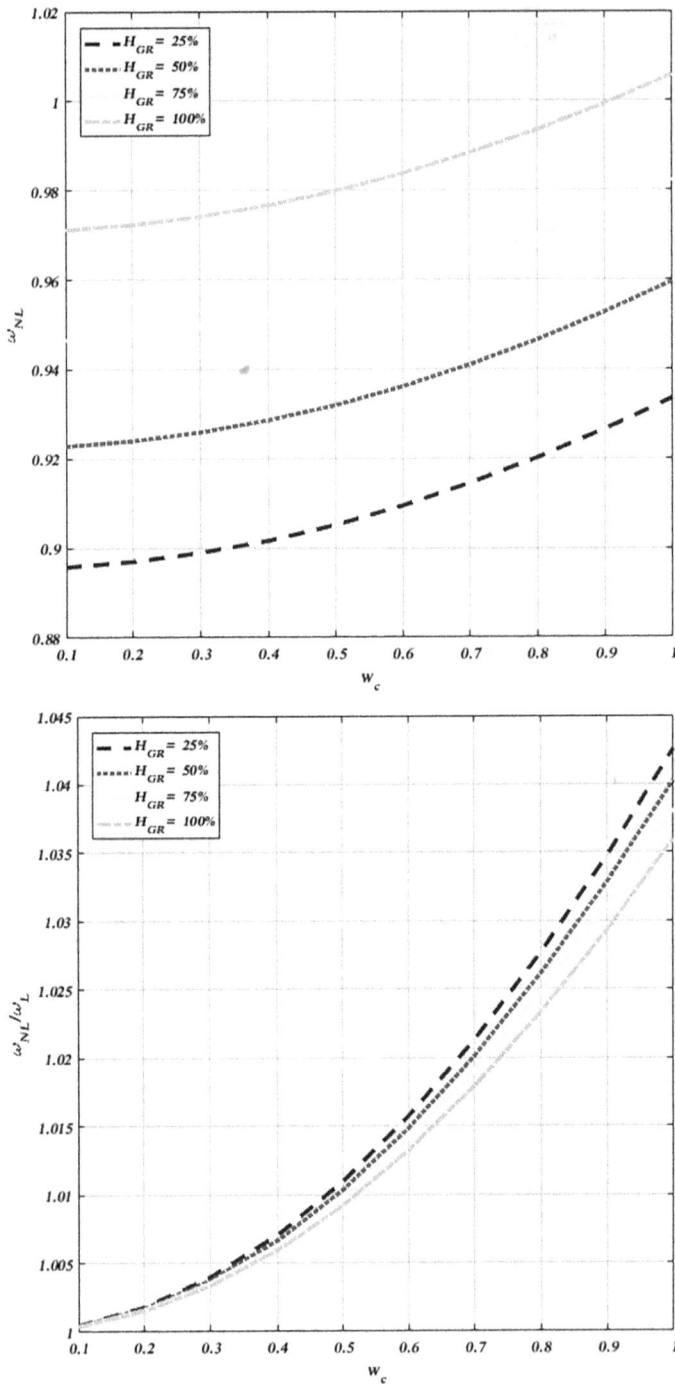

FIGURE 3.23 Influence of angle of GOri folding degree on dimensionless ω_{nl} of the sandwich plate in different B.C. (a) CCCC and (b) SSSS.

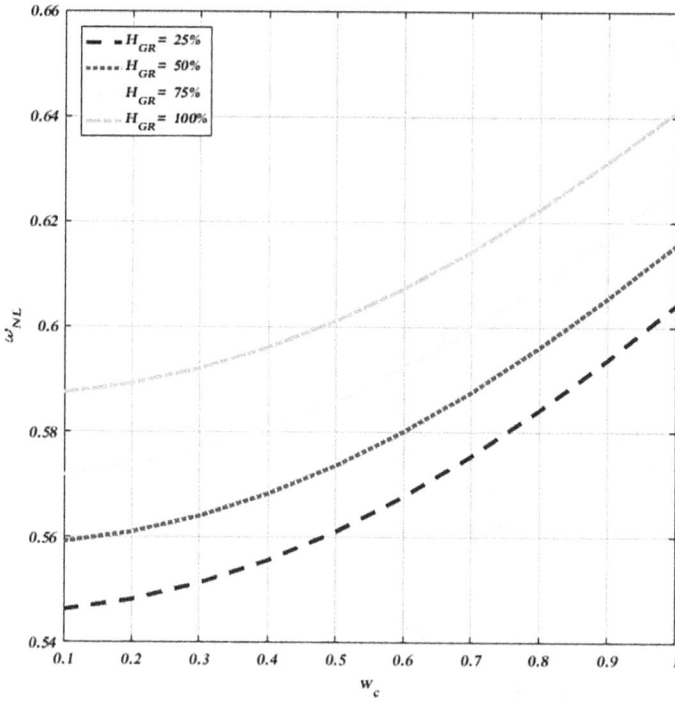

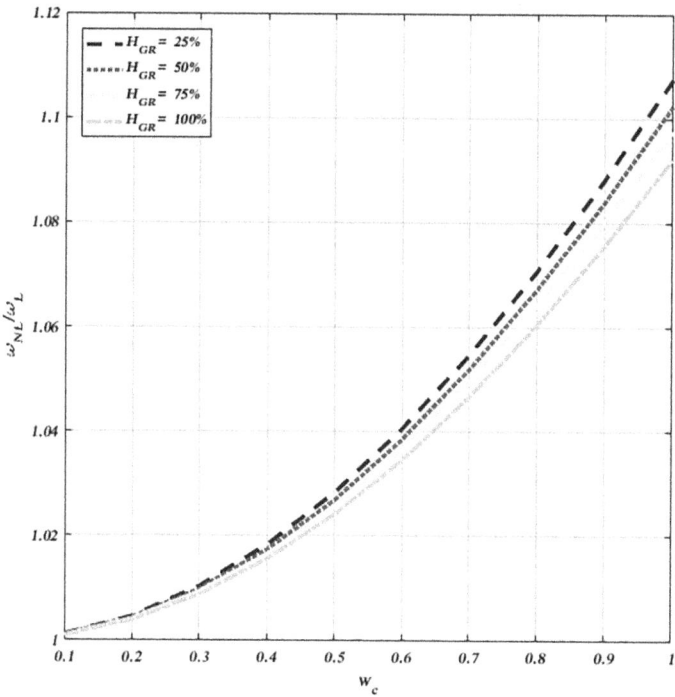

FIGURE 3.23   (Continued)

the information shown in this figure, it can be seen that increasing the maximum magnitude value from 0.1 to 1 leads to an elevation in both the frequency ratio and the nonlinear frequency. Furthermore, the reduction of the GOA folding angle inclination leads to a decrease in the nonlinear frequency. It is evident that the frequency ratio in SSSS B.C. is lower compared to that in CCCC B.C.

The influence of the carbon fiber volume percentage on the frequency ratio and nonlinear frequency of the smart hybrid plate is shown in Figure 3.24. As seen in this figure, it is evident that both the frequency ratio and the nonlinear frequency have exhibited an augmentation in magnitude due to the increment in $wmax$ from 0.1 to 1. The observed trend indicates a significant rise in the ratio of natural frequency to fundamental frequency ($\omega nl/\omega l$) when the carbon fiber value is reduced from 0.3 to 0.1 in both the CCCC and SSSS boundary conditions. The empirical evidence clearly demonstrates that the nonlinear frequency ratio in CCCC is comparatively smaller than that seen in SSSS B.C.

The influence of the pore size constant, $e0$, on both the ratio (ω_{nl}/ω_l) and the dimensionless ω_{nl} of the smart GOA plate is evident upon examination of Figure 3.25. As seen in this figure, the nonlinear frequency ratio experiences an increase when the

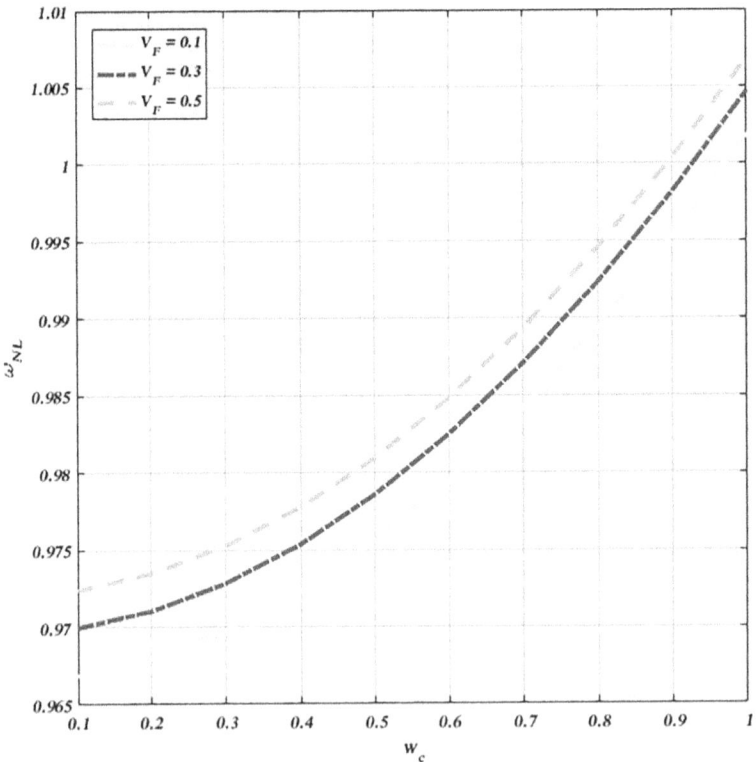

FIGURE 3.24 Effect of the volume fraction of fibers (V_F) on (w_{nl}/ω_l) and dimensionless ω_{nl} for the smart plate in several B.C. (a) CCCC and (b) SSSS.

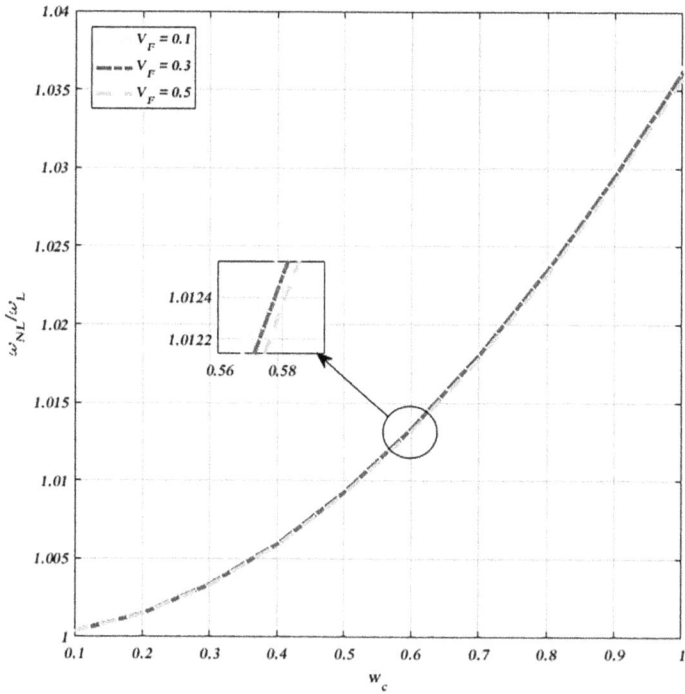

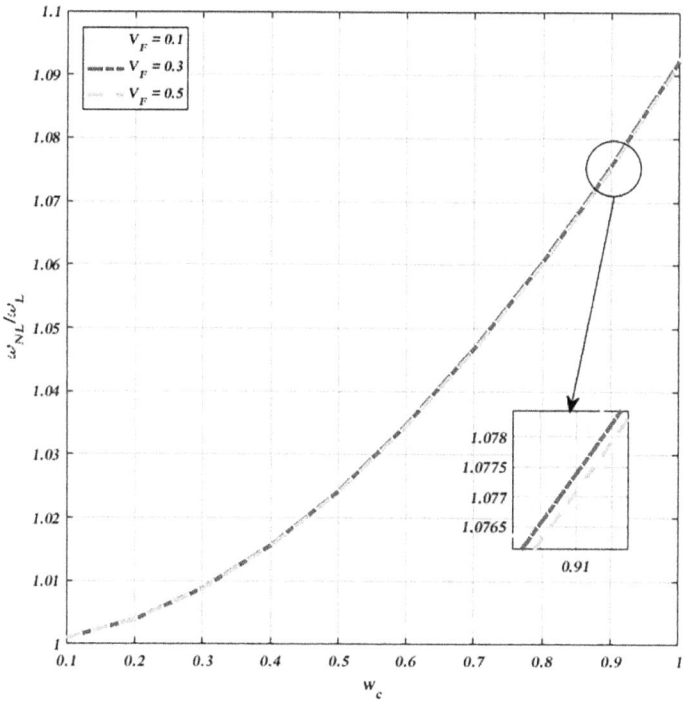

FIGURE 3.24 (Continued)

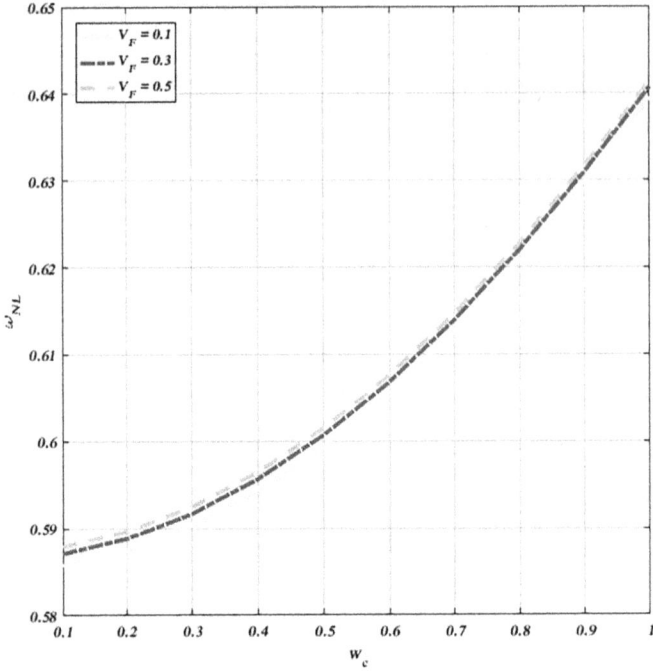

FIGURE 3.24 (Continued)

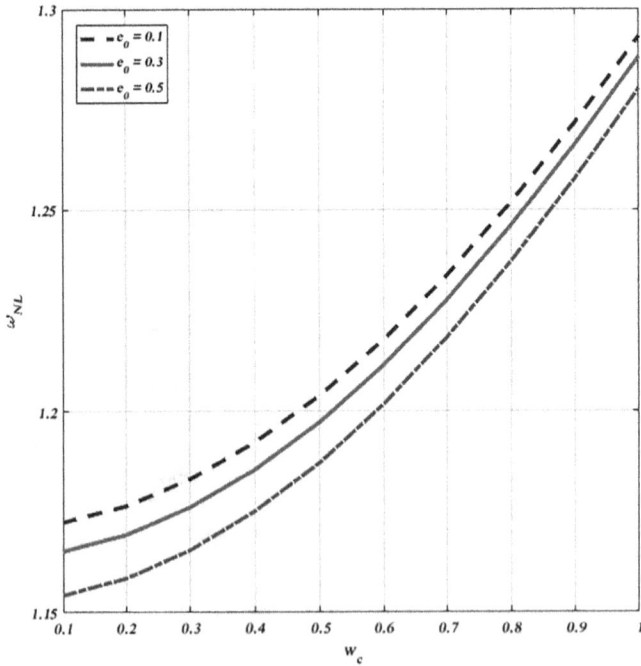

FIGURE 3.25 Influence of various porosity constant (e_0) on the (w_{nl}/ω_l) and dimensionless ω_{nl} of the sandwich plate in several B.C. (a) CCCC, (b) SSSS.

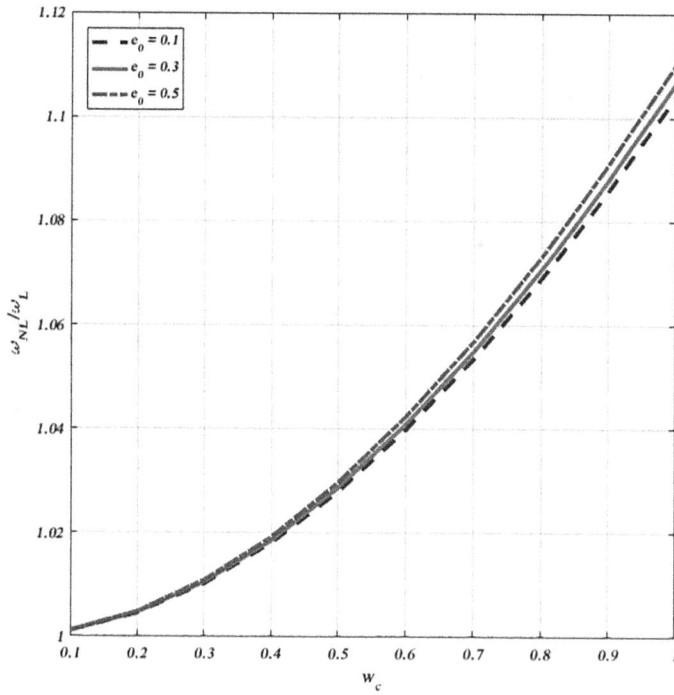

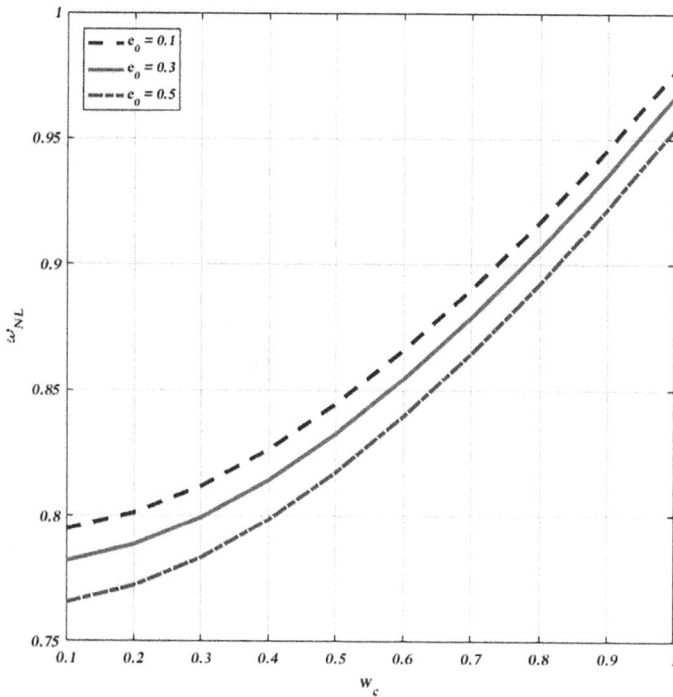

FIGURE 3.25   (Continued)

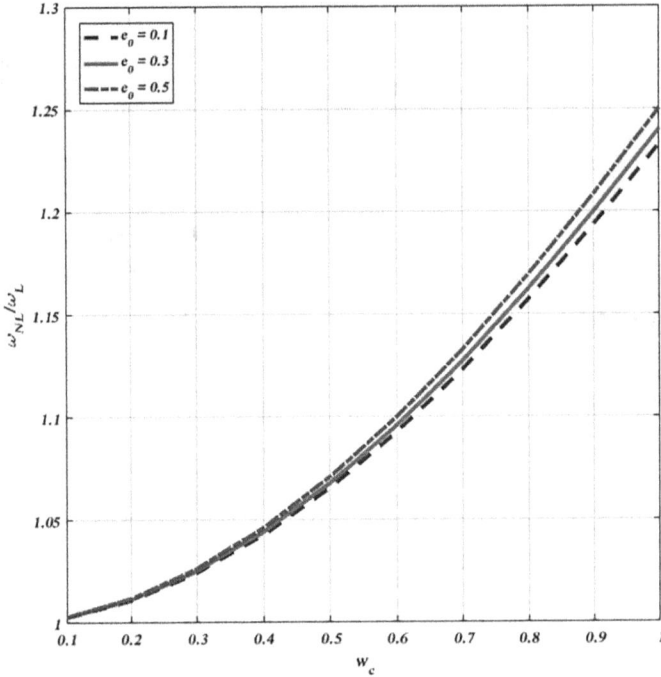

FIGURE 3.25 (Continued)

maximum magnitude value, wmax, is adjusted from 0.1 to 1. When the porosity constant ($e0$) is increased from 0.1 to 0.5 for both CCCC and SSSS boundary conditions, there is an observable increase in the frequency ratio value. The observation has been made that the frequency ratio in CCCC is notably lower compared to that of SSSS B.C.

The nonlinear frequency ratio of the smart hybrid FG-GOA plate is shown in Figure 3.26, illustrating its variation with respect to changes in the FG power index. The data shown in this figure indicates that the nonlinear frequency ratio exhibits an increasing trend when the amplitude (wmax) ranges from 0.1 to 1. Furthermore, it was observed that the augmentation of the FG-GOA power index, ranging from 0 to 1, resulted in a reduction of the frequency ratio under both CCCC and SSSS boundary conditions. The nonlinear frequency ratio in CCCC has a significantly reduced magnitude compared to that seen in SSSS B.C.

Figure 3.27 provides a clear illustration of the variation in the dimensionless natural frequency, denoted as ω_{nl} of the smart hybrid plate in response to an external voltage, denoted as V0. The figure clearly demonstrates that increasing the value of wmax from 0.1 to 1 leads to a significant enhancement in the dimensionless parameter ω_{nl}. When the external voltage (V0) decreases from +10 to −10 under CCCC and SSSS boundary conditions, it is evident that the dimensionless nonlinear natural frequency is greater than its previous value. It is evident that the dimensionless nonlinear natural frequency in SSSS exhibits a lower value compared to that seen in the historical period of CCCC B.C.

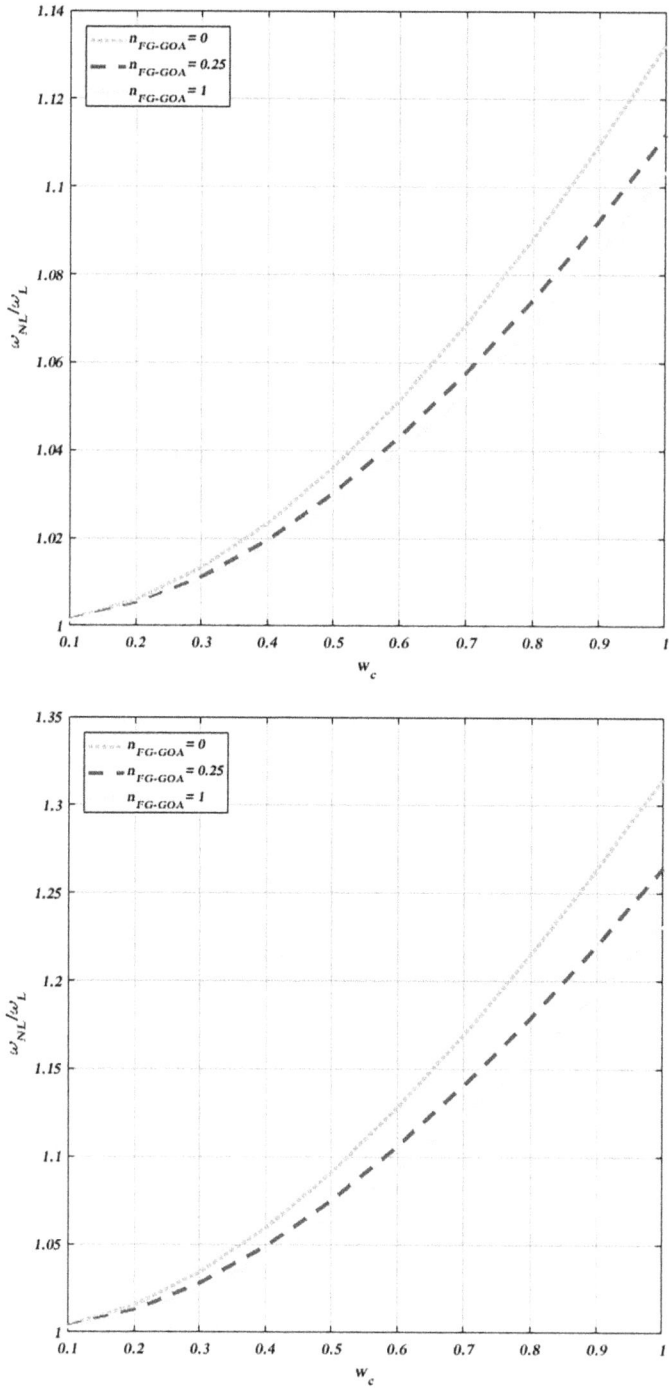

FIGURE 3.26   Influence of various FG power index (n) on the (ω_{nl}/ω_l) of the sandwich plate in several B.C. (a) CCCC, (b) SSSS.

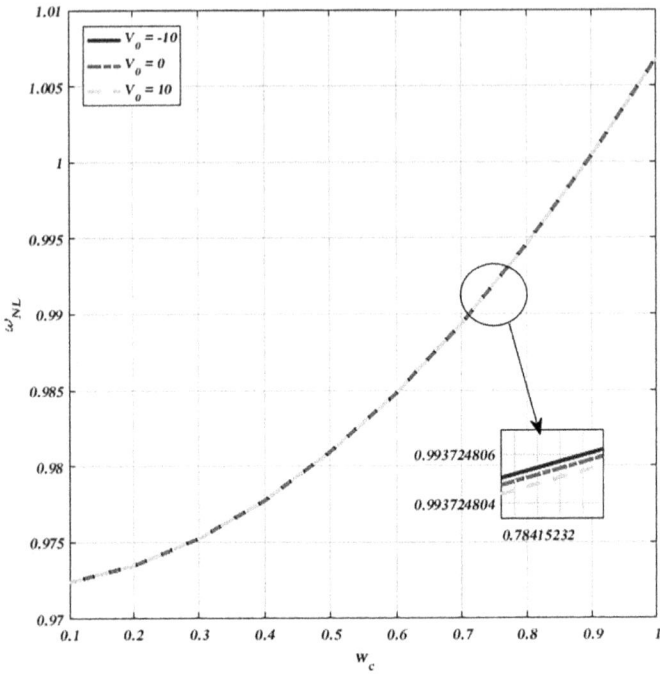

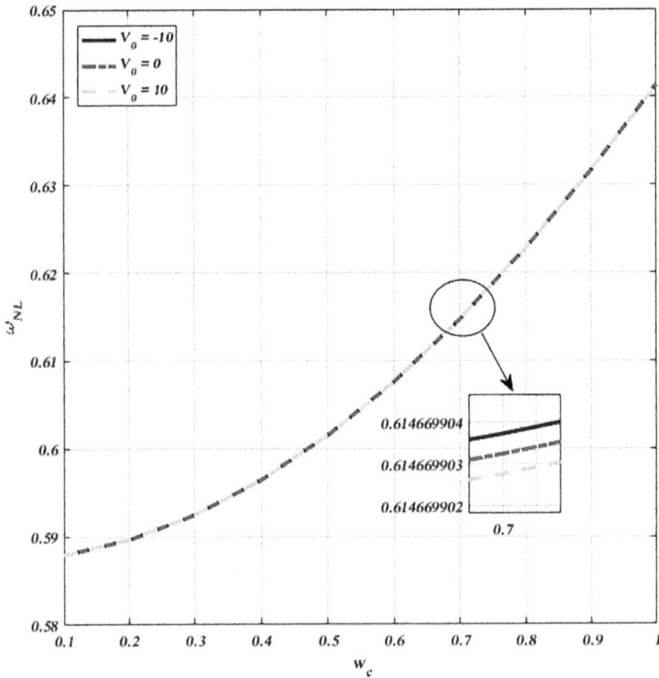

FIGURE 3.27 Influence of various exterior voltage on the dimensionless ω_{nl} of the sandwich plate in several B.C. (a) CCCC, (b) SSSS.

3.4   AUXETIC SHELLS: VIBRATION ANALYSIS

3.4.1   VIBRATION ANALYSIS OF CYLINDRICAL COMPOSITE SHELLS WITH AUXETIC CORE

This section examines a three-layer cylindrical shell with aluminum upper and bottom outer layers and auxetic materials with a negative Poisson's ratio $E_{Al} = 69GPa$, $\rho_{Al} = 2700 \, kg/m^3$, $v_{Al} = 0.3$. The auxetic core's effective shear modulus is and the aluminum material's characteristics are $G^c = 26GPa$. The top and bottom auxetic layers might be equal or differing thicknesses $3 \times 10^{-3} \, m$ (both are aluminum). Auxetic core thickness and top and bottom layer thickness assumed for primary analyses with equal thicknesses $10^{-3} \, m$.

Figure 3.28 depicts a θ three-layered, long, and thick composite cylindrical shell. The cylindrical coordinate system (x, θ, z) defines each shell point by three parameters, x, θ, and z, where x and θ are in the axial and circumferential directions and z is perpendicular to the surface.

Figure 3.29 shows the unit cell of the auxetic material and its geometric parameters, which affect the mechanical properties of the whole structure. (l), (e), and (t)

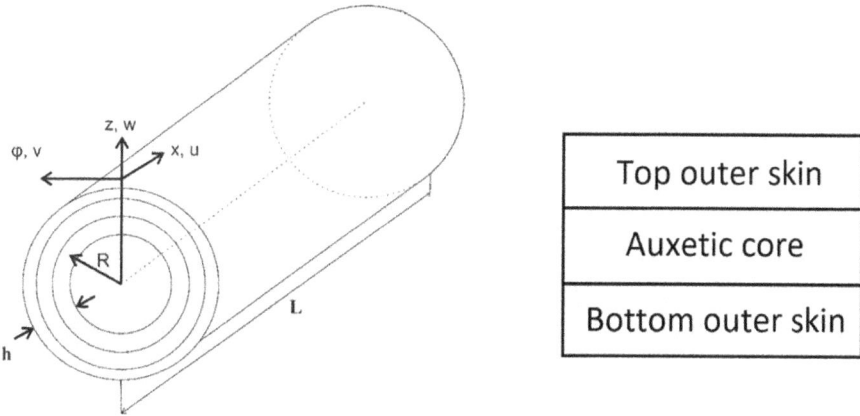

FIGURE 3.28   Geometry and schematic of a composite cylindrical shell.

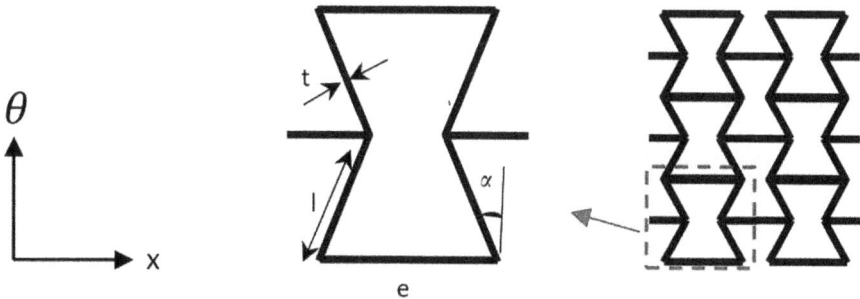

FIGURE 3.29   The unit cell of the auxetic material and its geometric parameters.

are the lengths of the inclined, vertical cell rib, and rib thickness, respectively, and (α) is the inclined angle.

As shown in Figure 3.29, the auxetic material's unit cell and geometric parameters impact the structure's mechanical characteristics. The inclined angle is (α) and the lengths of the inclined, vertical, and rib thickness are (l), (e), and (t), respectively.

The material properties of the auxetic structure can be written in the following form:

$$E_1^{(c)} = E^c \frac{\lambda_3^3 \left(\lambda_1 - \sin\alpha\right)}{\cos\alpha \left[1+\left(\tan^2\alpha + \lambda_1 \sec^2\alpha\right)\lambda_3^2\right]}$$

$$E_2^{(c)} = E^c \frac{\lambda_3^3}{\cos\alpha \left(\lambda_1 - \sin\alpha\right)\left(\tan^2\alpha + \lambda_3^2\right)}$$

$$G_{12}^{(c)} = G^c \frac{\lambda_3^3}{\lambda_1 \left(1+2\lambda_1\right)\cos\alpha}$$

$$G_{23}^{(c)} = G^c \frac{\lambda_3 \cos\alpha}{\lambda_1 - \sin\alpha}$$

$$G_{13}^{(c)} = G^c \frac{\lambda_3}{2\cos\alpha} \left[\frac{\lambda_1 - \sin\alpha}{1+2\lambda_1} + \frac{\lambda_1 + 2\sin^2\alpha}{2\left(\lambda_1 - \sin\alpha\right)}\right]$$

$$v_{12}^{(c)} = -\frac{\sin\alpha \left(1 - \lambda_3^2\right)\left(\lambda_1 - \sin\alpha\right)}{\cos^2\alpha \left[1+\left(\tan^2\alpha + \sec^2\alpha\lambda_1\right)\lambda_3^2\right]} \qquad (3.180)$$

$$v_{21}^{(c)} = \frac{-\sin\alpha \left(1 - \lambda_3^2\right)}{\left(\tan^2\alpha + \lambda_3^2\right)\left(\lambda_1 - \sin\alpha\right)}$$

$$v_{21}^{(2)} = \frac{\sin\alpha \left(1 - \lambda_3^2\right)}{\left(\tan^2\alpha + \lambda_3^2\right)\left(\lambda_1 - \sin\alpha\right)}$$

$$\rho^c = \rho \frac{\lambda_3 \left(\lambda_1 + 2\right)}{2\cos\alpha \left(\lambda_1 - \sin\alpha\right)}$$

$$\lambda_1 = d/l$$

$$\lambda_3 = t/l$$

where $E_1^{(c)}$, $E_2^{(c)}$, $G_{12}^{(c)}$, $G_{23}^{(c)}$, $v_{12}^{(c)}$, and $v_{21}^{(c)}$ are the effective Young's modulus of auxetic, effective shear modulus of auxetic, and effective Poisson's ratio of auxetic core, respectively. Furthermore, $v^{(c)}$, $E^{(c)}$, and $G^{(c)}$ specify, respectively, the Young's modulus, shear modulus, and Poisson's ratio of auxetic material. Also, ρ^c and ρ represents the effective density of the auxetic core and density of the auxetic material. Moreover, λ_1, λ_3 are the auxetic rib length ratio. It should be noted that symbol c represents core material. According to Donnell's improved shell theory for the non-shallow cylindrical shell, the relationship of nonlinear strain and displacement for a cylindrical shell can be guided as follows [2,4]:

$$\varepsilon_x = \varepsilon_x^0 - z\chi_x$$
$$\varepsilon_\theta = \varepsilon_\theta^0 - z\chi_\theta \qquad (3.181)$$
$$\gamma_{x\theta} = \gamma_{x\theta}^0 - 2z\chi_{x\theta}$$

where

$$\varepsilon_x^0 = \frac{\partial u}{\partial x} + \frac{1}{2}\left(\frac{\partial w}{\partial x}\right)^2$$
$$\varepsilon_\theta^0 = \frac{\partial v}{\partial \theta} - \frac{w}{R} + \frac{1}{2}\left(\frac{\partial w}{\partial \theta}\right)^2 \qquad (3.182a)$$
$$\gamma_{x\theta}^0 = \frac{\partial v}{\partial x} + \frac{\partial u}{\partial \theta} + \frac{\partial w}{\partial x}\frac{\partial w}{\partial \theta}$$

and

$$\chi_x = \frac{\partial^2 w}{\partial x^2}$$
$$\chi_\theta = \frac{1}{R}\frac{\partial v}{\partial \theta} + \frac{\partial^2 w}{\partial \theta^2} \qquad (3.182b)$$
$$\chi_{x\theta} = \frac{1}{2R}\frac{\partial v}{\partial x} + \frac{\partial^2 w}{\partial x\partial \theta}$$

in which $\varepsilon_x^0, \varepsilon_\theta^0$ denote the normal strains and $\gamma_{x\theta}^0$ represents the shear strain at the middle surface of the shell. Also, χ_x, χ_θ, and $\chi_{x\theta}$ specify the curvatures and twist, respectively. The following equations are based on Hook's law for a cylindrical shell:

$$\sigma_x = \frac{E}{1-v^2}\left(\varepsilon_x + v\varepsilon_\theta\right)$$
$$\sigma_\theta = \frac{E}{1-v^2}\left(\varepsilon_\theta + v\varepsilon_x\right) \qquad (3.183)$$
$$\sigma_{x\theta} = \frac{E}{2(1+v)}\gamma_{x\theta}$$

The internal force and momentum generated by the shell are determined by the stressed components confined inside the thickness, as expressed by the following equation:

$$(N_i, M_i) = \int_{-h/2}^{-h/2+h_{al}^b} \sigma_i^{al}(1,z)dz + \int_{-h/2+h_{al}^b}^{-h/2+h_{al}^b+h_a} \sigma_i^{a}(1,z)dz$$
$$+ \int_{-h/2+h_{al}^b+h_a}^{-h/2+h_{al}^b+h_a+h_{al}^t} \sigma_i^{al}(1,z)dz, \quad i = x, \theta, x\theta \qquad (3.184)$$

In the aforementioned equations, the variables h_{al}^b, h_a, and h_{al}^t represent, respectively, the thickness of the bottom outer skin, the thickness of the auxetic core, and the thickness of the top outer skin, as seen in Figure 3.28. By substituting the aforementioned equations into equation (3.184), the relationships may be expressed as follows:

$$
\begin{bmatrix} N_x \\ N_\theta \\ N_{x\theta} \\ M_x \\ M_\theta \\ M_{x\theta} \end{bmatrix} = \begin{bmatrix} A_{11} & A_{12} & 0 & B_{11} & B_{12} & 0 \\ A_{12} & A_{22} & 0 & B_{12} & B_{22} & 0 \\ 0 & 0 & A_{66} & 0 & 0 & B_{66} \\ B_{11} & B_{12} & 0 & D_{11} & D_{12} & 0 \\ B_{12} & B_{22} & 0 & D_{12} & D_{22} & 0 \\ 0 & 0 & B_{66} & 0 & 0 & D_{66} \end{bmatrix} \begin{bmatrix} \varepsilon_x^0 \\ \varepsilon_\theta^0 \\ \gamma_{x\theta}^0 \\ -\chi_x \\ -\chi_\theta \\ -2\chi_{x\theta} \end{bmatrix}
\tag{3.185}
$$

in which:

$$
\begin{bmatrix} A_{ij} \\ B_{ij} \\ D_{ij} \end{bmatrix} = \int_{-h/2}^{-h/2+h_{al}^b} Q_{ij}^{al} \begin{bmatrix} 1 \\ z \\ z^2 \end{bmatrix} dz + \int_{-h/2+h_{al}^b}^{-h/2+h_{al}^b+h_a} Q_{ij}^{a} \begin{bmatrix} 1 \\ z \\ z^2 \end{bmatrix} dz + \int_{-h/2+h_{al}^b+h_a}^{-h/2+h_{al}^b+h_a+h_{al}^t} Q_{ij}^{al} \begin{bmatrix} 1 \\ z \\ z^2 \end{bmatrix} dz \tag{3.186}
$$

where:

$$
Q_{11}^{al} = Q_{22}^{al} = \frac{E_{al}}{1-v_{al}^2}, \quad Q_{12}^{al} = \frac{E_{al}v_{al}}{1-v_{al}^2}, \quad Q_{66}^{al} = \frac{E_{al}}{2(1+v_{al})},
$$

$$
Q_{11}^{a} = Q_{22}^{a} = \frac{E_1^{(c)}}{1-v_{12}^c v_{21}^c}, \quad Q_{12}^{a} = \frac{E_2^{(c)}v_{12}^{(c)}}{1-v_{12}^c v_{21}^c}, \quad Q_{66}^{a} = \frac{E_2^{(c)}}{2(1+v_{12}^{(c)})}. \tag{3.187}
$$

Based on Donnell's enhanced shell theory for the non-shallow cylindrical shell, the equations of motion may be expressed in a nonlinear form.

$$
\frac{\partial N_x}{\partial x} + \frac{\partial N_{x\theta}}{\partial \theta} = \rho_1 \frac{\partial^2 u}{\partial t^2}
$$

$$
\frac{\partial N_{x\theta}}{\partial x} + \frac{\partial N_\theta}{\partial \theta} - \frac{1}{R}\left(\frac{\partial M_{x\theta}}{\partial x} + \frac{\partial M_\theta}{\partial \theta}\right) = \rho_1 \frac{\partial^2 v}{\partial t^2} \frac{\partial^2 M_x}{\partial x^2} + 2\frac{\partial^2 M_{x\theta}}{\partial x \partial \theta} + \frac{\partial^2 M_\theta}{\partial \theta^2} + \frac{N_x}{R}
$$

$$
+\frac{\partial}{\partial x}\left(N_x \frac{\partial w}{\partial x} + N_{x\theta}\frac{\partial w}{\partial \theta}\right) + \frac{\partial}{\partial \theta}\left(N_{x\theta}\frac{\partial w}{\partial x} + N_\theta \frac{\partial w}{\partial \theta}\right) - ph\frac{\partial^2 w}{\partial x^2} + q
$$

$$
= \rho_1 \frac{\partial^2 w}{\partial t^2} + 2\varepsilon\rho_1 \frac{\partial w}{\partial t} \tag{3.188}
$$

in which:

$$
\rho_1 = \int_{-h/2}^{h/2} \rho dz \tag{3.189}
$$

The variables denoted as p, q, and ε in equation (3.188) represent an axial compressive force applied to both ends of the shell, an external pressure uniformly distributed throughout the shell's surface, and a damping coefficient, respectively. By putting equations (3.182a) and (3.182b) into equation (3.185) and subsequently into equation (3.188), the nonlinear equations governing the motion of the cylindrical shell in terms of displacement components may be reformulated as follows:

$$l_{11}(u) + l_{12}(v) + l_{13}(w) + P_1(w) = \rho_1 \frac{\partial^2 u}{\partial t^2}$$

$$l_{21}(u) + l_{22}(v) + l_{23}(w) + P_2(w) = \rho_1 \frac{\partial^2 v}{\partial t^2}$$

$$l_{31}(u) + l_{32}(v) + l_{33}(w) + P_3(w) + Q_3(u,w)$$

$$+ R_3(v,w) - ph\frac{\partial^2 w}{\partial x^2} + q = \rho_1 \frac{\partial^2 w}{\partial t^2} + 2\varepsilon\rho_1 \frac{\partial w}{\partial t}$$

(3.190)

The terms of equation (3.190) are defined in Appendix 3.3. To perform a nonlinear dynamic analysis of the non-shallow cylindrical shell, the equations of the motion system (3.190) are utilized in conjunction with the appropriate boundary conditions and initial conditions. The boundary conditions for this analysis assume a simply supported configuration at a specific location $x = 0$ and $x = L$:

$$w = 0 \qquad v = 0 \qquad M_x = 0 \qquad (3.191)$$

By considering the boundary conditions, the system of equations (3.190) may be expressed in the form of approximate solutions as follows:

$$u = U(t)\cos\frac{m\pi x}{L}\sin\frac{n\theta}{R} \qquad v = V(t)\sin\frac{m\pi x}{L}\cos\frac{n\theta}{R}$$

$$w = W(t)\sin\frac{m\pi x}{L}\sin\frac{n\theta}{R} \qquad (3.192)$$

The amplitudes of vibration are shown in the instances where U, V, and W. m and n represent the quantities of half-waves in the axial direction and waves in the circumferential direction, respectively. By substituting equation (3.192) into equation (3.190) and then using the Galerkin technique, the equations may be reformulated as follows:

$$l_{11}U + l_{12}V + l_{13}W + n_1 W^2 = \rho_1 \frac{d^2 U}{dt^2}$$

$$l_{21}U + l_{22}V + l_{23}W + n_2 W^2 = \rho_1 \frac{d^2 V}{dt^2}$$

$$l_{31}U + l_{32}V + l_{33}W + n_3 W^2 + n_4 W^3 + n_5 UW$$

$$+ n_6 VW + \frac{16q}{\pi^2 mn} = \rho_1 \frac{d^2 W}{dt^2} + 2\varepsilon\rho_1 \frac{dW}{dt} \qquad (3.193)$$

The terms of equation (3.193) are described in Appendix 3.4.

Based on the Volmir's assumption [2,4] $\left(u \ll w, v \ll w, \rho_1\left(\partial^2 U/\partial t^2\right)\to 0\right.$ and $\left.\rho_1\left(\partial^2 V/\partial t^2\right)\to 0\right)$, the system equations (3.193) can be written as follows:

$$l_{11}U + l_{12}V + l_{13}W + n_1 W^2 = 0$$
$$l_{21}U + l_{22}V + l_{23}W + n_2 W^2 = 0$$
$$l_{31}U + l_{32}V + l_{33}W + n_3 W^2 + n_4 W^3 + n_5 UW$$

$$+n_6 VW + \frac{16q}{\pi^2 mn} = \rho_1 \frac{d^2 W}{dt^2} + 2\varepsilon\rho_1 \frac{dW}{dt} \tag{3.194}$$

By calculating the first and second equations of equation (3.194) with regard to variables U and V, and afterwards inserting the obtained values into the final equation of equation (3.194), an equation may be derived as follows:

$$\rho_1 \frac{d^2 W}{dt^2} + 2\varepsilon\rho_1 \frac{dW}{dt} + a_1 W - a_2 W^2 + a_3 W^3 = \frac{16q}{\pi^2 mn} \tag{3.195}$$

in which:

$$a_1 = -l_{33} - \frac{l_{31}\left(l_{12}l_{23} - l_{22}l_{13}\right) + l_{32}\left(l_{21}l_{13} - l_{11}l_{23}\right)}{l_{11}l_{22} - l_{12}^2}$$

$$a_2 = n_3 + \frac{l_{31}\left(l_{12}n_2 - l_{22}n_1\right) + l_{32}\left(l_{12}n_1 - l_{11}n_2\right) + n_5\left(l_{12}l_{23} - l_{22}l_{13}\right) + n_6\left(l_{21}l_{13} - l_{11}l_{23}\right)}{l_{11}l_{22} - l_{12}^2}$$

$$a_3 = -n_4 - \frac{n_5\left(l_{12}n_2 - l_{22}n_1\right) + n_6\left(l_{21}n_1 - l_{11}n_2\right)}{l_{11}l_{22} - l_{12}^2} \tag{3.196}$$

In due course, either the comprehensive set of equations (3.193) or the approximation equation (3.195) including Volmir's assumption may be used for conducting nonlinear dynamic analysis on a non-shallow cylindrical shell. The determination of the natural frequencies of a cylindrical shell may be achieved by calculating the equation under the assumption that there is no evenly distributed external pressure acting on the surface of the construction.

$$\begin{vmatrix} l_{11} + \rho_1\omega^2 & l_{12} & l_{13} \\ l_{21} & l_{22} + \rho_1\omega^2 & l_{23} \\ l_{31} & l_{32} & l_{33} + \rho_1\omega^2 \end{vmatrix} = 0 \tag{3.197}$$

By solving the determinant of the system, we are able to determine three angular frequencies associated with the cylindrical shell in the axial, circumferential, and radial directions. Among these frequencies, the least one is selected for consideration. The precise answer for the fundamental frequency of the shell is provided in equation (3.197), although in an implicit form. Furthermore, the explicit expression for the estimated fundamental frequencies of the shell may be derived from equation (3.195) and can be presented in the following format:

$$\omega_{mn} = \sqrt{\frac{a_1}{\rho_1}} \tag{3.198}$$

This section examines the nonlinear vibration of a cylindrical shell subjected to a transverse force that is equally distributed. It assumes the presence of pre-loaded compression, denoted by variables $q = Q \sin \Omega t$ and P. By following a derivation process, equation (3.195) may be obtained.

$$\rho_1 \frac{d^2 W}{dt^2} + 2\varepsilon \rho_1 \frac{dW}{dt} + a_1 W - a_2 W^2 + a_3 W^3 - \frac{16q}{\pi^2 mn} \sin \Omega t = 0 \tag{3.199}$$

which can be stated as follows:

$$\frac{d^2 W}{dt^2} + 2\varepsilon \frac{dW}{dt} + \omega_{mn}^2 \left(W - HW^2 + KW^3 \right) - F \sin \Omega t = 0 \tag{3.200}$$

The fundamental frequency of the linear vibration of cylindrical shells is denoted by ω_{mn}. In the above formula, the variables H, K, and F indicate specific quantities as follows:

$$H = \frac{a_2}{a_1} \qquad K = \frac{a_3}{a_1} \qquad F = \frac{16Q}{\rho_1 \pi^2 mn} \tag{3.201}$$

To get the amplitude-frequency characteristics of nonlinear vibration, one may use the following equation:

$$W = A \sin \Omega t \tag{3.201}$$

The introduction of equation (3.201) into equation (3.199) results in:

$$X \equiv A \left(\omega_{mn}^2 - \Omega^2 \right) \sin \Omega t + 2\varepsilon A \Omega \cos \Omega t - \omega_{mn}^2 HA^2 \sin^2 \Omega t$$
$$+ K \omega_{mn}^2 A^3 \sin^3 \Omega t - F \sin \Omega t = 0 \tag{3.202}$$

When doing the integration, it is necessary to consider a fourth of the vibration time.

$$\int_0^{\pi/2\Omega} X \sin \Omega t dt = 0 \tag{3.203}$$

The relationship between frequency and amplitude in nonlinear vibration is determined.

$$\Omega^2 - \frac{4\varepsilon}{\pi} \Omega = \omega_{mn}^2 \left(1 - \frac{8}{3\pi} HA + \frac{3K}{4} A^2 \right) - \frac{F}{A} \tag{3.204}$$

By defining $\alpha^2 = \Omega^2 / \omega_{mn}^2$, equation (3.204) can be rewritten as:

$$\alpha^2 - \frac{4\varepsilon}{\pi \omega_{mn}} \alpha = 1 - \frac{8}{3\pi} HA + \frac{3K}{4} A^2 - \frac{F}{A\omega_{mn}^2} \qquad (3.205)$$

The aforementioned relationship undergoes modification in the case of nonlinear vibration of the shell in the absence of dampening.

$$\Omega^2 = \omega_{mn}^2 \left(1 - \frac{8}{3\pi} HA + \frac{3K}{4} A^2 \right) - \frac{F}{A}$$

or

$$\alpha^2 = 1 - \frac{8}{3\pi} HA + \frac{3K}{4} A^2 - \frac{F}{A\omega_{mn}^2} \qquad (3.206)$$

Hence, in the absence of any harmonic excitation forces exerted on the shell, the equation describing the relationship between frequency and amplitude in free non-linear vibration may be derived as follows:

$$\omega_{NL}^2 = \omega_{mn}^2 \left(1 - \frac{8}{3\pi} HA + \frac{3K}{4} A^2 \right) \qquad (3.207)$$

The variable ω_{NL} is used to denote the representation of the nonlinear vibration frequency of the cylindrical shell.

In this section, we will examine the behavior of a circular cylindrical shell subjected to an equally distributed transverse force. Additionally, we will assume that the shell is under pre-loaded compression, denoted by symbols $q = Q\sin\Omega t$ and p. Under these conditions, the equations governing the motion of the shell are modified as follows:

$$l_{11}U + l_{12}V + l_{13}W + n_1 W^2 = \rho_1 \frac{d^2 U}{dt^2}$$

$$l_{21}U + l_{22}V + l_{23}W + n_2 W^2 = \rho_1 \frac{d^2 V}{dt^2}$$

$$l_{31}U + l_{32}V + l_{33}W + n_3 W^2 + n_4 W^3 + n_5 UW + n_6 VW$$

$$= \rho_1 \frac{d^2 W}{dt^2} + 2\varepsilon\rho_1 \frac{dW}{dt} - \frac{16Q\sin\Omega t}{\pi^2 mn} \qquad (3.208)$$

and the equation of motion using the Volmir assumption becomes:

$$\rho_1 \frac{d^2 W}{dt^2} + 2\varepsilon\rho_1 \frac{dW}{dt} + a_1 W - a_2 W^2 + a_3 W^3 = \frac{16Q\sin\Omega t}{\pi^2 mn} \qquad (3.209)$$

The determination of the nonlinear dynamic responses of cylindrical shells may be achieved by using the fourth-order Runge-Kutta technique, as described in equation

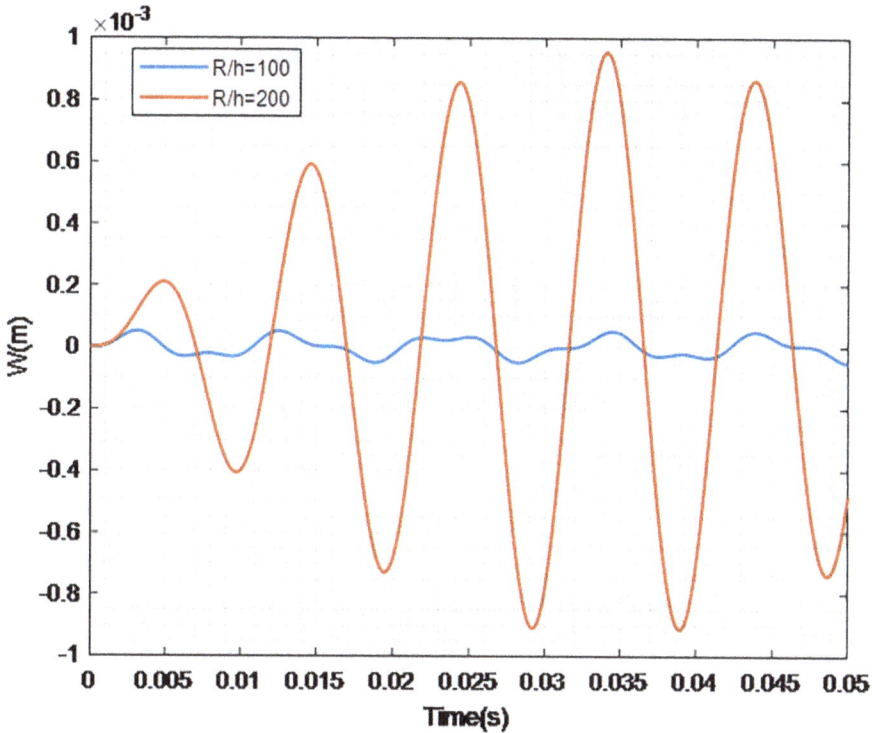

FIGURE 3.30 Variations of the nonlinear response of the cylindrical sandwich shell for different radius-to-thickness ratios R/h_{max} ($L/R = 2$, $P = 0$, $Q = 1500$, $\Omega = 600$, $(m, n) = (1, 3)$).

(3.208) or equation (3.209), in conjunction with the appropriate beginning conditions. The nonlinear reactions of cylindrical shells are shown in Figure 3.30, showcasing the impact of the radius/thickness ratio. The fluctuation of W has a discernible pattern over time, with a steady amplification of its amplitude. The findings indicate a significant rise in the vibratory amplitude of the shell as the ratio increases. This finding is consistent with the inherent characteristic of the construction, namely, that a thicker shell tends to exhibit more vibration damping compared to a thinner shell. Moreover, the amplitude of the shell oscillations will be reduced based on the relative sizes of the shell radii.

The effects of the length-to-radius ratio on the W–time curve are shown in Figure 3.31. As illustrated in the preceding figure, the current figure also demonstrates the occurrence of vibrations in W over time. It is evident from the figure that the amplitude of these vibrations escalates in proportion to the ratio L/R. This observation can be attributed to the structural characteristics of the composite cylindrical shell, wherein a shorter length and larger radius contribute to enhanced vibration absorption capabilities.

Figure 3.32 depicts the impact of varying degrees of inclination of the composite auxetic-core cylindrical shell on the amplitude, shown as a function of the nonlinear

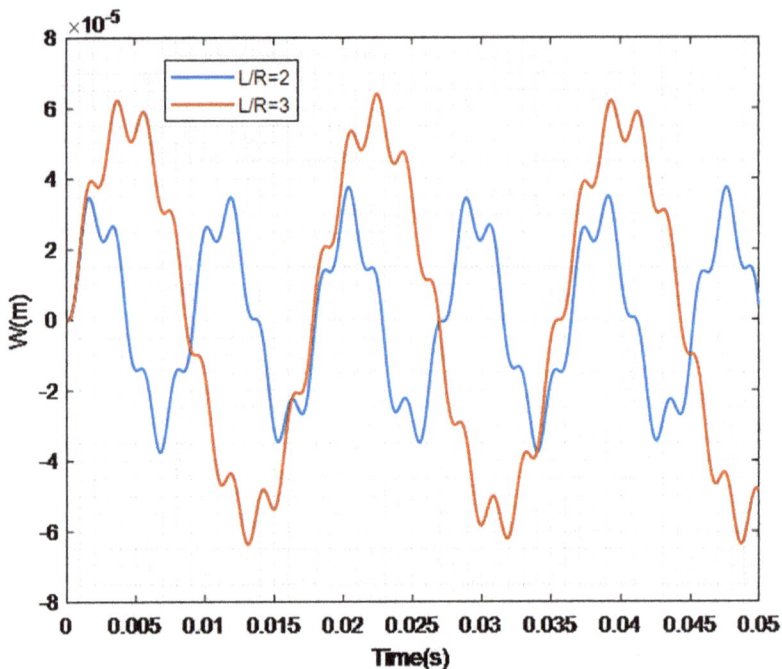

FIGURE 3.31 Variations of the nonlinear response of the cylindrical sandwich shell for different length-to-radius L/R ($R/h = 200$, $P = 0$, $Q = 3000$, $\Omega = 3000$, $(m, n) = (1, 3)$).

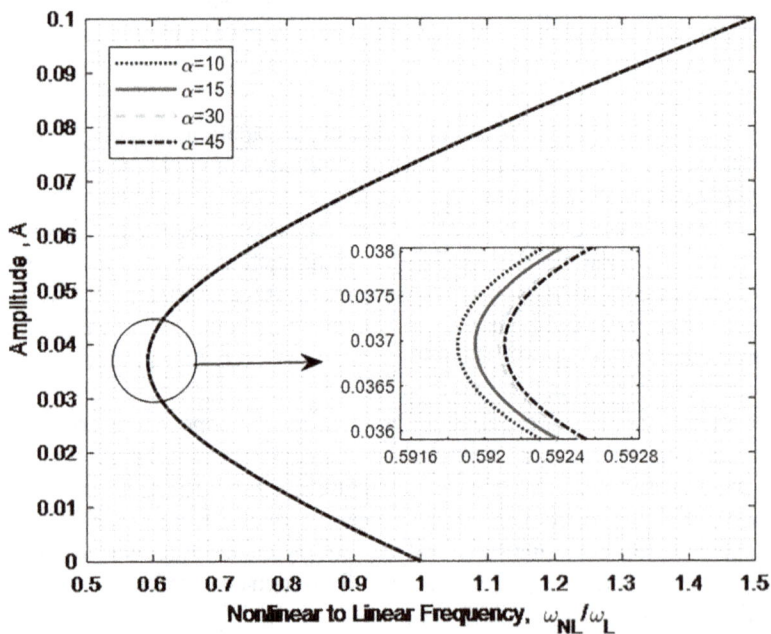

FIGURE 3.32 Changes of the amplitude as a function of nonlinear to the linear frequency with different values of inclination angle (α) ($R/h = 200$, $L/R = 2$, $P = 0$, $(m, n) = (1, 3)$).

to linear frequency ratio. The amplitude of vibration of the auxetic cylindrical shell is influenced by the inclination angle, which exhibits a diminishing effect. As the angle of inclination is increased, the amplitude of the oscillation diminishes. Moreover, when the amplitude is augmented, the frequency transitions from nonlinear to linear at an earlier stage, and subsequent to reaching its maximum point, it starts an upward trend with further increments in amplitude. Furthermore, it can be seen that the minimum frequency at which nonlinearity transitions to linearity has an upward trend as the inclination angle increases.

The graphic shown illustrates the impact of different lambdas on the nonlinear frequency fluctuation in relation to the linear frequency of the auxetic cylindrical shell. Figure 3.33 illustrates the impact of amplitude compared to nonlinear frequency on the linear frequency of the cylindrical composite shell, considering four lambda values. In contrast to the preceding Figure 3.32, when the amplitude is shown to decrease as a result of a certain factor, the lambda variable exhibits a positive correlation with the amplitude. Specifically, an increase in the number of lambdas corresponds to an increase in the amplitude. Furthermore, it can be shown that the lowest frequency at which nonlinearity transitions to linearity rises as the variable under consideration increases. This trend is consistent with the findings presented in Figure 3.46, which have been thoroughly analyzed. The figure illustrates a

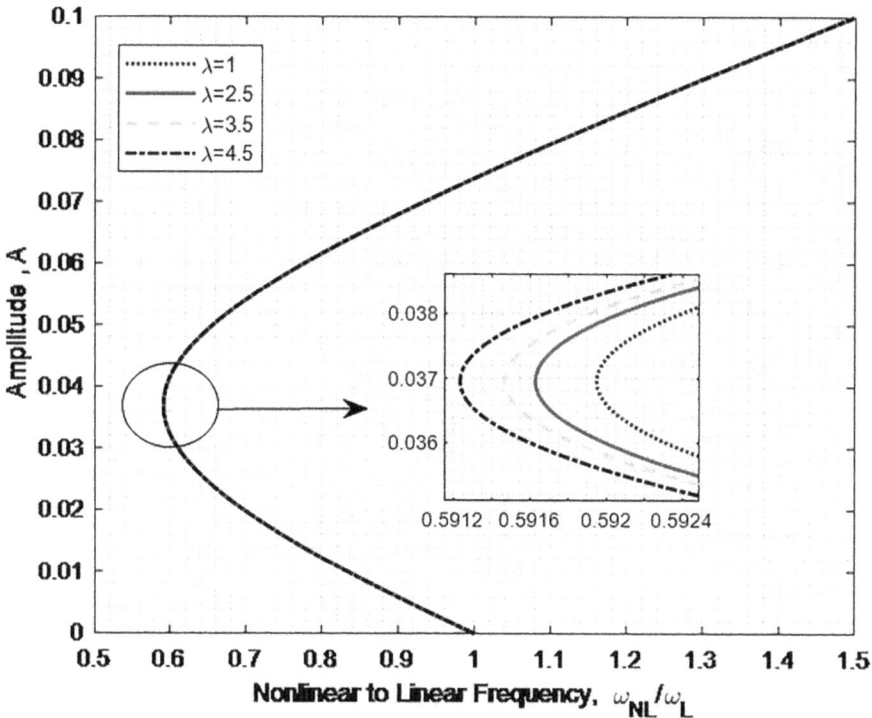

FIGURE 3.33 Effect of various values of lambda (λ) on the frequency-amplitude curve of cylindrical shell with the auxetic core ($R/h = 200$, $L/R = 2$, $P = 0$, $(m, n) = (1, 3)$).

relationship between amplitude and frequency. It is seen that initially, as the amplitude grows, the frequency falls. However, after a certain point, referred to as the
peak, the frequency starts to increase with further increases in amplitude.

To enhance visual observation, the impact of varying lambda values on the
temporal vibration amplitude curve is shown in the following sequence. As seen
in Figure 3.34a, the magnitude of the oscillations progressively escalates as time
elapses. Furthermore, a specific portion of Figure 3.34a has been magnified and
is shown as a distinct entity in Figure 3.34b. The figure provides a clear depiction
of the inverse relationship between the lambda value and the vibration amplitude,
indicating that a rise in lambda leads to a decrease in amplitude, resulting in fewer
oscillations in amplitude.

The figure serves to illustrate the benefits associated with the use of an auxetic
core in our approach. The impact of the inclined angle of the auxetic structure on
the nonlinear change in linear frequency per lambda is prominently seen in Figure
3.35. The function of the inclined angle becomes more significant in the transition
of the cylindrical shell with an auxetic core from a nonlinear to a linear frequency.
This transition continues until the inclination angle approaches 45 degrees, after
which the magnitude of the nonlinear to linear frequency drops as the alpha parameter increases. The findings of this study indicate that determining a universal rule
for forecasting the impact of the inclination angle on the nonlinear to linear frequency is a challenging task. Nevertheless, this observation illustrates that the use

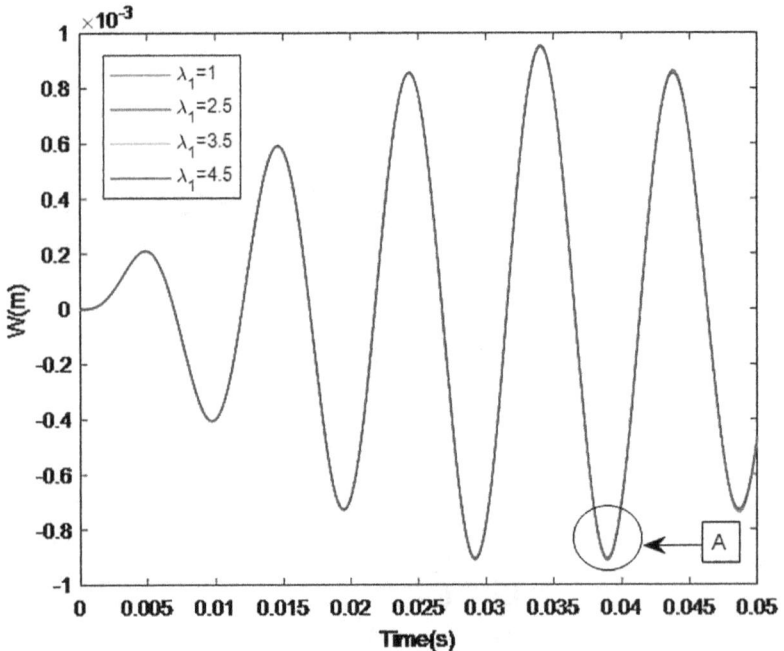

FIGURE 3.34　Effect of the various values of lambda (λ) on nonlinear responses of cylindrical shells ($R/h = 200$, $L/R = 2$, $P = 0$, $Q = 1500$, $\Omega = 600$, $(m, n) = (1, 3)$).

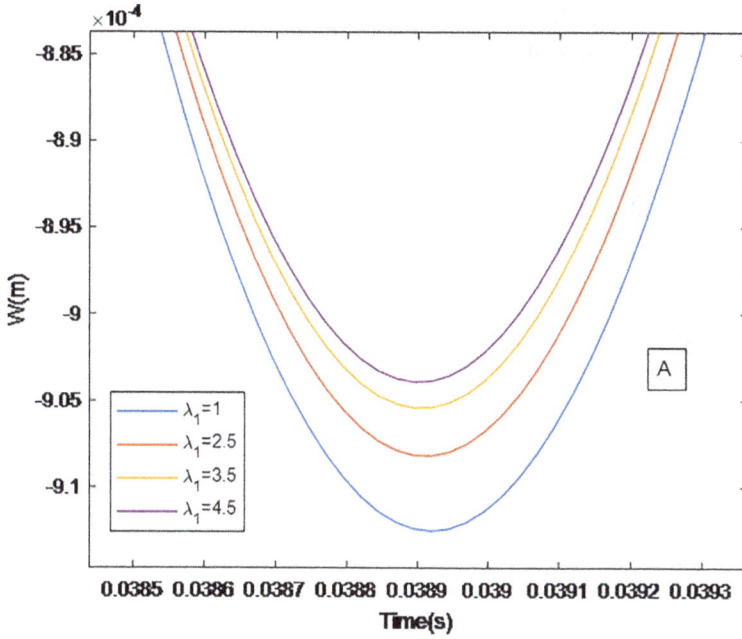

FIGURE 3.34   (Continued)

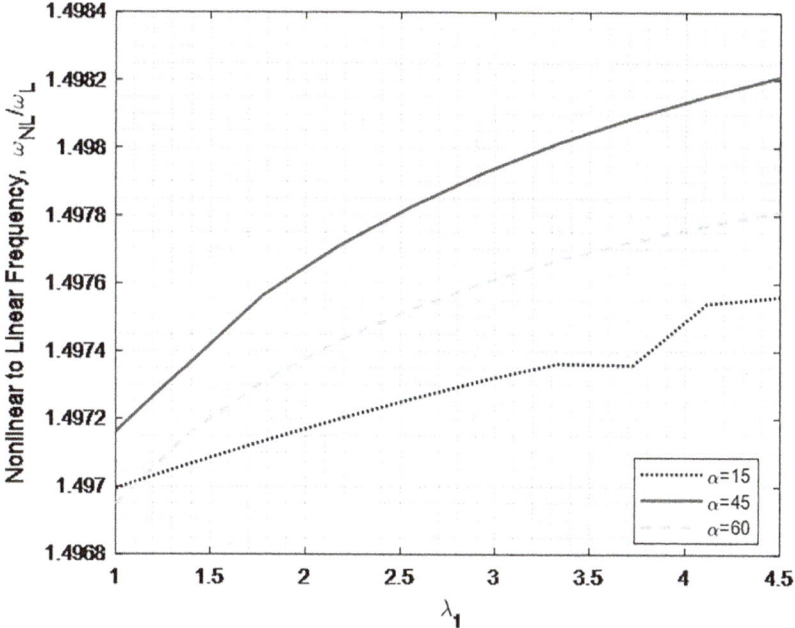

FIGURE 3.35   Variation of nonlinear to linear frequency versus rib length ratio of cylindrical shell with the auxetic core concerning different values of inclination angle ($R/h = 200$, $L/R = 2$, $P = 0$, $A = 0.1$, $(m, n) = (1, 3)$).

of auxetic material has the potential to decrease the weight of the structure without necessitating any explicit alterations in the frequency, transitioning from a nonlinear to a linear state.

The impact of various arrangements of the cylindrical shell with an auxetic core on the frequency-amplitude curve, namely the transition from nonlinear to linear behavior, is seen in Figure 3.36. It is evident that an increase in the thickness of the auxetic core results in a decrease in the nonlinear to linear frequency. This phenomenon may be attributed to the distinctive quality of the auxetic material and its high hardness. Moreover, it can be inferred that the use of auxetic material leads to a decrease in the nonlinear frequencies. In the given structural configuration, for example, in the structure 1–2–1, where the thickness of the auxetic material is one-half of the total thickness, we see a lower nonlinear frequency than in the 1–1–1, where the thickness of the auxetic represents one-third of the total thickness.

The data shown in Figure 3.37 demonstrates a positive correlation between the magnitude of the applied force and the observed amplitude at the beginning of work. This implies that the force assumes a progressively significant role in determining the amplitude. Furthermore, when the amplitude is augmented, the relationship between amplitude and the frequency ratio of nonlinear to linear behavior approaches that of the free vibration curve.

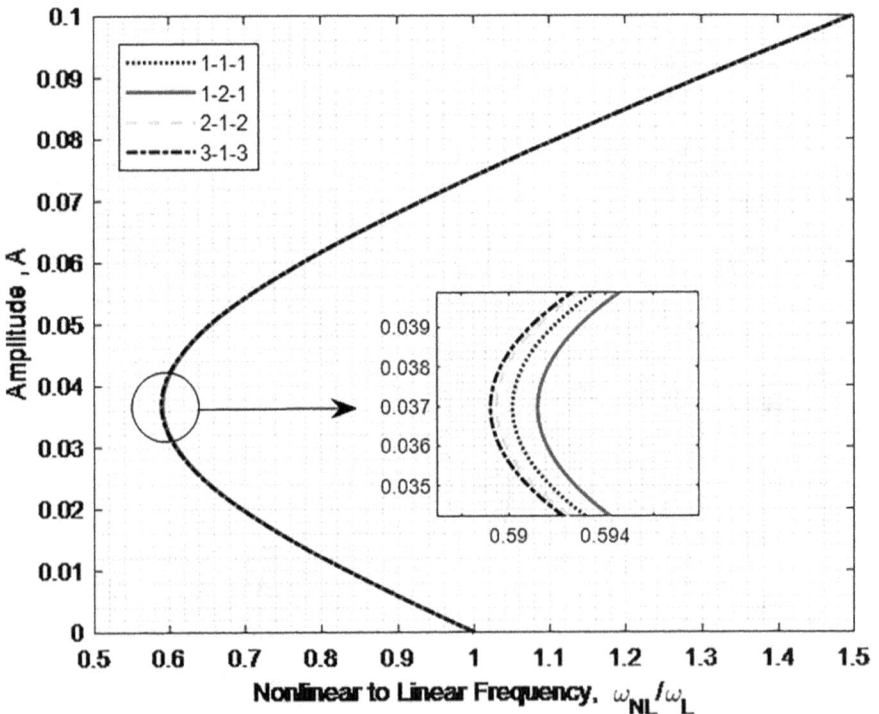

FIGURE 3.36 Effect of various thicknesses on the nonlinear to the linear frequency-amplitude curve of cylindrical shell with the auxetic core ($R/h = 200$, $L/R = 2$, $P = 0$, $(m, n) = (1, 3)$).

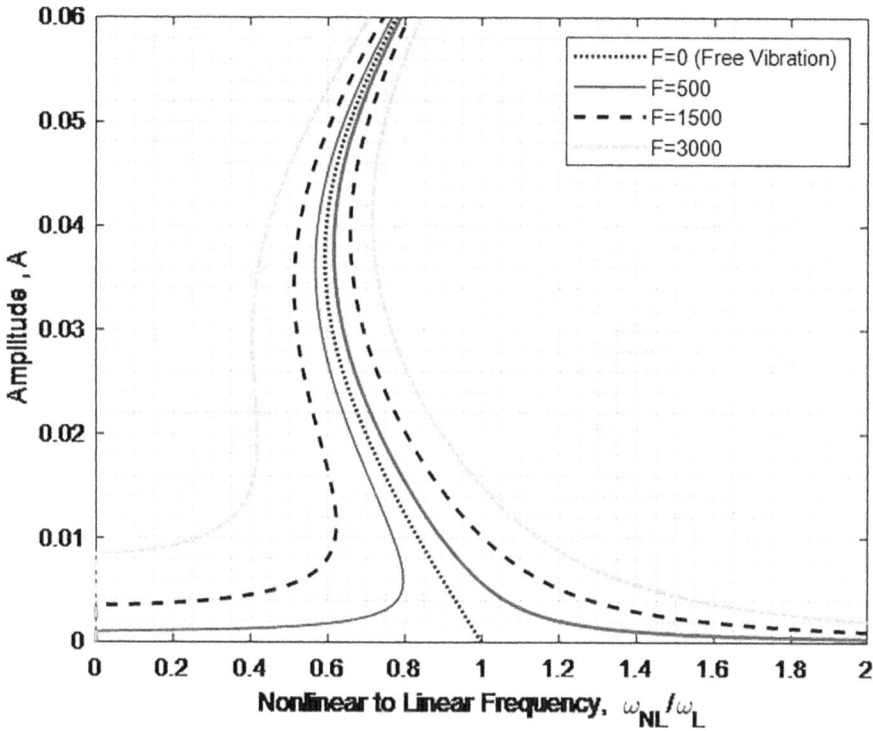

FIGURE 3.37 Variation of amplitude versus nonlinear to linear frequency for various levels of the harmonic excitation forces ($R/h = 200$, $L/R = 2$, $P = 0$, $(m, n) = (1, 3)$).

Figure 3.38 illustrates the frequency-amplitude curves of the cylindrical shell, showcasing the transition from nonlinear to linear behavior. Additionally, the impact of pre-loaded axial compression on these curves is examined. The findings obtained indicate a positive correlation between the amplitude of vibration of the shells and the magnitude of the applied load. Moreover, it can be said that when the amplitude is augmented, the frequency value transitions from being nonlinear to linear, first exhibiting a reduction and thereafter commencing a climb after reaching its maximum point. Furthermore, it can be seen that an increase in the pre-loaded axial compression leads to a drop in the lowest frequency at which the nonlinear behavior transitions to linear behavior.

Furthermore, Figure 3.39 investigates the impact of pre-loaded axial compression on the W curve with respect to time. The variability of omega is observable over time, and it is apparent that a rise in value p leads to greater amplitude in these variations.

The present study examines the impact of various pre-loaded axial compression values on the nonlinear to linear frequency diagram over time, as shown in Figure 3.40. It is evident from the figure that an increase in the value of p leads to a substantial rise in the nonlinear to linear frequency. Additionally, assuming continuous pre-loaded axial compression, the introduction of lambda modifications results in little alteration in the quantity of $\dfrac{\omega_{NL}}{\omega_L}$.

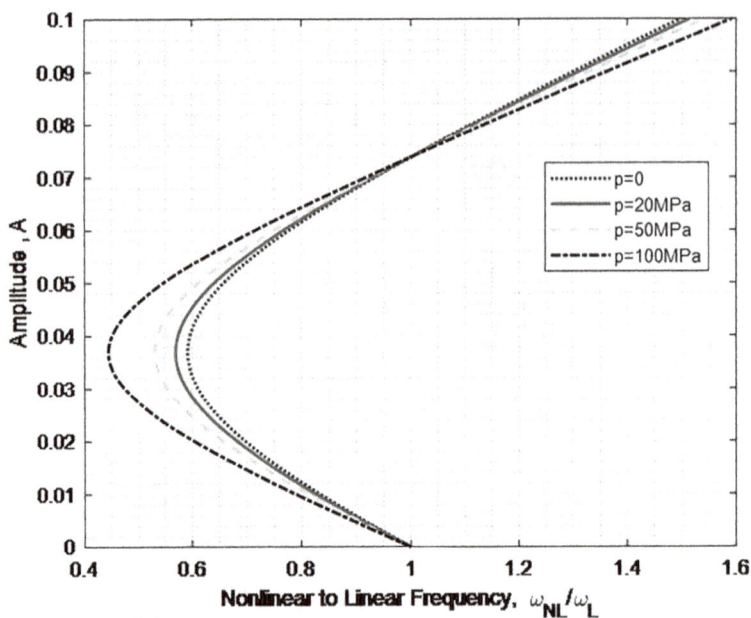

FIGURE 3.38 Effect of pre-loaded axial compression on frequency–amplitude curve of sandwich cylindrical shells ($R/h = 200$, $L/R = 2$, $(m, n) = (1, 3)$).

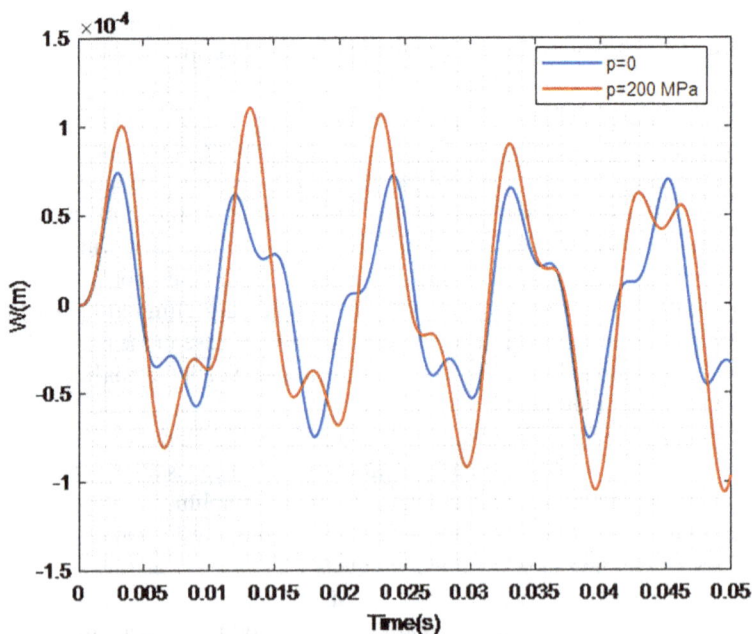

FIGURE 3.39 Effect of pre-loaded axial compression on nonlinear responses of cylindrical shells ($R/h = 200$, $L/R = 2$, $Q = 1500$, $\Omega = 600$, $(m, n) = (1, 3)$).

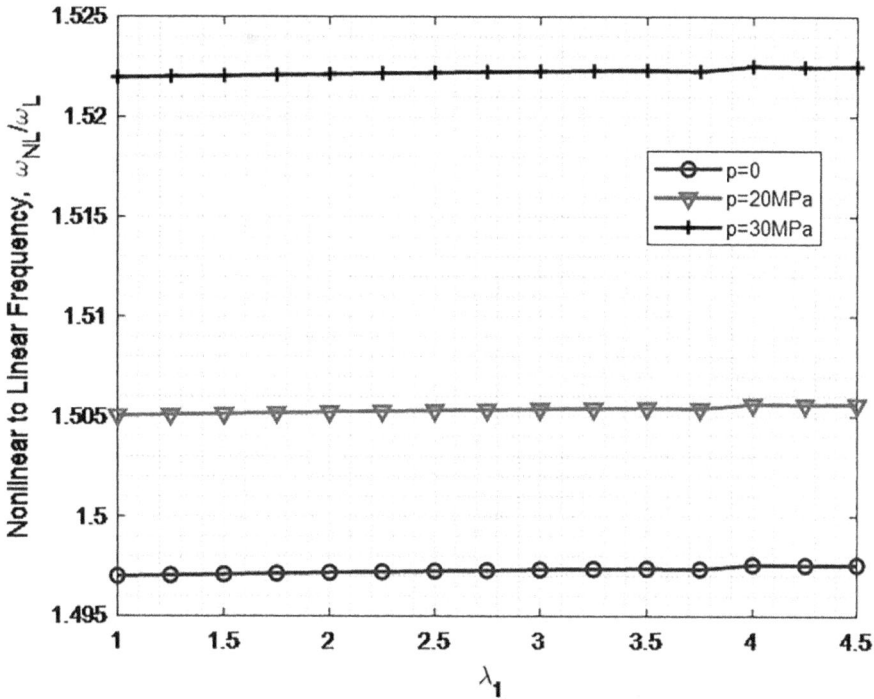

FIGURE 3.40 Effect of pre-loaded axial compression on the nonlinear to linear frequency function of rib length ratio of sandwich cylindrical shells with the auxetic core ($R/h = 200$, $L/R = 2$, $P = 0$, $A = 0.1$, $(m, n) = (1, 3)$).

Figure 3.41 illustrates the temporal fluctuations of the variable W across several values of n. The presented figure demonstrates that the quantity of waves in the circumferential direction significantly influences both the maximum amplitude of vibration and the number of fluctuations. A higher number of n leads to a larger amplitude of vibration, whereas a lower value of n results in a smaller frequency amplitude but more variations.

The preceding study examines the influence of variable m on the curve W over time, as depicted in Figure 3.42. In contrast, Figure 3.41 illustrates that the variable n exhibits a positive correlation with the amplitude of the oscillations. Specifically, as the number of half-waves in the axial direction increases, the amplitude of the oscillations decreases. Hence, as seen in the figure, when the variable m attains a value of 3, the occurrence of oscillations diminishes significantly, resulting in the transformation of the temporal curve W into a linear trajectory.

3.4.2 VIBRATION ANALYSIS OF SMART DOUBLY CURVED AUXETIC SHELLS

This section presents a novel smart origami auxetic sandwich panel, which combines a functionally graded origami auxetic core with multi-scale hybrid nanocomposite layers consisting of graphene platelet (GPL), carbon fiber (CF), and polyvinylidene

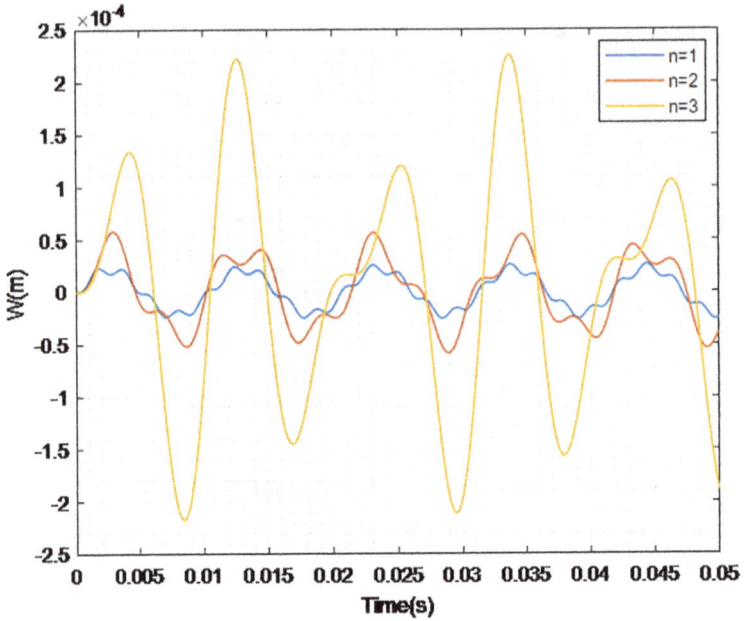

FIGURE 3.41 The comparison of the W in terms of time with different numbers of waves in the circumferential direction (n) ($R/h = 200$, $L/R = 2$, $P = 0$, $Q = 1500$, $\Omega = 600$, $m = 1$).

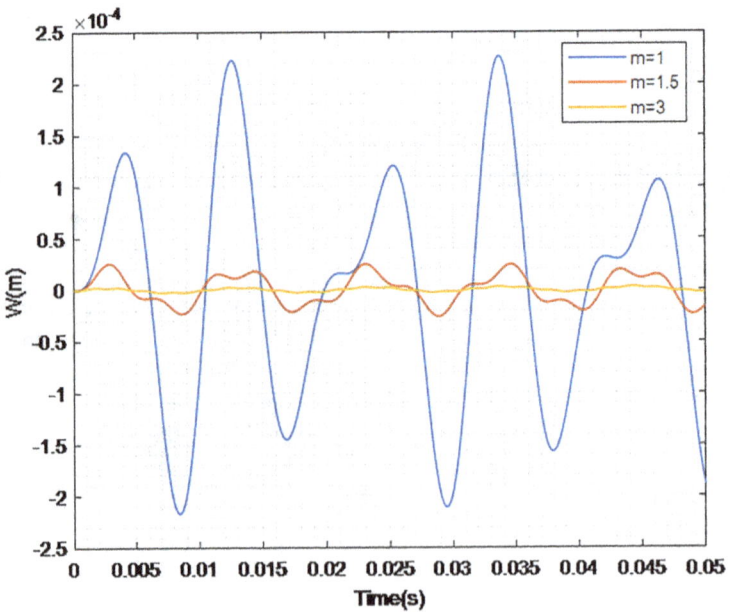

FIGURE 3.42 Effect of the various number of half-waves in the axial direction (m) on non-linear responses of cylindrical shells ($R/h = 200$, $L/R = 2$, $P = 0$, $Q = 1500$, $\Omega = 600$, $n = 3$).

fluoride polymer (PVDF). The chapter also examines the nonlinear vibrational characteristics of the intelligent doubly curved shell. The equations of motion are derived by using a Timoshenko shear deformation theory (TSDT) and including a Von-Karman geometric nonlinearity. The homotopy perturbation approach is used to solve the equations governing nonlinear motion. The accuracy of HPM has been shown to be superior to other perturbation methods throughout both nonlinear and linear frequency ranges. Additionally, this study examines the impacts of modifying various parameters, such as the radius of doubly curved surfaces, the thickness of the smart hybrid nanocomposite, the thickness of the auxetic metamaterial, the weight fraction of GOri nanofillers, the weight fraction of graphene nanoplatelets, the degree of GO folding angle, and the volume fraction of carbon fibers.

The efficiency characteristics of the GPL are assessed via the use of the HT and the rule of mixing (ROM). This chapter posits that the conventional structure of the GPL may be described as follows. Figure 3.43 displays the schematic representation of the smart multi-scale hybrid FG-graphene origami auxetic sandwich doubly curved panel, as described in references [2,4].

$$W_{GPL} = 100W_C W_{GPL}^t \tag{3.210}$$

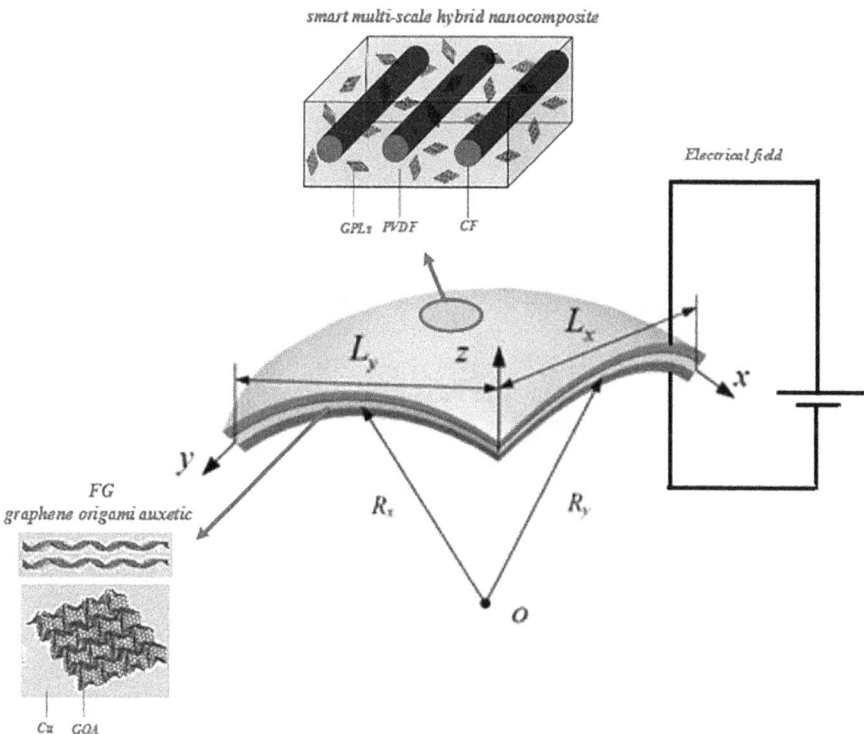

FIGURE 3.43 The schematic of the smart multi-scale hybrid FG-graphene origami auxetic sandwich doubly curved panel.

The exceptional quality and relative abundance of graphene nanoplatelets are denoted by W_C and W_{GPL}^t, respectively. The determination of the volume fraction of graphene nanoplatelets may be achieved by the following calculation:

$$V_{GPL} = \frac{W_{GPL}}{W_{GPL} + \left(\dfrac{\rho_{GPL}}{\rho_M}\right)\left(1 - W_{GPL}\right)} \tag{3.211}$$

The concentrations of graphene nanoplatelets and their PVDF matrix are denoted as ρ_{GPL} and ρ_M in equation (3.211). The determination of the true modulus of the smart nanocomposite (SNC) may be achieved by the use of the Halpin-Tsai theory.

$$E_{SNCM} = \left(\frac{3}{8}E_L + \frac{5}{8}E_T\right) \tag{3.212}$$

The longitudinal and transverse moduli of the SNC are denoted by E_L and E_T, respectively.

$$\{E_L, E_T\} = \left\{\frac{1 + \xi_L^{GPL}\eta_L^{GPL}V_{GPL}}{1 - \eta_L^{GPL}V_{GPL}}, \frac{1 + \xi_W^{GPL}\eta_W^{GPL}V_{GPL}}{1 - \eta_W^{GPL}V_{GPL}}\right\}E_M$$

$$\{\eta_L^{GPL}, \eta_W^{GPL}\} = \left\{\frac{\left(\dfrac{E_{GPL}}{E_M}\right) - 1}{\left(\dfrac{E_{GPL}}{E_M}\right) + \xi_L^{GPL}}, \frac{\left(\dfrac{E_{GPL}}{E_M}\right) - 1}{\left(\dfrac{E_{GPL}}{E_M}\right) + \xi_W^{GPL}}\right\} \tag{3.213}$$

$$\{\xi_L^{GPL}, \xi_W^{GPL}\} = \left\{\frac{2l_{GPL}}{t_{GPL}}, \frac{2w_{GPL}}{t_{GPL}}\right\}$$

The variables l_{GPL}, w_{GPL}, and t_{GPL} represent the dimensions of the graphene. By using the ROM, it is possible to compute the Poisson's ratio, density, dielectric, and piezo-electric constants of the SNC layer in the following manner:

$$V_{GPL} + V_M = 1 \tag{3.214a}$$

$$\rho_{SNC} = \left(V_M\rho^M + V_{GPL}\rho^{GPL}\right) \tag{3.214b}$$

$$v_{SNC} = V_M v^M + V_{GPL}v^{GPL} \tag{3.214c}$$

$$G_{SNC} = \left(\frac{E_{SNC}}{2\left(1 + v_{SNC}\right)}\right) \tag{3.214d}$$

$$e_{ij}^{SNC} = V_M e_{ij}^M + V_{GPL}e_{ij}^{GPL} \tag{3.214e}$$

$$s_{ij}^{SNC} = V_M s_{ij}^M + V_{GPL} s_{ij}^{GPL} \tag{3.214f0}$$

The smart hybrid nanocomposite (SHNC) layer is strengthened by the incorporation of carbon fiber and graphene. The effective qualities of the SHNC layer are expressed in the following manner:

$$V_{SNC} + V_F = 1 \tag{3.215a}$$

$$E_{SHNC} = \left(V_{SNC} E^{SNC} + V_F E^F \right) \tag{3.215b}$$

$$\rho_{SHNC} = \left(V_{SNC} \rho^{SNC} + V_F \rho^F \right) \tag{3.215c}$$

$$v_{SHNC} = \left(V_{SNC} v^{SNC} + V_F v^F \right) \tag{3.215d}$$

$$G_{SHNC} = \left(\frac{E_{SHNC}}{2\left(1 + v_{SHNC}\right)} \right) \tag{3.215e}$$

$$e_{ij}^{SHNC} = \left(V_{SNCM} e_{ij}^{SNC} + V_F e_{ij}^F \right) \tag{3.215f}$$

$$s_{ij}^{SHNC} = \left(V_{SNCM} s_{ij}^{SNC} + V_F s_{ij}^F \right) \tag{3.215g}$$

Furthermore, the specific characteristics of the GOA core are obtained by molecular dynamics (MD) simulation in the following manner:

$$E^{Auxetic} = E_{Cu} \frac{1 + \xi \eta V_{Gr}}{1 - \eta V_{Gr}} f_E \left(H_{Gr}, V_{Gr}, T \right) \tag{3.216a}$$

$$v^{Auxetic} = \left(V_{Gr} v_{Gr} + V_{Cu} v_{Cu} \right) f_v \left(H_{Gr}, V_{Gr}, T \right) \tag{3.216b}$$

$$\alpha^{Auxetic} = \left(V_{Gr} \alpha_{Gr} + V_{Cu} \alpha_{Cu} \right) f_\alpha \left(V_{Gr}, T \right) \tag{3.216c}$$

$$\rho^{Auxetic} = \left(V_{Gr} \rho_{Gr} + V_{Cu} \rho_{Cu} \right) f_\rho \left(V_{Gr}, T \right) \tag{3.216d}$$

$$f_E \left(H_{Gr}, V_{Gr}, T \right) = \begin{vmatrix} 1.11 - 1.22 V_{Gr} - 0.134 \left(\dfrac{T}{T_0} \right) \\ +0.559 V_{Gr} \left(\dfrac{T}{T_0} \right) - 5.5 H_{Gr} V_{Gr} \\ +38 H_{Gr} V_{Gr}^2 - 20.6 H_{Gr}^2 V_{Gr}^2 \end{vmatrix}$$

$$f_v\left(H_{Gr},V_{Gr},T\right)=\begin{pmatrix}1.01-1.43V_{Gr}+0.165\left(\dfrac{T}{T_0}\right)\\[2mm]-16.8H_{Gr}V_{Gr}-1.1H_{Gr}V_{Gr}\left(\dfrac{T}{T_0}\right)\\[2mm]+16H_{Gr}^2V_{Gr}^2\end{pmatrix}$$

$$f_\alpha\left(V_{Gr},T\right)=\begin{pmatrix}0.794-16.8V_{Gr}^2-0.0279\left(\dfrac{T}{T_0}\right)^2\\[2mm]+0.182\left(\dfrac{T}{T_0}\right)\left(V_{Gr}+1\right)\end{pmatrix}$$

$$f_\rho\left(V_{Gr},T\right)=\left(1.01-2.01V_{Gr}^2-0.0131\left(\dfrac{T}{T_0}\right)\right)\qquad(3.216e)$$

W_{Gr} represents the graphene origami weight percentage, and also $V_{Gr}=\dfrac{\rho_{Cu}W_{Gr}}{\rho_{Cu}W_{Gr}+\rho_{Gr}\left(1-W_{Gr}\right)}$, $V_{Cu}=1-V_{Gr}$ signifies the volume fractions of graphene origami and Cu, respectively. Moreover, undefined function $f_{E,v,\alpha,\rho}\left(H_{Gr},V_{Gr},T\right)$ and dimension constants $\eta=\dfrac{\dfrac{E_{Gr}}{E_{Cu}}-1}{\dfrac{E_{Gr}}{E_{Cu}}+\xi}$, $\xi=2\left(\dfrac{l_{Gr}}{t_{Gr}}\right)$ are detectable with MD.

Furthermore, the aforementioned produced shape may be used to characterize the FG distributions inside the graphene origami auxetic core (FG-GOA).

$$V_{Gr}\left(z\right)=V_{Gr}\left|\frac{4z}{h_{FG-GOA}}\right|^n\qquad(3.217)$$

The displacement fields at any specific spot inside the smart origami auxetic sandwich panel may be expressed as:

$$u=u_0+z\varphi_x-\frac{4}{3h^2}z^3\left(\varphi_x+\frac{\partial w_0}{\partial x}\right)\qquad(3.218a)$$

$$v = v_0 + z\varphi_y - \frac{4}{3h^2}z^3\left(\varphi_y + \frac{\partial w_0}{\partial y}\right) \tag{3.218b}$$

$$w = w_0 \tag{3.218c}$$

The symbols u_0, v_0, and w_0 represent the starting displacements of the smart origami auxetic sandwich panel along the x and y axes. Furthermore, the symbols φ_x and φ_y are used to represent the rotations of the transverse normal at the mid-plane along the x and y axes, respectively. The displacements of the smart origami auxetic sandwich panel are characterized by the functions. The variables $u_0, v_0, w_0, \varphi_x, \varphi_y$ are derived from the original set of five variables. In the scenario including geometric nonlinearity of the Von Karman type, the strain components ε_{xx}, ε_{yy}, and γ_{xy} may be shown in the following manner:

$$\begin{Bmatrix}\varepsilon_{xx}\\\varepsilon_{yy}\\\gamma_{xy}\end{Bmatrix} = \begin{Bmatrix}\varepsilon_{xx}^0\\\varepsilon_{yy}^0\\\gamma_{xy}^0\end{Bmatrix} + z\begin{Bmatrix}k_{xx}^0\\k_{yy}^0\\k_{xy}^0\end{Bmatrix} + z^3\begin{Bmatrix}k_{xx}^2\\k_{yy}^2\\k_{xy}^2\end{Bmatrix}$$

$$= \begin{Bmatrix}\dfrac{\partial u_0}{\partial x}+\dfrac{w_0}{R_1}+\dfrac{1}{2}(\dfrac{\partial w_0}{\partial x})^2\\[2mm]\dfrac{\partial v_0}{\partial y}+\dfrac{1}{2}(\dfrac{\partial w_0}{\partial y})^2+\dfrac{w_0}{R_2}\\[2mm]\dfrac{\partial u_0}{\partial y}+\dfrac{\partial w}{\partial x}\dfrac{\partial w}{\partial y}+\dfrac{\partial v_0}{\partial x}\end{Bmatrix} + z\begin{Bmatrix}\dfrac{\partial \varphi_x}{\partial x}\\[2mm]\dfrac{\partial \varphi_y}{\partial y}\\[2mm]\dfrac{\partial \varphi_x}{\partial y}+\dfrac{\partial \varphi_y}{\partial x}\end{Bmatrix}$$

$$-\frac{4}{3h^2}z^3\begin{Bmatrix}\dfrac{\partial \varphi_x}{\partial x}+\left(\dfrac{\partial^2 w_0}{\partial x^2}\right)\\[2mm]\dfrac{\partial \varphi_y}{\partial y}+\left(\dfrac{\partial^2 w_0}{\partial y^2}\right)\\[2mm]\dfrac{\partial \varphi_x}{\partial y}+\dfrac{\partial \varphi_y}{\partial x}+2\left(\dfrac{\partial^2 w_0}{\partial x\partial y}\right)\end{Bmatrix} \tag{3.219a}$$

$$\begin{Bmatrix}\tau_{yz}\\\tau_{xz}\end{Bmatrix} = \begin{Bmatrix}\gamma_{yz}^0\\\gamma_{xz}^0\end{Bmatrix} + z^2\begin{Bmatrix}k_{yz}^1\\k_{xz}^1\end{Bmatrix} \tag{3.219b}$$

$$= \begin{Bmatrix}\varphi_y+\dfrac{\partial w_0}{\partial y}\\[2mm]\varphi_x+\dfrac{\partial w_0}{\partial x}\end{Bmatrix} - \frac{4}{h^2}z^3\begin{Bmatrix}\varphi_y+\dfrac{\partial w_0}{\partial y}\\[2mm]\varphi_x+\dfrac{\partial w_0}{\partial x}\end{Bmatrix}$$

The electric potential of the smart sandwich panel may be determined in accordance with Maxwell's law.

$$\Phi(x,z,t) = -\cos(\xi z)\phi(x,t) + \frac{2z}{h}V \tag{3.220}$$

where $\xi = \pi / h$. As an additional note, V is the electrical voltage supplied to the intelligent sandwich panel. The smart sandwich plate's electrical field (E_x, E_y, E_z) may be calculated using Maxwell's laws as follows:

$$E_x = -\phi_{,x} = \cos(\xi z)\frac{\partial \phi}{\partial x} \tag{3.221a}$$

$$E_y = -\phi_{,y} = \cos(\xi z)\frac{\partial \phi}{\partial y} \tag{3.221b}$$

$$E_z = -\phi_{,z} = \xi \sin(\xi z) - \frac{2v}{h} \tag{3.221c}$$

Let's assume that the symbol denotes the angle formed by the fiber with the geometric horizontal axis. Then, the smart origami auxetic sandwich panel's basic connection may be expressed in geometric terms as follows:

$$\begin{Bmatrix} \sigma_{xx} \\ \sigma_{yy} \\ \tau_{yz} \\ \tau_{xy} \\ \tau_{xz} \end{Bmatrix}_T = \begin{bmatrix} \bar{Q}_{11}^n & \bar{Q}_{12}^n & 0 & 0 & 0 \\ \bar{Q}_{12}^n & \bar{Q}_{22}^n & 0 & 0 & 0 \\ 0 & 0 & \bar{Q}_{44}^n & 0 & 0 \\ 0 & 0 & 0 & \bar{Q}_{55}^n & 0 \\ 0 & 0 & 0 & 0 & \bar{Q}_{66}^n \end{bmatrix}_T \begin{Bmatrix} \varepsilon_{xx} - \alpha_{11}\left(T(z) - T_0\right) \\ \varepsilon_{yy} - \alpha_{22}\left(T(z) - T_0\right) \\ \gamma_{xz} \\ \gamma_{xy} \\ \gamma_{xz} \end{Bmatrix}_T$$

$$- \begin{bmatrix} 0 & 0 & e_{31} & 0 & 0 \\ 0 & 0 & e_{31} & 0 & 0 \\ 0 & 0 & 0 & 0 & 0 \\ 0 & e_{15} & 0 & 0 & 0 \\ e_{15} & 0 & 0 & 0 & 0 \end{bmatrix} \begin{Bmatrix} E_x \\ E_y \\ 0 \\ E_z \\ 0 \end{Bmatrix}_T \tag{3.222a}$$

$$
\begin{Bmatrix} D_x \\ D_y \\ 0 \\ D_z \\ 0 \end{Bmatrix}_T = \begin{bmatrix} 0 & 0 & 0 & 0 & e_{15} \\ 0 & 0 & e_{15} & 0 & 0 \\ 0 & 0 & 0 & 0 & 0 \\ e_{31} & e_{31} & 0 & 0 & 0 \\ 0 & 0 & 0 & 0 & 0 \end{bmatrix}_T \begin{Bmatrix} \varepsilon_{xx} - \alpha_{11}\left(T(z)-T_0\right) \\ \varepsilon_{yy} - \alpha_{22}\left(T(z)-T_0\right) \\ \gamma_{xz} \\ \gamma_{xy} \\ \gamma_{xz} \end{Bmatrix}_T
$$

$$
- \begin{bmatrix} S_{11} & 0 & 0 & 0 & 0 \\ 0 & S_{11} & 0 & 0 & 0 \\ 0 & 0 & 0 & 0 & 0 \\ 0 & 0 & 0 & S_{33} & 0 \\ 0 & 0 & 0 & 0 & 0 \end{bmatrix} \begin{Bmatrix} E_x \\ E_y \\ 0 \\ E_z \\ 0 \end{Bmatrix}_T
\qquad (3.222b)
$$

The stress-strain was represented by the symbols $\_\sigma_{ij} - \varepsilon_{ij}$, the electric potential was denoted by the symbol E_n, and the displacement was denoted by the symbol D_{ij}. Similarly, s_{ij} and e_{ij} determine the dielectric and piezoelectric constants. Reduced stiffness modulus of the smart sandwich may be expressed as:

$$
Q_{11} = \frac{E_{11}}{1-v_{12}v_{21}}, Q_{12} = \frac{v_{12}E_{22}}{1-v_{12}v_{21}}, Q_{22} = \frac{E_{22}}{1-v_{12}v_{21}},
$$
$$
Q_{44} = G_{23}, Q_{55} = G_{13}, Q_{66} = G_{12} \qquad (3.223)
$$

Appendix 3.5 details the key components of the converted smart origami auxetic sandwich panel.

The following is one way to state Hamilton's principle:

$$
\int_0^t \delta\left(V+U-T\right)dt = 0 \qquad (3.224)
$$

where (V) is the external work, (U) is the strain energy, and (T) is the kinetic energy. The strain energy is represented by the following formula:

$$
U = \frac{1}{2}\sum_{n=1}^{N}\int_0^a\int_0^b\int_{h_{n-1}}^{h_n} [(\sigma_{xx}\varepsilon_{xx} + \sigma_{yy}\varepsilon_{yy} + \tau_{yz}\gamma_{yz} + \tau_{xz}\gamma_{xz} + \tau_{xy}\gamma_{xy} - D_x\delta E_x
$$
$$
- D_y\delta E_y - D_z\delta E_z]\times\left(1+\frac{z}{R_1}\right)\left(1+\frac{z}{R_2}\right)dc_1dc_2dz \qquad (3.225)
$$

The variation of strain energy may be derived as follows:

$$\delta U = \sum_{n=1}^{N}\int_0^a\int_0^b [(N_{xx}\delta\varepsilon_{xx}^0 + M_{xx}\delta k_{xx}^0 + P_{xx}\delta k_{xx}^2 + N_{yy}\delta\varepsilon_{yy}^0$$

$$+ M_{yy}\delta k_{yy}^2 + P_{yy}\delta k_{yy}^2 + N_{xy}\delta\varepsilon_{xy}^0 + M_{xy}\delta k_{xy}^2$$

$$+ P_{xy}\delta k_{xy}^2 + K_{yy}\delta\gamma_{yz}^0 + R_{yy}k_{yz}^1 + K_{xx}\delta\gamma_{xz}^0 + R_{xx}k_{xz}^1 \times \left(1+\frac{z}{R_1}\right)\left(1+\frac{z}{R_2}\right) dc_1 dc_2$$

$$+ \int_0^a\int_0^b\int_{-h/2}^{h/2}\left[-D_x\cos(\xi z)\frac{\partial\delta\phi}{\partial x} - D_y\cos(\xi z)\frac{\partial\delta\phi}{\partial y} + D_z\xi\sin(\xi z)\delta\phi\right]dzdxdy \quad (3.226)$$

It is assumed that the double-curved panel has dimensions of a and b in the c_1 and c_2 planes. The Lame coefficients for the doubly curved panel are also $q_1 = c_1\left(1+\frac{z}{R_1}\right), q_2 = c_{.2}\left(1+\frac{z}{R_2}\right)$. Additionally, R_1 and R_2 stand for the curvature radii in the q_1 and q_1 directions, respectively.

It is possible to express kinetic energy in the following ways:

$$T = \frac{1}{2}\rho^n\sum_{n=1}^{N}\int_0^a\int_0^b\int_{h_{n-1}}^{h_n}\left(\dot{u}_0^2 + \dot{v}_0^2 + \dot{w}_0^2\right)\times\left(1+\frac{z}{R_1}\right)\left(1+\frac{z}{R_2}\right)dc_1 dc_2 dz \quad (3.227)$$

Possible smart origami auxetic sandwich panel discoveries that simplify kinetic energy coupling include:

$$T = \frac{1}{2}\rho^n\sum_{n=1}^{N}\int_0^a\int_0^b\left[\dot{u}_0^2 + \dot{v}_0^2 + \dot{w}_0^2 + h^2\left[\frac{17}{315}\left(\dot{\phi}_x^2 + \dot{\phi}_y^2\right) + \dot{\phi}_x\dot{u}_0\left(\frac{41}{120R_1}+\frac{2}{15R_2}\right)\right.\right.$$

$$+ \dot{\phi}_y\dot{v}_0\left(\frac{41}{120R_2}+\frac{2}{15R_1}\right) + \frac{1}{4}\left(\frac{\dot{u}_0^2}{R_1}+\frac{\dot{v}_0^2}{R_2}\right) + \frac{\dot{w}_0^2}{12R_1R_2} + \frac{\partial\dot{w}_0}{q_1\partial c_1}\left(\frac{\partial\dot{w}_0}{252q_1\partial c_1}+\frac{\dot{u}_0}{120R_1}\right.$$

$$\left.-\frac{\dot{u}_0}{30R_2}-8\frac{\dot{\phi}_x}{315}\right) + \frac{\partial\dot{w}_0}{q_2\partial c_2}\left(\frac{\partial\dot{w}_0}{252q_2\partial c_1}+\frac{\dot{v}_0}{120R_2}-\frac{\dot{v}_0}{30R_1}-8\frac{\dot{\phi}_y}{315}\right)\right]\right]dc_1 dc_2 \quad (3.228)$$

The external work variation may be calculated as follows:

$$\delta\Pi_w = \int_0^{2\pi}\int_0^L\left[N_x^0\frac{\partial w_0}{\partial x}\frac{\partial\delta w_0}{\partial x}\right]dc_1 dc_2 \quad (3.229)$$

The smart origami auxetic sandwich panel is assumed to be exposed to electric voltage V from the outside. In order to calculate N_x^0, we have:

$$N_x^0 = N^E = -\int_{-h/2}^{h/2} e_{31}\frac{2V}{h}dz \quad (3.230)$$

If the δu, δv, δw, $\delta\varphi_x$, $\delta\varphi_y$, $\delta\phi$ and $\delta\psi$ are all zero in equation (9), then (16–1) may be constructed using equations (10), (12), and (14).

$$\frac{\partial N_x}{\partial x}+\frac{\partial N_{xy}}{\partial y}=\bar{I}_0\left(\frac{\partial^2 u_0}{\partial t^2}\right)-\bar{J}_1\left(\frac{\partial^2 \varphi_x}{\partial t^2}\right)+s_1\bar{I}_3\frac{\partial^2}{\partial t^2}\left(\frac{\partial w_0}{\partial x}\right) \tag{3.231a}$$

$$\frac{\partial N_{xy}}{\partial x}+\frac{\partial N_y}{\partial y}=\bar{I}_0\left(\frac{\partial^2 v_0}{\partial t^2}\right)-\bar{J}_1\left(\frac{\partial^2 \varphi_y}{\partial t^2}\right)+s_1\bar{I}_3\frac{\partial^2}{\partial t^2}\left(\frac{\partial w_0}{\partial y}\right) \tag{3.231b}$$

$$\frac{\partial \bar{K}_x}{\partial x}+\frac{\partial \bar{K}_y}{\partial y}+\frac{\partial}{\partial x}\left[N_x\left(\frac{\partial w_0}{\partial x}-\frac{u_0}{R_1}\right)+N_{xy}\left(\frac{\partial w_0}{\partial y}-\frac{v_0}{R_2}\right)\right]$$

$$+\frac{\partial}{\partial y}\left[N_{xy}\left(\frac{\partial w_0}{\partial x}-\frac{u_0}{R_1}\right)+N_y\left(\frac{\partial w_0}{\partial y}-\frac{v_0}{R_2}\right)\right]+$$

$$s_1+\frac{\partial}{\partial y}\left[N_{xy}\left(\frac{\partial w_0}{\partial x}-\frac{u_0}{R_1}\right)+N_y\left(\frac{\partial w_0}{\partial y}-\frac{v_0}{R_2}\right)\right]$$

$$+s_1\left(\frac{\partial^2 P_y}{\partial x^2}+2\frac{\partial^2 P_{xy}}{\partial x\partial y}+\frac{\partial^2 P_y}{\partial y^2}\right)-\frac{N_x}{R_1}-\frac{N_y}{R_2}$$

$$=I_0\left(\frac{\partial^2 w_0}{\partial t^2}\right)-s_1^2 I_6\frac{\partial^2}{\partial t^2}\left(\frac{\partial^2 w_0}{\partial x^2}+\frac{\partial^2 u_0}{\partial y^2}\right)$$

$$+s_1[I_3\frac{\partial^2}{\partial t^2}\left(\frac{\partial u_0}{\partial x}\right)+I_3\frac{\partial^2}{\partial t^2}\left(\frac{\partial v_0}{\partial y}\right)+J_4\frac{\partial^2}{\partial t^2}\left(\frac{\partial \varphi_x}{\partial x}+\frac{\partial^2 \varphi_y}{\partial y}\right)] \tag{3.231c}$$

$$\frac{\partial \bar{M}_x}{\partial x}+\frac{\partial \bar{M}_{xy}}{\partial y}-\bar{K}_x=J_1\left(\frac{\partial^2 u_0}{\partial t^2}\right)+k_2\left(\frac{\partial^2 \varphi_x}{\partial t^2}\right)$$

$$-s_1 J_4\frac{\partial^2}{\partial t^2}\left(\frac{\partial w_0}{\partial x}\right) \tag{3.231d}$$

$$\frac{\partial \bar{M}_{xy}}{\partial x}+\frac{\partial \bar{M}_y}{\partial y}-\bar{K}_y=J_1\left(\frac{\partial^2 v_0}{\partial t^2}\right)+k_2\left(\frac{\partial^2 \varphi_y}{\partial t^2}\right)$$

$$-s_1 J_4\frac{\partial^2}{\partial t^2}\left(\frac{\partial w_0}{\partial y}\right) \tag{3.231e}$$

$$\int_{-h_n-1}^{h_n}\left[\cos(\xi z)\frac{\partial D_x}{\partial x}+\cos(\xi z)\frac{\partial D_y}{\partial y}+\xi\sin(\xi z)D_z\right]dz=0 \tag{3.231f}$$

where:

$$\bar{M}_i = M_i - s_1 P_i \ (i = 1,2,6), \ s_1 = \frac{4}{3h^2}, s_2 = 3s_1 \tag{3.232a}$$

$$\bar{K}_j = K_j - s_2 R_j \ (j = 1,2) \tag{3.232b}$$

Smart origami auxetic sandwich panels' moment of inertias may be represented using the following formula:

$$I_i = \sum_{n=1}^{N} \int_{J_n}^{n+1} \rho^n z_i dz, \quad (i = 0,...,6) \tag{3.233a}$$

$$J_i = I_i - s_1 I_{i+2}, \quad (i = 1,4) \tag{3.233b}$$

$$\bar{K}_2 = I_2 - 2s_1 I_4 + s_1^2 I_6 \tag{3.233c}$$

$$\bar{I}_0 = I_0 + 2\frac{s_1}{R_1} I_3 + \left(\frac{s_1}{R_1}\right)^2 I_6 \tag{3.233d}$$

$$\bar{J}_1 = J_1 + \frac{s_1}{R_1} I_4 \tag{3.233e}$$

$$\bar{I}_3 = I_3 + \frac{s_1}{R_1} I_5 \tag{3.233f}$$

It has been assumed that the boundary conditions of the smart auxetic sandwich panel are S-S-S-S.

$$u_0(x,0,t) = u_0(x,b,t) = 0, \ v_0(0,y,t) = v_0(x,b,t) = 0 \tag{3.234a}$$

$$w_0(x,0,t) = w_0(x,b,t) = 0 \tag{3.234b}$$

$$\varphi_y(0,y,t) = \varphi_y(a,y,t) = 0, \ \varphi_x(x,0,t) = \varphi_x(x,b,t) = 0 \tag{3.234c}$$

$$N_x(x,0,t) = N_x(x,b,t) = 0, N_y(0,y,t)$$
$$= N_y(a,y,t) = 0 \tag{3.234d}$$

$$\bar{M}_x(x,0,t) = \bar{M}_x(x,b,t) = 0, \bar{M}_y(0,y,t)$$
$$= \bar{M}_y(a,y,t) = 0 \tag{3.234e}$$

Furthermore, the boundary condition is attained by determining the displacement of the smart auxetic sandwich panel in the following way:

$$u(x,y,t) = \sum_{n=1}^{\infty}\sum_{m=1}^{\infty} U_{mn}(t)\cos\left(\frac{n\pi x}{b}\right)\sin(my) \tag{3.235a}$$

$$v_0(x,y,t) = \sum_{n=1}^{\infty}\sum_{m=1}^{\infty} V_{mn}(t)\cos\left(\frac{n\pi x}{b}\right)\sin(my) \tag{3.235b}$$

$$w_0(x,y,t) = \sum_{n=1}^{\infty}\sum_{m=1}^{\infty} W_{mn}(t)\cos\left(\frac{n\pi x}{b}\right)\sin(my) \tag{3.235c}$$

$$\varphi_x(x,y,t) = \sum_{n=1}^{\infty}\sum_{m=1}^{\infty} \varphi_{xmn}(t)\cos\left(\frac{n\pi x}{b}\right)\sin(my) \tag{3.235d}$$

$$\varphi_y(x,y,t) = \sum_{n=1}^{\infty}\sum_{m=1}^{\infty} \varphi_{ymn}(t)\cos\left(\frac{n\pi x}{b}\right)\sin(my) \tag{3.235e}$$

$$\phi(x,y,t) = \sum_{n=1}^{\infty}\sum_{m=1}^{\infty} \phi_{mn}(t)\cos\left(\frac{n\pi x}{b}\right)\sin(my) \tag{3.235f}$$

where n and m are the number of frequency modes in the forward and backward directions of motion, respectively, and $U_{mn}(t), V_{mn}(t), W_{mn}(t), \varphi_{xmn}(t), \varphi_{ymn}(t)$, and $\phi(t)$ are time-dependent unknowns. It is assumed here that $\dfrac{n\pi x}{b} =$. By substituting equation (3.235) into equation (3.231) and performing the Navier process, we get the following formulae:

$$a_{11}U_{mn}(t) + a_{12}V_{mn}(t) + a_{13}W_{mn}(t) + a_{14}\varphi_{xmn}(t) + a_{15}\varphi_{ymn}(t)$$
$$+ a_{16}\phi_{mn}(t) = M_{11}\ddot{U}_{mn}(t) + M_{13}\ddot{W}_{mn} + M_{14}\ddot{\varphi}_{xmn}(t) \tag{3.236a}$$

$$a_{21}U_{mn}(t) + a_{22}V_{mn}(t) + a_{23}W_{mn}(t) + a_{24}\varphi_{xmn}(t) + a_{25}\varphi_{ymn}(t)$$
$$+ a_{26}\phi_{mn}(t) = M_{22}\ddot{V}_{mn}(t) + M_{23}\ddot{W}_{mn} + M_{25}\ddot{\varphi}_{ymn}(t) \tag{3.236b}$$

$$a_{31}U_{mn(t)} + a_{32}V_{mn}(t) + a_{33}W_{mn}(t) + a_{34}W_{mn}^3(t) + a_{35}\varphi_{xmn}(t)$$
$$+ a_{36}\varphi_{ymn}(t) + a_{37}\phi_{mn}(t) = M_{33}\ddot{W}_{mn} + M_{34}\ddot{\varphi}_{xmn}(t) + M_{35}\ddot{\varphi}_{ymn}(t) \tag{3.236c}$$

$$a_{41}U_{mn}(t) + a_{42}V_{mn}(t) + a_{43}W_{mn}(t) + a_{44}\varphi_{xmn}(t) + a_{45}\varphi_{ymn}(t)$$
$$+ a_{46}\phi_{mn}(t) = M_{44}\ddot{\varphi}_{xmn}(t) \tag{3.236d}$$

$$a_{51}U_{mn}(t) + a_{52}V_{mn}(t) + a_{53}W_{mn}(t) + a_{54}\varphi_{xmn}(t)$$
$$+ a_{55}\varphi_{ymn}(t) + a_{56}\phi_{mn}(t) = 0 \tag{3.236e}$$

$$a_{61}U_{mn}(t) + a_{62}V_{mn}(t) + a_{63}W_{mn}(t) + a_{64}\varphi_{xmn}(t)$$
$$+ a_{65}\varphi_{ymn}(t) + a_{66}\phi_{mn}(t) = 0 \tag{3.236f}$$

where a_{ij} and M_{ij} represent the dispersion stiffness and mass matrix of the smart auxetic sandwich panel. In addition, Appendix 3.5 defines these terms (stiffness and mass matrix). The nonlinear differential equation needed to run the smart sandwich is as follows:

$$\frac{d^2 W_{mn}(t)}{dt^2} + P_1 W_{mn}(t) + P_2 W_{mn}{}^2(t) + P_3 W_{mn}{}^3(t) = 0 \tag{3.237}$$

where:

$$P_3 = -\frac{M_{33} + M_{34} + M_{35}}{a_{34}} \tag{3.238}$$

Also, the smart auxetic sandwich panel's linear frequency is written as:

$$\omega_l = \sqrt{P_1} \tag{3.239}$$

for which initial circumstances it is shown by means of:

$$W_{mn}(0) = \frac{\bar{W}}{h}, \frac{dW_{mn}(t)}{dt}\bigg|_{t=0} = 0 \tag{3.240}$$

The unidentified function $W_{mn}(t)$ is replaced by $g(t)$, and equations (3.236) and (3.239) to utilize a straightforward and clear method:

$$\frac{d^2 g(t)}{dt^2} + P_1\{g(t) + \varsigma g^3(t)\} = 0 \tag{3.241}$$

where:

$$\varsigma = \frac{P_1}{P_2} \tag{}$$

The equation (3.240) is solved using the homotopy perturbation approach as described in references [45] and [46]. The concept of being able to acquire or achieve something might be described as attainable.

$$\frac{d^2 g(t)}{dt^2} + \omega^2 g(t) + \xi\{(P_1 - \omega^2)g(t) + P_1 \varsigma g^3(t)\} = 0 \tag{3.242}$$

where the ξ parameter is from [0,1]. Substituting 0 into equation (3.241) yields the linear differential equation.

$$\frac{d^2 g(t)}{dt^2} + \omega^2 g(t) = 0 \tag{3.243}$$

where:

$$g(t) = g_0(t) + \xi g_1(t) + \xi^2 g_2(t) + \dots \tag{3.244}$$

The result is derived by substituting equation (3.244) into equation (3.243).

$$\xi^0 : \frac{d^2 g_0(t)}{dt^2} + \omega^2 g_0(t) = 0 \,, g_0(0) = \frac{\bar{W}}{h} \,, \left. \frac{dg_0(t)}{dt} \right|_{t=0} = 0 \tag{3.245a}$$

$$\xi^1 : \frac{d^2 g_1(t)}{dt^2} + \omega^2 g_1(t) + \left\{ (P_1 - \omega^2) g_0(t) + P_1 \varsigma g_0^3(t) \right\} = 0,$$

$$g_1(0) = \frac{\bar{W}}{h} \,, \left. \frac{dg_1(t)}{dt} \right|_{t=0} = 0 \tag{3.245b}$$

The result of equation (3.245) is shown below:

$$g_0(t) = \frac{\bar{W}}{h} \cos(\omega t)$$

$$a = \frac{\bar{W}}{h} \tag{3.246}$$

The variable "a" denotes a dimensionless amplitude. The nonlinear frequency of the smart auxetic sandwich panel is represented by:

$$\frac{d^2 g_1(t)}{dt^2} + P_1 g_1(t) + (P_1 - \omega^2 + \frac{3}{4} a^2 \varsigma P_1) a \cos(\omega t) + \frac{1}{4} P_1 a^3 \varsigma \cos(3\omega t) = 0 \tag{3.247}$$

after removing $g_0(t)$ terms, the equation leads to:

$$P_1 - \omega^2 + \frac{3}{4} a^2 \varsigma P_1 = 0 \tag{3.248}$$

in which the nonlinear frequency of the smart auxetic sandwich panel may be written as:

$$\omega_{nl} = \omega_l \sqrt{1 + \frac{3}{4} a^2 \varsigma} = 0 \tag{3.249}$$

where:

$$\hat{a} = \frac{\overline{W}}{h^2}$$

$$\omega_{nl} = \omega_l \sqrt{1 + \frac{3}{4} h^2 \varsigma (a^*)^2} = 0$$

(3.250)

This section presents the mathematical findings pertaining to the nonlinear vibration characteristics shown by a smart origami auxetic sandwich panel. The geometrical and electromechanical properties of the FG-GOA metamaterial and SHNC sheets are shown in Table 3.2. Furthermore, Table 3.3 provides a comprehensive overview of the parameters used in this investigation.

The impact of the doubly curved radius on the nonlinear frequency ratio of the smart sandwich doubly curved panel is seen in Figure 3.44. This study illustrates the effects of varying nonlinear vibration amplitudes, ranging from 0 to 1, on the result of doubly curved radii. Specifically, doubly curved radii within the range of $\frac{b}{0.02}$ to $\frac{b}{0.08}$ are considered. The investigation reveals that the rise of the radius leads to an increase in the nonlinear frequency ratio of the smart sandwich doubly curved panel. As shown in this figure, it is evident that the magnitude of the nonlinear frequency ratio increment far surpasses all other values presented in this figure. Additionally, the minimum value of the nonlinear frequency ratio increment is seen in this figure. $\frac{b}{0.02}$ (The circumstances under consideration are governed by intricate and precise layers, which are influenced by geometric nonlinear effects, namely within the range of R = b/0.02 to b/0.08. The graphic shown demonstrates the phenomenon of hardening behavior in the resonance dynamics of the intelligent structure.

TABLE 3.3

Details of Parameters Used in the Present Study

Parameter	Assumed values
V_0	5
W_{GPL}^t	4%
h_{SHNC}	20
h_a	15
$\dfrac{b}{R_2}$	0.02
V_F	0.25
H_{GR}	50%
W_{GR}	100%

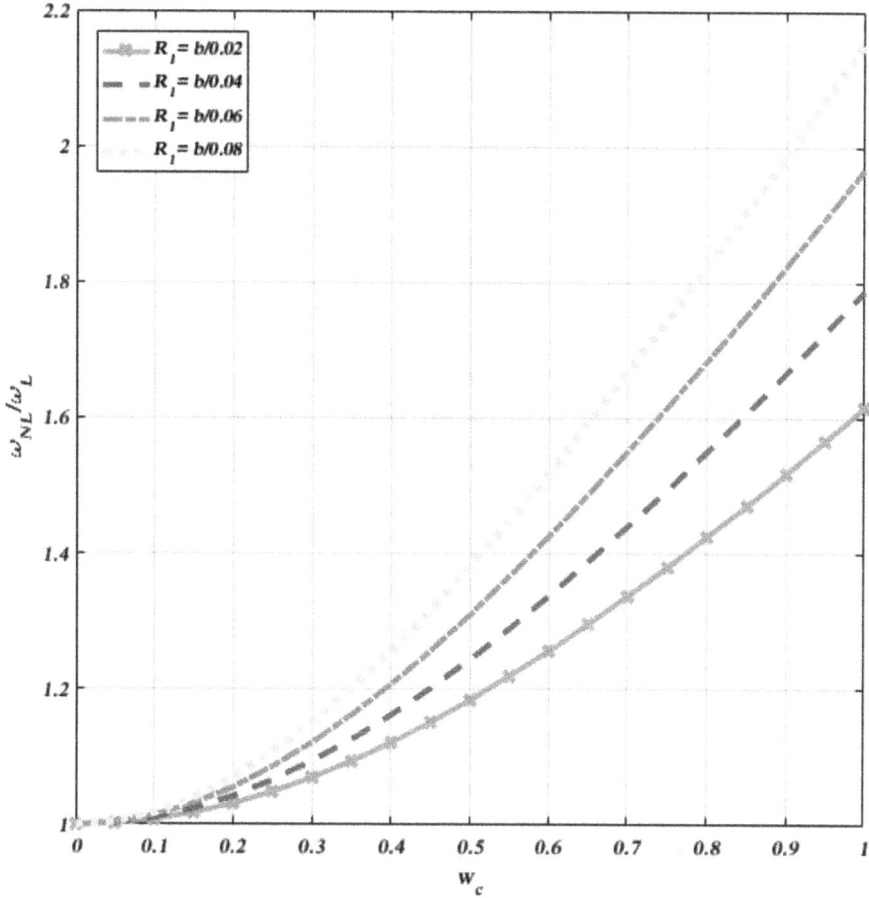

FIGURE 3.44 Effect of radius on nonlinear frequency ratio of smart multi-scale hybrid FG-graphene origami auxetic sandwich doubly curved panel.

The impact of the thickness of the smart hybrid nanocomposite (h_{SHNC}) on the nonlinear frequency ratio of the sandwich doubly curved panel made of the smart hybrid nanocomposite is clearly illustrated in Figure 3.45. This figure presents the results of varying the thickness of the doubly curved smart hybrid nanocomposite panel, ranging from 0 to 1, with assumed values of 3 to 6. The investigation reveals that the rise of smart thickness contributes to an increase in the nonlinear frequency ratio of the smart sandwich doubly curved panel. As seen in Figure 3.45, it is evident that the intensity of rise in the nonlinear frequency ratio is highest for $h_{SHNC} = 6$ compared to all other values presented in the figure. Conversely, the lowest value of the nonlinear frequency ratio is associated with $h_{SHNC} = 3$.

The aforementioned criteria (h_{SHN}=3:6) are regulated by intricate and precise layers under the influence of geometric nonlinear effects. As seen in this figure, the resonance dynamics of the smart structure are observed to exhibit hardening

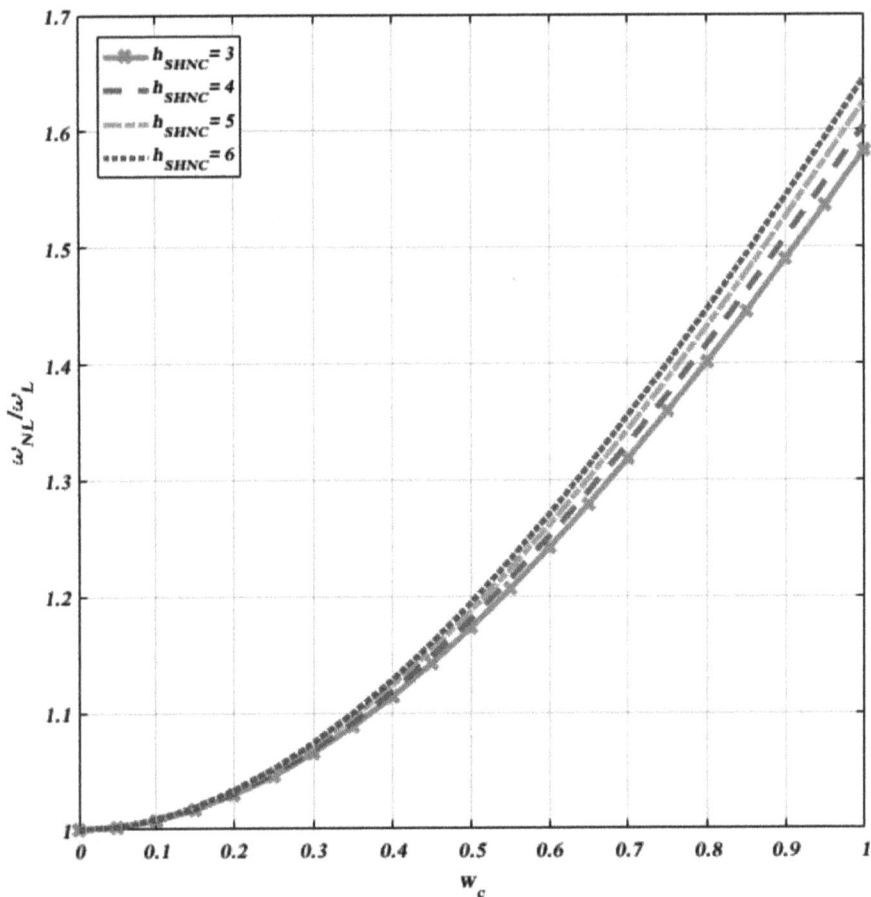

FIGURE 3.45   Effect of smart hybrid nanocomposite thickness on nonlinear frequency ratio of smart multi-scale hybrid FG-graphene origami auxetic sandwich doubly curved panel.

behavior. The impact of the thickness (h_a) of the auxetic metamaterial on the nonlinear frequency ratio of the smart sandwich doubly curved panel is clearly seen in Figure 3.46. The presented figure illustrates the results of a study on the behavior of doubly curved smart hybrid nanocomposite thickness under different nonlinear vibration amplitudes ranging from 0 to 1. The thickness of the doubly curved auxetic metamaterial is assumed to be within the range of 10 to 20.

It can be investigated that auxetic metamaterial thickness development leading up in growing nonlinear frequency ratio of the smart sandwich doubly curved panel. As can be understood in this figure, the intensity of growth in the nonlinear frequency ratio in $h_a = 20$ is greater than all the values available in this figure, and its lowest value is related to ha = 10. These complex and specific layers govern these conditions via geometric nonlinear effects. As shown in this figure, it causes hardening behavior in the resonance dynamics of the smart structure. The influence of the fiber

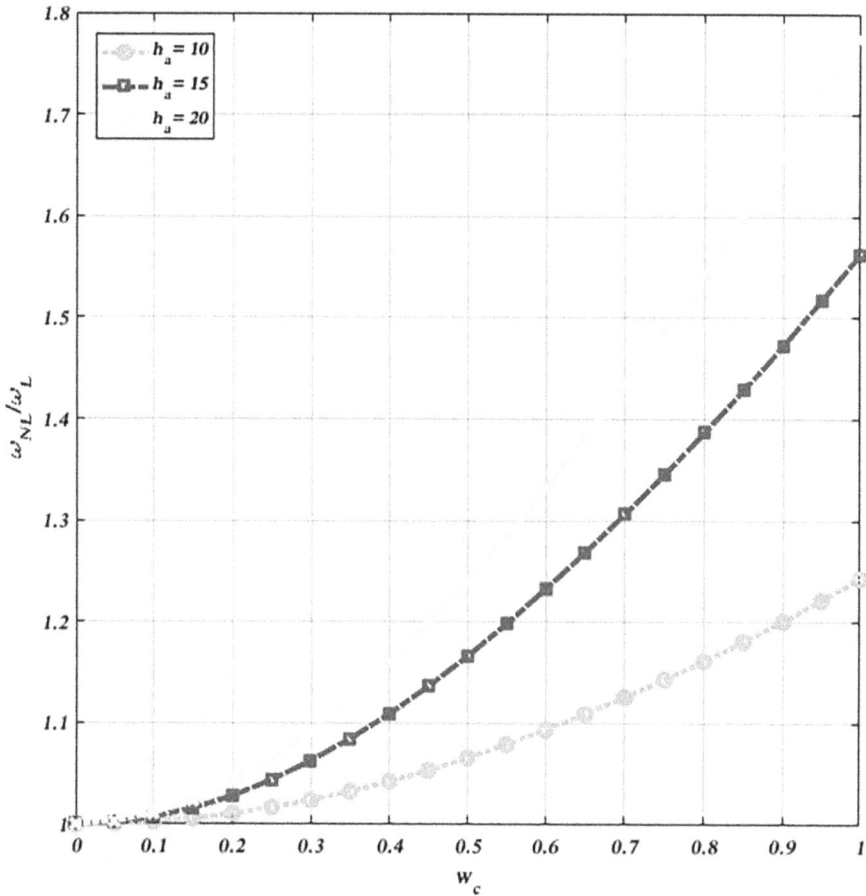

FIGURE 3.46 Effect of auxetic metamaterial thickness on nonlinear frequency ratio of smart multi-scale hybrid FG-graphene origami auxetic sandwich doubly curved panel.

angle of the smart layer (θ1) on the nonlinear frequency ratio of the smart sandwich doubly curved panel is evident in Figure 3.47. This figure, demonstrating the outcome of the fiber angle of the smart layer (θ1) through various nonlinear vibration amplitude from 0 to 1, fiber angle of the smart layer (θ1) such as 0 to $\pi/2$ are assumed. The fiber angle of smart layer (θ1) development can be investigated leading up to decreasing nonlinear frequency ratio of the smart sandwich doubly curved panel. As can be understood in this figure, the intensity of growth in the nonlinear frequency ratio in θ1 = 0 is greater than all the values available in this figure, and its lowest value is related to $\theta_1 = \dfrac{\pi}{2}$. These conditions $\left(\theta_1 = 0 : \dfrac{\pi}{2}\right)$ that is governed by these complex and specific layers via geometric nonlinear effects. As shown in this figure, it causes hardening behavior in the resonance dynamics of the smart structure. The influence of the fiber angle of the auxetic core (θ_2) on the nonlinear frequency ratio

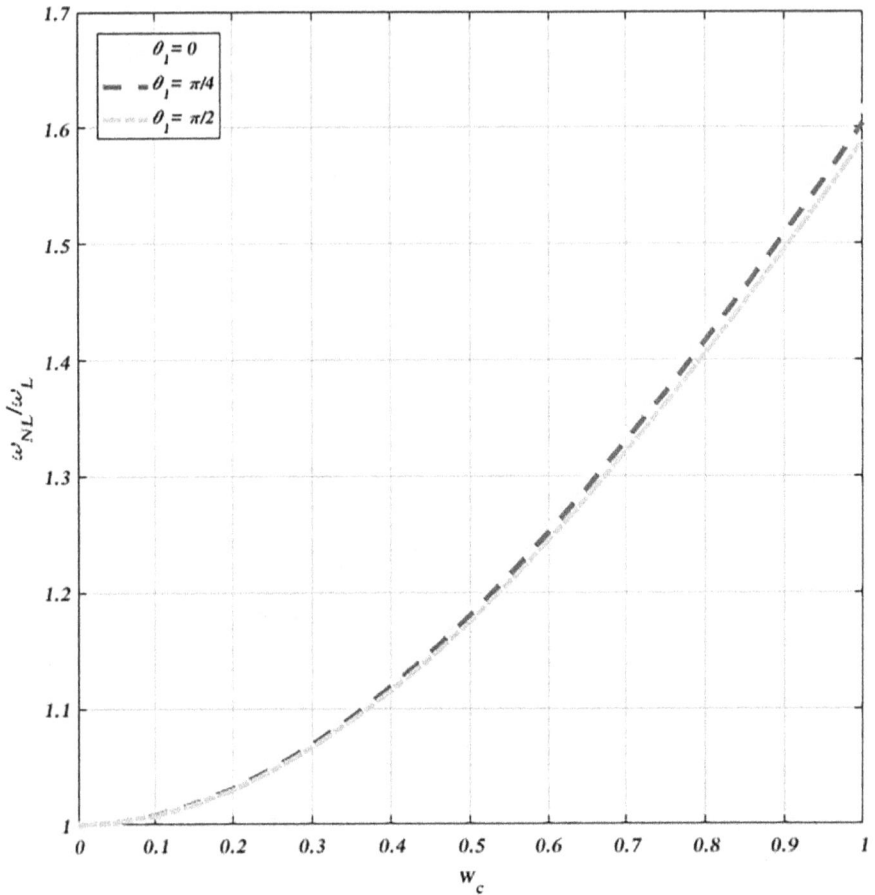

FIGURE 3.47 Effect of fiber angle of the smart layer on nonlinear frequency ratio of smart multi-scale hybrid FG-graphene origami auxetic sandwich doubly curved panel.

of the smart sandwich doubly curved panel is evident in Figure 3.48. This figure demonstrating the outcome of the fiber angle of auxetic core (θ_2) through various nonlinear vibration amplitude from 0 to 1, fiber angle of auxetic core (θ_2) such as $\pi/6$ to $\pi/3$ are assumed. Fiber angle of auxetic core (θ_2) development can be investigated it may leading up in growing nonlinear frequency ratio of the smart sandwich doubly curved panel.

The figure demonstrates that the nonlinear frequency ratio has a higher level of increase at $\theta_2 = \pi/6$ compared to all other values shown. Additionally, the smallest value of the nonlinear frequency ratio is associated with $\theta_2 = \pi/3$. The aforementioned circumstances, characterized by the range of θ_2 values from $\pi/6$ to $\pi/3$, are subject to the influence of intricate and distinct layers, which exhibit geometric nonlinear effects. Additionally, it induces a phenomenon of increased stiffness in the vibrational dynamics of the intelligent structure.

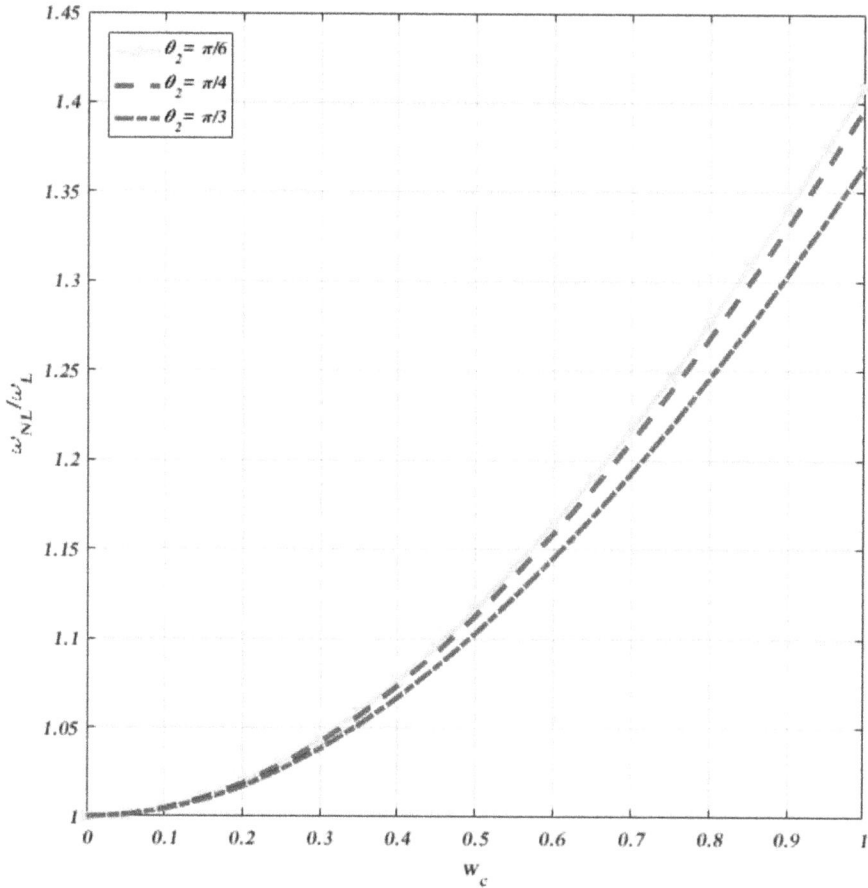

FIGURE 3.48 Effect of fiber angle of the auxetic core on nonlinear frequency ratio of smart multi-scale hybrid FG-graphene origami auxetic sandwich doubly curved panel.

The impact of the weight fraction of GOri nanofillers on the nonlinear frequency ratio of the smart sandwich doubly curved panel is clearly seen in Figure 3.49. The study presented the findings of the impact of different nonlinear vibration amplitudes ranging from 0 to 1 on the doubly weight fraction of GOri nanofillers. The weight fractions of GOri nanofillers included in the analysis were assumed to be within the range of 0.25 to 1. The investigation focuses on the analysis of the weight fraction of GOri nanofillers and its impact on the nonlinear frequency ratio of the smart sandwich doubly curved panel. The magnitude of evolutionary intensity in the nonlinear frequency ratio at/= 1 exceeds all the values shown in this figure, with its minimum value being associated with $W_{GR} = 0.25$. The circumstances are governed by intricate and precise layers, which are influenced by geometric nonlinear effects at a ratio of 0.25:1. The phenomenon induces a stiffening response in the resonance dynamics of the intelligent structure.

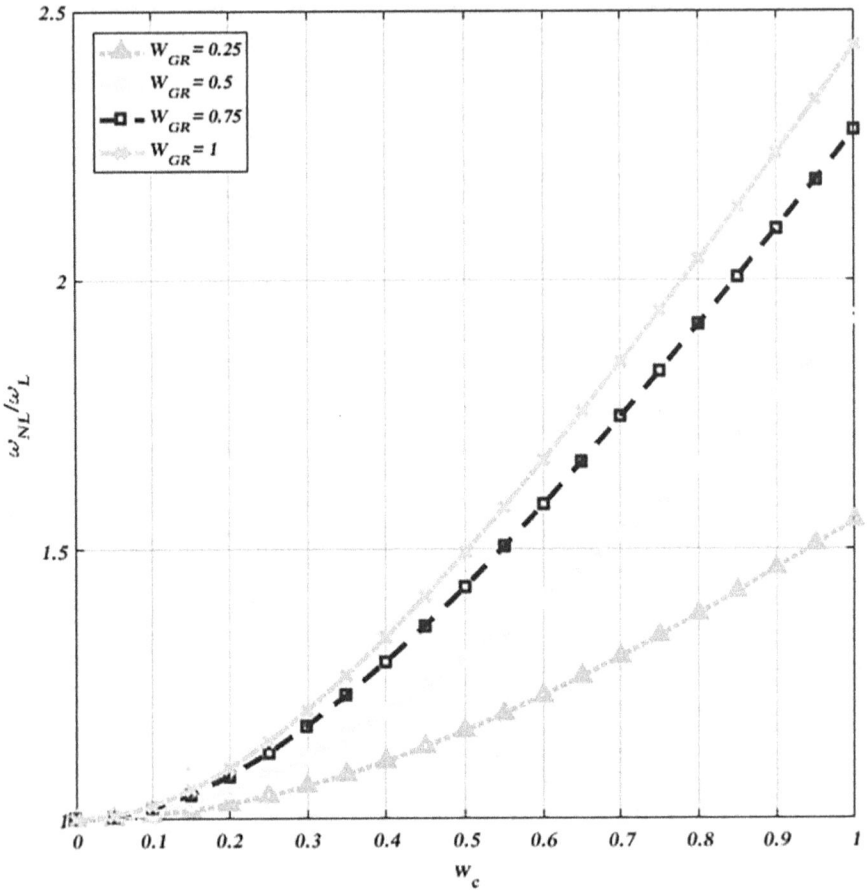

FIGURE 3.49 Effect of weight fraction of graphene origami nanofillers on nonlinear frequency ratio of smart multi-scale hybrid FG-graphene origami auxetic sandwich doubly curved panel.

The impact of the folding angle (θ) on the nonlinear frequency ratio of the smart sandwich doubly curved panel is clearly seen in Figure 3.50. The study presented the findings of investigating the impact of nonlinear vibration amplitudes ranging from 0 to 1 on the folding angle of GO. Specifically, the study focused on the doubling of the angle of GO folding degree. The range of folding angles considered in the study was between 0.5 and 0.9. The progressive increase in the folding degree $\left(H_{GR}\right)$ of the GO angle results in an escalation of the nonlinear frequency ratio of the smart sandwich doubly curved panel. The magnitude of evolutionary change in the nonlinear frequency ratio at/= 0.9 surpasses all the values shown in this figure, with its minimum value being associated with to $H_{GR} = 0.5$. The circumstances under consideration are governed by intricate and precise layers, which are influenced by geometric nonlinear phenomena. As seen in this figure, the resonance dynamics of the smart structure are influenced by the phenomenon of hardening behavior.

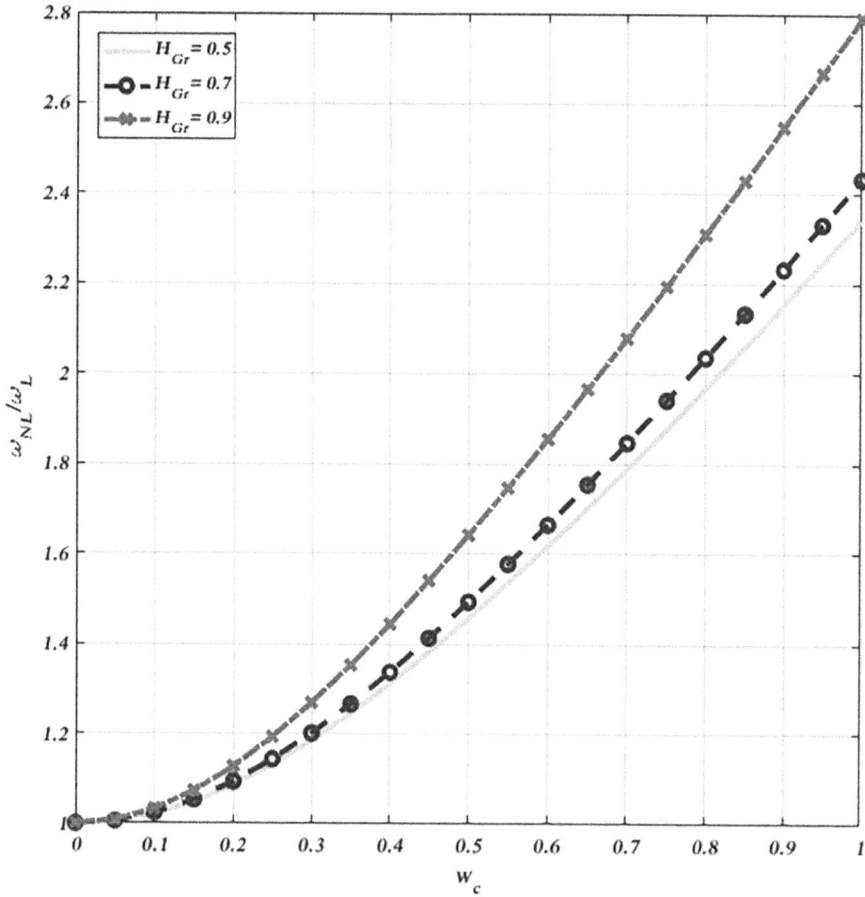

FIGURE 3.50 Effect of angle of graphene origami folding degree on nonlinear frequency ratio of smart multi-scale hybrid FG-graphene origami auxetic sandwich doubly curved panel.

The impact of the weight percentage of graphene nanoplatelets $\left(W_{GPL}^{t}\right)$ on the nonlinear frequency ratio of the smart sandwich doubly curved panel is evident in Figure 3.51. The study presented the findings of investigating the impact of nonlinear vibration amplitudes ranging from 0 to 1 on the weight fraction of graphene nanoplatelets. Specifically, weight fractions of graphene nanoplatelets ranging from 1% to 4% were considered. The relationship between the weight percentage of graphene nanoplatelets and the nonlinear frequency ratio of the smart sandwich doubly curved panel exhibits a rising trend. The magnitude of evolutionary changes in the nonlinear frequency ratio at 4% exceeds all other values shown in this figure, with its minimum value corresponding to 1%. The circumstances are governed by intricate and precise layers, which are influenced by geometric nonlinear effects at a rate of 1% to 4%. The image shown illustrates the phenomenon of hardening behavior seen in the resonance dynamics of the intelligent structure.

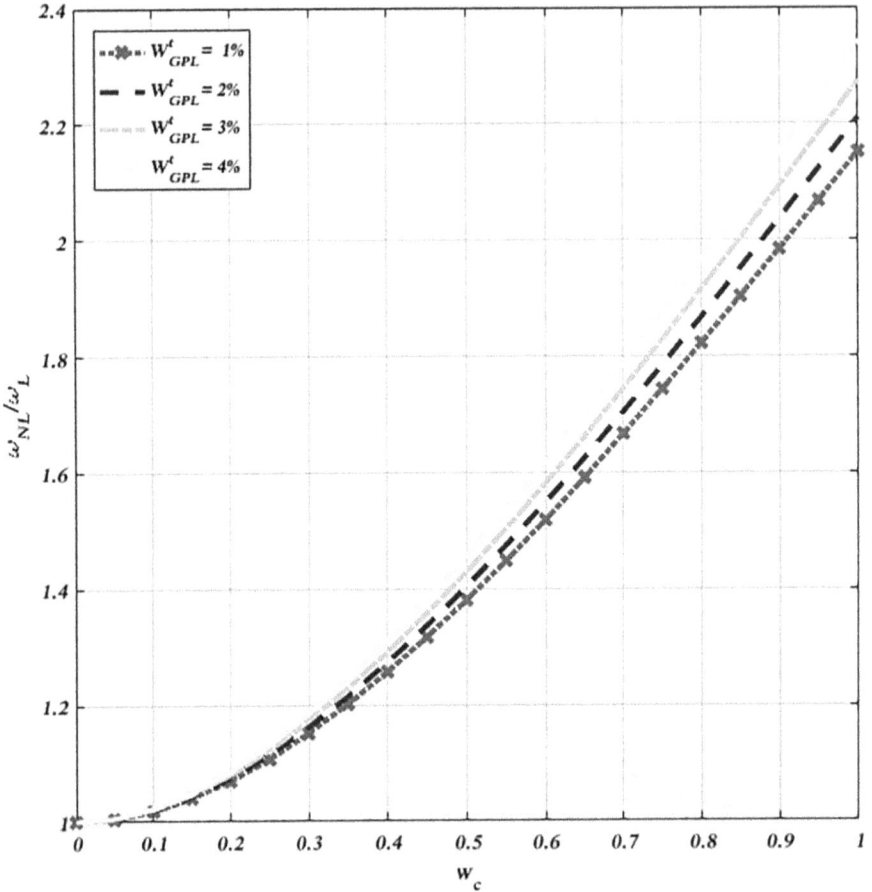

FIGURE 3.51 Effect of percent of weight fraction of graphene nanoplatelets on nonlinear frequency ratio of smart multi-scale hybrid FG-graphene Origami auxetic sandwich doubly curved panel.

The impact of the carbon fiber volume percentage (V_F) on the nonlinear frequency ratio of the smart sandwich doubly curved panel is clearly seen in Figure 3.52. The study presented findings on the relationship between the volume percent of carbon fibers and nonlinear vibration amplitude. The investigation explored a range of amplitudes from 0 to 1, while considering volume fractions of carbon fibers ranging from 0 to 0.5. The relationship between the volume percentage of carbon fibers and the nonlinear frequency ratio of the smart sandwich doubly curved panel exhibits a declining trend. As seen in this figure, the intensity of evolution in the nonlinear frequency ratio at $V_F = 0$ surpasses all other values presented in the figure. The minimum value of this intensity is associated with = 0.5. The circumstances are governed by intricate and precise layers, which are influenced by geometric nonlinear effects. The resonance dynamics of the smart structure exhibit a phenomenon of hardening behavior.

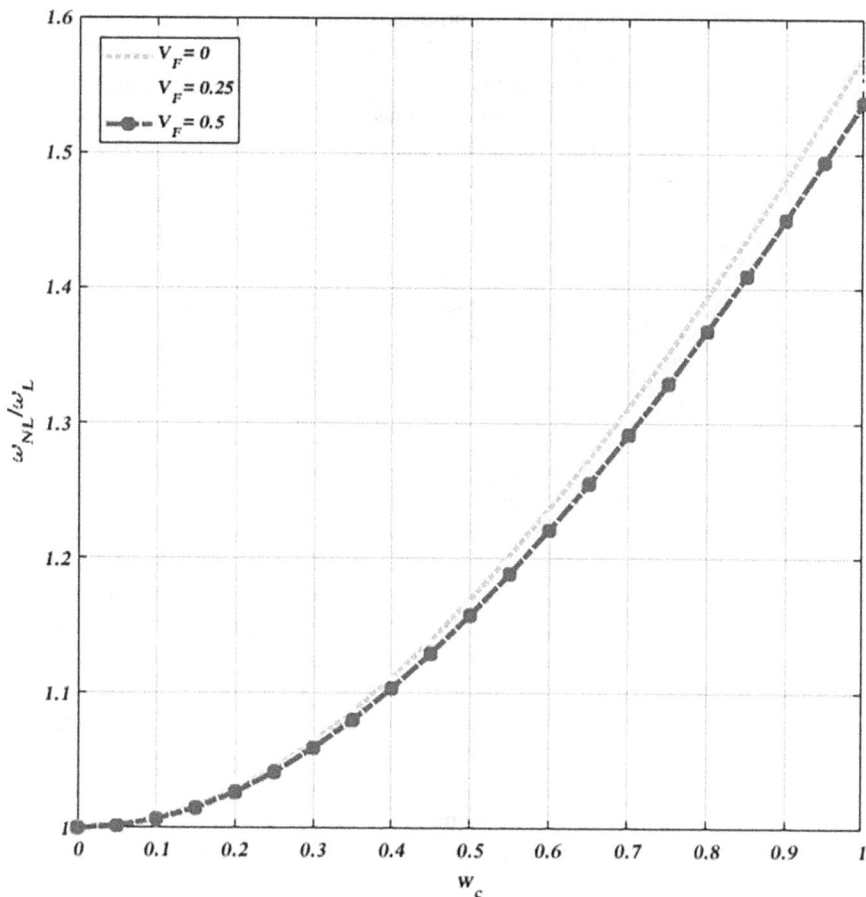

FIGURE 3.52 Effect of volume fraction of carbon fibers on nonlinear frequency ratio of smart multi-scale hybrid FG-graphene origami auxetic sandwich doubly curved panel.

3.5 SUMMARY

This chapter presents an investigation on the vibration and dynamic features of auxetic structures. A variety of auxetic structures, including non-circular nanorods, plates, shells, and sandwich doubly curved panels, are assessed in this study. The primary objective of this chapter was to examine the impact of auxeticity and other factors on the nonlinear vibration characteristics responses of metamaterial constructions. The investigation of vibration and dynamic features of composite structures with origami auxetic cores allowed by graphene is also explored. The nonlinear equations regulating the sandwich structure and the corresponding boundary conditions are obtained by the application of the Hamiltonian principle. The findings of this chapter have great significance in the realm of sensitive structure design and have the potential for application across diverse materials and varying cross-sectional configurations. The present study focuses on conducting vibrational assessments of circular composite cylindrical shells. These shells are composed of an auxetic core positioned as a middle layer, and are exposed to both axial compressive load and uniform external pressure. The objective is to assess the behavior of these shells when supported by elastic foundations. The governing equations for the dynamic analysis of shells, considering both linear and nonlinear aspects, are derived using Galerkin's method, Volmir's assumption, and Runge-Kutta's fourth-order method. These equations are based on Donnell's improved shell theory, which does not consider the shallowness of cylindrical shells but does account for kinematic nonlinearity. This study provides a comprehensive analysis of the many factors that affect the vibration characteristics of a circular composite cylindrical shell. These factors include the geometrical parameters of the shell and auxetic core, the level of pre-loaded axial compression, the external excitation force, and the distinct modes of vibration. The effect of each parameter is examined in depth.

Furthermore, this chapter explores the nonlinear vibration characteristics of a smart multi-scale hybrid functionally graded (FG) graphene origami auxetic sandwich doubly curved panel. The nonlinear model is achieved inside the theoretical framework by using the third-order shear deformation theory and Von Karman's geometrical nonlinearity factors. By using the Hamiltonian principle as a foundational concept, one may derive the equation of motion. Subsequently, the nonlinear equations are examined using the Homotopy Perturbation Method technique. This section considers multiple parameters, including the increase in the radius of curvature, the thickness of the smart hybrid nanocomposite, the thickness of the auxetic metamaterial, the weight fraction of GOri nanofillers, the weight fraction of graphene nanoplatelets, the degree of GO folding angle, and the volume fraction of carbon fibers. These parameters have been shown to have a significant impact on the frequency-amplitude curve that was acquired.

The salient points that may be seen are as follows, as shown by this comprehensive investigation [110–115]:

- The impact of nonlocal factors on the fluctuation of natural frequencies of nanorods with triangular and elliptical cross-sections was investigated, revealing a diminishing effect. This influence was visually represented using several diagrams.
- It has been shown that, alongside the nonlocal parameter, the inclined angle of the auxetic honeycomb structure has a diminishing influence on the natural frequency in both types of nanorods.
- Additionally, an investigation was conducted to examine the impact of the edge length of a nanorod triangle. The findings revealed that selecting triangles with lower edge lengths leads to higher natural frequencies.
- Furthermore, the findings indicate that the aspect ratio of an elliptical nanorod has a substantial influence on the fluctuation of its natural frequency. Notably, the maximum natural frequency is seen when the aspect ratio is equal to 1, at which point the elliptical shape of the nanorod transforms into a circular shape.
- In the case of low aspect ratios, an increase in the ratio leads to an increase in the natural frequency. However, when considering a vertical ellipse

(where the minor axis is greater than the major axis), the natural frequency decreases as the aspect ratio increases. This reduction in natural frequency can be attributed to a decrease in the moment of inertia, similar to what is observed in a horizontal ellipse (where the major axis is greater than the minor axis).

- It has been shown that, in all instances, nanorods exhibiting C-C boundary conditions exhibit greater natural frequencies when compared to nanorods having C-F boundary conditions.

- The frequency ratio (ω_{nl}/w_l) for auxetic sandwich plates may be decreased by increasing the thickness of the magneto-electrical layer (h_{SC}), the thickness of the auxetic core, the geometrical parameter (e/l), the external voltage, and the vibration amplitude at a given value of $\theta = -55$.

- The frequency ratio of auxetic sandwich plates is determined by the installation of the magnetic potential and power index at a certain angle of -55 degrees.

- The ratio of (ω_{nl}/w_l) for auxetic sandwich plates has increased due to the increase in the thicknesses of the auxetic core) and the vibration amplitude for $\theta = -15$.

- Increasing the weight ratio of graphene platelets and their corresponding resonance amplitude results in an elevation of the nonlinear frequency ratio of auxetic sandwich plates.

- The enhancement of the frequency ratio in auxetic sandwich plates may be achieved by increasing the porosity and the thickness of the porous SPHNC layer.

- The frequency ratio in CCCC and SSSS boundary conditions is reduced when the FG-GOA power index and volume percentage of fibers in auxetic sandwich plates are increased. The nonlinear frequency ratio in CCCC is seen to be much lower compared to SSSS B.C. in auxetic sandwich plates.

- Among the three kinds of plates studied for this investigation, it is seen that the solitary magnetostrictive plate exhibits the highest values of dimensionless natural frequency.

- The dimensionless natural frequencies decrease as the inclination angle increases.

- The dimensionless natural frequency exhibits an upward trend when the length of the rib is increased.

- The behavior of the plate is influenced by the presence of magnetostrictive material, which exerts a regulating effect.

- The addition of a graphene-origami auxetic layer onto a magnetostrictive plate results in an increase in the dimensionless natural frequency.

- An increase in temperature results in a decrease in the natural frequencies.

- Increasing the weight percentage of graphene leads to an increase in the natural frequencies until it reaches around 1 weight percent of graphene, after which there is a noticeable decrease.

- The geometric dimensions of auxetics have a crucial role in determining the behavior of composite cylindrical shells under various situations.

- In general, the manipulation of the auxetic rib length ratio has the potential to effectively regulate the vibrations of the cylindrical shell, resulting in a reduction in the amplitude of system fluctuation.
- In addition, it is worth noting that larger levels of auxetic inclination angles are likewise correlated with increased frequencies of nonlinearity relative to linearity.
- The increase in core thickness leads to a drop in the ratio of nonlinear to linear frequencies due to the distinctive characteristics of auxetic structure, such as its notable hardness. Additionally, auxetic structures contribute to a reduction in weight of the overall structure.
- The findings of this study demonstrate that auxetic materials have the potential to serve as a core component in delicate structures, hence enhancing overall efficiency.
- The auxetic core's exceptional lightweight and excellent strength-to-weight ratio make it suitable for designing various materials and sections.
- The fundamental frequency ratios will increase as the thickness of the smart hybrid nanocomposite and the auxetic metamaterial increase.
- The frequency ratio transition in a smart sandwich doubly curved panel is influenced by the rise in the GO folding degree parameter angle and the reduction in the volume fraction of carbon fibers.
- The weight fraction of graphene oxide-reduced nanofillers and its influence on the nonlinear frequency are quantified by determining the percentage of weight fraction of graphene nanoplatelets that exhibit a favorable effect.

REFERENCES

1. Seyfi, A., Teimouri, A. & Ebrahimi, F., 2021. Scale-dependent torsional vibration response of non-circular nanoscale auxetic rods. *Waves in Random and Complex Media*, pp. 1–17.
2. Ebrahimi, F. & Dabbagh, A., 2020. *Mechanics of Nanocomposites: Homogenization and Analysis.* 1st ed. Boca Raton, FL: CRC Press.
3. Ebrahimi, F. & Dabbagh, A., 2022. Vibration analysis of multi-scale hybrid nanocomposite shells by considering nanofillers' aggregation. *Waves in Random and Complex Media*, 32(3), pp. 1060–1078.
4. Ebrahimi, F. & Dabbagh, A., 2022. *Mechanics of Multiscale Hybrid Nanocomposites.* Duxford: Elsevier. ISBN: 9780128196144
5. Ebrahimi, F., Nopour, R. & Dabbagh, A., 2022. Effect of viscoelastic properties of polymer and wavy shape of the CNTs on the vibrational behaviors of CNT/glass fiber/polymer plates. *Engineering with Computers*, 38(5), pp. 4113–4126.
6. Ebrahimi, F., Nopour, R. & Dabbagh, A., 2022. Effects of polymer's viscoelastic properties and curved shape of the CNTs on the dynamic response of hybrid nanocomposite beams. *Waves in Random and Complex Media*, 2022, pp. 1–18.
7. Nopour, R., Ebrahimi, F., Dabbagh, A. & Aghdam, M. M., 2023. Nonlinear forced vibrations of three-phase nanocomposite shells considering matrix rheological behavior and nano-fiber waviness. *Engineering with Computers*, 39(1), pp. 557–574.
8. Ebrahimi, F., Nouraei, M., Dabbagh, A. & Rabczuk, T., 2019. Thermal buckling analysis of embedded graphene-oxide powder-reinforced nanocomposite plates. *Advances in Nano Research*, 7(5), pp. 293–310.

9. Ebrahimi, F., Dabbagh, A., Rastgoo, A. & Rabczuk, T., 2020. Agglomeration effects on static stability analysis of multi-scale hybrid nanocomposite plates. *Computers, Materials & Continua*, 63(1), pp. 41–64.

10. Dabbagh, A., Rastgoo, A. & Ebrahimi, F., 2021. Static stability analysis of agglomerated multi-scale hybrid nanocomposites via a refined theory. *Engineering with Computers*, 37(3), pp. 2225–2244.

11. Ebrahimi, F. & Dabbagh, A., 2021. An analytical solution for static stability of multi-scale hybrid nanocomposite plates. *Engineering with Computers*, 37(1), pp. 545–559.

12. Dabbagh, A. & Rastgoo, A. & Ebrahimi, F., 2022. Post-buckling analysis of imperfect multi-scale hybrid nanocomposite beams rested on a nonlinear stiff substrate. *Engineering with Computers*, 38(1), pp. 301–314.

13. Ebrahimi, F., Dabbagh, A. & Rastgoo, A., 2023. Static stability analysis of multi-scale hybrid agglomerated nanocomposite shells. *Mechanics Based Design of Structures and Machines*, 51(1), pp. 501–517.

14. Ebrahimi, F., Dabbagh, A. & Taheri, M., 2021. Vibration analysis of porous metal foam plates rested on viscoelastic substrate. *Engineering with Computers*, 37(4), pp. 3727–3739.

15. Ebrahimi, F. & Dabbagh, A., 2019. *Wave Propagation Analysis of Smart Nanostructures*. Boca Raton, FL: CRC Press.

16. Dabbagh, A. & Ebrahimi, F., 2021. Postbuckling analysis of meta-nanocomposite beams by considering the CNTs' agglomeration. *The European Physical Journal Plus*, 136(11), pp. 1168.

17. Eringen, A. C., 1972. Linear theory of nonlocal elasticity and dispersion of plane waves. *International Journal of Engineering Science*, 10(5), pp. 425–435.

18. Ebrahimi, F. & Barati, M., 2016e. Static stability analysis of smart magneto-electro-elastic heterogeneous nanoplates embedded in an elastic medium based on a four-variable refined plate theory. *Smart Materials and Structures*, 25(10).

19. Ebrahimi, F. & Barati, M. R., 2016a. Vibration analysis of nonlocal beams made of functionally graded material in thermal environment. *The European Physical Journal Plus*, 131(8), p. 279.

20. Ebrahimi, F. & Barati, M. R., 2016b. A unified formulation for dynamic analysis of nonlocal heterogeneous nanobeams in hygro-thermal environment. *Applied Physics A*, 122(9), p. 792.

21. Ebrahimi, F. & Barati, M. R., 2016c. A nonlocal higher-order refined magneto-electro-viscoelastic beam model for dynamic analysis of smart nanostructures. *International Journal of Engineering Science*, 107, pp. 183–196.

22. Ebrahimi, F. & Barati, M. R., 2016d. Hygrothermal buckling analysis of magnetically actuated embedded higher order functionally graded nanoscale beams considering the neutral surface position. *Journal of Thermal Stresses*, 39(10), pp. 1210–1229.

23. Ebrahimi, F. & Barati, M. R., 2016f. Wave propagation analysis of quasi-3D FG nanobeams in thermal environment based on nonlocal strain gradient theory. *Applied Physics A*, 122(9), p. 843.

24. Ebrahimi, F. & Barati, M. R., 2016g. A nonlocal higher-order shear deformation beam theory for vibration analysis of size-dependent functionally graded nanobeams. *Arabian Journal for Science and Engineering*, 41(5), pp. 1679–1690.

25. Ebrahimi, F. & Barati, M. R., 2016h. Thermal buckling analysis of size-dependent FG nanobeams based on the third-order shear deformation beam theory. *Acta Mechanica Solida Sinica*, 29(5), pp. 547–554.

26. Ebrahimi, F. & Barati, M. R., 2016i. Magneto-electro-elastic buckling analysis of nonlocal curved nanobeams. *The European Physical Journal Plus*, 131(9), p. 346.

27. Ebrahimi, F. & Barati, M. R., 2016j. Analytical solution for nonlocal buckling characteristics of higher-order inhomogeneous nanosize beams embedded in elastic medium. *Advances in Nano Research*, 4(3), pp. 229–249.

28. Ebrahimi, F. & Barati, M. R., 2016k. Electromechanical buckling behavior of smart piezoelectrically actuated higher-order size-dependent graded nanoscale beams in thermal environment. *International Journal of Smart and Nano Materials*, 7(2), pp. 69–90.

29. Ebrahimi, F. & Barati, M. R., 2016l. Nonlocal thermal buckling analysis of embedded magneto-electro-thermo-elastic nonhomogeneous nanoplates. *Iranian Journal of Science and Technology, Transactions of Mechanical Engineering*, 40(4), pp. 243–264.

30. Ebrahimi, F. & Barati, M. R., 2016m. Size-dependent thermal stability analysis of graded piezomagnetic nanoplates on elastic medium subjected to various thermal environments. *Applied Physics A*, 122(10), p. 910.

31. Ebrahimi, F. & Barati, M. R., 2017a. Hygrothermal effects on vibration characteristics of viscoelastic FG nanobeams based on nonlocal strain gradient theory. *Composite Structures*, 159, pp. 433–444.

32. Ebrahimi, F. & Barati, M. R., 2017b. A nonlocal strain gradient refined beam model for buckling analysis of size-dependent shear-deformable curved FG nanobeams. *Composite Structures*, 159, pp. 174–182.

33. Ebrahimi, F. & Barati, M. R., 2017c. Vibration analysis of nonlocal strain gradient embedded single-layer graphene sheets under nonuniform in-plane loads. *Journal of Vibration and Control*, 24(20), pp. 4751–4763.

34. Ebrahimi, F. & Barati, M. R., 2017d. A nonlocal strain gradient mass sensor based on vibrating hygro-thermally affected graphene nanosheets. *Iranian Journal of Science and Technology, Transactions of Mechanical Engineering*, pp. 1–16.

35. Ebrahimi, F. & Barati, M. R., 2017e. Damping vibration analysis of smart piezoelectric polymeric nanoplates on viscoelastic substrate based on nonlocal strain gradient theory. *Smart Materials and Structures*, 26(6), p. 065018.

36. Ebrahimi, F. & Barati, M. R., 2017f. Vibration analysis of viscoelastic inhomogeneous nanobeams resting on a viscoelastic foundation based on nonlocal strain gradient theory incorporating surface and thermal effects. *Acta Mechanica*, 328(3), pp. 1197–1210.

37. Ebrahimi, F. & Barati, M. R., 2017g. Flexural wave propagation analysis of embedded S-FGM nanobeams under longitudinal magnetic field based on nonlocal strain gradient theory. *Arabian Journal for Science and Engineering*, 42(5), pp. 1715–1726.

38. Ebrahimi, F. & Barati, M. R., 2017h. Vibration analysis of viscoelastic inhomogeneous nanobeams incorporating surface and thermal effects. *Applied Physics A*, 123(1), p. 5.

39. Ebrahimi, F. & Barati, M. R., 2017i. Buckling analysis of smart size-dependent higher order magneto-electro-thermo-elastic functionally graded nanosize beams. *Journal of Mechanics*, 33(1), pp. 23–33.

40. Ebrahimi, F. & Barati, M. R., 2017j. Small-scale effects on hygro-thermo-mechanical vibration of temperature-dependent nonhomogeneous nanoscale beams. *Mechanics of Advanced Materials and Structures*, 24(11), pp. 924–936.

41. Ebrahimi, F. & Barati, M. R., 2017k. Through-the-length temperature distribution effects on thermal vibration analysis of nonlocal strain-gradient axially graded nanobeams subjected to nonuniform magnetic field. *Journal of Thermal Stresses*, 40(5), pp. 548–563.

42. Ebrahimi, F. & Barati, M. R., 2017l. Size-dependent vibration analysis of viscoelastic nanocrystalline silicon nanobeams with porosities based on a higher order refined beam theory. *Composite Structures*, 166, pp. 256–267.

43. Ebrahimi, F. & Barati, M. R., 2017m. Dynamic modeling of preloaded size-dependent nano-crystalline nano-structures. *Applied Mathematics and Mechanics*, 38(12), pp. 1753–1772.

44. Ebrahimi, F. & Barati, M. R., 2017n. Static stability analysis of embedded flexoelectric nanoplates considering surface effects. *Applied Physics A*, 123(10), p. 123.

45. Ebrahimi, F. & Barati, M. R., 2017o. Vibration analysis of embedded size dependent FG nanobeams based on third-order shear deformation beam theory. *Structural Engineering And Mechanics*, 61(6), pp. 721–736.

46. Ebrahimi, F. & Barati, M. R., 2017p. Vibration analysis of embedded biaxially loaded magneto-electrically actuated inhomogeneous nanoscale plates. *Journal of Vibration and Control*, p. 1077546317708105.

47. Ebrahimi, F. & Barati, M. R., 2017q. A third-order parabolic shear deformation beam theory for nonlocal vibration analysis of magneto-electro-elastic nanobeams embedded in two-parameter elastic foundation. *Advances in Nano Research*, 5(4), pp. 313–336.

48. Ebrahimi, F. & Barati, M. R., 2017r. Dynamic modeling of magneto-electrically actuated compositionally graded nanosize plates lying on elastic foundation. *Arabian Journal for Science and Engineering*, 42(5), pp. 1977–1997.

49. Ebrahimi, F. & Barati, M. R., 2017. Size-dependent dynamic modeling of inhomogeneous curved nanobeams embedded in elastic medium based on nonlocal strain gradient theory. *Proceedings of the Institution of Mechanical Engineers, Part C: Journal of Mechanical Engineering Science*, 231(23), pp. 4457–4469.

50. Ebrahimi, F. & Barati, M. R., 2018. A nonlocal strain gradient refined plate model for thermal vibration analysis of embedded graphene sheets via DQM. *Structural Engineering and Mechanics*, 66(6), pp. 693–701.

51. Ebrahimi, F. & Barati, M. R., 2018a. Vibration analysis of smart piezoelectrically actuated nanobeams subjected to magneto-electrical field in thermal environment. *Journal of Vibration and Control*, 24(3), pp. 549–564.

52. Ebrahimi, F. & Barati, M. R., 2018b. Damping vibration analysis of graphene sheets on viscoelastic medium incorporating hygro-thermal effects employing nonlocal strain gradient theory. *Composite Structures*, 185, pp. 241–253.

53. Ebrahimi, F. & Barati, M. R., 2018c. Vibration analysis of graphene sheets resting on the orthotropic elastic medium subjected to hygro-thermal and in-plane magnetic fields based on the nonlocal strain gradient theory. *Proceedings of the Institution of Mechanical Engineers, Part C: Journal of Mechanical Engineering Science*, 232(13), pp. 2469–2481.

54. Ebrahimi, F. & Barati, M. R., 2018d. Buckling analysis of nonlocal strain gradient axially functionally graded nanobeams resting on variable elastic medium. *Proceedings of the Institution of Mechanical Engineers, Part C: Journal of Mechanical Engineering Science*, 232(11), pp. 2067–2078.

55. Ebrahimi, F. & Barati, M. R., 2018e. Effect of three-parameter viscoelastic medium on vibration behavior of temperature-dependent non-homogeneous viscoelastic nanobeams in a hygro-thermal environment. *Mechanics of Advanced Materials and Structures*, 25(5), pp. 361–374.

56. Ebrahimi, F. & Barati, M. R., 2018f. Vibration analysis of piezoelectrically actuated curved nanosize FG beams via a nonlocal strain-electric field gradient theory. *Mechanics of Advanced Materials and Structures*, 25(4), pp. 350–359.

57. Ebrahimi, F. & Barati, M. R., 2018g. Static stability analysis of double-layer graphene sheet system in hygro-thermal environment. *Microsystem Technologies*, 24(9), pp. 3713–3727.

58. Ebrahimi, F. & Barati, M. R., 2018h. Damping vibration behavior of visco-elastically coupled double-layered graphene sheets based on nonlocal strain gradient theory. *Microsystem Technologies*, 24(3), pp. 1643–1658.

59. Ebrahimi, F. & Barati, M. R., 2018i. Vibration analysis of biaxially compressed double-layered graphene sheets based on nonlocal strain gradient theory. *Mechanics of Advanced Materials and Structures*, 26(10), pp. 854–865.

60. Ebrahimi, F. & Barati, M. R., 2018j. Vibration analysis of size-dependent flexoelectric nanoplates incorporating surface and thermal effects. *Mechanics of Advanced Materials and Structures*, 25(7), pp. 611–621.

61. Ebrahimi, F. & Barati, M. R., 2018l. Vibration analysis of graphene sheets resting on the orthotropic elastic medium subjected to hygro-thermal and in-plane magnetic fields based on the nonlocal strain gradient theory. *Proceedings of the Institution of Mechanical Engineers, Part C: Journal of Mechanical Engineering Science*, 232(13), pp. 2469–2481.

62. Ebrahimi, F. & Barati, M. R., 2018m. Nonlocal and surface effects on vibration behavior of axially loaded flexoelectric nanobeams subjected to in-plane magnetic field. *Arabian Journal for Science and Engineering*, 43(3), pp. 1423–1433.

63. Ebrahimi, F. & Barati, M. R., 2018n. Longitudinal varying elastic foundation effects on vibration behavior of axially graded nanobeams via nonlocal strain gradient elasticity theory. *Mechanics of Advanced Materials and Structures*, 25(11), pp. 953–963.

64. Ebrahimi, F., Barati, M. R. & Dabbagh, A., 2016. A nonlocal strain gradient theory for wave propagation analysis in temperature-dependent inhomogeneous nanoplates. *International Journal of Engineering Science*, 107, pp. 169–182.

65. Ebrahimi, F., Barati, M. R. & Dabbagh, A., 2016d. Wave dispersion characteristics of axially loaded magneto-electro-elastic nanobeams. *Applied Physics A*, 122(11), p. 949.

66. Ebrahimi, F., Barati, M. R. & Haghi, P., 2017. Thermal effects on wave propagation characteristics of rotating strain gradient temperature-dependent functionally graded nanoscale beams. *Journal of Thermal Stresses*, 40(5), pp. 535–547.

67. Ebrahimi, F., Barati, M. R. & Haghi, P., 2016c. Nonlocal thermo-elastic wave propagation in temperature-dependent embedded small-scaled nonhomogeneous beams. *The European Physical Journal Plus*, 131(11), p. 383.

68. Ebrahimi, F., Barati, M. R. & Haghi, P., 2017a. Wave propagation analysis of size-dependent rotating inhomogeneous nanobeams based on nonlocal elasticity theory. *Journal of Vibration and Control*, p. 1077546317711537.

69. Ebrahimi, F. & Barati, M. R., 2016. Temperature distribution effects on buckling behavior of smart heterogeneous nanosize plates based on nonlocal four-variable refined plate theory. *International Journal of Smart and Nano Materials*, 7(3), pp. 119–143.

70. Ebrahimi, F. & Barati, M. R., 2017. Buckling analysis of piezoelectrically actuated smart nanoscale plates subjected to magnetic field. *Journal of Intelligent Material Systems and Structures*, 28(11), pp. 1472–1490.

71. Ebrahimi, F. & Barati, M., 2017d. Surface effects on the vibration behavior of flexoelectric nanobeams based on nonlocal elasticity theory. *The European Physical Journal Plus*, 132(1), p. 19.

72. Ebrahimi, F. & Barati, M. R., 2016e. Dynamic modeling of a thermo-piezo-electrically actuated nanosize beam subjected to a magnetic field. *Applied Physics A*, 122(4), p. 451.

73. Ebrahimi, F. & Barati, M. R., 2016f. Magnetic field effects on buckling behavior of smart size-dependent graded nanoscale beams. *The European Physical Journal Plus*, 131(7), p. 238.

74. Ebrahimi, F. & Barati, M. R., 2018d. Magnetic field effects on buckling characteristics of smart flexoelectrically actuated piezoelectric nanobeams based on nonlocal and surface elasticity theories. *Microsystem Technologies*, pp. 1–11.

75. Ebrahimi, F. & Dabbagh, A., 2017a. Nonlocal strain gradient based wave dispersion behavior of smart rotating magneto-electro-elastic nanoplates. *Materials Research Express*, 4(2), p. 025003.

76. Ebrahimi, F. & Dabbagh, A., 2017b. Wave propagation analysis of smart rotating porous heterogeneous piezo-electric nanobeams. *The European Physical Journal Plus*, 132(4), p. 153.

77. Ebrahimi, F. & Dabbagh, A., 2017c. On flexural wave propagation responses of smart FG magneto-electro-elastic nanoplates via nonlocal strain gradient theory. *Composite Structures*, 162, pp. 281–293.

78. Ebrahimi, F. & Dabbagh, A., 2017d. Wave propagation analysis of embedded nanoplates based on a nonlocal strain gradient-based surface piezoelectricity theory. *The European Physical Journal Plus*, 132(11), p. 449.

79. Ebrahimi, F. & Dabbagh, A., 2018a. Viscoelastic wave propagation analysis of axially motivated double-layered graphene sheets via nonlocal strain gradient theory. *Waves in Random and Complex Media*, pp. 1–20.

80. Ebrahimi, F. & Dabbagh, A., 2018b. Wave dispersion characteristics of nonlocal strain gradient double-layered graphene sheets in hygro-thermal environments. *Structural Engineering and Mechanics*, 65(6), pp. 645–656.

81. Ebrahimi, F. & Dabbagh, A., 2018c. Thermo-magnetic field effects on the wave propagation behavior of smart magnetostrictive sandwich nanoplates. *The European Physical Journal Plus*, 133(3), p. 97.

82. Ebrahimi, F. & Dabbagh, A., 2018. Effect of humid-thermal environment on wave dispersion characteristics of single-layered graphene sheets. *Applied Physics A*, 124(4), p. 301.

83. Ebrahimi, F., Dabbagh, A. & Barati, M. R., 2016a. Wave propagation analysis of a size-dependent magneto-electro-elastic heterogeneous nanoplate. *The European Physical Journal Plus*, 131(12), p. 433.

84. Ebrahimi, F. & Dabbagh, A. 2019. Thermo-mechanical wave dispersion analysis of nonlocal strain gradient single-layered graphene sheet rested on elastic medium. *Microsystem Technologies*, 25, pp. 587–597.

85. Ebrahimi, F. & Dashti, S., 2015. Free vibration analysis of a rotating non-uniform functionally graded beam. *Steel and Composite Structures*, 19(5), pp. 1279–1298.

86. Ebrahimi, F., Dabbagh, A., Rastgoo, A. & Rabczuk, T., 2020. Agglomeration effects on static stability analysis of multi-scale hybrid nanocomposite plates. *Computers, Materials & Continua*, 63(1), pp. 41–64.

87. Ebrahimi, F. & Dabbagh, A., 2020. A brief review on the influences of nanotubes' entanglement and waviness on the mechanical behaviors of CNTR polymer nanocomposites. *Journal of Computational Applied Mechanics*, 51(1), pp. 247–252.

88. Ebrahimi, F. & Dabbagh, A., 2019. A comprehensive review on modeling of nanocomposite materials and structures. *Journal of Computational Applied Mechanics*, 50(1), pp. 197–209.

89. Ebrahimi, F., Ehyaei, J. & Babaei, R., 2016g. Thermal buckling of FGM nanoplates subjected to linear and nonlinear varying loads on Pasternak foundation. *Advances in Materials Research-An International Journal*, 5(4), pp. 245–261.

90. Ebrahimi, F. & Fardshad, R. E., 2018. Analytical solution for scale-dependent static stability analysis of temperature-dependent nanobeams subjected to uniform temperature distributions. *Wind and Structures*, 26(4), pp. 205–214.

91. Ebrahimi, F., Ghadiri, M., Salari, E., Hoseini, S. A. H. & Shaghaghi, G. R., 2015. Application of the differential transformation method for nonlocal vibration analysis of functionally graded nanobeams. *Journal of Mechanical Science and Technology*, 29(3), pp. 1207–1215.

92. Ebrahimi, F. & Haghi, P., 2018. Wave dispersion analysis of rotating heterogeneous nanobeams in thermal environment. *Advances in Nano Research*, 6(1), pp. 21–37.

93. Ebrahimi, F. & Hosseini, S., 2016a. Nonlinear electroelastic vibration analysis of NEMS consisting of double-viscoelastic nanoplates. *Applied Physics A*, 122(10), p. 922.

94. Ebrahimi, F. & Hosseini, S., 2016. Thermal effects on nonlinear vibration behavior of viscoelastic nanosize plates. *Journal of Thermal Stresses*, 39(5), pp. 606–625.

95. Ebrahimi, F. & Hosseini, S., 2017. Surface effects on nonlinear dynamics of NEMS consisting of double-layered viscoelastic nanoplates. *The European Physical Journal Plus*, 132(4), p. 172.

96. Ebrahimi, F. & Shafiei, N., 2017. Influence of initial shear stress on the vibration behavior of single-layered graphene sheets embedded in an elastic medium based on Reddy's higher-order shear deformation plate theory. *Mechanics of Advanced Materials and Structures*, 24(9), pp. 761–772.

97. Ebrahimi, F., Shaghaghi, G. R. & Boreiry, M., 2016e. An investigation into the influence of thermal loading and surface effects on mechanical characteristics of nanotubes. *Structural Engineering and Mechanics*, 57(1), pp. 179–200.

98. Ebrahimi, F. & Haghi, P., 2017. Wave propagation analysis of rotating thermoelastically-actuated nanobeams based on nonlocal strain gradient theory. *Acta Mechanica Solida Sinica*, 30(6), pp. 647–657.

99. Ebrahimi, F., Dehghan, M. & Seyfi, A., 2019. Eringen's nonlocal elasticity theory for wave propagation analysis of magneto-electro-elastic nanotubes. *Advances in Nano Research*, 7(1), p. 1.

100. Ebrahimi, F., Seyfi, A. & Dabbagh, A., 2019. Dispersion of waves in FG porous nanoscale plates based on NSGT in thermal environment. *Advances in Nano Research*, 7(5), pp. 325–335.

101. Ebrahimi, F., Seyfi, A. & Dabbagh, A., 2019. A novel porosity-dependent homogenization procedure for wave dispersion in nonlocal strain gradient inhomogeneous nanobeams. *The European Physical Journal Plus*, 134(5), p. 226.

102. Eringen, A. C., 1984. Plane waves in nonlocal micropolar elasticity. *International Journal of Engineering Science*, 22(8–10), pp. 1113–1121.

103. Ebrahimi, F., Seyfi, A., Dabbagh, A. & Tornabene, F., 2019. Wave dispersion characteristics of porous graphene platelet-reinforced composite shells. *Structural Engineering and Mechanics, An Int'l Journal*, 71(1), pp. 99–107.

104. Qaderi, S., Ebrahimi, F. & Mahesh, V., 2019. Free vibration analysis of graphene platelets—reinforced composites plates in thermal environment based on higher-order shear deformation plate theory. *International Journal of Aeronautical and Space Sciences*, 20, pp. 902–912.

105. Qaderi, S. & Ebrahimi, F., 2022. Vibration analysis of polymer composite plates reinforced with graphene platelets resting on two-parameter viscoelastic foundation. *Engineering with Computers*, pp. 1–17.

106. Qaderi, S., Ebrahimi, F. & Vinyas, M., 2019. Dynamic analysis of multi-layered composite beams reinforced with graphene platelets resting on two-parameter viscoelastic foundation. *The European Physical Journal Plus*, 134, pp. 1–11.

107. Ebrahimi, F. & Qaderi, S., 2019. Stability analysis of embedded graphene platelets reinforced composite plates in thermal environment. *The European Physical Journal Plus*, 134(7), p. 349.
108. Ebrahimi, F. & Dabbagh, A., 2018. Wave dispersion characteristics of embedded graphene platelets-reinforced composite microplates. *The European Physical Journal Plus*, 133, pp. 1–13.
109. Ebrahimi, F., Seyfi, A. & Teimouri, A., 2021. Torsional vibration analysis of scale-dependent non-circular graphene oxide powder-strengthened nanocomposite nanorods. *Engineering with Computers*, pp. 1–12.
110. Mahinzare, M., Rastgoo, A. & Ebrahimi, F., 2023. On nonlinear vibration of piezo-electrically multi-scale hybrid nanocomposite sandwich plate including an auxetic core based on HSDT. *International Journal of Structural Stability and Dynamics*. https://doi.org/10.1142/S021945542450069X
111. Mahinzare, M., Rastgoo, A. & Ebrahimi, F., 2023. Magnetic field effects on wave dispersion of piezo-electrically actuated auxetic sandwich shell via GPL reinforcement.
112. Ebrahimi, F., Nopour, R. & Dabbagh, A., 2021. Smart laminates with an auxetic ply rested on visco-Pasternak medium: Active control of the system's oscillation. *Engineering with Computers*, 1–11.
113. Ebrahimi, F., Dabbagh, A. & Rabczuk, T., 2021. On wave dispersion characteristics of magnetostrictive sandwich nanoplates in thermal environments. *European Journal of Mechanics-A/Solids*, 85, p. 104130.
114. Ebrahimi, F., et al., 2019. Hygro-thermal effects on wave dispersion responses of magnetostrictive sandwich nanoplates. *Advances in Nano Research*, 7(3), p. 157.
115. Ebrahimi, F. & Ahari, M. F., 2021. Magnetostriction-assisted active control of the multi-layered nanoplates: Effect of the porous functionally graded facesheets on the system's behavior. *Engineering with Computers*, pp. 1–15.

APPENDICES

APPENDIX 3.1

$$L_{11}(u) = A_{11}\frac{\partial^2 u}{\partial x^2} + A_{66}\frac{\partial^2 u}{\partial t^2}$$

$$L_{12}(v) = \left(A_{12} + A_{66} - \frac{B_{12}+B_{66}}{R}\right)\frac{\partial^2 v}{\partial x\partial\theta}$$

$$L_{13}(w) = -\frac{A_{12}}{R}\frac{\partial w}{\partial x} - B_{11}\frac{\partial^3 w}{\partial x^3} - (B_{12}+2B_{66})\frac{\partial^3 w}{\partial x\partial\theta^2}$$

$$P_1(w) = A_{11}\frac{\partial w}{\partial x}\frac{\partial^2 w}{\partial x^2} + (A_{12}+A_{66})\frac{\partial w}{\partial\theta}\frac{\partial^2 w}{\partial x\partial\theta} + A_{66}\frac{\partial w}{\partial x}\frac{\partial^2 w}{\partial\theta^2}$$

$$L_{21}(u) = \left(A_{12}+A_{66} - \frac{B_{12}+B_{66}}{R}\right)\frac{\partial^2 u}{\partial x\partial\theta}$$

$$L_{22}(v) = \left(A_{66} - \frac{2B_{66}}{R} + \frac{D_{66}}{R^2}\right)\frac{\partial^2 v}{\partial x^2} + \left(A_{11} - \frac{2B_{11}}{R} + \frac{D_{11}}{R^2}\right)\frac{\partial^2 v}{\partial\theta^2}$$

$$L_{23}(w) = -\left(\frac{A_{11}}{R} - \frac{B_{11}}{R^2}\right)\frac{\partial w}{\partial \theta} - \left(B_{11} - \frac{D_{11}}{R}\right)\frac{\partial^3 w}{\partial \theta^3}$$

$$+ \left(B_{12} + 2B_{66} - \frac{D_{12} + 2D_{66}}{R}\right)\frac{\partial^3 w}{\partial x^2 \partial \theta}$$

$$P_2(w) = \left(A_{66} - \frac{B_{66}}{R}\right)\frac{\partial^2 w}{\partial x^2}\frac{\partial w}{\partial \theta} + \left(A_{11} - \frac{B_{11}}{R}\right)\frac{\partial w}{\partial \theta}\frac{\partial^2 w}{\partial \theta^2}$$

$$+ \left(A_{12} + A_{66} - \frac{B_{12} + B_{66}}{R}\right)\frac{\partial w}{\partial x}\frac{\partial^2 w}{\partial x \partial \theta}$$

$$L_{31}(u) = \frac{A_{12}}{R}\frac{\partial u}{\partial x} + B_{11}\frac{\partial^3 u}{\partial x^3} + \left(B_{12} + 2B_{66}\right)\frac{\partial^3 u}{\partial x \partial \theta^2}$$

$$L_{32}(v) = \left(\frac{A_{11}}{R} - \frac{B_{11}}{R^2}\right)\frac{\partial v}{\partial \theta} + \left(B_{11} - \frac{D_{11}}{R}\right)\frac{\partial^3 v}{\partial \theta^3} + \left(B_{12} + 2B_{66} - \frac{D_{12} + 2D_{66}}{R}\right)\frac{\partial^3 v}{\partial x^2 \partial \theta}$$

$$L_{33}(w) = -\frac{A_{11}}{R^2}w - 2\frac{B_{11}}{R^2}\frac{\partial^2 w}{\partial \theta^2} - D_{11}\left(\frac{\partial^2 w}{\partial x^4} + \frac{\partial^2 w}{\partial \theta^4}\right) - 2\left(D_{12} + 2D_{66}\right)\frac{\partial^2 w}{\partial x^2 \partial \theta^2}$$

$$P_3(w) = +2\left(A_{12} + 2A_{66}\right)\frac{\partial w}{\partial x}\frac{\partial w}{\partial \theta}\frac{\partial^2 w}{\partial x \partial \theta} - \frac{w}{R}\left(A_{12}\frac{\partial^2 w}{\partial x^2} + A_{11}\frac{\partial^2 w}{\partial \theta^2}\right)$$

$$+2\left(B_{66} - B_{12}\right)\frac{\partial^2 w}{\partial x^2}\frac{\partial^2 w}{\partial \theta^2} + 2\left(B_{12} - B_{66}\right)\left(\frac{\partial^2 w}{\partial x \partial \theta}\right)^2$$

$$-\frac{A_{12}}{2R}\left(\frac{\partial w}{\partial x}\right)^2 - \frac{A_{11}}{2R}\left(\frac{\partial w}{\partial \theta}\right)^2 + \frac{3A_{11}}{2}\left[\frac{\partial^2 w}{\partial x^2}\left(\frac{\partial w}{\partial x}\right)^2 + \frac{\partial^2 w}{\partial \theta^2}\left(\frac{\partial w}{\partial \theta}\right)^2\right]$$

$$+ \left(\frac{A_{12}}{2} + A_{66}\right)\left[\frac{\partial^2 w}{\partial \theta^2}\left(\frac{\partial w}{\partial x}\right)^2 + \frac{\partial^2 w}{\partial x^2}\left(\frac{\partial w}{\partial \theta}\right)^2\right]$$

$$R_3(v,w) = \left(A_{12} - \frac{B_{12}}{R}\right)\frac{\partial v}{\partial \theta}\frac{\partial^2 w}{\partial x^2} + 2\left(A_{66} - \frac{B_{66}}{R}\right)\frac{\partial v}{\partial x}\frac{\partial^2 w}{\partial x \partial \theta}$$

$$+ \left(A_{11} - \frac{B_{11}}{R}\right)\left(\frac{\partial v}{\partial \theta}\frac{\partial^2 w}{\partial \theta^2} + \frac{\partial^2 v}{\partial \theta^2}\frac{\partial w}{\partial \theta}\right)$$

$$+ \left(A_{66} - \frac{B_{66}}{R}\right)\frac{\partial^2 v}{\partial x^2}\frac{\partial w}{\partial \theta} + \left(A_{12} + A_{66} - \frac{B_{12} + B_{66}}{R}\right)\frac{\partial^2 v}{\partial x \partial u}\frac{\partial w}{\partial x}$$

Appendix 3.2

$$l_{11} = -A_{11}\frac{\pi^2 m^2}{L^2} - A_{66}\frac{n^2}{R^2}$$

$$l_{12} = l_{21} = \left(-A_{12} - A_{66} + \frac{B_{12}+B_{66}}{R}\right)\frac{\pi mn}{LR}$$

$$l_{13} = l_{31} = -A_{12}\frac{\pi m}{LR} + B_{11}\frac{\pi^3 m^3}{L^3} + (B_{12}+2B_{66})\frac{\pi mn^2}{LR^2}$$

$$l_{22} = \left(-A_{66} + \frac{2B_{66}}{R} - \frac{D_{66}}{R^2}\right)\frac{\pi^2 m^2}{L^2} + \left(-A_{11}+\frac{2B_{11}}{R}-\frac{D_{11}}{R^2}\right)\frac{n^2}{R^2}$$

$$l_{23} = l_{32} = \left(-\frac{A_{11}}{R}+\frac{B_{11}}{R^2}\right)\frac{n}{R}+\left(B_{11}-\frac{D_{11}}{R}\right)\frac{n^3}{R^3}+\left(B_{12}+2B_{66}-\frac{D_{12}+2D_{66}}{R}\right)\frac{\pi^2 m^2 n}{L^2 R}$$

$$l_{33} = 2B_{12}\frac{\pi^2 m^2}{L^2 R}+2B_{11}\frac{n^2}{R^3}-D_{11}\frac{\pi^4 m^4}{L^4}-D_{11}\frac{n^4}{R^4}-2(D_{12}+2D_{66})\frac{\pi^2 m^2 n^2}{L^2 R^2}$$
$$-\frac{A_{11}}{R^2}+\frac{ph\pi^2 m^2}{L^2}$$

$$n_1 = -32A_{11}\frac{\pi m^2}{9L^3 n}+16(A_{12}-A_{66})\frac{n}{9\pi LR^2}$$

$$n_2 = \left(-A_{66}+A_{12}+\frac{B_{66}-B_{12}}{R}\right)\frac{16m}{9L^2 R}+\left(-A_{11}+\frac{B_{11}}{R}\right)\frac{32n^2}{9\pi^2 R^3 m}$$

$$n_3 = 16A_{12}\frac{m}{3L^2 Rn}+16A_{11}\frac{n}{3\pi^2 R^3 m}+32(B_{66}-B_{12})\frac{mn}{3L^2 R^2}$$

$$n_4 = -9A_{11}\frac{\pi^4 m^4}{32L^4}-(A_{12}+2A_{66})\frac{\pi^2 m^2 n^2}{16L^2 R^2}-9A_{11}\frac{n^4}{32R^4}$$

$$n_5 = 32A_{11}\frac{\pi m^2}{9L^3 n}+32(A_{12}-A_{66})\frac{n}{9\pi LR^2}$$

$$n_6 = \left(A_{12}-A_{66}+\frac{B_{66}-B_{12}}{R}\right)\frac{32m}{9L^2 R}+\left(A_{11}-\frac{B_{11}}{R}\right)\frac{32n^2}{9\pi^2 R^3 m}$$

Appendix 3.3

Transformed smart multi-scale hybrid FG-graphene origami auxetic sandwich doubly curved panel principle coordinate can be presented by:

$$\bar{Q}_{11}^n = Q_{11}^n \cos^4\theta + 2\left(Q_{12}^n + 2Q_{66}^n\right)\sin^2\theta\,\cos^2\theta + Q_{22}^n \sin^4\theta$$

$$\bar{Q}_{12}^n = (Q_{11}^n + Q_{22}^n - 4Q_{66}^n)\sin^2\theta\,\cos^2\theta + Q_{12}^n\left(\sin^4\theta + \cos^4\theta\right)$$

$$\bar{Q}_{22}^n = Q_{11}^n \sin^4\theta + 2\left(Q_{12}^n + 2Q_{66}^n\right)\sin^2\theta\,\cos^2\theta + Q_{22}^n \cos^4\theta$$

$$\bar{Q}_{66}^{n} = \left(Q_{11}^{n} + Q_{22}^{n} - 2Q_{12}^{n} - 2Q_{66}^{n} \right) \sin^2\theta \, \cos^2\theta + Q_{66}^{n} \left(\sin^4\theta + \cos^4\theta \right) \tag{A.1}$$

$$\bar{Q}_{44}^{n} = Q_{44}^{n} \cos^2\theta + Q_{55}^{n} \sin^2\theta$$

$$\bar{Q}_{55}^{n} = Q_{55}^{n} \cos^2\theta + Q_{44}^{n} \sin^2\theta$$

where $\bar{Q}_{ij} \left(i, j = 1, 2, 3, 4, 5, 6 \right)$ offered the transformed reduced stiffness modulus.

$$
\begin{Bmatrix} Nx \\ Ny \\ Nxy \end{Bmatrix} =
$$

$$
\begin{pmatrix} A11 & A12 & 0 \\ A21 & A22 & 0 \\ 0 & 0 & A66 \end{pmatrix}
\begin{bmatrix} Nx \\ Ny \\ Nxy \end{bmatrix}
\begin{Bmatrix} \dfrac{\partial u_0}{\partial x} \\[2mm] \dfrac{\partial v_0}{\partial y} \\[2mm] \dfrac{\partial v_0}{\partial x} + \dfrac{\partial u_0}{\partial y} \end{Bmatrix} +
$$

$$
\begin{pmatrix} B11 & B12 & 0 \\ B21 & B22 & 0 \\ 0 & 0 & B66 \end{pmatrix}
\begin{Bmatrix} \dfrac{w_0}{R_1} + \dfrac{1}{2}(\dfrac{\partial w_0}{\partial x})^2 \\[2mm] \dfrac{1}{2}(\dfrac{\partial w_0}{\partial y})^2 + \dfrac{w_0}{R_2} \\[2mm] \dfrac{\partial w}{\partial x} \dfrac{\partial w}{\partial y} \end{Bmatrix} +
\begin{bmatrix} A_{31} \\ A_{32} \\ 0 \end{bmatrix} \phi +
\begin{bmatrix} F_{31} \\ F_{32} \\ 0 \end{bmatrix} \psi -
\begin{Bmatrix} N_x^0 \\ 0 \\ 0 \end{Bmatrix} \tag{A.2}
$$

$$
\begin{Bmatrix} Mx \\ My \\ Mxy \end{Bmatrix} =
$$

$$
\begin{pmatrix} B11 & B12 & 0 \\ B21 & B22 & 0 \\ 0 & 0 & B66 \end{pmatrix}
\begin{bmatrix} Nx \\ Ny \\ Nxy \end{bmatrix}
\begin{Bmatrix} \dfrac{\partial u_0}{\partial x} \\[2mm] \dfrac{\partial v_0}{\partial y} \\[2mm] \dfrac{\partial v_0}{\partial x} + \dfrac{\partial u_0}{\partial y} \end{Bmatrix} +
$$

$$
\begin{pmatrix} D11 & D12 & 0 \\ D21 & D22 & 0 \\ 0 & 0 & D66 \end{pmatrix}
\begin{Bmatrix} \dfrac{w_0}{R_1} + \dfrac{1}{2}(\dfrac{\partial w_0}{\partial x})^2 \\[2mm] \dfrac{1}{2}(\dfrac{\partial w_0}{\partial y})^2 + \dfrac{w_0}{R_2} \\[2mm] \dfrac{\partial w}{\partial x} \dfrac{\partial w}{\partial y} \end{Bmatrix} +
\begin{bmatrix} E_{31} \\ E_{32} \\ 0 \end{bmatrix} \phi +
\begin{bmatrix} Q_{31} \\ Q_{32} \\ 0 \end{bmatrix} \psi -
\begin{Bmatrix} M_x \\ M_y \\ 0 \end{Bmatrix} \tag{A.3}
$$

Here, N_x, N_y, N_{xy} and M_x, M_y, M_{xy} expressed the total in-plane forced and moment resultants and P_x, P_y, P_{xy} and R_x, R_y are the third-order stresses resultants can be defined by:

$$P_x = \int_{-h/2}^{h/2} \left(\sigma_x, \sigma_y \right) z^3 dz \tag{A.4}$$

$$R_x, R_y = \int_{-h/2}^{h/2} \left(\tau_{xz}, \tau_{yz} \right) z^3 dz \tag{A.5}$$

Also, normal forces and moments can be written as:

$$M_x = \int_{-h/2}^{h/2} \left(e_{31} \frac{2V}{h} + q_{31} \frac{2\psi}{h} \right) z dz, \quad M_y = \int_{-h/2}^{h/2} \left(e_{32} \frac{2V}{h} + q_{32} \frac{2\psi}{h} \right) z dz \tag{A.6}$$

$$\int_{-h_{n-1}}^{h_n} \begin{Bmatrix} D_x \\ D_y \end{Bmatrix} \cos(\xi z) dz = F_{11}^e \begin{Bmatrix} \dfrac{\partial \phi}{\partial x} \\ \dfrac{\partial \phi}{\partial y} \end{Bmatrix} \tag{A.7}$$

$$\int_{-h_{n-1}}^{h_n} \xi \sin(\xi z) D_z dz = \left[A_{31} \left(\frac{\partial u_0}{\partial x} + \frac{\partial v_0}{\partial y} \right) - E_{31} \nabla^2 w \right] \tag{A.8}$$

in which:

$$(A_{31}, E_{31}) = \int_{-h/2}^{h/2} e_{31} \xi \sin(\xi z) \{1, z\}_z \, dz, (A_{32}, E_{32}) = \int_{-h/2}^{h/2} e_{32} \xi \sin(\xi z) \{1, z\}_z \, dz \tag{A.9}$$

$$(F_{31}, F_{33}) = \int_{-h/2}^{h/2} \left\{ s_{11} \cos^2(\xi z), s_{33} \xi^2 \sin^2(\xi z) \right\} dz \tag{A.10}$$

where the cross-sectional rigidities of smart multi-scale hybrid FG-graphene origami auxetic sandwich doubly curved panel can be expressed as follows:

$$(A_{ij}, B_{ij}, D_{ij}, E_{ij}, F_{ij}) = \int_{h_k}^{h_{k+1}} \bar{Q}_{ij}^n \left(1, z, z^2, z^3, z^4, z^6 \right) dz, \quad (i, j = 1, 2, 6) \tag{A.11}$$

and:

$$\bar{A}_{ij} = A_{ij} - 3s_1 D_{ij}, \quad \bar{D}_{ij} = A_{ij} - 3s_1 F_{ij}, \quad \hat{A}_{ij} = \bar{A}_{ij} - 3s_1 \bar{D}_{ij} \ (i, j = 4, 5)$$
$$\hat{B}_{ij} = B_{ij} - s_1 E_{ij}, \quad \hat{F}_{ij} = F_{ij} - s_1 H_{ij}, \quad \hat{D}_{ij} = D_{ij} - s_1 F_{ij} \tag{A.11}$$

4 Active Vibration Control of Auxetic Structures

4.1 BACKGROUND

The main objective of this chapter is to provide a set of principles for effectively controlling the vibrational characteristics of different auxetic structures. More specifically, the focus is on structures that include an auxetic core or a core made of graphene-origami auxetics. The governing equations and corresponding boundary conditions for sandwich plates are obtained by including geometric nonlinearity via the use of von Karman's theory, Maxwell's law, and Hamilton's principle. The governing equations of a sandwich plate are discretized using the generalized differential quadrature technique in a subsequent manner. The investigation focuses on the time-dependent viscoelastic deflection properties of laminated composite structures. These structures are composed of an auxetic core, which is surrounded by layers of piezoelectric material and gold. The micromechanical technique is used to get the material characteristics of the auxetic ply. The derivation of the kinematic motion equation for a thin-walled plate, which is placed on a viscoelastic medium with three parameters, involves entering the displacement field of the Kirchhoff-Love plate theorem into the dynamic form of the principle of virtual work. The ultimate controlling equation of the dynamic issue is derived by combining the constitutive equations of the laminate with the motion equation of the plate. This study investigates the influence of the thicknesses of piezoelectric and auxetic layers on the determination of the viscoelastic deflection-time curve of the system. The use of a proportional active control system has been shown to effectively regulate the fluctuation of the system. Additionally, it has been claimed that the damping coefficient of the visco-Pasternak substrate may serve as an efficient passive damper. In addition to its primary objective, the present work also seeks to regulate the vibrational characteristics of an auxetic plate that has been covered with a magnetostrictive substance. The kinematic relationships governing the behavior of a plate supported by a Winkler-Pasternak medium are formulated using the first-order shear deformation theory. The governing equations are solved using the analytical approach of using Navier's technique. A comprehensive analysis is conducted to observe the impact of many factors, including auxetic inclination angle, auxetic rib length, and feedback gain, on the control dynamics of the system. The findings of the study suggest that the incorporation of an auxetic core into a magnetostrictive plate leads to an augmentation in the dimensionless natural frequency. The discoveries made in this study have the potential to contribute to advancements in the field of sensor technology, actuator design, and vibration cancellation techniques. Additionally, the obtained findings might serve as a foundational basis for subsequent investigations. This chapter presents an inquiry aimed at regulating the vibrational features of a graphene-origami auxetic (GOA) plate by the use of magnetostrictive

 DOI: 10.1201/9781003289296-4

chemicals. The research attentively observes the influence of several factors, such as weight fraction, temperature, geometrical parameters, and feedback gain, on the control dynamics of the system. The results indicate that the integration of a graphene-origami auxetic core into a magnetostrictive plate results in an augmentation of the dimensionless natural frequency. The findings derived from this investigation have the potential to provide valuable contributions to the progress of sensor and actuator technology, as well as the use of techniques aimed at mitigating vibrations.

4.2   ACTIVE VIBRATION CONTROL OF SMART LAMINATES WITH AN AUXETIC PLY

In this section, we will address the free vibration issue of a continuous system composed of a three-layered laminate. This laminate comprises an auxetic core, with piezoelectric and gold layers situated at the bottom and top, respectively. The derivation of the governing equations for simulating the motion of the sandwich construction will be conducted using the Kirchhoff-Love plate theory. Subsequently, the final equation will be derived, the analytical solution of which will provide the time-dependent response [2]. In this section, the material parameters of each ply will be determined separately. To begin with, the matter of determining the attributes of the auxetic layer will be afterwards addressed. The schematic representation of an auxetic lattice and its corresponding unit cell is shown in Figure 4.1.

The observation of orthotropic behavior in aluminum is a consequence of auxeticity. The corresponding features of the auxetic core may be determined using the formulae provided in reference [2].

$$E_{11}^a = E_{Al}\left(t/l\right)^3 \frac{\cos\theta}{\left(e/l+\sin\theta\right)\sin^2\theta} \tag{4.1a}$$

$$E_{22}^a = E_{Al}\left(t/l\right)^3 \frac{\left(e/l+\sin\theta\right)}{\cos^3\theta} \tag{4.1b}$$

$$G_{12}^a = E_{Al}\left(t/l\right)^3 \frac{\left(e/l+\sin\theta\right)}{\left(e/l\right)^2\left(1+2e/l\right)\cos\theta} \tag{4.1c}$$

$$v_{12}^a = \frac{\cos^2\theta}{\left(e/l+\sin\theta\right)\sin\theta} \tag{4.1d}$$

$$\rho_a = \rho_{Al} \frac{\left(t/l\right)\left(e/l+2\right)}{2\cos\theta\left(e/l+\sin\theta\right)} \tag{4.1e}$$

The subscripts and superscripts denoted by a in equations (4.1a–e) are associated with the auxetic core. The given equations involve the initial inputs, which consist of the material properties of aluminum and the geometrical parameters of the auxetic lattice. These parameters include the thickness-to-length ratio of the inclined cell rib (e/l), the ratio of the length of the vertical cell rib to the length of the inclined cell rib (e/l), and the trigonometric functions of the inclined angle ($\sin\theta$ and $\cos\theta$).

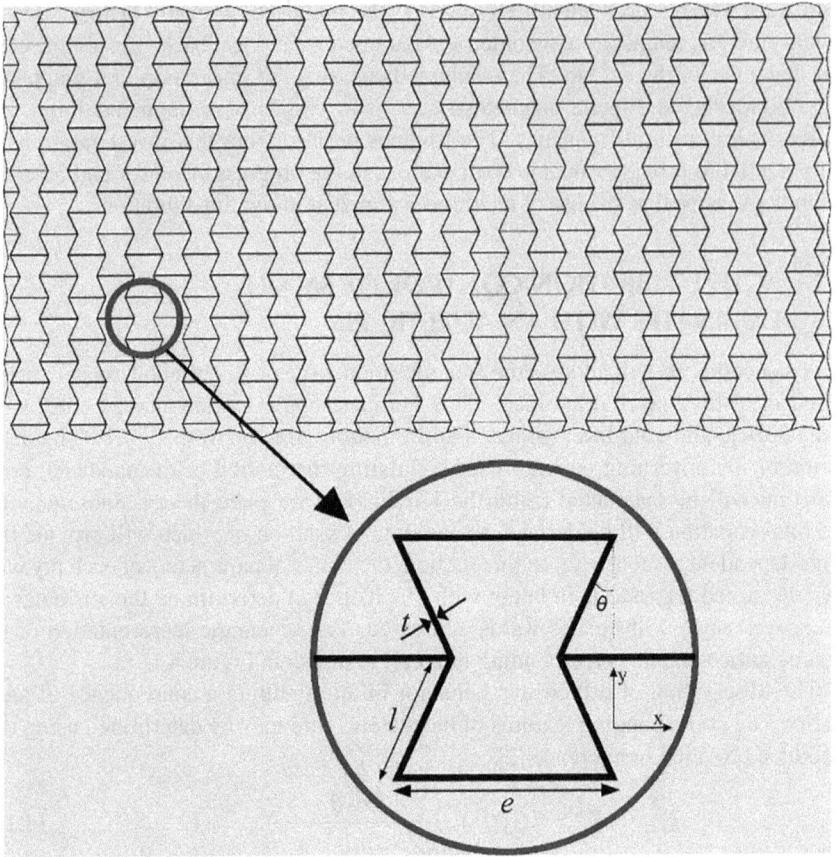

FIGURE 4.1 The structure of an auxetic lattice. The enlarged figure depicts the geometric parameters used in the process of homogenizing the auxetic ply.

In this section, we shall elucidate the process by which the motion equation of the plate is derived. This chapter discusses the thin-walled nature of a plate-type element due to its much larger length and breadth compared to its thickness. Therefore, it may be concluded that there would be little disparity between the mechanical responses derived from classical theory and those acquired from higher-order shear deformation theories (HSDT). The use of the Kirchhoff-Love plate theory will be employed in this study to mitigate excessive computing expenses, given the aforementioned rationale. The components of the displacement field of the plate may be represented as follows [2]:

$$u_x(x,y,z,t) = -z\frac{\partial w(x,y,t)}{\partial x},$$

$$u_y(x,y,z,t) = -z\frac{\partial w(x,y,t)}{\partial y}, \qquad (4.2)$$

$$u_z(x,y,z,t) = w(x,y,t)$$

In the aforementioned definition, w refers to the flexural deformation of the plate. Based on the notion of infinitesimal strains in the field of continuum mechanics, it is possible to enhance the non-zero stresses of the plate by using the following relation [2]:

$$\varepsilon_{ij} = \frac{1}{2}\left(u_{i,j} + u_{j,i}\right) \tag{4.3}$$

The variable ε_{ij} represents the constituent elements of the strain tensor. Upon substituting the displacement field vector from equation (4.2) into equation (4.3), it can be shown that the strain tensor [2] exhibits the following non-zero components:

$$\left\{\varepsilon_{xx} \quad \varepsilon_{yy} \quad \gamma_{xy}\right\} = -z\left\{\frac{\partial^2 w}{\partial x^2} \quad \frac{\partial^2 w}{\partial y^2} \quad 2\frac{\partial^2 w}{\partial x \partial y}\right\} \tag{4.4}$$

The non-zero stresses of the plate have now been attained, enabling the commencement of the process of obtaining the motion equations. In order to achieve this objective, the use of the dynamic version of the concept of virtual labor, often referred to as Hamilton's principle, will be applied. In accordance with this fundamental concept, it is essential that the alteration in the aggregate energy of the system during a certain time frame be shown as zero. The mathematical formulation of the aforementioned concept may be represented as follows [2]:

$$\int_0^t \delta(U - T + V)dt = 0 \tag{4.5}$$

The symbols U, T, and V represent the strain energy, kinetic energy, and work done by external loading, respectively. Now, it is necessary to compute the fluctuations of each of the aforementioned energies in a corresponding manner. The expression for the fluctuation of the strain energy may be represented as follows [2]:

$$
\begin{aligned}
\delta U &= \int_\forall \left(\sigma_{xx}\delta\varepsilon_{xx} + \sigma_{yy}\delta\varepsilon_{yy} + \sigma_{xy}\delta\gamma_{xy}\right)d\forall \\
&= -\int_0^b\int_0^a \left(M_{xx}\frac{\partial^2 \delta w}{\partial x^2} + 2M_{xy}\frac{\partial^2 \delta w}{\partial x \partial y} + M_{yy}\frac{\partial^2 \delta w}{\partial y^2}\right)dxdy
\end{aligned}
\tag{4.6}
$$

The entire volume of the continuous system is denoted by variable $\forall$. Additionally, M_{ij} represents the constituent elements of the bending moment, which may be determined by using the definition provided below:

$$M_{ij} = \int_{-h/2}^{-h/2+h_p} z\sigma_{ij}^p dz + \int_{-h/2+h_p}^{-h/2+h_p+h_a} z\sigma_{ij}^a dz + \int_{-h/2+h_p+h_a}^{-h/2+h_p+h_a+h_g} z\sigma_{ij}^g dz, \quad (i,j=x,y) \tag{4.7}$$

The components of the Cauchy stress tensor in the piezoelectric, auxetic, and gold layers are denoted by σ_{ij}^p, σ_{ij}^a, and σ_{ij}^g respectively. Additionally, the thicknesses of the aforementioned plies are denoted as h_p, h_a, and h_g accordingly. However, the fluctuation in the system's kinetic energy may be determined using the following equation [2]:

$$\delta T = \int\limits_{\forall} \left(\frac{\partial u_x}{\partial t} \frac{\partial \delta u_x}{\partial t} + \frac{\partial u_y}{\partial t} \frac{\partial \delta u_y}{\partial t} + \frac{\partial u_z}{\partial t} \frac{\partial \delta u_z}{\partial t} \right) \rho d\forall$$

$$= \int\limits_{0}^{b} \int\limits_{0}^{a} \left(I_0 \frac{\partial w}{\partial t} \frac{\partial \delta w}{\partial t} + I_2 \left[\frac{\partial^2 w}{\partial x \partial t} \frac{\partial^2 \delta w}{\partial x \partial t} + \frac{\partial^2 w}{\partial y \partial t} \frac{\partial^2 \delta w}{\partial y \partial t} \right] \right) dxdy \qquad (4.8)$$

In the aforementioned definition, the mass moments of inertia may be enhanced by using the following formula:

$$\begin{aligned}
[I_0, I_2] &= \int\limits_{-h/2}^{-h/2+h_p} \left[1, z^2 \right] \rho_p dz + \int\limits_{-h/2+h_p}^{-h/2+h_p+h_a} \left[1, z^2 \right] \rho_a dz \\
&\quad + \int\limits_{-h/2+h_p+h_a}^{-h/2+h_p+h_a+h_g} \left[1, z^2 \right] \rho_g dz, \quad (i,j = x, y)
\end{aligned} \qquad (4.9)$$

The calculation of the variance in the work performed by external loading is now required. The assumed configuration consists of a structural system supported by a viscoelastic foundation characterized by three parameters. This foundation incorporates both linear and shear springs, along with a viscous damper. Furthermore, the impact of the applied electric voltage on the continuous system will manifest via the application of a bi-axial electrical force on the plate. Therefore, the expression for the change in work performed by external loading may be formulated as follows [2]:

$$\delta V = \int\limits_{\forall} \left[N_x^e \frac{\partial^2 w}{\partial x^2} + N_y^e \frac{\partial^2 w}{\partial y^2} - k_w w + k_p \left(\frac{\partial^2 w}{\partial x^2} + \frac{\partial^2 w}{\partial y^2} \right) - c_d \frac{\partial w}{\partial t} \right] \delta w d\forall \qquad (4.10)$$

The variables denoted as Winkler, Pasternak, and damping coefficients, respectively, are represented by k_w, k_p, and c_d. Additionally, the electric force in the axial and transverse axes is denoted by N_x^e and N_y^e, correspondingly. Now, it is necessary to elucidate the theoretical derivation of the electric force. This study will use an active control mechanism to regulate the fluctuations of the system. In the present active control system, which is also referred to as a proportional feedback control system, the determination of the electric voltage delivered to the structure may be computed using the following definition [2]:

$$V_e = K_c \frac{\partial u_z}{\partial t} = K_c \frac{\partial w}{\partial t} \qquad (4.11)$$

The control gain of the proportional system K_c is located at a certain point. In relation to the characterization of the electric field resulting from the application of a voltage on a structure with a certain thickness, denoted as $E = V_e/h$, the magnitude of the electric force exerted on the structure in both the axial and transverse directions may be determined using the formula shown below [2]:

$$\begin{Bmatrix} N_x^e \\ N_y^e \end{Bmatrix} = 2 \int\limits_{-h/2}^{h/2+h_p} \begin{Bmatrix} e_{31} \\ e_{32} \end{Bmatrix} E dz = 2K_c \begin{Bmatrix} e_{31} \\ e_{32} \end{Bmatrix} \frac{\partial w}{\partial t} \qquad (4.12)$$

The piezoelectric coefficients in the axial and transverse axes are denoted by e_{31} and e_{32}, respectively. The equation of motion for the thin sandwich plate may be derived by inserting the changes of strain energy, kinetic energy, and work done by external loading from equations (4.6), (4.8), and (4.10) into equation (4.5) as follows:

$$
\frac{\partial^2 M_{xx}}{\partial x^2} + 2\frac{\partial^2 M_{xy}}{\partial x \partial y} + \frac{\partial^2 M_{yy}}{\partial y^2} + N_x^e \frac{\partial^2 w}{\partial x^2} + N_y^e \frac{\partial^2 w}{\partial y^2} - k_w w +
$$
$$
k_p \left[\frac{\partial^2 w}{\partial x^2} + \frac{\partial^2 w}{\partial y^2} \right] - c_d \frac{\partial w}{\partial t} = I_0 \frac{\partial^2 w}{\partial t^2} - I_2 \left(\frac{\partial^4 w}{\partial x^2 \partial t^2} + \frac{\partial^4 w}{\partial y^2 \partial t^2} \right)
$$

(4.13)

This section will provide the constitutive behaviors of the constituent plies with the aim of deriving the components of bending moment based on the displacement field of the plate. The first focus of this study will be on examining the constitutive characteristics of the piezoelectric layer. As stated by reference [2], the following constitutive behavior may be regarded as applicable to the piezoelectric layer:

$$
\begin{Bmatrix} \sigma_{xx}^P \\ \sigma_{yy}^P \\ \sigma_{xy}^P \end{Bmatrix} = \begin{bmatrix} Q_{11}^P & Q_{12}^P & 0 \\ Q_{12}^P & Q_{22}^P & 0 \\ 0 & 0 & Q_{66}^P \end{bmatrix} \begin{Bmatrix} \varepsilon_{xx} \\ \varepsilon_{yy} \\ \gamma_{xy} \end{Bmatrix} - \begin{bmatrix} 0 & 0 & e_{31} \\ 0 & 0 & e_{32} \\ 0 & 0 & 0 \end{bmatrix} \begin{Bmatrix} E_x \\ E_y \\ E_z \end{Bmatrix}
$$

(4.14)

The expression for the components of the elasticity tensor for the piezoelectric layer may be represented in the following manner [2]:

$$
Q_{11}^P = \frac{E_{11}^P}{1 - v_{12}^P v_{21}^P}, \quad Q_{12}^P = \frac{v_{12}^P E_{22}^P}{1 - v_{12}^P v_{21}^P}, \quad Q_{22}^P = \frac{E_{22}^P}{1 - v_{12}^P v_{21}^P}, \quad Q_{66}^P = G_{12}^P
$$

(4.15)

The longitudinal modulus, transverse modulus, and shear modulus of the piezoelectric layer in the specified plane are denoted as E_{11}^P, E_{22}^P, and G_{12}^P, respectively. Additionally, the Poisson's ratios are denoted by v_{12}^P and v_{21}^P. Furthermore, the relationship between the piezoelectric coefficients and the dielectric coefficients may be established based on the definition provided in reference [2].

$$
e_{31} = Q_{11}^P d_{31} + Q_{12}^P d_{32}
$$

(4.16a)

$$
e_{32} = Q_{12}^P d_{31} + Q_{22}^P d_{32}
$$

(4.16b)

The introduction of the constitutive equations pertaining to the auxetic layer is now necessary. The auxetic core, which is composed of aluminum, exhibits orthotropic behavior. Hence, the constitutive equation pertaining to this particular ply may be represented as follows [2]:

$$
\begin{Bmatrix} \sigma_{xx}^a \\ \sigma_{yy}^a \\ \sigma_{xy}^a \end{Bmatrix} = \begin{bmatrix} Q_{11}^a & Q_{12}^a & 0 \\ Q_{12}^a & Q_{22}^a & 0 \\ 0 & 0 & Q_{66}^a \end{bmatrix} \begin{Bmatrix} \varepsilon_{xx} \\ \varepsilon_{yy} \\ \gamma_{xy} \end{Bmatrix}
$$

(4.17a)

where

$$Q_{11}^a = \frac{E_{11}^a}{1-v_{12}^a v_{21}^a}, \quad Q_{12}^a = \frac{v_{12}^a E_{22}^a}{1-v_{12}^a v_{21}^a}, \quad Q_{22}^a = \frac{E_{22}^a}{1-v_{12}^a v_{21}^a}, \quad Q_{66}^a = G_{12}^a \qquad (4.17b)$$

The constitutive equations of the top layer, which is composed of gold, may be expressed as follows [2]:

$$\begin{Bmatrix} \sigma_{xx}^g \\ \sigma_{yy}^g \\ \sigma_{xy}^g \end{Bmatrix} = \begin{bmatrix} Q_{11}^g & Q_{12}^g & 0 \\ Q_{12}^g & Q_{22}^g & 0 \\ 0 & 0 & Q_{66}^g \end{bmatrix} \begin{Bmatrix} \varepsilon_{xx} \\ \varepsilon_{yy} \\ \gamma_{xy} \end{Bmatrix} \qquad (4.18)$$

in which

$$Q_{11}^g = Q_{22}^g = \frac{E_g}{1-v_g^2}, \quad Q_{12}^g = \frac{v_g E_g}{1-v_g^2}, \quad Q_{66}^g = \frac{E_g}{2(1+v_g)} \qquad (4.19)$$

The constitutive equations for each of the plies in the sandwich structure have been obtained, allowing for the expression of bending moments in relation to the displacement field of the plate. In order to achieve this objective, it is necessary to use equations (4.12), (4.16), and (4.18), and include them into equation (4.7). Upon performing the aforementioned mathematical operation, the following connection may be attained:

$$\begin{Bmatrix} M_{xx} \\ M_{yy} \\ M_{xy} \end{Bmatrix} = \begin{bmatrix} D_{11} & D_{12} & 0 \\ D_{12} & D_{22} & 0 \\ 0 & 0 & D_{66} \end{bmatrix} \begin{Bmatrix} -\dfrac{\partial^2 w}{\partial x^2} \\ -\dfrac{\partial^2 w}{\partial y^2} \\ -2\dfrac{\partial^2 w}{\partial x \partial y} \end{Bmatrix} - \begin{Bmatrix} M_x^e \\ M_y^e \\ 0 \end{Bmatrix} \qquad (4.20)$$

The bending moments resulting from the electric force are denoted by the variables M_x^e and M_y^e. The calculation of the through-the-thickness rigidities in the aforementioned equation may be determined using the following definition:

$$\begin{Bmatrix} D_{11} \\ D_{22} \\ D_{12} \\ D_{66} \end{Bmatrix} = \int_{-h/2}^{-h/2+h_p} \begin{Bmatrix} Q_{11}^p \\ Q_{22}^p \\ Q_{12}^p \\ Q_{66}^p \end{Bmatrix} z^2 dz + \int_{-h/2+h_p}^{-h/2+h_p+h_a} \begin{Bmatrix} Q_{11}^a \\ Q_{22}^a \\ Q_{12}^a \\ Q_{66}^a \end{Bmatrix} z^2 dz + \int_{-h/2+h_p+h_a}^{-h/2+h_p+h_a+h_g} \begin{Bmatrix} Q_{11}^g \\ Q_{22}^g \\ Q_{12}^g \\ Q_{66}^g \end{Bmatrix} z^2 dz \qquad (4.21)$$

This section aims to derive the governing equation of the plate by using the Euler-Lagrange equation as provided in equation (4.11). Upon substitution of equation

(4.20) into equation (4.11), the governing equation for the smart sandwich plate may be formulated as follows:

$$
\begin{aligned}
&-D_{11}\frac{\partial^4 w}{\partial x^4} - 2\left(D_{12}+2D_{66}\right)\frac{\partial^4 w}{\partial x^2 \partial y^2} - D_{22}\frac{\partial^4 w}{\partial y^4} - \left(N_x^e\frac{\partial^2 w}{\partial x^2}+N_y^e\frac{\partial^2 w}{\partial y^2}\right) - \\
&k_w w + k_p\left(\frac{\partial^2 w}{\partial x^2}+\frac{\partial^2 w}{\partial y^2}\right) - c_d\frac{\partial w}{\partial t} - I_0\frac{\partial^2 w}{\partial t^2} + I_2\left(\frac{\partial^4 w}{\partial x^2\partial t^2}+\frac{\partial^4 w}{\partial y^2\partial t^2}\right) = 0
\end{aligned}
$$

(4.22)

In this section, the use of the separation of variables technique will be applied to solve the governing equation of the issue. It is important to note that the boundary conditions at all edges of the structure are considered to be of the simply supported kind. Hence, it is necessary to establish that both bending deflection and moments at each edge are equal to zero, namely at the boundaries (i.e., $w = \partial^2 w/\partial x^2 = \partial^2 w/\partial y^2 = 0$ at $x = 0, a$ and $y = 0, b$). After implementing the aforementioned constraints on the spatial component of the solution, the following approach will be used to calculate the deflection of the plate:

$$
w(x,y,t) = W(t)\sin\alpha x\sin\beta y \tag{4.23a}
$$

In the aforementioned approach, the functions $\alpha = \dfrac{m\pi}{a}$ and $\beta = \dfrac{n\pi}{b}$ are used. Additionally, the magnitude of the bending deflection is shown by the transient amplitude $W(t)$. By incorporating the bending deflection of the plate from equation (4.23a) into equation (4.22), the following relationship may be derived:

$$
M\frac{d^2 W(t)}{dt^2} + C\frac{dW(t)}{dt} + KW(t) = F \tag{4.23b}
$$

In the aforementioned equation, the variables M, C, and K represent the corresponding values of the mass, damping coefficient, and stiffness of the continuous system. Additionally, a steady distributed load F is uniformly delivered throughout the whole surface of the plate. Nevertheless, the problem-solving process remains incomplete at this particular level. In order to get an expression for the deflection of the structure in the frequency domain, it is necessary to apply the Laplace transformation method to the above equation. Subsequently, the acquired equation has to be converted into the time domain by the process of Laplace inversion.

$$
\mathcal{L}\{f(t)\} = F(s) = \int_0^\infty e^{-st}f(t)dt \tag{4.24a}
$$

The function $f(t)$ denotes a representation in the time domain, whereas $F(s)$ signifies the same representation in the frequency domain. By using the approach outlined in equation (4.23b) and obtaining the Laplace domain transformation, the displacement vector in the frequency domain may be calculated as follows:

$$
\{\bar{W}\} = ([K]+s[C]+s^2[M])^{-1}\{\bar{F}\} \tag{4.24b}
$$

The bar symbol represents the converted variations in the frequency domain. It is evident that the determination of the unknown may be achieved by the computation of the inverse Laplace transformation inside the region of convergence (ROC) of the system. The computation of the inverse Laplace transformation may be achieved by the use of the well-recognized Bromwich integral, as seen below:

$$\mathcal{L}^{-1}\{F(s)\} = f(t) = \frac{1}{2\pi i} \int_{c-i\infty}^{c+i\infty} e^{st} F(s) ds \tag{4.24c}$$

The calculation of the integral is necessary for each value of c that is inside the region of convergence along the route from $c-i\infty$ to $c+i\infty$. In order to streamline the discussion and acknowledge the prevalent use of this approach in existing scholarly works, more elaboration will not be provided in this section. Additional information on this strategy may be obtained by consulting reference [2].

This section will provide a series of examples to facilitate a discussion on the factors influencing the transient behaviors of the intelligent continuous system. In each of the numerical examples provided, the length-to-thickness ratio of the plate-type element is assumed to be 100. This assumption is made in order to accurately represent the behavior of thin-walled plates. Additionally, this research focuses on the examination of square plates. Furthermore, it is assumed that the inclination angle is $\theta = 55$ degrees, and dimensionless ratios of $e/l = 4$ and $t/l = 0.0138571$ will be used in the next numerical case studies. The auxetic ply is often fabricated using aluminum as the basis material, with its material characteristics being as follows: $E_{Al} = 70$ GPa, $\rho_{Al} = 2705$ kg / m^3, and $v_{Al} = 0.3$ [2]. Furthermore, the characteristics of G-1195N, which are considered to represent the piezoelectric ply, include: $E_p = 63$ GPa, $\rho_p = 7600$ kg / m^3, and $v_p = 0.3$. The material parameters of the gold layer are shown as follows: $E_g = 79$ GPa, $\rho_g = 19300$ kg / m^3, and $v_p = 0.4$ [2].

This section will examine the time-dependent mechanical responses of smart composite plates when a laminated composite structure is exposed to a tensile load of 10 kPa. For the purpose of simplicity, the dimensionless form of the foundation parameters and deflection amplitude will be used in all subsequent demonstrations.

$$K_w = \frac{k_w a^4}{D_{11}}, \quad K_p = \frac{k_p a^2}{D_{11}}, \quad C_d = \frac{c_d a^2}{\sqrt{\rho_{Al} h D_{11}}}, \quad W(t) = \frac{w(a/2, b/2, t) E_{Al} h}{P_0 a^2} \tag{4.24d}$$

The continuous traction $P_0 = 10$ kPa is uniformly imparted to the plate over the whole time domain. The dimensionless deflection-time curve of the plate is shown in Figure 4.2, showcasing the impact of different values provided to the thickness of the auxetic layer. Research has shown that variations in the thickness of the auxetic layer have a significant impact on the center deflection of the plate. It is evident that plates with thicker auxetic cores exhibit greater stability consistently throughout time. The amplitude of their fluctuation is comparatively reduced, resulting in a lower steady-state deflection reaction compared to the other entities. The primary factor contributing to this phenomenon is the remarkable rigidity of the auxetic core, which may significantly enhance the equivalent stiffness of the structure when a greater

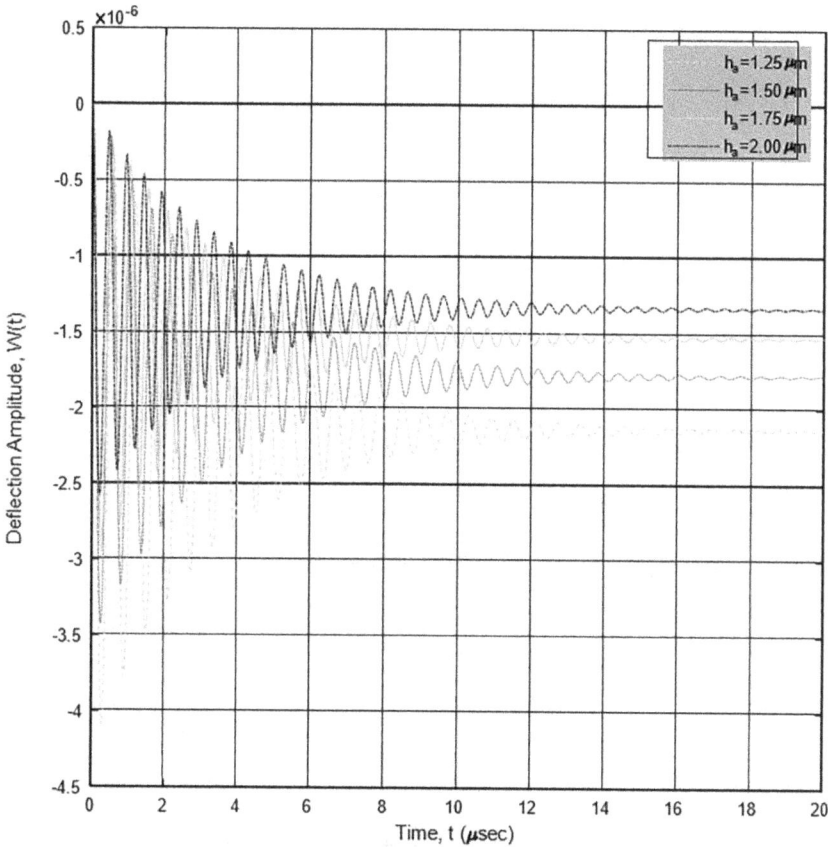

FIGURE 4.2 Variation of the dimensionless deflection of the plate versus time for various amounts of the auxetic layer's thickness ($h_g=1$ μm, $h_p=2$ mm, $K_c=400$, $K_w=K_p=C_d=0$).

thickness of this material is used. It is common to see increased rigidity, shown by less deflection, when using thicker cores during the production of laminated panels.

Figure 4.3 depicts the examination of the impact of the auxetic layer's thickness on the deflection response of the laminated composite plate. It is evident that plates including thinner piezoelectric plies demonstrate less deflections. The impact of the thickness of the piezo layer on the dynamic response of the plate differs significantly from that of the auxetic layer. The observed phenomenon may be ascribed to the comparatively reduced bending stiffness of the piezo layer in contrast to the stiffness of the auxetic core. Consequently, the inclusion of a thicker piezoelectric layer might lead to the creation of structures with reduced effective stiffness, resulting in increased deflection amplitudes. Furthermore, the temporal point at which the system's deflection reaches its steady-state value may be altered by modifying the thickness of the piezo layer. The experimental findings demonstrate that the inclusion of the additional thickness of this layer leads to a noticeable prolongation in the time required for the composite plate to reach a stable state.

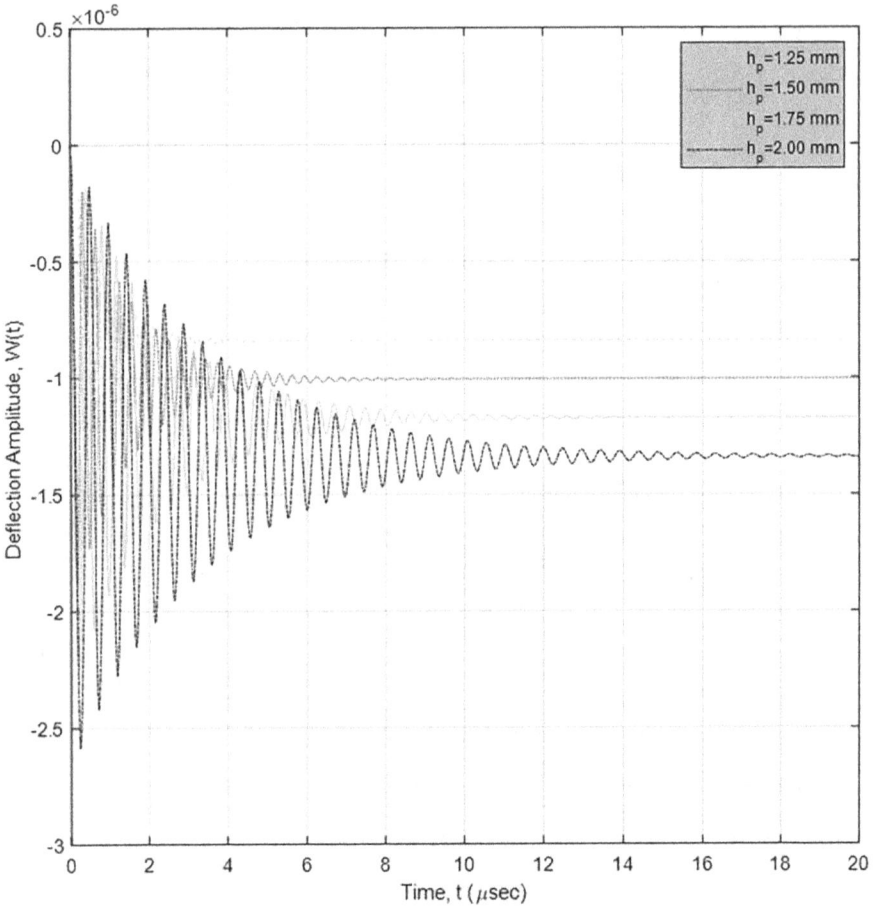

FIGURE 4.3 Variation of the dimensionless deflection of the plate versus time for various amounts of the piezoelectric layer's thickness (h_a=2 µm, h_g=1 µm, K_c=400, K_w=K_p=C_d=0).

Figure 4.4 depicts the relationship between the dimensional deflection of the plate and time across different levels of control gain. Based on the diagram shown, it can be seen that the dynamic deflection of the laminate exhibits a more rapid attenuation when the gain of the proportional control system is augmented. The inclusion of the control gains results in an augmentation of the damping component. This is attributed to the linear relationship between the electrical force and the first derivative of the system. It is evident that both the magnitude of deflection and the time required for the system to attain its steady-state response are reduced, which is a notable feature.

The examination of the influence of the foundation parameters on the controlled and free vibrations of the laminated plate is shown within the context of Figure 4.5. It is evident that systems equipped with an active controller would achieve their

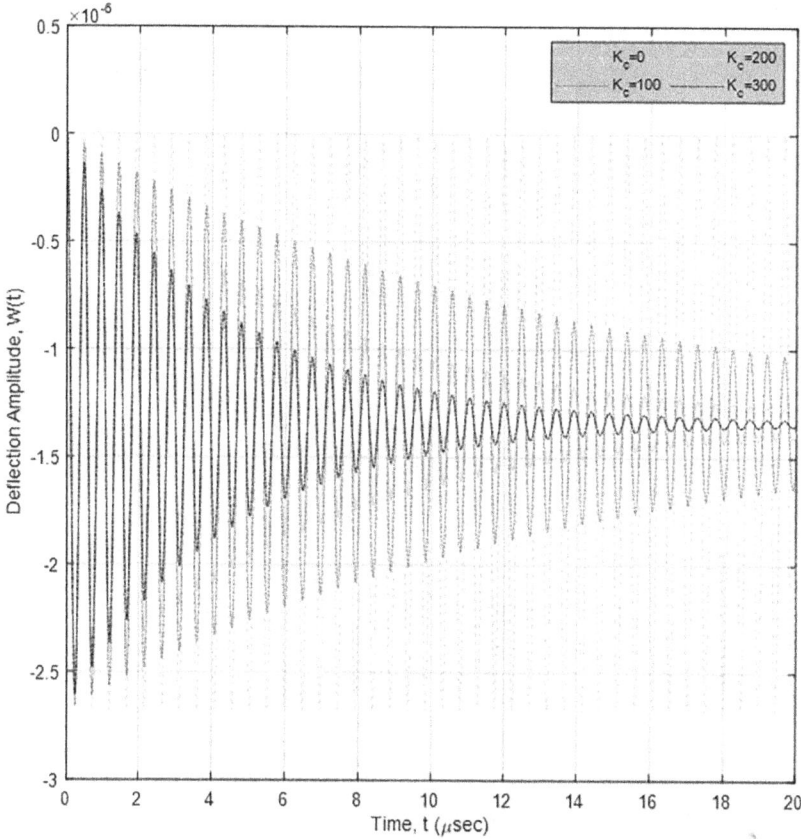

FIGURE 4.4 Variation of the dimensionless deflection of the plate versus time for various amounts of the control gain of the feedback control system (h_a=2 μm, h_p=2 mm, h_g=1 μm, K_w=K_p=C_d=0).

steady-state response in a shorter duration. Conversely, it is evident that including a higher level of rigidity in the foundation results in reduced deflection amplitudes. The underlying cause of this phenomenon may be attributed to the favorable influence exerted by the foundation's stringent specifications on the overall stiffness of the continuous system. It is important to highlight that the impact of the Pasternak spring is greater in magnitude compared to that of the Winkler spring. Furthermore, it is evident that the presence of a viscous damper in the foundation would lead to a quick attenuation of the system's oscillation, resulting in the steady-state response occurring earlier compared to scenarios where viscous losses are disregarded. Therefore, it can be inferred that the incorporation of passive-type dampers in the foundation of a chosen structure may lead to a significant decrease in the vibrational movement of the system.

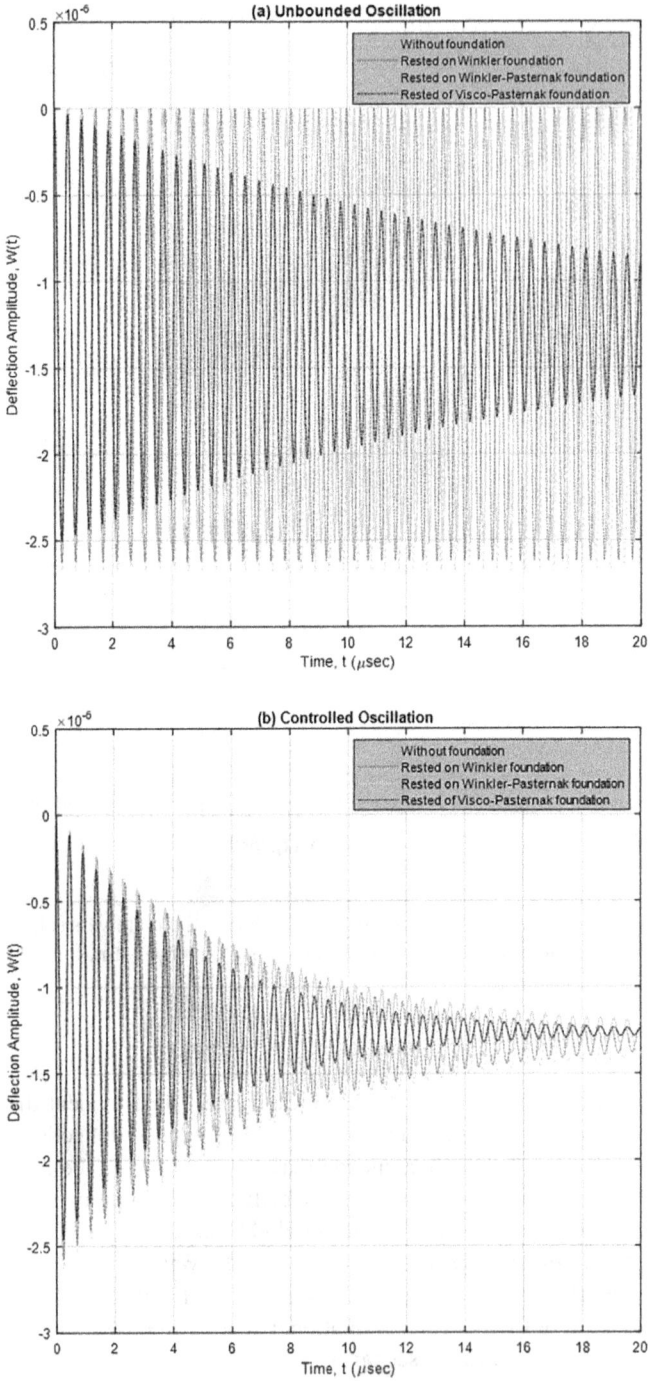

FIGURE 4.5 Variation of the dimensionless deflection of the plate versus time for various types of seats used to support the structure (h_a=2 μm, h_p=2 mm, h_g=1 μm).

4.3 ACTIVE VIBRATION CONTROL OF COMPOSITE AUXETIC PLATES WITH MAGNETOSTRICTIVE COATINGS

The objective of this section is to analyze the use of plates with auxetic core and coated magnetostrictive materials in the context of vibration control. The Cartesian coordinate system is used to establish the coordinates (x, y, and z) in this context. The origin of this system is positioned at one corner of the plate, as seen in Figure 4.6. The Winkler-Pasternak foundation has been used to quantify the impact of said foundation. The FSDT is used for the purpose of characterizing the motion of a composite plate.

The FSDT will be utilized to frame the issue in this chapter. The related FSDT displacement field can be expressed as [3–5]:

$$U_x(x,y,z,t) = u_0(x,y,t) + \phi_x(x,y,t) \tag{4.25a}$$

$$V_y(x,y,z,t) = v_0(x,y,t) + \phi_y(x,y,t) \tag{4.25b}$$

$$W_z(x,y,z,t) = w_0(x,y,t) \tag{4.25c}$$

where the complete displacement fields in the Cartesian x, y, and z axes are denoted by U_x, V_y, and W_z.

The auxetic core layer is composed of cells that are organized in a honeycomb pattern. The mechanical properties of the honeycomb auxetic core layer, such as Young's modulus, shear modulus, negative Poisson's ratio, and thermal expansion coefficients, are believed to be influenced by the geometrical parameters of the individual unit cell. The potential consequences for the effective mechanical features of the auxetic core are outlined as follows [3]:

$$E_{11}^a = E_{Al}\left(\frac{t}{l}\right)^3 \frac{\cos\theta}{\left(\dfrac{e}{l}+\sin\theta\right)\sin^2\theta} \tag{4.26a}$$

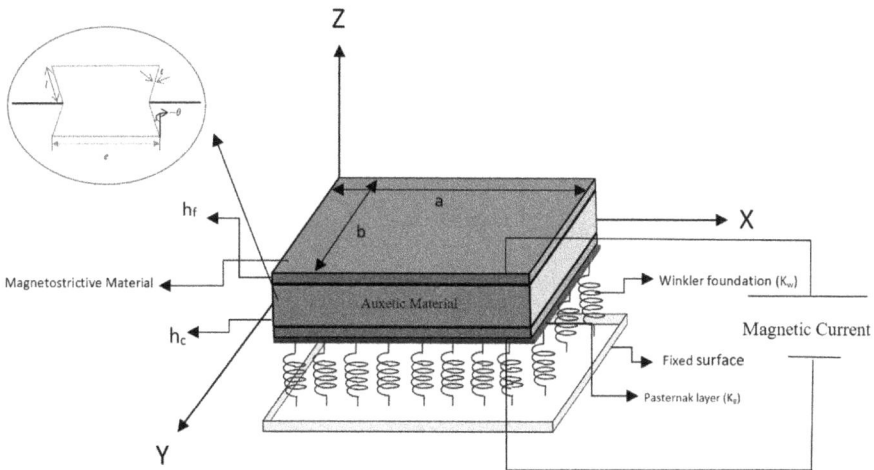

FIGURE 4.6 Plan view of the structure.

$$E_{22}^a = E_{Al} \left(\frac{t}{l}\right)^3 \frac{\left(\dfrac{e}{l} + \sin\theta\right)}{\cos^3\theta} \qquad (4.26b)$$

$$G_{12}^a = E_{Al} \left(\frac{t}{l}\right)^3 \frac{\left(\dfrac{e}{l} + \sin\theta\right)}{\left(\dfrac{e}{l}\right)^2 \left(1 + \dfrac{2e}{l}\right)\cos\theta} \qquad (4.26c)$$

$$v_{12}^a = \frac{\cos^2\theta}{\left(\dfrac{e}{l} + \sin\theta\right)\sin\theta} \qquad (4.26d)$$

$$\rho_a = \rho_{Al} \frac{\left(\dfrac{t}{l}\right)\left(\dfrac{e}{l} + 2\right)}{2\cos\theta\left(\dfrac{e}{l} + \sin\theta\right)} \qquad (4.26e)$$

The equations (4.26a–e) provide the necessary information on the mechanical characteristics of the used auxetic core. The rib thickness of the auxetic cell is represented by the variable t, while its horizontal rib length is marked by e, and its vertical inclined rib length is indicated by l. There are two efficacious auxetic core moduli, namely the Young's modulus E_{11}^a and E_{22}^a, as well as the shear modulus G_{11}^a and G_{22}^a. Furthermore, both v_{11}^a and v_{22}^a may be seen as effective auxetic variations of Poisson's ratio. In the context of auxetic materials, the correlation between stress and strain may be expressed as shown in reference [3].

$$\sigma_{xx}^a = C_{11}^a \varepsilon_{xx} + C_{12}^a \varepsilon_{yy} \qquad (4.27a)$$

$$\sigma_{yy}^a = C_{21}^a \varepsilon_{xx} + C_{22}^a \varepsilon_{yy} \qquad (4.27b)$$

$$\tau_{yz}^a = C_{44}^a \gamma_{yz} \qquad (4.27c)$$

$$\tau_{xz}^a = C_{55}^a \gamma_{xz} \qquad (4.27d)$$

$$\tau_{xy}^a = C_{66}^a \gamma_{xy} \qquad (4.27e)$$

Auxetic layer non-zero stresses are defined as:

$$\begin{Bmatrix} \varepsilon_{xx}^a \\ \varepsilon_{yy}^a \\ \varepsilon_{yz}^a \\ \varepsilon_{xz}^a \\ \varepsilon_{xy}^a \end{Bmatrix} = \begin{Bmatrix} \epsilon_{xx}^a \\ \epsilon_{yy}^a \\ \epsilon_{yz}^a \\ \epsilon_{xz}^a \\ \epsilon_{xy}^a \end{Bmatrix} + z \begin{Bmatrix} \kappa_{xx}^a \\ \kappa_{yy}^a \\ \kappa_{yz}^a \\ \kappa_{xz}^a \\ \kappa_{xy}^a \end{Bmatrix} \qquad (4.28)$$

in which

$$
\begin{Bmatrix} \epsilon_{xx}^a \\ \epsilon_{yy}^a \\ \epsilon_{yz}^a \\ \epsilon_{xz}^a \\ \epsilon_{xy}^a \end{Bmatrix} = \begin{Bmatrix} \dfrac{\partial u_0}{\partial x} \\ \dfrac{\partial v_0}{\partial y} \\ \left(\phi_y + \dfrac{\partial w_0}{\partial y}\right) \\ \left(\phi_x + \dfrac{\partial w_0}{\partial x}\right) \\ \left(\dfrac{\partial u_0}{\partial y} + \dfrac{\partial v_0}{\partial x}\right) \end{Bmatrix} \quad \begin{Bmatrix} \kappa_{xx}^a \\ \kappa_{yy}^a \\ \kappa_{yz}^a \\ \kappa_{xz}^a \\ \kappa_{xy}^a \end{Bmatrix} = \begin{Bmatrix} \dfrac{\partial \phi_x}{\partial x} \\ \dfrac{\partial \phi_y}{\partial y} \\ 0 \\ 0 \\ \dfrac{\partial \phi_x}{\partial y} + \dfrac{\partial \phi_y}{\partial x} \end{Bmatrix}
\tag{4.29}
$$

The phenomenon referred to as the Joule effect, sometimes known as magnetostriction, pertains to the alteration experienced by a ferromagnetic substance when subjected to an externally applied magnetic field. The ramifications of this phenomenon may be seen in a diverse range of applications. The phenomenon of magnetostriction was originally described by James Joule, an English physicist, in the year 1842. The researcher observed that a control sample composed of a ferromagnetic substance, such as iron, exhibits a change in length when exposed to a magnetic field. Terfenol-D is often regarded as the most optimal magnetostrictive material due to its superior strain capabilities and high Curie temperature. The investigation and advancement of magnetostrictive supermaterials were started throughout the 1960s by researchers, including Clark et al. Terfenol-D was discovered during the 1970s at the Naval Ordinance Laboratory under the leadership of a research group led by Keller. Keller was attributed with the responsibility of bestowing the laboratory with its name. The formula [3] indicates the relationship between stress, strain, and magnetic susceptibility in the magnetostrictive layer.

$$
\sigma_{xx}^m = C_{11}^m \varepsilon_{xx} + C_{12}^m \varepsilon_{yy} - e_{31} H_{zz}
\tag{4.30a}
$$

$$
\sigma_{yy}^m = C_{21}^m \varepsilon_{xx} + C_{22}^m \varepsilon_{yy} - e_{32} H_{zz}
\tag{4.30b}
$$

$$
\tau_{yz}^m = C_{44}^m \gamma_{yz}
\tag{4.30c}
$$

$$
\tau_{xz}^m = C_{55}^m \gamma_{xz}
\tag{4.30d}
$$

$$
\tau_{xy}^m = C_{66}^m \gamma_{xy}
\tag{4.30e}
$$

The stress and strain tensors are denoted by arrays in the form of $\sigma_{ij}, \varepsilon_{ij}$ $(i, j = x, y, z)$. In addition, the magnetic field is denoted as H_{zz}. Furthermore, it is important to note that the symbol C_{ij} denotes the arrays of elasticity tensors. Assuming a negligible thickness stretching, denoted as $\varepsilon_{zz} = 0$, the elastic constants inside the stress plane may be expressed as shown in reference [1].

$$C_{11}^m = \frac{E_m}{1-\vartheta_m^2} \quad C_{12}^m = \frac{\vartheta_m E_m}{1-\vartheta_m^2} \quad C_{66}^m = \frac{E_m}{2(1+\vartheta_m)} \quad C_{55}^m = C_{44}^m = k_s C_{66}^m \qquad (4.31)$$

where E_m and v_m stand for the magnetostrictive layers' elasticity modulus and Poisson's ratio, respectively. It's important to note that the shear correction parameter, symbolized by the constant value k_s, has been assigned different values by various theories. Using the following connection, magnetic field (H_{zz}) can be calculated as [5]:

$$H_{zz} = k_c I(x,y,t) \qquad (4.32a)$$

$$I(x,y,t) = C(t)\frac{\partial w_0(x,y,t)}{\partial t} \qquad (4.32b)$$

The coil constant, represented as k_c, is dependent on the physical attributes of the coil, including its width, radius, and number of turns, as well as the composition of the material used. Furthermore, the variables (x,y,t) and $C(t)$ represent the current flowing through the coil and the gain of the coil, respectively. It is important to acknowledge that our assumption is based on a continual control gain. The gain of the velocity feedback is represented by the product of the gain constant Kc and the transfer function $K_c C(t)$. The relationship between coil characteristics and the coil constant may be described as follows:

$$K_c = \frac{n_c}{\sqrt{b_c^2 + 4r_c^2}} \qquad (4.33)$$

The aforementioned equation utilizes the variables n_c, b_c, and r_c to denote the number of turns, width, and radius of the coil, correspondingly. The non-zero stresses of the magnetostrictive layer may be characterized as:

$$\begin{Bmatrix} \varepsilon_{xx}^m \\ \varepsilon_{yy}^m \\ \varepsilon_{yz}^m \\ \varepsilon_{xz}^m \\ \varepsilon_{xy}^m \end{Bmatrix} = \begin{Bmatrix} \epsilon_{xx}^m \\ \epsilon_{yy}^m \\ \epsilon_{yz}^m \\ \epsilon_{xz}^m \\ \epsilon_{xy}^m \end{Bmatrix} + z \begin{Bmatrix} \kappa_{xx}^m \\ \kappa_{yy}^m \\ \kappa_{yz}^m \\ \kappa_{xz}^m \\ \kappa_{xy}^m \end{Bmatrix} \qquad (4.34)$$

where

$$\begin{Bmatrix} \epsilon_{xx}^m \\ \epsilon_{yy}^m \\ \epsilon_{yz}^m \\ \epsilon_{xz}^m \\ \epsilon_{xy}^m \end{Bmatrix} = \begin{Bmatrix} \dfrac{\partial u_0}{\partial x} \\ \dfrac{\partial v_0}{\partial y} \\ \left(\phi_y + \dfrac{\partial w_0}{\partial y}\right) \\ \left(\phi_x + \dfrac{\partial w_0}{\partial x}\right) \\ \left(\dfrac{\partial u_0}{\partial y} + \dfrac{\partial v_0}{\partial x}\right) \end{Bmatrix} \qquad \begin{Bmatrix} \kappa_{xx}^m \\ \kappa_{yy}^m \\ \kappa_{yz}^m \\ \kappa_{xz}^m \\ \kappa_{xy}^m \end{Bmatrix} = \begin{Bmatrix} \dfrac{\partial \phi_x}{\partial x} \\ \dfrac{\partial \phi_y}{\partial y} \\ 0 \\ 0 \\ \dfrac{\partial \phi_x}{\partial y} + \dfrac{\partial \phi_y}{\partial x} \end{Bmatrix} \qquad (4.35)$$

The significant properties of magnetostrictive layer auxetic materials are as follows [3–5].

Properties of magnetostrictive material:

$$E_m \ (\text{GPa})= 30, \ \rho_m \ (^{kg}\!/_{m^3}) = 9.25*10e3, \ e_{31} = e_{32} \left(^{N}\!/_{mA}\right) = 442.55, \ \vartheta_c = 0.25$$

Properties of auxetic materials:

$$E_a \ (\text{GPa})= 69, \ G_a \ (\text{GPa})= 26, \ \rho_m \ (^{kg}\!/_{m^3})= 2700$$

In this part, the motion equations are obtained by the use of the energy technique. The notion developed by Hamilton will be used for this objective. One potential formulation of this concept is as follows [5]:

$$\int_0^t \delta(U - T + V)\,dt = 0 \tag{4.36}$$

In this context, U, T, and V represent the strain energy, kinetic energy, and external work, respectively. The first variation of strain energy may be expressed as:

$$\delta U^a = \int_0^a \int_0^b \begin{pmatrix} N_{xx}^a \left(\delta\epsilon_{xx}^a\right) + M_{xx}^a \left(\delta\kappa_{xx}^a\right) + N_{yy}^a \left(\delta\epsilon_{yy}^a\right) + M_{yy}^a \left(\delta\kappa_{yy}^a\right) + \\ N_{xz}^a \left(\delta\epsilon_{xz}^a\right) + M_{xz}^a \left(\delta\kappa_{xz}^a\right) + N_{yz}^a \left(\delta\epsilon_{yz}^a\right) + M_{yz}^a \left(\delta\kappa_{yz}^a\right) + \\ N_{xy}^a \left(\delta\epsilon_{xy}^a\right) + M_{xy}^a \left(\delta\kappa_{xy}^a\right) \end{pmatrix} dA \tag{4.37a}$$

$$\delta U^m = \int_0^a \int_0^b \begin{pmatrix} N_{xx}^m \left(\delta\epsilon_{xx}^m\right) + M_{xx}^m \left(\delta\kappa_{xx}^m\right) + N_{yy}^m \left(\delta\epsilon_{yy}^m\right) + M_{yy}^m \left(\delta\kappa_{yy}^m\right) + \\ N_{xz}^m \left(\delta\epsilon_{xz}^m\right) + M_{xz}^m \left(\delta\kappa_{xz}^m\right) + N_{yz}^m \left(\delta\epsilon_{yz}^m\right) + M_{yz}^m \left(\delta\kappa_{yz}^m\right) + \\ N_{xy}^m \left(\delta\epsilon_{xy}^m\right) + M_{xy}^m \left(\delta\kappa_{xy}^m\right) \end{pmatrix} dA \tag{4.37b}$$

where

$$\left(N_{ij}^a, M_{ij}^a\right) = \int_{-\frac{h_a}{2}}^{\frac{h_a}{2}} \sigma_{ij}^a (1, z)\,dz, \ (i, j = x, \ y, \ z) \tag{4.38a}$$

$$\left(N_{ij}^m, M_{ij}^m\right) = \int_{\frac{h_a}{2}}^{\frac{h_a}{2}+h_m} \sigma_{ij}^m (1, z)\,dz + \int_{-\left(\frac{h_a}{2}+h_m\right)}^{2} \sigma_{ij}^m (1, z)\,dz, \ (i, j = x, \ y, z) \tag{4.38b}$$

The equations stated above designate the auxetic and magnetostrictive layers with the superscripts a and m, respectively. Furthermore, the forces $N_{i.j}$ and the moments $M_{i.j}$ are designated. The interconnections between forces and bending moments may be described as follows:

$$
\begin{bmatrix} N_{xx} \\ N_{yy} \\ N_{yz} \\ N_{xz} \\ N_{xy} \end{bmatrix} = \left(\begin{bmatrix} A_{11} & A_{12} & 0 & 0 & 0 \\ A_{21} & A_{22} & 0 & 0 & 0 \\ 0 & 0 & A_{44} & 0 & 0 \\ 0 & 0 & 0 & A_{55} & 0 \\ 0 & 0 & 0 & 0 & A_{66} \end{bmatrix} \begin{bmatrix} \dfrac{\partial u_0}{\partial x} \\[2mm] \dfrac{\partial v_0}{\partial y} \\[2mm] \phi_y + \dfrac{\partial w_0}{\partial y} \\[2mm] \phi_x + \dfrac{\partial w_0}{\partial x} \\[2mm] \dfrac{\partial u_0}{\partial y} + \dfrac{\partial v_0}{\partial x} \end{bmatrix} + \begin{bmatrix} B_{11} & B_{12} & 0 & 0 & 0 \\ B_{21} & B_{22} & 0 & 0 & 0 \\ 0 & 0 & B_{44} & 0 & 0 \\ 0 & 0 & 0 & B_{55} & 0 \\ 0 & 0 & 0 & 0 & B_{66} \end{bmatrix} \begin{bmatrix} \dfrac{\partial \phi_x}{\partial x} \\[2mm] \dfrac{\partial \phi_y}{\partial y} \\[2mm] 0 \\[2mm] 0 \\[2mm] \dfrac{\partial \phi_x}{\partial y} + \dfrac{\partial \phi_y}{\partial x} \end{bmatrix} \right) \tag{4.39a}
$$

$$
\begin{bmatrix} M_{xx} \\ M_{yy} \\ M_{yz} \\ M_{xz} \\ M_{xy} \end{bmatrix} = \left(\begin{bmatrix} B_{11} & B_{12} & 0 & 0 & 0 \\ B_{21} & B_{22} & 0 & 0 & 0 \\ 0 & 0 & B_{44} & 0 & 0 \\ 0 & 0 & 0 & B_{55} & 0 \\ 0 & 0 & 0 & 0 & B_{66} \end{bmatrix} \begin{bmatrix} \dfrac{\partial u_0}{\partial x} \\[2mm] \dfrac{\partial v_0}{\partial y} \\[2mm] \phi_y + \dfrac{\partial w_0}{\partial y} \\[2mm] \phi_x + \dfrac{\partial w_0}{\partial x} \\[2mm] \dfrac{\partial u_0}{\partial y} + \dfrac{\partial v_0}{\partial x} \end{bmatrix} + \begin{bmatrix} D_{11} & D_{12} & 0 & 0 & 0 \\ D_{21} & D_{22} & 0 & 0 & 0 \\ 0 & 0 & D_{44} & 0 & 0 \\ 0 & 0 & 0 & D_{55} & 0 \\ 0 & 0 & 0 & 0 & D_{66} \end{bmatrix} \begin{bmatrix} \dfrac{\partial \phi_x}{\partial x} \\[2mm] \dfrac{\partial \phi_y}{\partial y} \\[2mm] 0 \\[2mm] 0 \\[2mm] \dfrac{\partial \phi_x}{\partial y} + \dfrac{\partial \phi_y}{\partial x} \end{bmatrix} \right) \tag{4.39b}
$$

where

$$
\begin{bmatrix} A_{11} & B_{11} & D_{11} \\ A_{22} & B_{22} & D_{22} \\ A_{12} & B_{12} & D_{12} \\ A_{44} & B_{44} & D_{44} \\ A_{55} & B_{55} & D_{55} \\ A_{66} & B_{66} & D_{66} \end{bmatrix} = \int_{-\frac{h_a}{2}}^{\frac{h_a}{2}} \begin{bmatrix} 1 & z & z^2 \end{bmatrix} \begin{bmatrix} C_{11}^a \\ C_{22}^a \\ C_{12}^a \\ C_{44}^a \\ C_{55}^a \\ C_{66}^a \end{bmatrix} dz + \int_{\frac{h_a}{2}}^{\frac{h_a}{2}+h_m} \begin{bmatrix} 1 & z & z^2 \end{bmatrix} \begin{bmatrix} C_{11}^m \\ C_{22}^m \\ C_{12}^m \\ C_{44}^m \\ C_{55}^m \\ C_{66}^m \end{bmatrix} dz + \int_{-\left(\frac{h_a}{2}+h_m\right)}^{-\frac{h_a}{2}} \begin{bmatrix} 1 & z & z^2 \end{bmatrix} \begin{bmatrix} C_{11}^m \\ C_{22}^m \\ C_{12}^m \\ C_{44}^m \\ C_{55}^m \\ C_{66}^m \end{bmatrix} dz
$$

(4.40)

The external work's first variation can be expressed as

$$
\delta V = k_w . w - k_g \left(\frac{\partial^2 w}{\partial x^2} + \frac{\partial^2 w}{\partial y^2} \right) \tag{4.41}
$$

where k_w and k_g are the Winkler stiffness and shear layer stiffness, respectively. The kinetic energy's first variation can be written as:

$$\delta T = \frac{1}{2}\iint \begin{bmatrix} I_0\left(\dfrac{\partial^2 u_0}{\partial t^2}\right)\delta u_0 + I_1\left(\dfrac{\partial^2 u_0}{\partial t^2}\right)\delta\phi_x + I_1\left(\dfrac{\partial^2 \phi_x}{\partial t^2}\right)\delta u_0 + I_2\left(\dfrac{\partial^2 \phi_x}{\partial t^2}\right)\delta\phi_x + \\ I_0\left(\dfrac{\partial^2 v_0}{\partial t^2}\right)\delta v_0 + I_1\left(\dfrac{\partial^2 v_0}{\partial t^2}\right)\delta\phi_y + I_1\left(\dfrac{\partial^2 \phi_y}{\partial t^2}\right)\delta v_0 + I_2\left(\dfrac{\partial^2 \phi_y}{\partial t^2}\right)\delta\phi_y + \\ I_0\left(\dfrac{\partial^2 w_0}{\partial t^2}\right)\delta w_0 \end{bmatrix} dA \tag{4.42}$$

in which, the mass moments of inertia are defined as:

$$I_0 = \int_{\frac{h_a}{2}}^{h_a} \rho_a dz + \left(\int_{-\frac{(h_a+h_m)}{2}}^{-\frac{h_a}{2}} \rho_m(z)\,dz + \int_{\frac{h_a}{2}}^{\frac{(h_a+h_m)}{2}} \rho_m(z)\,dz \right) \tag{4.43a}$$

$$I_1 = \int_{-h_a/2}^{h_a/2} z\rho_a dz + \left(\int_{-\frac{(h_a+h_m)}{2}}^{-\frac{h_a}{2}} z\rho_m(z)\,dz + \int_{\frac{h_a}{2}}^{\frac{(h_a+h_m)}{2}} z\rho_m(z)\,dz \right) \tag{4.43b}$$

$$I_2 = \int_{-h_a/2}^{h_a/2} z^2\rho_a dz + \left(\int_{-\frac{(h_a+h_m)}{2}}^{-\frac{h_a}{2}} z^2\rho_m(z)\,dz + \int_{\frac{h_a}{2}}^{\frac{(h_a+h_m)}{2}} z^2\rho_m(z)\,dz \right) \tag{4.43c}$$

and thus:

$$\delta u_0 : \frac{\partial N_{xx}}{\partial x} + \frac{\partial N_{xy}}{\partial y} - e_{31}h_m K_c C(t)\frac{\partial^2 w_0}{\partial x\partial t} = I_0\frac{\partial^2 u}{\partial t^2} + I_1\frac{\partial^2 \phi_x}{\partial t^2} \tag{4.44a}$$

$$\delta v_0 : \frac{\partial N_{xx}}{\partial y} + \frac{\partial N_{xy}}{\partial x} - e_{32}h_m K_c C(t)\frac{\partial^2 w_0}{\partial y\partial t} = I_0\frac{\partial^2 v}{\partial t^2} + I_1\frac{\partial^2 \phi_y}{\partial t^2} \tag{4.44b}$$

$$\delta w_0 : \frac{\partial N_{xz}}{\partial x} + \frac{\partial N_{yz}}{\partial y} - k_w . w + k_g\left(\frac{\partial^2 w}{\partial x^2} + \frac{\partial^2 w}{\partial y^2}\right)$$
$$-e_{31}h_m K_c C(t)\frac{\partial^2 U_x}{\partial x\partial t} - e_{32}h_m K_c C(t)\frac{\partial^2 V_y}{\partial y\partial t} = I_0\frac{\partial^2 w}{\partial t^2} \tag{4.44c}$$

$$\delta\phi_x : \frac{\partial M_{xx}}{\partial x} + \frac{\partial M_{xy}}{\partial y} - N_{xz} = I_1\frac{\partial^2 u}{\partial t^2} + I_2\frac{\partial^2 \phi_x}{\partial t^2} \tag{4.44d}$$

$$\delta\phi_y : \frac{\partial M_{yy}}{\partial y} + \frac{\partial M_{xy}}{\partial x} - N_{yz} = I_1\frac{\partial^2 v}{\partial t^2} + I_2\frac{\partial^2 \phi_y}{\partial t^2} \tag{4.44e}$$

The analytical solution proposed by Navier was used to provide a resolution for the governing equation. In order for the proposed solutions to Navier's problem to be effective, they must adhere to the specified boundary conditions of the problem. To continue with the calculation of the natural frequency, it is necessary to substitute the

indicated response into equations (4.44a–e), which represents the governing equation of the system. The Navier methodology has been effectively used in the determination of many issues in the field of vibration analysis. It is a definitive method
capable of accurately solving partial differential equations, exhibiting high levels of
convergence and performance. Furthermore, this methodology has been effectively
used to ascertain several additional issues within the domain of vibration analysis.
The displacements may be defined in line with Navier's solution by assuming that
the boundary conditions are simply supported.

$$
\begin{bmatrix} u_0(x,y,t) \\ v_0(x,y,t) \\ w_0(x,y,t) \\ \phi_x(x,y,t) \\ \phi_y(x,y,t) \end{bmatrix} = \begin{Bmatrix} \sum_{m=1}^{\infty}\sum_{n=1}^{\infty} U_{mn}\cos(\alpha x)\sin(\beta y)e^{i\omega t} \\ \sum_{m=1}^{\infty}\sum_{n=1}^{\infty} V_{mn}\sin(\alpha x)\cos(\beta y)e^{i\omega t} \\ \sum_{m=1}^{\infty}\sum_{n=1}^{\infty} W_{mn}\sin(\alpha x)\sin(\beta y)e^{i\omega t} \\ \sum_{m=1}^{\infty}\sum_{n=1}^{\infty} \phi_{xmn}\cos(\alpha x)\sin(\beta y)e^{i\omega t} \\ \sum_{m=1}^{\infty}\sum_{n=1}^{\infty} \phi_{ymn}\sin(\alpha x)\cos(\beta y)e^{i\omega t} \end{Bmatrix}
\qquad (4.44f)
$$

The amplitudes of the vibrations are denoted by U_{mn}, V_{mn}, W_{mn}, ϕ_{xmn}, and ϕ_{ymn}.
Additionally, the symbol ω is used to represent the fundamental frequency.
Furthermore, the symbols α and β are mathematically defined as $\beta = \dfrac{n\pi}{y}$ $\alpha = \dfrac{m\pi}{a}$.
The natural frequencies may be determined by putting equation (4.44f) into equations
(4.44a–e) and solving for the variables $u_0, v_0, w_0, \phi_x, \phi_y$. In order to accurately capture
the fundamental frequencies, it is necessary to ensure that the determinant of the
given equation is equal to zero.

$$
\left\| [K] - \omega^2 [M] + i\omega [C] \right\| = 0
\qquad (4.44g)
$$

The stiffness, mass, and damping matrices, denoted as K, M, and C correspondingly, are defined in Appendix 4.1. In addition, the dimensionless formulations of the
Winkler and Pasternak foundation models are regarded as:

$$
K_w = \frac{k_w a^4}{D_{11}} \quad K_g = \frac{k_g a^2}{D_{11}}
\qquad (4.44h)
$$

Table 4.1 has been used to monitor the first three mode forms of the plate in order
to investigate the impact of the auxetic inclination angle and velocity feedback gain
on the natural frequencies. Based on the findings shown in Table 4.1, it can be seen
that an increase in the auxetic inclination angle leads to a drop in the natural frequencies. This decrease may be attributed to the reduction in flexural stiffness. This phenomenon may mostly be attributed to the values of the Poisson's ratio. Furthermore,
as seen in Table 4.1, augmenting the feedback gain results in a decrease in the dimensionless natural frequencies of the system. This reduction facilitates the regulation of
the system's fluctuation frequency.

Tables 4.2 and 4.3 have been presented to analyze the impact of the vertical rib
length and thickness of the auxetic materials on the natural frequencies, considering

TABLE 4.1

Dimensionless Natural Frequencies of a Square Plate for the Different Values of Auxetic Inclination Angle and Velocity Feedback Gain for Three First Modes (t=1mm, l=4mm, e=8mm, a=b=60mm, hc=4mm, hf=0.1mm)

Dimensionless Frequency	Feedback gain	$\theta = -75°$	$\theta = -60°$	$\theta = -30°$	$\theta = -15°$	$\theta = +15°$	$\theta = +30°$	$\theta = +60°$	$\theta = +75°$
Ω_{11}	$k_c C(t) = 0$	2.8911	2.88791	2.88325	2.88314	2.88416	2.88518	2.88432	2.88343
Ω_{11}	$k_c C(t) = 1*10^5$	2.8883	2.88511	2.88046	2.88034	2.88136	2.88238	2.88153	2.88063
Ω_{11}	$k_c C(t) = 5*10^5$	2.87713	2.87395	2.8693	2.86919	2.87021	2.87123	2.87037	2.86948
Ω_{11}	$k_c C(t) = 9*10^5$	2.866	2.86283	2.85819	2.85808	2.8591	2.86012	2.85927	2.85836
Ω_{22}	$k_c C(t) = 0$	10.4732	10.4518	10.4263	10.4254	10.4283	10.4323	10.4279	10.428
Ω_{22}	$k_c C(t) = 1*10^5$	10.4639	10.4426	10.4171	10.4162	10.4191	10.4231	10.4187	10.4189
Ω_{22}	$k_c C(t) = 5*10^5$	10.4271	10.4058	10.3805	10.3795	10.3824	10.3864	10.382	10.3821
Ω_{22}	$k_c C(t) = 9*10^5$	10.3903	10.3692	10.3439	10.3429	10.3459	10.3498	10.3456	10.3456
Ω_{33}	$k_c C(t) = 0$	20.7403	20.6778	20.6095	20.6063	20.6105	20.619	20.6077	20.6151
Ω_{33}	$k_c C(t) = 1*10^5$	20.724	20.6616	20.5933	20.5902	20.5944	20.6028	20.5916	20.5989
Ω_{33}	$k_c C(t) = 5*10^5$	20.6589	20.5967	20.5287	20.5256	20.5298	20.5383	20.5271	20.5343
Ω_{33}	$k_c C(t) = 9*10^5$	20.594	20.5321	20.4644	20.4613	20.4655	20.474	20.4628	20.4699

TABLE 4.2

Dimensionless Natural Frequencies of a Square Plate for the Different Values of Rib Length and Velocity Feedback Gain (t=0.25mm, e=8mm, a=b=60mm, hc=4mm, hf=0.1mm, teta=60)

Natural Frequency	Feedback gain	$l = 8$mm	$l = 4$mm	$l = 2$mm	$l = 1$mm	$l = 0.5$mm
Ω_{11}	$k_c C(t) = 0*10^5$	2.88417	2.88418	2.88381	2.88315	2.87827
Ω_{11}	$k_c C(t) = 1*10^5$	2.88138	2.88138	2.88102	2.88035	2.87547
Ω_{11}	$k_c C(t) = 5*10^5$	2.87022	2.87023	2.86986	2.8692	2.86432
Ω_{12}	$k_c C(t) = 0*10^5$	6.83553	6.83553	6.83475	6.83343	6.82466
Ω_{12}	$k_c C(t) = 1*10^5$	6.82923	6.82924	6.82845	6.82714	6.81836
Ω_{12}	$k_c C(t) = 5*10^5$	6.80411	6.80411	6.80332	6.80201	6.79324
Ω_{22}	$k_c C(t) = 0*10^5$	10.4291	10.4291	10.4279	10.4253	10.4079
Ω_{22}	$k_c C(t) = 1*10^5$	10.4199	10.4199	10.4187	10.4161	10.3987
Ω_{22}	$k_c C(t) = 5*10^5$	10.3832	10.3832	10.382	10.3794	10.362

TABLE 4.3

Dimensionless Natural Frequencies of a Square Plate for the Different Values of the Auxetic Thickness (l=2mm, e=8mm, a=b=60mm, hc=4mm, hf=0.1mm, teta=60)

Natural Frequency	Feedback gain	$t = 1$mm	$t = 0.5$mm	$t = 0.25$mm	$t = 0.1$mm	$t = 0.01$mm
Ω_{11}	$k_c C(t) = 0*10^5$	2.86089	2.88128	2.88381	2.88415	2.88417
Ω_{11}	$k_c C(t) = 1*10^5$	2.85809	2.87848	2.88102	2.88136	2.88138
Ω_{11}	$k_c C(t) = 5*10^5$	2.84694	2.86732	2.86986	2.8702	2.87023
Ω_{12}	$k_c C(t) = 0*10^5$	6.78499	6.82924	6.83475	6.83548	6.83553
Ω_{12}	$k_c C(t) = 1*10^5$	6.77869	6.82294	6.82845	6.82919	6.82924
Ω_{12}	$k_c C(t) = 5*10^5$	6.75335	6.79781	6.80332	6.80406	6.80411
Ω_{22}	$k_c C(t) = 0*10^5$	10.3491	10.4191	10.4279	10.429	10.4291
Ω_{22}	$k_c C(t) = 1*10^5$	10.3399	10.41	10.4187	10.4198	10.4199
Ω_{22}	$k_c C(t) = 5*10^5$	10.3032	10.3733	10.382	10.3831	10.3832

various values of velocity feedback gain. The statistics clearly demonstrate that an increase in the vertical rib length results in a corresponding rise in the natural frequencies. Additionally, the data shown in these tables demonstrates that when the feedback gain is increased, there is a corresponding decrease in the natural frequency.

Table 4.4 provides a comprehensive depiction of the influence exerted by the thickness of the magnetostrictive layer on the active vibrational control characteristics of

TABLE 4.4

Dimensionless Natural Frequency of the Plate for Different Values of Auxetic Inclination Angle and Magnetostrictive Thickness (l=2mm, e=8mm, t=0.25mm, a=b=60mm, ha=4mm)

Magnetostrictive Thickness	Velocity Feedback Gain	$\theta = +15°$	$\theta = +30°$	$\theta = +60°$	$\theta = +75°$
$h_m=0.15$mm	$k_c C(t) = 0*10^5$	3.4475	3.44758	3.44721	3.44761
	$k_c C(t) = 1*10^5$	3.44351	3.44359	3.44321	3.44362
	$k_c C(t) = 5*10^5$	3.42758	3.42766	3.42728	3.42769
$h_c=0.1$mm	$k_c C(t) = 0*10^5$	2.88419	2.8843	2.88381	2.88433
	$k_c C(t) = 1*10^5$	2.8814	2.8815	2.88102	2.88154
	$k_c C(t) = 5*10^5$	2.87024	2.87035	2.86986	2.87038
$h_c=0.01$mm	$k_c C(t) = 0*10^5$	0.959487	0.959846	0.958167	0.959975
	$k_c C(t) = 1*10^5$	0.959177	0.959537	0.957858	0.959666
	$k_c C(t) = 5*10^5$	0.957941	0.9583	0.956621	0.95843

the continuous system. It is evident that an increase in the thickness of the magneto-strictive layer leads to a reduction in non-dimensional frequencies, hence establishing stability within the system.

Figure 4.7 is shown to investigate the impact of the inclination angle on the natural frequency of the plate. As seen in this figure, it is evident that an increase in the inclination angle results in a reduction in the dimensionless natural frequencies. This correlation has been previously substantiated in the preceding paragraphs from a physics perspective.

Furthermore, Figure 4.8 is shown to illustrate the impact of the feedback gain parameter on the dimensionless natural frequency. The figure demonstrates that the system's natural frequencies experience a notable drop when the feedback gain is increased. This decrease may be attributed to the damping impact caused by the magnetostriction phenomena on the system's behavior. The significance of the velocity feedback gain in governing the stability of the system is evident.

The impact of the vertical rib length on the stability of the system has been examined in Figure 4.8. The figure demonstrates that augmenting the vertical rib length results in a corresponding improvement in the dimensionless natural frequencies of the system.

Figure 4.9 provides a clear representation of the impact of the auxetic layer's thickness on the behavior of the continuous system. It is evident that an increase in the thickness of the core layer leads to a reduction in non-dimensional frequencies, hence establishing stability within the system.

Figure 4.10 has been generated to provide a comparative analysis of the impact of various plate components on two distinct velocity feedback gain values. The

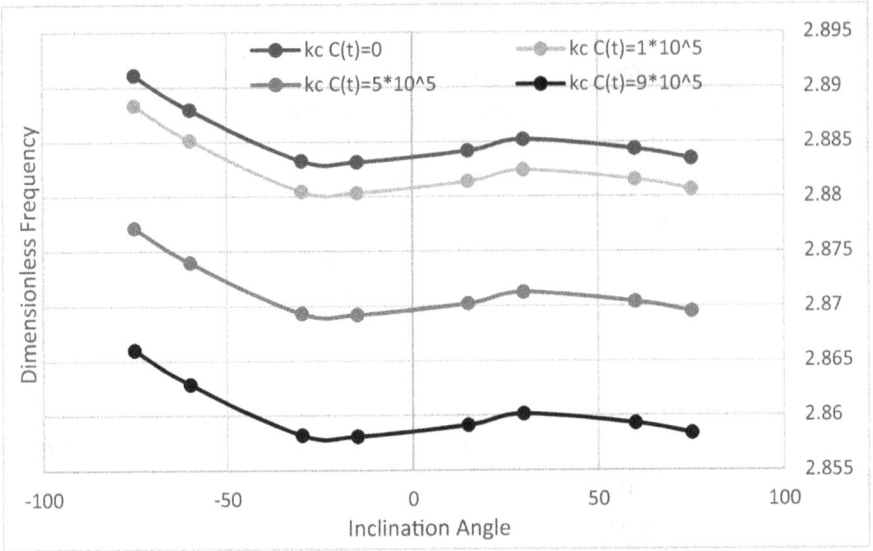

FIGURE 4.7 Dimensionless natural frequencies versus inclination angle for various amounts of velocity feedback gain (t=1mm, l=4mm, e=8mm, a=b=60mm, hc=4mm, hf=0.1mm).

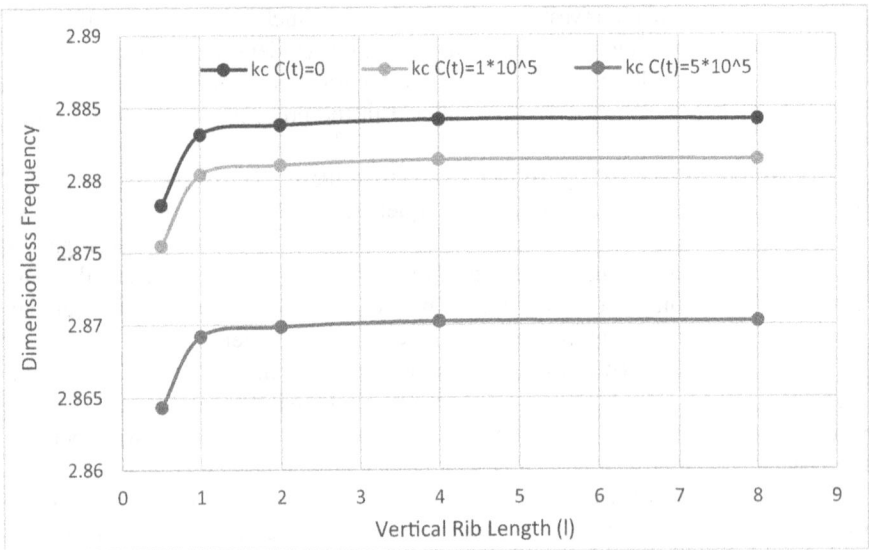

FIGURE 4.8 Dimensionless natural frequencies versus vertical rib length for various amounts of velocity feedback gain.

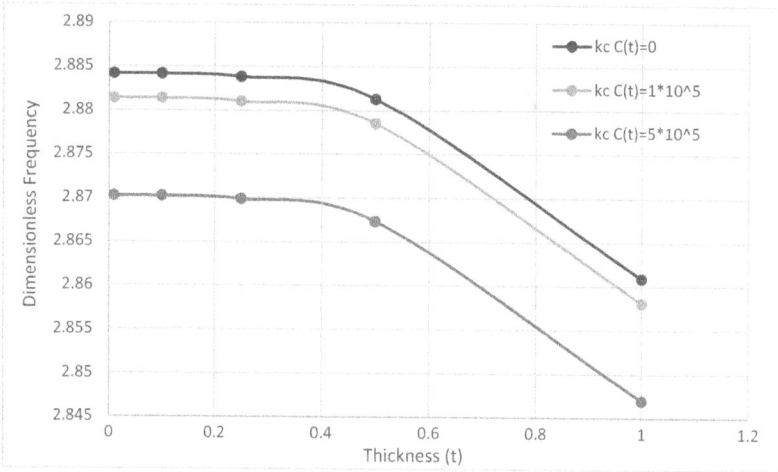

FIGURE 4.9 Dimensionless natural frequencies versus the thickness (t) for various amounts of velocity feedback gain.

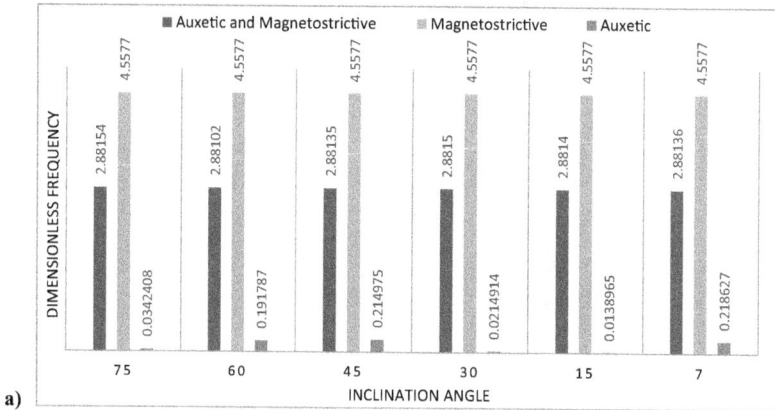

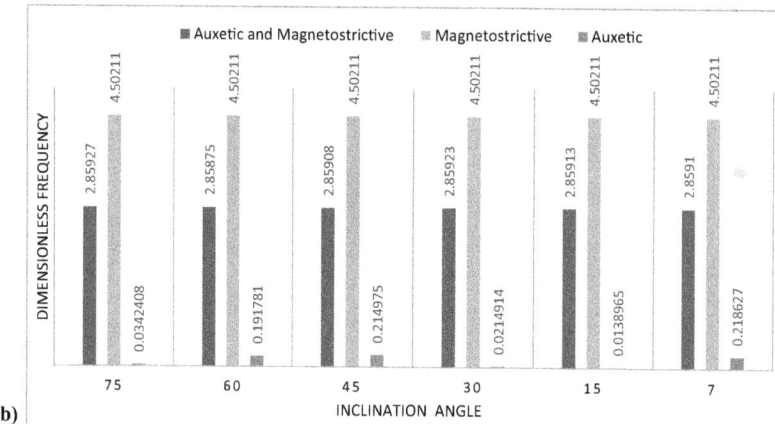

FIGURE 4.10 Comparison of dimensionless natural frequency for three different types of plate: (a) velocity feedback gain = 5*10^5; (b) velocity feedback gain = 9*10^5.

findings shown clearly demonstrate that the inclusion of a magnetostrictive layer in the plate leads to a significant augmentation in the dimensionless natural frequency. Moreover, it is worth noting that the pure magnetostrictive plate has the highest natural frequency when compared to the other two kinds mentioned.

4.4 ACTIVE VIBRATION CONTROL OF GRAPHENE-ORIGAMI-ENABLED AUXETIC PLATES WITH MAGNETOSTRICTIVE COATINGS

The objective of this research is to examine the use of plates including a graphene-origami-enabled auxetic core and coated magnetostrictive materials for the purpose of vibration control. The determination of the coordinates (x, y, and z) occurs inside a Cartesian coordinate system, where the origin is located at one of the corners of the plate. The illustration may be seen in Figure 4.11. The Winkler-Pasternak foundation has been implemented to integrate the influence of said foundation. The finite strip displacement technique (FSDT) is used to analyze and describe the movement of the composite plate.

The first-order shear deformation theory (FSDT) will be used to provide a contextual framework for the subject matter discussed in this publication. The mathematical representation of the displacement field associated with the FSDT may be found in the previous section, specifically in equations (4.44a–c). The displacement fields along the Cartesian x, y, and z axes are denoted as $U_x, V_y,$ and W_z, respectively. The derivation process would provide the features of graphene-origami auxetic (GOA).

$$E^{Auxetic} = E_{Cu} \frac{1 + \xi \eta V_{Gr}}{1 - \eta V_{Gr}} f_E \left(H_{Gr}, V_{Gr}, T \right) \tag{4.45a}$$

$$v^{Auxetic} = \left(V_{Gr} v_{Gr} + V_{Cu} v_{Cu} \right) f_v \left(H_{Gr}, V_{Gr}, T \right) \tag{4.45b}$$

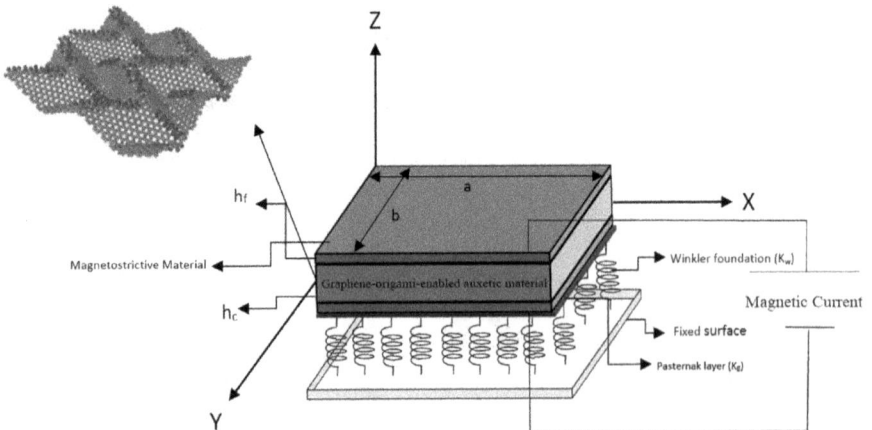

FIGURE 4.11 Plan view of the structure.

$$\alpha^{Auxetic} = \left(V_{Gr}\alpha_{Gr} + V_{Cu}\alpha_{Cu}\right)f_{\alpha}\left(V_{Gr},T\right) \tag{4.45c}$$

$$\rho^{Auxetic} = \left(V_{Gr}\rho_{Gr} + V_{Cu}\rho_{Cu}\right)f_{\rho}\left(V_{Gr},T\right) \tag{4.45d}$$

$$f_E\left(H_{Gr},V_{Gr},T\right) = \begin{pmatrix} 1.11 - 1.22V_{Gr} - 0.134\left(\dfrac{T}{T_0}\right) \\[2mm] +0.559V_{Gr}\left(\dfrac{T}{T_0}\right) - 5.5H_{Gr}V_{Gr} \\[2mm] +38H_{Gr}V_{Gr}^2 - 20.6H_{Gr}^2V_{Gr}^2 \end{pmatrix} \tag{4.45e}$$

$$f_v\left(H_{Gr},V_{Gr},T\right) = \begin{pmatrix} 1.01 - 1.43V_{Gr} + 0.165\left(\dfrac{T}{T_0}\right) \\[2mm] -16.8H_{Gr}V_{Gr} - 1.1H_{Gr}V_{Gr}\left(\dfrac{T}{T_0}\right) \\[2mm] +16H_{Gr}^2V_{Gr}^2 \end{pmatrix} \tag{4.45f}$$

$$f_\alpha\left(V_{Gr},T\right) = \begin{pmatrix} 0.794 - 16.8V_{Gr}^2 - 0.0279\left(\dfrac{T}{T_0}\right)^2 \\[2mm] +0.182\left(\dfrac{T}{T_0}\right)\left(V_{Gr}+1\right) \end{pmatrix} \tag{4.45g}$$

$$f_\rho\left(V_{Gr},T\right) = \left(1.01 - 2.01V_{Gr}^2 - 0.0131\left(\dfrac{T}{T_0}\right)\right) \tag{4.45h}$$

The value of T_0 is equal to 300K, whereas T represents the ambient temperature. The volume fraction (V_{Gr}) of the GOA material with a uniform distribution is formally defined as:

$$V_{Gr} = \frac{\rho_{Cu}W_{Gr}}{\rho_{Cu}W_{Gr} + \rho_{Gr}\left(1 - W_{Gr}\right)}$$

$$V_{Cu} = 1 - V_{Gr} \tag{4.46}$$

The symbols ρ_{Cu} and ρ_{Gr} represent the corresponding densities of pure copper and graphene. Moreover, the symbol V_{Cu} represents the volume proportion of copper. Additionally, it is expected that the degree of folding in graphene-origami, as measured by the hydrogen atom coverage (H_{Gr}) in the crease, varies gradually along the direction of thickness. Additionally, the material coefficient (η) and size coefficient (ξ) are explicitly specified as:

$$\eta = \frac{\dfrac{E_{Gr}}{E_{Cu}} - 1}{\dfrac{E_{Gr}}{E_{Cu}} + \xi}, \xi = 2\left(\frac{l_{Gr}}{t_{Gr}}\right) \tag{4.47}$$

The variables l_{Gr} and t_{Gr} represent the length and thickness of graphene, respectively. In the realm of auxetic materials facilitated by graphene origami, the functional relationship between stress and strain may be articulated in the following manner:

$$\sigma_{xx}^{a} = C_{11}^{a}\varepsilon_{xx} + C_{12}^{a}\varepsilon_{yy} \tag{4.48a}$$

$$\sigma_{yy}^{a} = C_{21}^{a}\varepsilon_{xx} + C_{22}^{a}\varepsilon_{yy} \tag{4.48b}$$

$$\tau_{yz}^{a} = C_{44}^{a}\gamma_{yz} \tag{4.48c}$$

$$\tau_{xz}^{a} = C_{55}^{a}\gamma_{xz} \tag{4.48d}$$

$$\tau_{xy}^{a} = C_{66}^{a}\gamma_{xy} \tag{4.48e}$$

The idea of "auxetic layer non-zero stresses" is a well-established phrase in academic discourse.

$$\begin{Bmatrix} \varepsilon_{xx}^{a} \\ \varepsilon_{yy}^{a} \\ \varepsilon_{yz}^{a} \\ \varepsilon_{xz}^{a} \\ \varepsilon_{xy}^{a} \end{Bmatrix} = \begin{Bmatrix} \epsilon_{xx}^{a} \\ \epsilon_{yy}^{a} \\ \epsilon_{yz}^{a} \\ \epsilon_{xz}^{a} \\ \epsilon_{xy}^{a} \end{Bmatrix} + z \begin{Bmatrix} \kappa_{xx}^{a} \\ \kappa_{yy}^{a} \\ \kappa_{yz}^{a} \\ \kappa_{xz}^{a} \\ \kappa_{xy}^{a} \end{Bmatrix} \tag{4.49}$$

in which,

$$\begin{Bmatrix} \epsilon_{xx}^{a} \\ \epsilon_{yy}^{a} \\ \epsilon_{yz}^{a} \\ \epsilon_{xz}^{a} \\ \epsilon_{xy}^{a} \end{Bmatrix} = \begin{Bmatrix} \dfrac{\partial u_0}{\partial x} \\ \dfrac{\partial v_0}{\partial y} \\ \left(\phi_y + \dfrac{\partial w_0}{\partial y}\right) \\ \left(\phi_x + \dfrac{\partial w_0}{\partial x}\right) \\ \left(\dfrac{\partial u_0}{\partial y} + \dfrac{\partial v_0}{\partial x}\right) \end{Bmatrix}, \begin{Bmatrix} \kappa_{xx}^{a} \\ \kappa_{yy}^{a} \\ \kappa_{yz}^{a} \\ \kappa_{xz}^{a} \\ \kappa_{xy}^{a} \end{Bmatrix} = \begin{Bmatrix} \dfrac{\partial \phi_x}{\partial x} \\ \dfrac{\partial \phi_y}{\partial y} \\ 0 \\ 0 \\ \dfrac{\partial \phi_x}{\partial y} + \dfrac{\partial \phi_y}{\partial x} \end{Bmatrix} \tag{4.50}$$

The stress, strain, and magnetic susceptibility of the magnetostrictive layer are expressed using the formula shown in equations (4.30a–e) in the preceding section.

The stress and strain tensors are represented as arrays using the notation $\sigma_{ij}, \varepsilon_{ij}$ $(i,j = x, y, z)$. Furthermore, the magnetic field is represented by the symbol H_{zz}. Under the assumption of zero-thickness stretching, denoted as $\varepsilon_{zz} = 0$, the elastic constants pertaining to the stress plane may be mathematically represented as follows:

$$C_{11}^m = \frac{E_m}{1-\vartheta_m^2} \quad C_{12}^m = \frac{\vartheta_m E_m}{1-\vartheta_m^2} \quad C_{66}^m = \frac{E_m}{2(1+\vartheta_m)} \quad C_{55}^m = C_{44}^m = k_s C_{66}^m \tag{4.51}$$

The variables E_m and v_m denote the elasticity modulus and Poisson's ratio of the magnetostrictive layers, respectively. It is noteworthy that the shear correction parameter, denoted by the constant k_s, has been attributed with distinct values by diverse theories. The magnetic field (H_{zz}) can be computed utilizing the subsequent connection:

$$H_{zz} = k_c I(x,y,t) \tag{4.52a}$$

$$I(x,y,t) = C(t)\frac{\partial w_0(x,y,t)}{\partial t} \tag{4.52b}$$

The coil constant, denoted as k_c, is determined by the physical characteristics of the coil, such as its width, radius, number of turns, and material composition. Moreover, the variables $I(x,y,t)$ and $C(t)$ denote the electric current traversing the coil and the amplification factor of the coil, respectively. It is crucial to recognize that a consistent control gain is being claimed. The gain of the velocity feedback is denoted as $K_c C(t)$. The correlation between the features of a coil and its coil constant may be mathematically represented as follows:

$$K_c = \frac{n_c}{\sqrt{b_c^2 + 4r_c^2}} \tag{4.53}$$

The equation described above includes the variables n_c, b_c, and r_c, which represent the number of turns, width, and radius of the coil, respectively. The non-zero stresses shown by the magnetostrictive layer may be described in the following manner:

$$\begin{Bmatrix} \varepsilon_{xx}^m \\ \varepsilon_{yy}^m \\ \varepsilon_{yz}^m \\ \varepsilon_{xz}^m \\ \varepsilon_{xy}^m \end{Bmatrix} = \begin{Bmatrix} \epsilon_{xx}^m \\ \epsilon_{yy}^m \\ \epsilon_{yz}^m \\ \epsilon_{xz}^m \\ \epsilon_{xy}^m \end{Bmatrix} + z \begin{Bmatrix} \kappa_{xx}^m \\ \kappa_{yy}^m \\ \kappa_{yz}^m \\ \kappa_{xz}^m \\ \kappa_{xy}^m \end{Bmatrix} \tag{4.54}$$

where

$$
\begin{Bmatrix} \epsilon_{xx}^{m} \\ \epsilon_{yy}^{m} \\ \epsilon_{yz}^{m} \\ \epsilon_{xz}^{m} \\ \epsilon_{xy}^{m} \end{Bmatrix} = \begin{Bmatrix} \dfrac{\partial u_0}{\partial x} \\ \dfrac{\partial v_0}{\partial y} \\ \left(\phi_y + \dfrac{\partial w_0}{\partial y} \right) \\ \left(\phi_x + \dfrac{\partial w_0}{\partial x} \right) \\ \left(\dfrac{\partial u_0}{\partial y} + \dfrac{\partial v_0}{\partial x} \right) \end{Bmatrix} \quad \begin{Bmatrix} \kappa_{xx}^{m} \\ \kappa_{yy}^{m} \\ \kappa_{yz}^{m} \\ \kappa_{xz}^{m} \\ \kappa_{xy}^{m} \end{Bmatrix} = \begin{Bmatrix} \dfrac{\partial \phi_x}{\partial x} \\ \dfrac{\partial \phi_y}{\partial y} \\ 0 \\ 0 \\ \dfrac{\partial \phi_x}{\partial y} + \dfrac{\partial \phi_y}{\partial x} \end{Bmatrix}
\tag{4.55}
$$

The properties of auxetic materials enabled by graphene origami are as follows:

$$
E_{gr} \text{ (GPa)} = 929.57, \ E_{Cu} \text{ (GPa)} = 65.79, \ \vartheta_{gr} = 0.22, \ \vartheta_{Cu} = 0.387,
$$

$$
\alpha_{gr} = -3.98 * 10^{-6} \left(\text{K}^{-1} \right), \ \alpha_{Cu} = 16.51 * 10^{-6} \left(\text{K}^{-1} \right), \ \rho_{gr} \left(\text{kg} \middle/ m^3 \right) = 1800,
$$

$$
\rho_{Cu} \left(\text{kg} \middle/ m^3 \right) = 8800
$$

The motion equations in this section are obtained by using the energy technique. The notion put out by Hamilton will be used for this purpose. One possible way to express this concept is as follows:

$$
\int_0^t \delta \left(U - T + V \right) dt = 0
\tag{4.56}
$$

The symbols U, T, and V represent the strain energy, kinetic energy, and external work, correspondingly. The first variation of strain energy may be expressed as:

$$
\delta U^a = \int_0^a \int_0^b \begin{pmatrix} N_{xx}^a \left(\delta \epsilon_{xx}^a \right) + M_{xx}^a \left(\delta \kappa_{xx}^a \right) + N_{yy}^a \left(\delta \epsilon_{yy}^a \right) + M_{yy}^a \left(\delta \kappa_{yy}^a \right) + \\ N_{xz}^a \left(\delta \epsilon_{xz}^a \right) + M_{xz}^a \left(\delta \kappa_{xz}^a \right) + N_{yz}^a \left(\delta \epsilon_{yz}^a \right) + M_{yz}^a \left(\delta \kappa_{yz}^a \right) + \\ N_{xy}^a \left(\delta \epsilon_{xy}^a \right) + M_{xy}^a \left(\delta \kappa_{xy}^a \right) \end{pmatrix} dA
\tag{4.57a}
$$

$$
\delta U^m = \int_0^a \int_0^b \begin{pmatrix} N_{xx}^m \left(\delta \epsilon_{xx}^m \right) + M_{xx}^m \left(\delta \kappa_{xx}^m \right) + N_{yy}^m \left(\delta \epsilon_{yy}^m \right) + M_{yy}^m \left(\delta \kappa_{yy}^m \right) + \\ N_{xz}^m \left(\delta \epsilon_{xz}^m \right) + M_{xz}^m \left(\delta \kappa_{xz}^m \right) + N_{yz}^m \left(\delta \epsilon_{yz}^m \right) + M_{yz}^m \left(\delta \kappa_{yz}^m \right) + \\ N_{xy}^m \left(\delta \epsilon_{xy}^m \right) + M_{xy}^m \left(\delta \kappa_{xy}^m \right) \end{pmatrix} dA
\tag{4.57b}
$$

where

$$
\left(N_{ij}^a, M_{ij}^a \right) = \int_{-\frac{h_a}{2}}^{\frac{h_a}{2}} \sigma_{ij}^a \left(1, z \right) dz, \ \left(i, j = x, \ y, \ z \right)
\tag{4.57c}
$$

$$
\left(N_{ij}^m, M_{ij}^m \right) = \int_{\frac{h_a}{2}}^{\frac{h_a}{2} + h_m} \sigma_{ij}^m \left(1, z \right) dz + \int_{-\left(\frac{h_a}{2} + h_m \right)}^{-\frac{h_a}{2}} \sigma_{ij}^m \left(1, z \right) dz, \ \left(i, j = x, \ y, z \right)
\tag{4.57d}
$$

The aforementioned equations use superscripts a and m to denote the auxetic and magnetostrictive layers, correspondingly. Additionally, the forces $N_{i,j}$ and moments $M_{i,j}$ are indicated. The relationship between forces and bending moments may be mathematically described by the following equations:

$$
\begin{bmatrix} N_{xx} \\ N_{yy} \\ N_{yz} \\ N_{xz} \\ N_{xy} \end{bmatrix} = \begin{bmatrix} A_{11} & A_{12} & 0 & 0 & 0 \\ A_{21} & A_{22} & 0 & 0 & 0 \\ 0 & 0 & A_{44} & 0 & 0 \\ 0 & 0 & 0 & A_{55} & 0 \\ 0 & 0 & 0 & 0 & A_{66} \end{bmatrix} \left(\begin{bmatrix} \dfrac{\partial u_0}{\partial x} \\ \dfrac{\partial v_0}{\partial y} \\ \phi_y + \dfrac{\partial w_0}{\partial y} \\ \phi_x + \dfrac{\partial w_0}{\partial x} \\ \dfrac{\partial u_0}{\partial y} + \dfrac{\partial v_0}{\partial x} \end{bmatrix} + \begin{bmatrix} B_{11} & B_{12} & 0 & 0 & 0 \\ B_{21} & B_{22} & 0 & 0 & 0 \\ 0 & 0 & B_{44} & 0 & 0 \\ 0 & 0 & 0 & B_{55} & 0 \\ 0 & 0 & 0 & 0 & B_{66} \end{bmatrix} \begin{bmatrix} \dfrac{\partial \phi_x}{\partial x} \\ \dfrac{\partial \phi_y}{\partial y} \\ 0 \\ 0 \\ \dfrac{\partial \phi_x}{\partial y} + \dfrac{\partial \phi_y}{\partial x} \end{bmatrix} \right)
$$

(4.58a)

$$
\begin{bmatrix} M_{xx} \\ M_{yy} \\ M_{yz} \\ M_{xz} \\ M_{xy} \end{bmatrix} = \begin{bmatrix} B_{11} & B_{12} & 0 & 0 & 0 \\ B_{21} & B_{22} & 0 & 0 & 0 \\ 0 & 0 & B_{44} & 0 & 0 \\ 0 & 0 & 0 & B_{55} & 0 \\ 0 & 0 & 0 & 0 & B_{66} \end{bmatrix} \left(\begin{bmatrix} \dfrac{\partial u_0}{\partial x} \\ \dfrac{\partial v_0}{\partial y} \\ \phi_y + \dfrac{\partial w_0}{\partial y} \\ \phi_x + \dfrac{\partial w_0}{\partial x} \\ \dfrac{\partial u_0}{\partial y} + \dfrac{\partial v_0}{\partial x} \end{bmatrix} + \begin{bmatrix} D_{11} & D_{12} & 0 & 0 & 0 \\ D_{21} & D_{22} & 0 & 0 & 0 \\ 0 & 0 & D_{44} & 0 & 0 \\ 0 & 0 & 0 & D_{55} & 0 \\ 0 & 0 & 0 & 0 & D_{66} \end{bmatrix} \begin{bmatrix} \dfrac{\partial \phi_x}{\partial x} \\ \dfrac{\partial \phi_y}{\partial y} \\ 0 \\ 0 \\ \dfrac{\partial \phi_x}{\partial y} + \dfrac{\partial \phi_y}{\partial x} \end{bmatrix} \right)
$$

(4.58b)

where

$$
\begin{bmatrix} A_{11} & B_{11} & D_{11} \\ A_{22} & B_{22} & D_{22} \\ A_{12} & B_{12} & D_{12} \\ A_{44} & B_{44} & D_{44} \\ A_{55} & B_{55} & D_{55} \\ A_{66} & B_{66} & D_{66} \end{bmatrix} = \int_{-\frac{h_a}{2}}^{\frac{h_a}{2}} \begin{bmatrix} 1 & z & z^2 \end{bmatrix} \begin{bmatrix} C_{11}^a \\ C_{22}^a \\ C_{12}^a \\ C_{44}^a \\ C_{55}^a \\ C_{66}^a \end{bmatrix} dz + \int_{\frac{h_a}{2}}^{\frac{h_a}{2}+h_m} \begin{bmatrix} 1 & z & z^2 \end{bmatrix} \begin{bmatrix} C_{11}^m \\ C_{22}^m \\ C_{12}^m \\ C_{44}^m \\ C_{55}^m \\ C_{66}^m \end{bmatrix} dz + \int_{-\left(\frac{h_a}{2}+h_m\right)}^{-\frac{h_a}{2}} \begin{bmatrix} 1 & z & z^2 \end{bmatrix} \begin{bmatrix} C_{11}^m \\ C_{22}^m \\ C_{12}^m \\ C_{44}^m \\ C_{55}^m \\ C_{66}^m \end{bmatrix} dz
$$

(4.58c)

The first variation of external work may be mathematically represented as:

$$\delta V = \int_0^a \int_0^b \left[k_w . \delta w_0 - k g \left(\frac{\partial^2 \delta w_0}{\partial x^2} + \frac{\partial^2 \delta w_0}{\partial y^2} \right) - N_{xx}^T \left(\frac{\partial^2 \delta w_0}{\partial x^2} \right) - N_{yy}^T \left(\frac{\partial^2 \delta w_0}{\partial y^2} \right) \right] \quad (4.59)$$

$$N_{ii}^T = -\sum_{l=1}^{n} \int_{z_k}^{z_{k+1}} \alpha_{ii} \left(T - T_0 \right), ii = xx, yy$$

The term N_{ii}^T represents the force that is generated as a result of changes in temperature. The values k_w and k_g denote the Winkler stiffness and shear layer stiffness, respectively. The initial change in kinetic energy may be mathematically represented as:

$$\delta T = \frac{1}{2} \int \int \begin{bmatrix} I_0 \left(\frac{\partial^2 u_0}{\partial t^2} \right) \delta u_0 + I_1 \left(\frac{\partial^2 u_0}{\partial t^2} \right) \delta \phi_x + I_1 \left(\frac{\partial^2 \phi_x}{\partial t^2} \right) \delta u_0 + I_2 \left(\frac{\partial^2 \phi_x}{\partial t^2} \right) \delta \phi_x + \\ I_0 \left(\frac{\partial^2 v_0}{\partial t^2} \right) \delta v_0 + I_1 \left(\frac{\partial^2 v_0}{\partial t^2} \right) \delta \phi_y + I_1 \left(\frac{\partial^2 \phi_y}{\partial t^2} \right) \delta v_0 + I_2 \left(\frac{\partial^2 \phi_y}{\partial t^2} \right) \delta \phi_y + \\ I_0 \left(\frac{\partial^2 w_0}{\partial t^2} \right) \delta w_0 \end{bmatrix} dA \quad (4.60)$$

in which, the mass moments of inertia are defined as:

$$I_0 = \int_{-\frac{h_a}{2}}^{\frac{h_a}{2}} \rho_a dz + \left(\int_{-\frac{(h_a + h_m)}{2}}^{-\frac{h_a}{2}} \rho_m (z) dz + \int_{\frac{h_a}{2}}^{\frac{(h_a + h_m)}{2}} \rho_m (z) dz \right) \quad (4.61a)$$

$$I_1 = \int_{-h_a/2}^{h_a/2} z \rho_a dz + \left(\int_{-\frac{(h_a + h_m)}{2}}^{-\frac{h_a}{2}} z \rho_m (z) dz + \int_{\frac{h_a}{2}}^{\frac{(h_a + h_m)}{2}} z \rho_m (z) dz \right) \quad (4.61b)$$

$$I_2 = \int_{-h_a/2}^{h_a/2} z^2 \rho_a dz + \left(\int_{-\frac{(h_a + h_m)}{2}}^{-\frac{h_a}{2}} z^2 \rho_m (z) dz + \int_{\frac{h_a}{2}}^{\frac{(h_a + h_m)}{2}} z^2 \rho_m (z) dz \right) \quad (4.61c)$$

And thus:

$$\delta u_0 : \frac{\partial N_{xx}}{\partial x} + \frac{\partial N_{xy}}{\partial y} - e_{31} h_m K_c C(t) \frac{\partial^2 w_0}{\partial x \partial t} = I_0 \frac{\partial^2 u}{\partial t^2} + I_1 \frac{\partial^2 \phi_x}{\partial t^2} \quad (4.62a)$$

$$\delta v_0 : \frac{\partial N_{xx}}{\partial y} + \frac{\partial N_{xy}}{\partial x} - e_{32} h_m K_c C(t) \frac{\partial^2 w_0}{\partial y \partial t} = I_0 \frac{\partial^2 v}{\partial t^2} + I_1 \frac{\partial^2 \phi_y}{\partial t^2} \quad (4.62b)$$

$$\delta w_0 : \frac{\partial N_{xz}}{\partial x} + \frac{\partial N_{yz}}{\partial y} - k_w . w + k_g \left(\frac{\partial^2 w}{\partial x^2} + \frac{\partial^2 w}{\partial y^2} \right) - e_{31} h_m K_c C(t) \frac{\partial^2 U_x}{\partial x \partial t}$$

$$- e_{32} h_m K_c C(t) \frac{\partial^2 V_y}{\partial y \partial t} = I_0 \frac{\partial^2 w}{\partial t^2} \quad (4.62c)$$

$$\delta\phi_x : \frac{\partial M_{xx}}{\partial x} + \frac{\partial M_{xy}}{\partial y} - N_{xz} = I_1 \frac{\partial^2 u}{\partial t^2} + I_2 \frac{\partial^2 \phi_x}{\partial t^2} \qquad (4.62d)$$

$$\delta\phi_y : \frac{\partial M_{yy}}{\partial y} + \frac{\partial M_{xy}}{\partial x} - N_{yz} = I_1 \frac{\partial^2 v}{\partial t^2} + I_2 \frac{\partial^2 \phi_y}{\partial t^2} \qquad (4.62e)$$

Once again, the analytical solution proposed by Navier was used to get a solution for the governing equation, as discussed in the preceding section. In order to accurately capture the fundamental frequencies, it is essential to ensure that the determinant of the given equation is equated to zero.

$$\left| [K] - \omega^2 [M] + i\omega [C] \right| = 0 \qquad (4.63)$$

The stiffness, mass, and damping matrices, denoted as K, M, and C correspondingly, are specified in Appendix 4.2. The objective of this part is to perform a comprehensive analysis into the effects of various factors on the control of vibrations in the proposed constructions. Table 4.5 has been developed in order to monitor the first three mode forms of the plate, with the aim of analyzing the influence of the graphene weight percentage and velocity feedback gain on the natural frequencies. The findings suggest that an augmentation in the weight percentage of graphene results in a corresponding augmentation and a significant decline in the natural frequencies, occurring at about $W_{Gr}(wt\%) = 1$. Furthermore, it can be seen that increasing the feedback gain value leads to a reduction in the dimensionless natural frequencies of the system. The aforementioned decrease enables the effective control of the frequency fluctuations inside the system.

Figure 4.12 shows how weight fraction affects plate natural frequency. Figure 4.12 shows that with increasing weight fraction, dimensionless natural frequencies drop.

Table 4.6 shows how increasing temperature affects natural frequencies throughout velocity feedback gain settings. The table shows that natural frequencies decrease with temperature. The table also shows that increasing feedback gain decreases natural frequency. A decreased Young's modulus and a higher absolute value of negative Poisson's ratio cause tiny changes of plane stress-reduced stiffness in the GOA composite plate, causing the observed phenomena.

Figure 4.13 shows how the feedback gain parameter affects dimensionless natural frequency. The magnetostriction phenomenon dampens the system's activity, lowering natural frequencies as the feedback gain increases. It's apparent that velocity feedback gain controls system stability.

Table 4.7 shows how the magnetostrictive layer thickness affects the continuous system's active vibrational control. An increase in magnetostrictive layer thickness decreases non-dimensional frequencies, stabilizing the system.

TABLE 4.5

Effect of the Graphene Weight Fraction and Velocity Feedback Gain for Three First Modes (a=b=0.1m, hc=0.003m, hf=0.001nm, T=300, Hgr=100%)

Dimensionless Frequency	Feedback gain	$W_{Gr}(wt\%)$ 0	0.5	1	1.5	2	2.5
Ω_{11}	$k_c C(t)=0$	0.227716	0.276133	0.125479	0.10813	0.0971478	0.0891444
Ω_{11}	$k_c C(t)=1*10^5$	0.227427	0.2761	0.125445	0.108102	0.0971239	0.0891236
Ω_{11}	$k_c C(t)=5*10^5$	0.226273	0.275968	0.125308	0.107989	0.0970284	0.0890405
Ω_{11}	$k_c C(t)=9*10^5$	0.225125	0.275836	0.125171	0.107875	0.0969329	0.0889575
Ω_{22}	$k_c C(t)=0$	0.890566	1.08644	0.492438	0.424247	0.381178	0.349798
Ω_{22}	$k_c C(t)=1*10^5$	0.889459	1.08631	0.492306	0.424138	0.381086	0.34717
Ω_{22}	$k_c C(t)=5*10^5$	0.885043	1.0858	0.491779	0.423701	0.380718	0.349397
Ω_{22}	$k_c C(t)=9*10^5$	0.880649	1.08529	0.491252	0.423266	0.380351	0.349078
Ω_{33}	$k_c C(t)=0$	1.93535	2.38151	1.0754	0.926191	0.832268	0.763843
Ω_{33}	$k_c C(t)=1*10^5$	1.93302	2.38123	1.07512	0.92596	0.83072	0.76673
Ω_{33}	$k_c C(t)=5*10^5$	1.92373	2.38015	1.074	0.925034	0.831291	0.762994
Ω_{33}	$k_c C(t)=9*10^5$	1.91448	2.37906	1.07289	0.92411	0.830511	0.762315

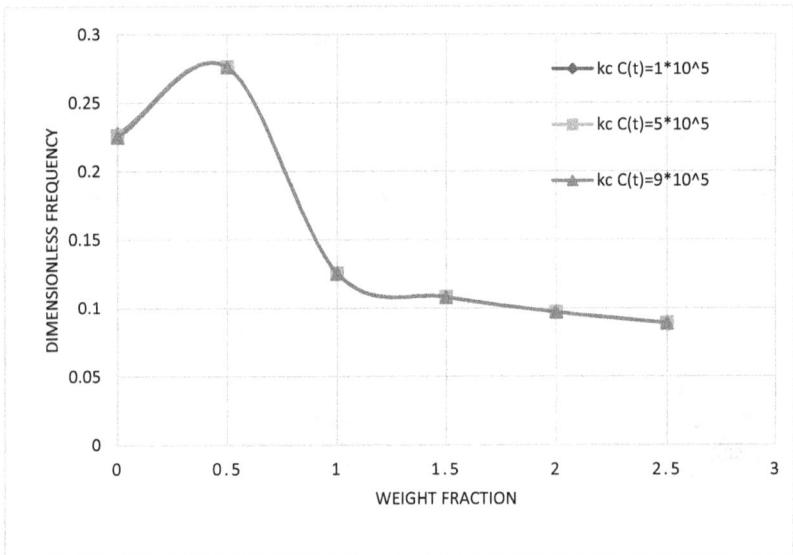

FIGURE 4.12 Dimensionless natural frequencies versus weight fraction for various amounts of velocity feedback gain.

TABLE 4.6

Effect of the Temperature and Velocity Feedback Gain (a=b=0.1m, hc=0.003m, hf=0.001nm, W_{Gr} ($wt\%$) =2.5, Hgr=100%)

Dimensionless Frequency	Feedback gain	T (K)						
		300	310	320	330	340	350	360
Ω_{11}	$k_c C(t) = 0*10^5$	0.0891444	0.0891226	0.0891059	0.0890937	0.0890851	0.0890797	0.0890769
Ω_{11}	$k_c C(t) = 1*10^5$	0.0891236	0.0891018	0.089085	0.0890729	0.0890643	0.0890589	0.0890561
Ω_{11}	$k_c C(t) = 5*10^5$	0.0890405	0.0890187	0.089002	0.0889871	0.0889811	0.0889757	0.0889729
Ω_{11}	$k_c C(t) = 9*10^5$	0.0889575	0.0889356	0.0889189	0.0889066	0.088898	0.0888926	0.0888898
Ω_{22}	$k_c C(t) = 0*10^5$	0.349798	0.349701	0.349625	0.349567	0.349524	0.349493	0.349474
Ω_{22}	$k_c C(t) = 1*10^5$	0.34717	0.349621	0.349545	0.349487	0.349444	0.349413	0.349394
Ω_{22}	$k_c C(t) = 5*10^5$	0.349397	0.349301	0.349225	0.349167	0.349123	0.349093	0.349073
Ω_{22}	$k_c C(t) = 9*10^5$	0.349078	0.348981	0.348905	0.348847	0.348803	0.348773	0.348753
Ω_{33}	$k_c C(t) = 0*10^5$	0.763843	0.763598	0.763401	0.763243	0.76312	0.763026	0.762957
Ω_{33}	$k_c C(t) = 1*10^5$	0.76673	0.763428	0.763231	0.763073	0.76295	0.762856	0.762787
Ω_{33}	$k_c C(t) = 5*10^5$	0.762994	0.762749	0.762551	0.762393	0.76227	0.762176	0.762107
Ω_{33}	$k_c C(t) = 9*10^5$	0.762315	0.76207	0.761872	0.761714	0.761591	0.761497	0.761428

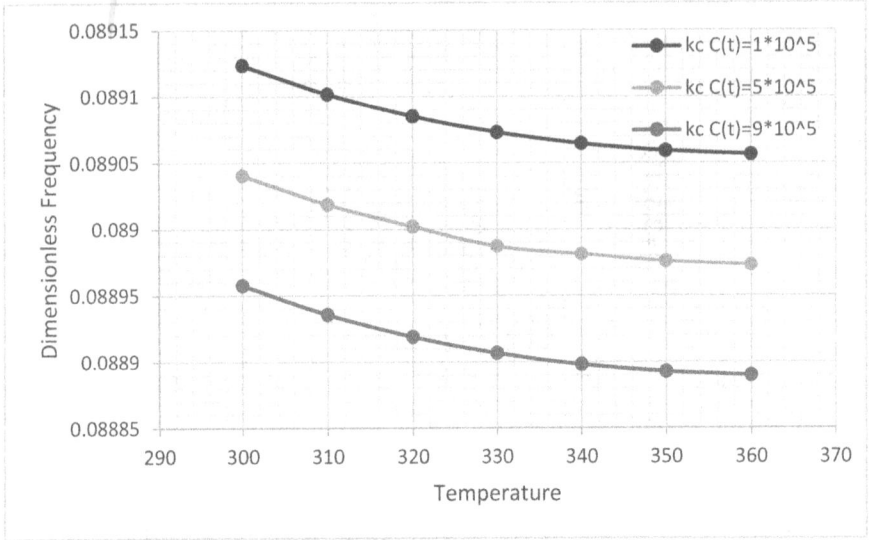

FIGURE 4.13 Dimensionless natural frequencies versus temperature for various amounts of velocity feedback gain.

TABLE 4.7

Effect of the Geometry ($W_{Gr}\left(wt\%\right)$=2.5, Hgr=100%, T=320, $k_c C\left(t\right) = 5*10^5$)

Magnetostrictive Thickness	Graphene-origami auxetic thickness	a/b 1	1/2	1/3
h_m=0.001m	h_a=0.003	0.089002	0.0557643	0.0495911
	h_a=0.002	0.0466667	0.0292039	0.0259652
	h_a=0.001	0.0161414	0.0100934	0.00897278
h_m=0.002m	h_a=0.003	0.0592357	0.0371448	0.0330379
	h_a=0.002	0.0321067	0.0201068	0.0178794
	h_a=0.001	0.0120426	0.0075347	0.00669884
h_m=0.003nm	h_a=0.003	0.0477746	0.0299917	0.0266814
	h_a=0.002	0.0265467	0.0166412	0.0148004
	h_a=0.001	0.0111465	0.00697971	0.00620636

As the last example, Table 4.8 compares plate component effects for two velocity feedback gain levels. This table shows that adding graphene-origami auxetic layer to the plate increases dimensionless natural frequency.

Figure 4.14 depicts a lucid representation of the impact of the auxetic layer's thickness on the behavior of the continuous system.

TABLE 4.8
Effect of the Adding Graphene-Origami Auxetic Layer

Wgr	$k_cC(t)=5*10e5$	Magnetostrictive	Graphene-origami Auxetic+ Magnetostrictive
0		0.0776514	0.225542
0.5		0.0776514	0.246503
1		0.0776514	0.124282
1.5		0.0776514	0.107597
2		0.0776514	0.0968695
2.5		0.0776514	0.089002
	$k_cC(t)=9*10e5$		
0		0.0774214	0.224399
0.5		0.0774214	0.246372
1		0.0774214	0.124145
1.5		0.0774214	0.107484
2		0.0774214	0.096774
2.5		0.0774214	0.0889189

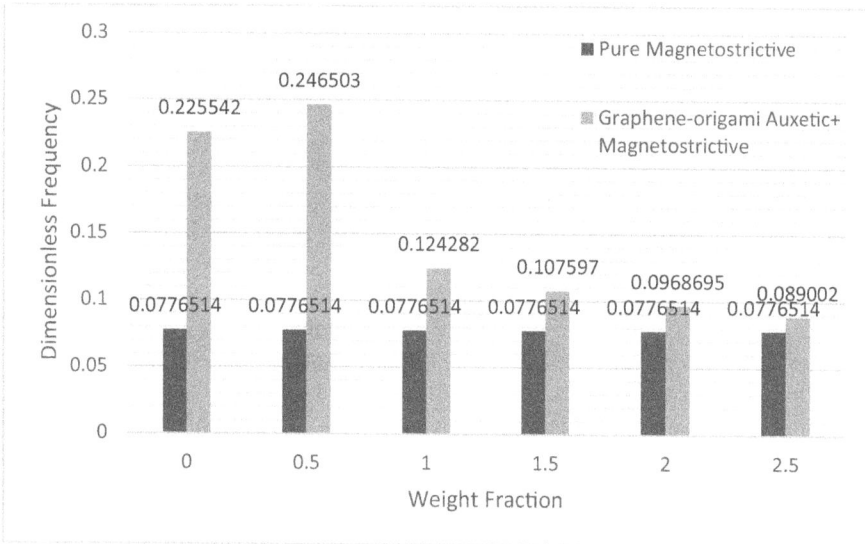

a)

FIGURE 4.14   Effect of the magnetostrictive layer: (a) velocity feedback gain = 5*10^5; (b) velocity feedback gain = 9*10^5 (a=b=0.1m, hc=0.003m, hf=0.001nm, Hgr=100%, T=320, $k_cC(t)=5*10^5$).

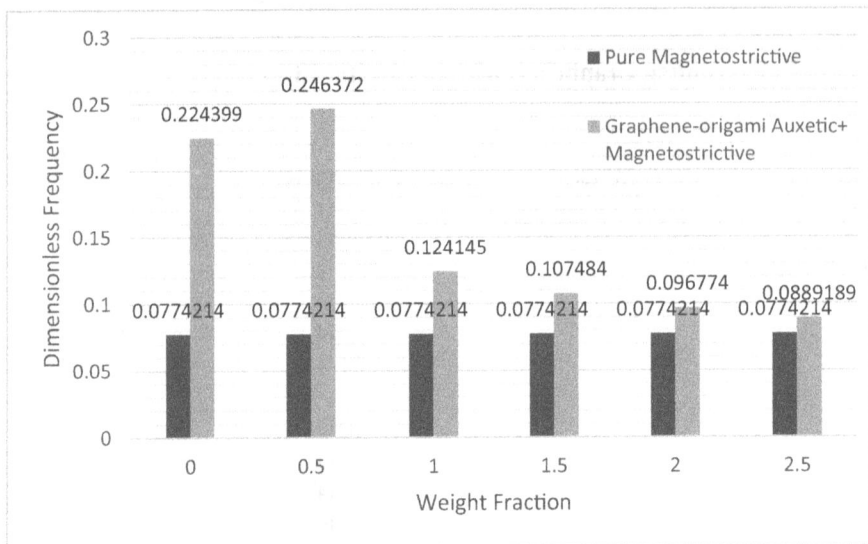

b)

FIGURE 4.14 (Continued)

4.5 SUMMARY

The primary objective of this chapter is to systematically investigate the transient behaviors shown by smart laminated composite constructions that include a smart piezoelectric layer. The current study also aimed to analyze the active control behavior of plates including an auxetic or graphene-origami auxetic core and covered with magnetostrictive materials. This action was undertaken with the purpose of enhancing comprehension of the phenomena. The current study's results have brought attention to numerous significant aspects, as recognized by the researchers:

- Laminates including thicker auxetic cores exhibit enhanced suitability for the objective of achieving a prompt steady-state response in the intended system applications.
- In order to effectively manage plate vibrations and minimize system fluctuations, it is advisable to use thin layers of the piezoelectric material.
- While increasing the control gain of the feedback control system may effectively decrease the vibration of the deflection, the incorporation of passive-type viscous dampers in the foundation of the system can lead to a more rapid attenuation of the deflection of the structure.
- Among the three kinds of plates studied for this investigation, it is seen that the solitary magnetostrictive plate exhibits the highest values of dimensionless natural frequency.

- The dimensionless natural frequencies decrease as the inclination angle increases.
- The dimensionless natural frequency exhibits an upward trend as the length of the rib grows.
- The behavior of the plate is influenced by the presence of magnetostrictive material, which exerts a regulating effect.
- The addition of a graphene-origami auxetic layer to a magnetostrictive plate results in an increase in the dimensionless natural frequency.
- An increase in temperature causes a decrease in the natural frequencies.
- Increasing the weight percentage of graphene leads to an increase in the natural frequencies until it reaches roughly 1 weight percent of graphene, after which there is a significant decrease.

REFERENCES

1. Ebrahimi, F. & Barati, M. R., 2016h. Thermal buckling analysis of size-dependent FG nanobeams based on the third-order shear deformation beam theory. *Acta Mechanica Solida Sinica*, 29(5), pp. 547–554.
2. Ebrahimi, F., Nopour, R. & Dabbagh, A., 2021. Smart laminates with an auxetic ply rested on visco-Pasternak medium: Active control of the system's oscillation. *Engineering with Computers*, pp. 1–11.
3. Ebrahimi, F., Dabbagh, A. & Rabczuk, T., 2021. On wave dispersion characteristics of magnetostrictive sandwich nanoplates in thermal environments. *European Journal of Mechanics-A/Solids*, 85, p. 104130.
4. Ebrahimi, F., et al., 2019. Hygro-thermal effects on wave dispersion responses of magnetostrictive sandwich nanoplates. *Advances in Nano Research*, 7(3), p. 157.
5. Ebrahimi, F. & Ahari, M. F., 2021. Magnetostriction-assisted active control of the multi-layered nanoplates: Effect of the porous functionally graded facesheets on the system's behavior. *Engineering with Computers*, pp. 1–15.

APPENDICES

APPENDIX 4.1

$$K_{11} = -A_{11}\left(\alpha^2\right) - A_{66}\left(\beta^2\right)$$

$$K_{12} = -A_{12}\left(\alpha\beta\right) - A_{66}\left(\alpha\beta\right)$$

$$K_{13} = 0$$

$$K_{14} = -B_{11}\left(\alpha^2\right) - B_{66}\left(\beta^2\right)$$

$$K_{15} = -B_{12}\left(\alpha\beta\right) - B_{66}\left(\alpha\beta\right)$$

$$K_{21} = -A_{21}\left(\alpha\beta\right) - A_{66}\left(\alpha\beta\right)$$

$$K_{22} = -A_{22}\left(\alpha^2\right) - A_{66}\left(\beta^2\right)$$

$$K_{23} = 0$$

$$K_{24} = -B_{21}\left(\alpha\beta\right) - B_{66}\left(\alpha\beta\right)$$

$$K_{25} = -B_{22}\left(\alpha^2\right) - B_{66}\left(\beta^2\right)$$

$$K_{31} = 0$$

$$K_{32} = 0$$

$$K_{33} = -A_{55}\left(\alpha^2\right) - A_{44}\left(\beta^2\right) - K_w - K_g\left(\alpha^2\right) - K_g\left(\beta^2\right) - K_w\left(\mu^2\alpha^2\right) -$$
$$K_w\left(\mu^2\beta^2\right) - K_g\left(\mu^2\alpha^4\right) - K_g\left(\mu^2\beta^4\right) - 2K_g\left(\mu^2\alpha^2\beta^2\right)$$

$$K_{34} = -A_{55}\left(\alpha\right) \qquad\qquad\qquad K_{35} = -A_{44}\left(\beta\right)$$

$$K_{41} = -B_{11}\left(\alpha^2\right) - B_{66}\left(\beta^2\right) \qquad\qquad K_{42} = -B_{21}\left(\alpha\beta\right) - B_{66}\left(\alpha\beta\right)$$

$$K_{43} = -A_{55}\left(\alpha\right) \qquad\qquad\qquad K_{44} = -D_{11}\left(\alpha^2\right) - D_{66}\left(\beta^2\right) - A_{55}$$

$$K_{45} = -D_{12}\left(\alpha\beta\right) - D_{66}\left(\alpha\beta\right)$$

$$K_{51} = -B_{12}\left(\alpha\beta\right) - B_{66}\left(\alpha\beta\right) \qquad\qquad K_{52} = -B_{22}\left(\alpha^2\right) - B_{66}\left(\beta^2\right)$$

$$K_{53} = -A_{44}\left(\beta\right) \qquad\qquad\qquad K_{54} = -D_{12}\left(\alpha\beta\right) - D_{66}\left(\alpha\beta\right)$$

$$K_{55} = -D_{22}\left(\beta^2\right) - D_{66}\left(\alpha^2\right) - A_{44}$$

$$M_{11} = I_0\left(1+\mu^2\alpha^2+\mu^2\beta^2\right) \qquad\qquad M_{14} = I_1\left(1+\mu^2\alpha^2+\mu^2\beta^2\right)$$

$$M_{22} = I_0\left(1+\mu^2\alpha^2+\mu^2\beta^2\right) \qquad\qquad M_{25} = I_1\left(1+\mu^2\alpha^2+\mu^2\beta^2\right)$$

$$M_{33} = I_0\left(1+\mu^2\alpha^2+\mu^2\beta^2\right) \qquad\qquad M_{41} = I_1\left(1+\mu^2\alpha^2+\mu^2\beta^2\right)$$

$$M_{44} = I_2\left(1+\mu^2\alpha^2+\mu^2\beta^2\right) \qquad\qquad M_{52} = I_1\left(1+\mu^2\alpha^2+\mu^2\beta^2\right)$$

$$M_{55} = I_2\left(1+\mu^2\alpha^2+\mu^2\beta^2\right)$$

$$C_{11} = C_{12} = C_{14} = C_{15} = 0$$

$$C_{13} = \int_{-\frac{(h_a+h_m)}{2}}^{\frac{-h_a}{2}} e_{31}\left(\alpha\right)K_c C(t)\,dz + \int_{\frac{h_a}{2}}^{\frac{(h_a+h_m)}{2}} e_{31}\left(\alpha\right)K_c C(t)\,dz$$

$$C_{21} = C_{22} = C_{24} = C_{25} = 0$$

$$C_{23} = \int_{-\frac{(h_a+h_m)}{2}}^{\frac{-h_a}{2}} e_{32}\left(\beta\right)K_c C(t)\,dz + \int_{\frac{h_a}{2}}^{\frac{(h_a+h_m)}{2}} e_{32}\left(\beta\right)K_c C(t)\,dz$$

$$C_{31} = C_{13} \qquad\qquad\qquad\qquad C_{32} = C_{23}$$

$$C_{33} = 0$$

$$C_{34} = \int_{-\frac{(h_a+h_m)}{2}}^{\frac{-h_a}{2}} z e_{31}\left(\alpha\right)K_c C(t)\,dz + \int_{\frac{h_a}{2}}^{\frac{(h_a+h_m)}{2}} z e_{31}\left(\alpha\right)K_c C(t)\,dz$$

$$C_{35} = \int_{-\frac{h_c}{2}}^{\frac{h_c}{2}} z e_{32}\left(\beta\right)K_c C(t)\,dz$$

$$C_{41} = C_{42} = C_{44} = C_{45} = 0 \qquad\qquad C_{43} = C_{34}$$

$$C_{51} = C_{52} = C_{54} = C_{55} = 0$$

$$C_{53} = \int_{-\frac{(h_a+h_m)}{2}}^{\frac{-h_a}{2}} ze_{32}(\beta)K_cC(t)dz + \int_{\frac{h_a}{2}}^{\frac{(h_a+h_m)}{2}} ze_{32}(\beta)K_cC(t)dz$$

$$A_{11} = \int_{-h_a/2}^{h_a/2} C_{11}^a dz + \int_{-\frac{(h_a+h_m)}{2}}^{\frac{-h_a}{2}} C_{11}^m dz + \int_{\frac{h_a}{2}}^{\frac{(h_a+h_m)}{2}} C_{11}^m dz$$

$$A_{12} = \int_{-h_a/2}^{h_a/2} C_{12}^a dz + \int_{-\frac{(h_a+h_m)}{2}}^{\frac{-h_a}{2}} C_{12}^m dz + \int_{h_c}^{\frac{(h_a+h_m)}{2}} C_{12}^m dz$$

$$A_{22} = \int_{-h_a/2}^{h_a/2} C_{22}^a dz + \int_{-\frac{(h_a+h_m)}{2}}^{\frac{-h_a}{2}} C_{22}^m dz + \int_{h_c}^{\frac{(h_a+h_m)}{2}} C_{22}^m dz$$

$$A_{44} = \int_{-h_a/2}^{h_a/2} C_{44}^a dz + \int_{-\frac{(h_a+h_m)}{2}}^{\frac{-h_a}{2}} C_{44}^m dz + \int_{h_c}^{\frac{(h_a+h_m)}{2}} C_{44}^m dz$$

$$A_{55} = \int_{-h_a/2}^{h_a/2} C_{55}^a dz + \int_{-\frac{(h_a+h_m)}{2}}^{\frac{-h_a}{2}} C_{55}^m dz + \int_{h_c}^{\frac{(h_a+h_m)}{2}} C_{55}^m dz$$

$$A_{66} = \int_{-h_a/2}^{h_a/2} C_{66}^a dz + \int_{-\frac{(h_a+h_m)}{2}}^{\frac{-h_a}{2}} C_{66}^m dz + \int_{h_c}^{\frac{(h_a+h_m)}{2}} C_{66}^m dz$$

$$B_{11} = \int_{-h_a/2}^{h_a/2} zC_{11}^a dz + \int_{-\frac{(h_a+h_m)}{2}}^{\frac{-h_a}{2}} zC_{11}^m dz + \int_{h_c}^{\frac{(h_a+h_m)}{2}} zC_{11}^m dz$$

$$B_{12} = \int_{-h_a/2}^{h_a/2} zC_{12}^a dz + \int_{-\frac{(h_a+h_m)}{2}}^{\frac{-h_a}{2}} zC_{12}^m dz + \int_{h_c}^{\frac{(h_a+h_m)}{2}} zC_{12}^m dz$$

$$B_{22} = \int_{-h_a/2}^{h_a/2} zC_{22}^a dz + \int_{-\frac{(h_a+h_m)}{2}}^{\frac{-h_a}{2}} zC_{22}^m dz + \int_{h_c}^{\frac{(h_a+h_m)}{2}} zC_{22}^m dz$$

$$B_{44} = \int_{-h_a/2}^{h_a/2} zC_{44}^a dz + \int_{-\frac{(h_a+h_m)}{2}}^{\frac{-h_a}{2}} zC_{44}^m dz + \int_{h_c}^{\frac{(h_a+h_m)}{2}} zC_{44}^m dz$$

$$B_{55} = \int_{-h_a/2}^{h_a/2} zC_{55}^a dz + \int_{-\frac{(h_a+h_m)}{2}}^{\frac{-h_a}{2}} zC_{55}^m dz + \int_{h_c}^{\frac{(h_a+h_m)}{2}} zC_{55}^m dz$$

$$B_{66} = \int_{-h_a/2}^{h_a/2} zC_{66}^a dz + \int_{-\frac{(h_a+h_m)}{2}}^{\frac{-h_a}{2}} zC_{66}^m dz + \int_{h_c}^{\frac{(h_a+h_m)}{2}} zC_{66}^m dz$$

$$D_{11} = \int_{-h_a/2}^{h_a/2} z^2C_{11}^a dz + \int_{-\frac{(h_a+h_m)}{2}}^{\frac{-h_a}{2}} z^2C_{11}^m dz + \int_{h_c}^{\frac{(h_a+h_m)}{2}} z^2C_{11}^m dz$$

$$D_{12} = \int_{-h_a/2}^{h_a/2} z^2C_{12}^a dz + \int_{-\frac{(h_a+h_m)}{2}}^{\frac{-h_a}{2}} z^2C_{12}^m dz + \int_{h_c}^{\frac{(h_a+h_m)}{2}} z^2C_{12}^m dz$$

$$D_{22} = \int_{-h_a/2}^{h_a/2} z^2C_{22}^a dz + \int_{-\frac{(h_a+h_m)}{2}}^{\frac{-h_a}{2}} z^2C_{22}^m dz + \int_{h_c}^{\frac{(h_a+h_m)}{2}} z^2C_{22}^m dz$$

$$D_{44} = \int_{-h_a/2}^{h_a/2} z^2 C_{44}^a dz + \int_{-\frac{(h_a+h_m)}{2}}^{-\frac{h_a}{2}} z^2 C_{44}^m dz + \int_{\frac{h_c}{2}}^{\frac{(h_a+h_m)}{2}} z^2 C_{44}^m dz$$

$$D_{55} = \int_{-h_a/2}^{h_a/2} z^2 C_{55}^a dz + \int_{-\frac{(h_a+h_m)}{2}}^{-\frac{h_a}{2}} z^2 C_{55}^m dz + \int_{\frac{h_c}{2}}^{\frac{(h_a+h_m)}{2}} z^2 C_{55}^m dz$$

$$D_{66} = \int_{-h_a/2}^{h_a/2} z^2 C_{66}^a dz + \int_{-\frac{(h_a+h_m)}{2}}^{-\frac{h_a}{2}} z^2 C_{66}^m dz + \int_{\frac{h_c}{2}}^{\frac{(h_a+h_m)}{2}} z^2 C_{66}^m dz$$

APPENDIX 4.2

$K_{11} = -A_{11}\left(\alpha^2\right) - A_{66}\left(\beta^2\right)$ $\qquad$ $K_{12} = -A_{12}\left(\alpha\beta\right) - A_{66}\left(\alpha\beta\right)$

$K_{13} = 0$ $\qquad$ $K_{14} = -B_{11}\left(\alpha^2\right) - B_{66}\left(\beta^2\right)$

$K_{15} = -B_{12}\left(\alpha\beta\right) - B_{66}\left(\alpha\beta\right)$

$K_{21} = -A_{21}\left(\alpha\beta\right) - A_{66}\left(\alpha\beta\right)$ $\qquad$ $K_{22} = -A_{22}\left(\alpha^2\right) - A_{66}\left(\beta^2\right)$

$K_{23} = 0$ $\qquad$ $K_{24} = -B_{21}\left(\alpha\beta\right) - B_{66}\left(\alpha\beta\right)$

$K_{25} = -B_{22}\left(\alpha^2\right) - B_{66}\left(\beta^2\right)$

$K_{31} = 0$ $\qquad$ $K_{32} = 0$

$K_{33} = -A_{55}\left(\alpha^2\right) - A_{44}\left(\beta^2\right) - K_w - K_g\left(\alpha^2\right) - K_g\left(\beta^2\right)$

$K_{34} = -A_{55}\left(\alpha\right)$ $\qquad$ $K_{35} = -A_{44}\left(\beta\right)$

$K_{41} = -B_{11}\left(\alpha^2\right) - B_{66}\left(\beta^2\right)$ $\qquad$ $K_{42} = -B_{21}\left(\alpha\beta\right) - B_{66}\left(\alpha\beta\right)$

$K_{43} = -A_{55}\left(\alpha\right)$ $\qquad$ $K_{44} = -D_{11}\left(\alpha^2\right) - D_{66}\left(\beta^2\right) - A_{55}$

$K_{45} = -D_{12}\left(\alpha\beta\right) - D_{66}\left(\alpha\beta\right)$

$K_{51} = -B_{12}\left(\alpha\beta\right) - B_{66}\left(\alpha\beta\right)$ $\qquad$ $K_{52} = -B_{22}\left(\alpha^2\right) - B_{66}\left(\beta^2\right)$

$K_{53} = -A_{44}\left(\beta\right)$ $\qquad$ $K_{54} = -D_{12}\left(\alpha\beta\right) - D_{66}\left(\alpha\beta\right)$

$K_{55} = -D_{22}\left(\beta^2\right) - D_{66}\left(\alpha^2\right) - A_{44}$

$M_{11} = I_0$ $\qquad$ $M_{14} = I_1$

$M_{22} = I_0$ $\qquad$ $M_{25} = I_1$

$$M_{33} = I_0 \qquad\qquad M_{41} = I_1$$

$$M_{44} = I_2 \qquad\qquad M_{52} = I_1$$

$$M_{55} = I_2$$

$$C_{11} = C_{12} = C_{14} = C_{15} = 0$$

$$C_{13} = \int_{-\frac{(h_a+h_m)}{2}}^{\frac{-h_a}{2}} e_{31}(\alpha) K_c C(t) dz + \int_{\frac{h_a}{2}}^{\frac{(h_a+h_m)}{2}} e_{31}(\alpha) K_c C(t) dz$$

$$C_{21} = C_{22} = C_{24} = C_{25} = 0$$

$$C_{23} = \int_{-\frac{(h_a+h_m)}{2}}^{\frac{-h_a}{2}} e_{32}(\beta) K_c C(t) dz + \int_{\frac{h_a}{2}}^{\frac{(h_a+h_m)}{2}} e_{32}(\beta) K_c C(t) dz$$

$$C_{31} = C_{13} \qquad\qquad C_{32} = C_{23}$$

$$C_{33} = 0$$

$$C_{34} = \int_{-\frac{(h_a+h_m)}{2}}^{\frac{-h_a}{2}} z e_{31}(\alpha) K_c C(t) dz + \int_{\frac{h_a}{2}}^{\frac{(h_a+h_m)}{2}} z e_{31}(\alpha) K_c C(t) dz$$

$$C_{35} = \int_{\frac{h_a}{2}}^{\frac{h_a}{2}} z e_{32}(\beta) K_c C(t) dz$$

$$C_{41} = C_{42} = C_{44} = C_{45} = 0 \qquad\qquad C_{43} = C_{34}$$

$$C_{51} = C_{52} = C_{54} = C_{55} = 0$$

$$C_{53} = \int_{-\frac{(h_a+h_m)}{2}}^{\frac{h_a}{2}} z e_{32}(\beta) K_c C(t) dz + \int_{\frac{h_a}{2}}^{\frac{(h_a+h_m)}{2}} z e_{32}(\beta) K_c C(t) dz$$

$$A_{11} = \int_{-h_a/2}^{h_a/2} C_{11}^a dz + \int_{-\frac{(h_a+h_m)}{2}}^{\frac{-h_a}{2}} C_{11}^m dz + \int_{\frac{h_a}{2}}^{\frac{(h_a+h_m)}{2}} C_{11}^m dz$$

$$A_{12} = \int_{-h_a/2}^{h_a/2} C_{12}^a dz + \int_{-\frac{(h_a+h_m)}{2}}^{\frac{-h_a}{2}} C_{12}^m dz + \int_{\frac{h_a}{2}}^{\frac{(h_a+h_m)}{2}} C_{12}^m dz$$

$$A_{22} = \int_{-h_a/2}^{h_a/2} C_{22}^a dz + \int_{-\frac{(h_a+h_m)}{2}}^{\frac{-h_a}{2}} C_{22}^m dz + \int_{\frac{h_a}{2}}^{\frac{(h_a+h_m)}{2}} C_{22}^m dz$$

$$A_{44} = \int_{-h_a/2}^{h_a/2} C_{44}^a dz + \int_{-\frac{(h_a+h_m)}{2}}^{\frac{-h_a}{2}} C_{44}^m dz + \int_{\frac{h_a}{2}}^{\frac{(h_a+h_m)}{2}} C_{44}^m dz$$

$$A_{55} = \int_{-h_a/2}^{h_a/2} C_{55}^a dz + \int_{-\frac{(h_a+h_m)}{2}}^{\frac{-h_a}{2}} C_{55}^m dz + \int_{\frac{h_a}{2}}^{\frac{(h_a+h_m)}{2}} C_{55}^m dz$$

$$A_{66} = \int_{-h_a/2}^{h_a/2} C_{66}^a dz + \int_{-\frac{(h_a+h_m)}{2}}^{\frac{-h_a}{2}} C_{66}^m dz + \int_{\frac{h_a}{2}}^{\frac{(h_a+h_m)}{2}} C_{66}^m dz$$

$$B_{11} = \int_{-h_a/2}^{h_a/2} zC_{11}^a dz + \int_{-\frac{(h_a+h_m)}{2}}^{\frac{-h_a}{2}} zC_{11}^m dz + \int_{\frac{h_a}{2}}^{\frac{(h_a+h_m)}{2}} zC_{11}^m dz$$

$$B_{12} = \int_{-h_a/2}^{h_a/2} zC_{12}^a dz + \int_{-\frac{(h_a+h_m)}{2}}^{\frac{-h_a}{2}} zC_{12}^m dz + \int_{\frac{h_a}{2}}^{\frac{(h_a+h_m)}{2}} zC_{12}^m dz$$

$$B_{22} = \int_{-h_a/2}^{h_a/2} zC_{22}^a dz + \int_{-\frac{(h_a+h_m)}{2}}^{\frac{-h_a}{2}} zC_{22}^m dz + \int_{\frac{h_a}{2}}^{\frac{(h_a+h_m)}{2}} zC_{22}^m dz$$

$$B_{44} = \int_{-h_a/2}^{h_a/2} zC_{44}^a dz + \int_{-\frac{(h_a+h_m)}{2}}^{\frac{-h_a}{2}} zC_{44}^m dz + \int_{\frac{h_a}{2}}^{\frac{(h_a+h_m)}{2}} zC_{44}^m dz$$

$$B_{55} = \int_{-h_a/2}^{h_a/2} zC_{55}^a dz + \int_{-\frac{(h_a+h_m)}{2}}^{\frac{-h_a}{2}} zC_{22}^m dz + \int_{\frac{h_a}{2}}^{\frac{(h_a+h_m)}{2}} zC_{55}^m dz$$

$$B_{66} = \int_{-h_a/2}^{h_a/2} zC_{66}^a dz + \int_{-\frac{(h_a+h_m)}{2}}^{\frac{-h_a}{2}} zC_{66}^m dz + \int_{\frac{h_a}{2}}^{\frac{(h_a+h_m)}{2}} zC_{66}^m dz$$

$$D_{11} = \int_{-h_a/2}^{h_a/2} z^2 C_{11}^a dz + \int_{-\frac{(h_a+h_m)}{2}}^{\frac{-h_a}{2}} z^2 C_{11}^m dz + \int_{\frac{h_a}{2}}^{\frac{(h_a+h_m)}{2}} z^2 C_{11}^m dz$$

$$D_{12} = \int_{-h_a/2}^{h_a/2} z^2 C_{12}^a dz + \int_{-\frac{(h_a+h_m)}{2}}^{\frac{-h_a}{2}} z^2 C_{12}^m dz + \int_{\frac{h_a}{2}}^{\frac{(h_a+h_m)}{2}} z^2 C_{12}^m dz$$

$$D_{22} = \int_{-h_a/2}^{h_a/2} z^2 C_{22}^a dz + \int_{-\frac{(h_a+h_m)}{2}}^{\frac{-h_a}{2}} z^2 C_{22}^m dz + \int_{\frac{h_a}{2}}^{\frac{(h_a+h_m)}{2}} z^2 C_{22}^m dz$$

$$D_{44} = \int_{-h_a/2}^{h_a/2} z^2 C_{44}^a dz + \int_{-\frac{(h_a+h_m)}{2}}^{\frac{-h_a}{2}} z^2 C_{44}^m dz + \int_{\frac{h_a}{2}}^{\frac{(h_a+h_m)}{2}} z^2 C_{44}^m dz$$

$$D_{55} = \int_{-h_a/2}^{h_a/2} z^2 C_{55}^a dz + \int_{-\frac{(h_a+h_m)}{2}}^{\frac{-h_a}{2}} z^2 C_{55}^m dz + \int_{\frac{h_a}{2}}^{\frac{(h_a+h_m)}{2}} z^2 C_{55}^m dz$$

$$D_{66} = \int_{-h_a/2}^{h_a/2} z^2 C_{66}^a dz + \int_{-\frac{(h_a+h_m)}{2}}^{\frac{-h_a}{2}} z^2 C_{66}^m dz + \int_{\frac{h_a}{2}}^{\frac{(h_a+h_m)}{2}} z^2 C_{66}^m dz$$

5 Wave Dispersion Characteristics of Auxetic Structures

5.1 BACKGROUND

A comprehensive comprehension of wave propagation is of utmost importance in the analysis of the mechanical attributes of structures comprising materials exhibiting diverse properties. Furthermore, this specific study may be used within the framework of structural health monitoring and non-destructive testing (NDT). The aforementioned criteria provide enough motivation for scientists to undertake an examination of wave propagation. Moreover, several inquiries have been conducted to examine the wave dispersion properties of diverse composite constructions [1–34]. On the other hand, after completing an extensive literature survey, it becomes apparent that there is a lack of study focused on investigating the behavior of wave propagation in auxetic metamaterial beams that are enhanced by the presence of functionally graded graphene origami on an elastic substrate. The aim of this chapter is to address the existing research gap. This chapter discusses the wave propagation characteristics of auxetic structures, specifically focusing on beams, plates, and shells. This work utilizes the Timoshenko beam theory to comprehensively examine the wave propagation characteristics of auxetic metamaterial beams that have been enhanced by the use of functionally graded graphene origami. The present investigation is being carried out with the assistance of the Winkler-Pasternak Foundation, which provides support to the beams. Furthermore, extensive investigations have been conducted on various graphene origami (GOri) distribution patterns, revealing consistent and graded properties across the thickness dimension. The use of Hamilton's principle allows for the capture of the kinetic equations pertaining to auxetic metamaterial beams. The analytical solution for the governing equations of the auxetic metamaterial beam is obtained using the provided information. Analyses have been conducted to examine the impact of various factors on the wave frequency and phase velocity of the auxetic metamaterial beam, leading to comparisons being drawn. The characteristics included in this study comprise the distribution pattern and content of graphene, the folding degree of GOri, the temperature, wave number, and the coefficients of the elastic foundation. The identification of critical elements in each figure has been accomplished by a comprehensive examination and comparison of the findings.

The second half of the study focuses on examining the wave dispersion properties of an auxetic plate composed of smart piezoelectric nanocomposite layers positioned at the top and bottom of the composite plate, while including an auxetic core. The equation of motion was derived in this section by using the improved

DOI: 10.1201/9781003289296-5

higher-order shear deformation theory (HSDT). Furthermore, the use of a micromechanical model is employed to ascertain the characteristics of the auxetic core and the piezoelectric smart nanocomposite layers. At that particular moment, Hamilton's principle was used to deduce the governing equations of intelligent composite plates. Subsequently, the governing equations of the smart composite plate are solved using an analytical approach based on the exponential form. The phase velocity of the smart composite plate is determined by the thicknesses of the auxetic core and the smart piezoelectric nanocomposite layers. Furthermore, the impact of the Winkler-Pasternak model, viscoelastic foundation, external electric voltage, and volume percentage of reinforcement in the matrix on the phase velocity is calculated and shown in each image. In the last section of this chapter, an examination was conducted on the influence of the auxetic layer on wave propagation in auxetic shells, as well as its impact on the effective features of cylindrical sandwich shells. A theoretical method is developed to characterize the transmission of waves by considering wave frequency and wave velocity based on the principles of first-order shear deformable shell theory. Furthermore, the Hamiltonian concept was used. Subsequently, the governing equation was solved by the use of an analytical approach. The study examined many characteristics, including the inclination of auxetic cell angles and diameters, in relation to wave behavior. The last phase entails the execution of a quantitative inquiry. The findings are presented in a series of graphics, which examine the impact of each parameter on wave propagation.

5.2 AUXETIC BEAMS: WAVE PROPAGATION ANALYSIS

This chapter examines the wave propagation properties of auxetic metamaterial beams, which are enhanced by the use of functionally graded graphene origami. These beams are further supported by an elastic basis. This research investigates four discrete distribution patterns of GOri. Furthermore, the beam has been subjected to analysis using the Timoshenko beam theory. Moreover, Hamilton's concept is utilized to develop equations of motion. The images offered in this research elucidate the influence of several factors. The current study considers a beam made of functionally graded graphene origami-enabled auxetic metamaterial (FG-GOEAM) that is placed on an elastic medium. The beam's position is defined using the supplied coordinate system, as shown in Figure 5.1. The beam is denoted as having a length L and a thickness h.

A metallic auxetic metamaterial stand has been fabricated by the integration of a carefully formulated combination of a copper matrix and GOri nanofillers. The material parameters of GOEAMs, including Young's modulus, Poisson's ratio, coefficient of thermal expansion, and density, are determined using the GP-assisted micromechanical models established by Zhao et al. [35].

$$E_c = \frac{1+\xi\eta\ V_{GR}}{1-\eta\ V_{GR}} E_{Cu} \times f_E(H_{GR},V_{GR},T) \qquad (5.1a)$$

$$v_c = (v_{GR}V_{GR} + v_{Cu}V_{Cu}) \times f_v(H_{GR},V_{GR},T) \qquad (5.1b)$$

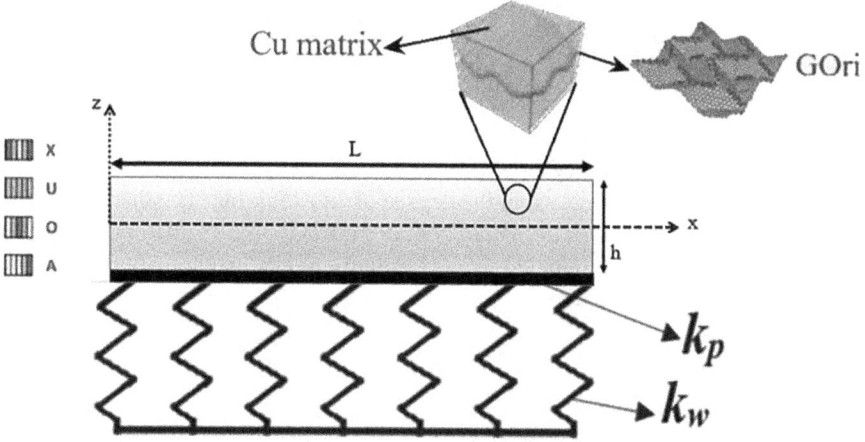

FIGURE 5.1 The shape of a functionally graded graphene origami-enabled auxetic meta-material beam embedded on an elastic medium.

$$\alpha_c = (\alpha_{GR} V_{GR} + \alpha_{Cu} V_{Cu}) \times f_\alpha (V_{GR}, T) \tag{5.1c}$$

$$\rho_c = (\rho_{GR} V_{GR} + \rho_{Cu} V_{Cu}) \times f_\rho (V_{GR}, T) \tag{5.1d}$$

The coefficients pertaining to material (η) and size (ξ) are delineated in the following manner:

$$\eta = \frac{(E_{GR}/E_{Cu}) - 1}{(E_{GR}/E_{Cu}) + \xi} \tag{5.2a}$$

$$\xi = 2(l_{GR}/t_{GR}) \tag{5.2b}$$

where l_{GR} and t_{GR} are the length and thickness of graphene, respectively. The transformation functions denoted by $f_{E,v,\alpha,\rho}(H_{GR}, V_{GR}, T)$ were produced by the use of the GP method, as developed by Zhao et al. [35].

$$f_E(H_{GR}, V_{GR}, T) = 1.11 - 1.22 V_{GR} - 0.134\left(\frac{T}{T_0}\right) + 0.559 V_{GR}\left(\frac{T}{T_0}\right)$$
$$- 5.5 H_{GR} V_{GR} + 38 H_{GR} V_{GR}^2 - 20.6 H_{GR}^2 V_{GR}^2 \tag{5.3a}$$

$$f_v(H_{GR}, V_{GR}, T) = 1.01 - 1.43 V_{GR} + 0.165\left(\frac{T}{T_0}\right)$$
$$- 16.8 H_{GR} V_{GR} - 1.1 H_{GR} V_{GR}\left(\frac{T}{T_0}\right) + 16 H_{GR}^2 V_{GR}^2 \tag{5.3b}$$

$$f_\alpha(V_{GR}, T) = 0.794 - 16.8 V_{GR}^2 - 0.0279\left(\frac{T}{T_0}\right)^2 + 0.182\left(\frac{T}{T_0}\right)(1 + V_{GR}) \tag{5.3c}$$

$$f_\rho(V_{GR},T) = 1.01 - 2.01\ V_{GR}^2 - 0.0131\left(\frac{T}{T_0}\right) \tag{5.3d}$$

Given that the temperature T_0 is fixed at 300 K, it may be said with confidence that the current ambient temperature can be represented by the T. The variable H_{GR} represents the quantity of hydrogen atoms that are engaged in bonding, serving as a criterion for assessing the degree of GOri folding. The distribution of GOri nanofillers exhibits variations in concentration that follow certain patterns $X-W_{GR}$, $U-W_{GR}$, $O-W_{GR}$, and $A-W_{GR}$, characterized by an even number of layers. The layer's volume fractions are determined using the following equations [35]:

Pattern $X-W_{GR}$: $V_{GR}(k) = 2\ V_{GR}\left|2k-N_L-1\right|/N_L$ (5.4a)

Pattern $U-W_{GR}$: $V_{GR}(k) = V_{GR}$ (5.4b)

Pattern $O-W_{GR}$: $V_{GR}(k) = 2\ V_{GR}(1-\left|2k-N_L-1\right|/N_L)$ (5.4c)

Pattern $A-W_{GR}$: $V_{GR}(k) = V_{GR}(2k-1)/N_L$ (5.4d)

in which $k = 1,2,...,N_L$. The symbol N_L is used to represent the entire number of layers. The calculation of the volume percentage of GOri in the beam, represented by V_{GR}, may be determined based on its weight fraction using the following formula [36]:

$$V_{GR} = \frac{\rho_{Cu}W_{GR}}{\rho_{Cu}W_{GR} + \rho_{GR}(1-W_{GR})} \tag{5.5a}$$

$$V_{Cu} = 1 - V_{GR} \tag{5.5b}$$

In the above equation, the variables ρ_{GR} and ρ_{Cu} represent the densities of graphene and pure copper, respectively. The variable V_{Cu} signifies the volume percentage of copper. The research done by Zhao et al. [35–37] shows that an escalation in the quantity of layers in the FG beams leads to the emergence of NPR features that are closely correlated with the increased GOri content and folding degree. The results of their study demonstrate that the $X-W_{GR}$ beam has a greater negative Poisson's ratio (NPR) in the outermost layers, whereas the interior layers show positive NPR values. The $U-W_{GR}$ beam has a consistent Poisson's ratio across all of its layers. The adherence to the Timoshenko beam theory is crucial for the derivation of kinetic relations. The displacement domain of the FG-GOEAM beam may be defined as follows, as shown by references [1,2,3]:

$$u_x(x,z,t) = u(x,t) + z\psi(x,t)$$
$$u_z(x,z,t) = w(x,t) \tag{5.6}$$

where u and w are displacement elements at the beam's middle surface, ψ is the section normal vector rotations about the y, and t denotes time. The non-zero stresses of the beam may be expressed as tails [1,2].

$$\varepsilon_{xx} = \frac{\partial u}{\partial x} + z\frac{\partial \psi}{\partial x}, \qquad \gamma_{xz} = \frac{\partial w}{\partial x} + \psi \tag{5.7}$$

The motion equations will be derived by using Hamilton's principle as the underlying theoretical framework. According to sources [1,2], this concept is articulated in the following manner:

$$\int_0^t \delta(U - T + V)dt = 0 \tag{5.8}$$

The symbols U, T, and V are used to denote the ideas of strain energy, kinetic energy, and work executed by external forces, in that order. The equation representing the variation in strain energy is provided in references [1,2].

$$\delta U = \int_V (\sigma_{xx}\delta\varepsilon_{xx} + \tau_{xz}\delta\gamma_{xz})dV \tag{5.9}$$

By substituting equations (5.6) through (5.7) into equation (5.8), we obtain:

$$\delta U = \int_0^L \left(N_x\delta\left(\frac{\partial u}{\partial x}\right) + M_x\delta\left(\frac{\partial \psi}{\partial x}\right) + Q_x\delta\left(\frac{\partial w}{\partial x} + \psi\right) \right)dx \tag{5.10}$$

The outcomes of stress, represented by the variables N_x, M_x, and Q_x, may be expressed in the following format:

$$[N_x, M_x, Q_x] = \int_{-h/2}^{h/2} [\sigma_{xx}, z\sigma_{xx}, \tau_{xz}]dz \tag{5.11}$$

Subsequently, an equation representing the fluctuations in the quantity of work achieved by external influences might be denoted as tails [1].

$$\delta V = \int_0^L \left(-k_w\delta w + (k_p + N^T)\delta\left(\frac{\partial^2 w}{\partial x^2}\right) \right)dx \tag{5.12}$$

The first variation of kinetic energy can be presented as [2]:

$$\delta T = \int_V \left(\dot{u}_x\delta\dot{u}_x + \dot{u}_z\delta\dot{u}_z \right)\rho(z)dV \tag{5.13}$$

$$= \int_0^L \left(\begin{array}{c} I_1\left(\frac{\partial u}{\partial t}\delta\frac{\partial u}{\partial t} + \frac{\partial w}{\partial t}\delta\frac{\partial w}{\partial t}\right) + I_2\left(\frac{\partial u}{\partial t}\delta\frac{\partial^2 \psi}{\partial t} + \frac{\partial^2 \psi}{\partial t}\delta\frac{\partial u}{\partial t}\right) \\ + I_3\frac{\partial^2 \psi}{\partial t}\delta\frac{\partial^2 \psi}{\partial t} \end{array} \right)dx$$

The dot superscript used in all equations denotes the derivative with respect to time. The mass inertias used in the aforementioned equations are obtained in the following style, from reference [3].

$$[I_1, I_2, I_3] = \int_{-h/2}^{h/2} \rho(z)(1, z, z^2)dz \tag{5.14}$$

The Euler-Lagrange equations of FG-GOEAM beams may be derived by substituting equations (5.10), (5.12), and (5.13) into equation (5.8) and setting the coefficients of δu, δw, and $\delta \psi$ to zero.

$$\frac{\partial N_x}{\partial x} = I_1 \frac{\partial^2 u}{\partial t^2} + I_2 \frac{\partial^2 \psi}{\partial t^2} \tag{5.15a}$$

$$\frac{\partial N_x}{\partial x}\frac{\partial w}{\partial x} + \left[N_x + N_x^T \right]\frac{\partial^2 w}{\partial x^2} - k_w w + k_p \frac{\partial^2 w}{\partial x^2} + \frac{\partial Q_x}{\partial x} = I_1 \frac{\partial^2 w}{\partial t^2} \tag{5.15b}$$

$$\frac{\partial M_x}{\partial x} - Q_x = I_2 \frac{\partial^2 u}{\partial t^2} + I_3 \frac{\partial^2 \psi}{\partial t^2} \tag{5.15c}$$

Furthermore, an examination is conducted on the elastic stress-strain relationships of FG-GOEAM with the aim of deducing the fundamental elastic equations pertaining to these materials. The constitutive equations presented in this context might be expressed as indicated in references [1,2].

$$\sigma_{ij} = C_{ijkl}(\varepsilon_{kl} - \alpha_{ij}\Delta T) \tag{5.16}$$

The letters σ_{ij}, ε_{kl}, and C_{ijkl} represent distinct components of the stress, strain, and elasticity tensors, respectively. Furthermore, it can be seen that ΔT denotes the rise in temperature. When analyzing a beam structure, it is possible to express the following equations with considerable simplicity:

$$\sigma_{xx} = Q_{11}\varepsilon_{xx} \tag{5.17a}$$

$$\sigma_{xx}^T = -Q_{11}\alpha_c \Delta T \tag{5.17b}$$

$$\tau_{xz} = Q_{55}\gamma_{xz} \tag{5.17c}$$

$$Q_{11} = \frac{E_c}{1-v_c^2}, \quad Q_{55} = \frac{E_c}{2(1+v_c)} \tag{5.17d}$$

Moreover, by performing the integration of equations (5.17a) and (5.17c) across the cross-sectional area of the beam, the resulting relationships may be expressed [1,2]:

$$N_x = A_{11}\frac{\partial u}{\partial x} + B_{11}\frac{\partial \psi}{\partial x} \tag{5.17e}$$

$$M_b = B_{11}\frac{\partial u}{\partial x} + D_{11}\frac{\partial \psi}{\partial x} \tag{5.17f}$$

$$Q_x = A_{55}\left(\frac{\partial w}{\partial x} + \psi\right) \tag{5.17g}$$

In instances when the cross-sectional stiffness may be mathematically represented as [1,2]:

$$\left[A_{11}, B_{11}, D_{11}\right] = \int_{-h/2}^{h/2} (1, z, z^2)Q_{11}dz \tag{5.18a}$$

$$A_{55} = \int_{-h/2}^{h/2} \kappa Q_{55} dz \tag{5.18b}$$

in which $\kappa = 5/6$. Furthermore, the thermal force may be mathematically represented by the following formula [1,2]:

$$[N_x^T] = -A_{11}\alpha_c \Delta T \tag{5.18c}$$

The derivation of the equations regulating FG-GOEAM beams may be accomplished by including equations (5.17e–g) into equations (5.15a–c) in the following way.

$$A_{11}\frac{\partial^2 u}{\partial x^2} + B_{11}\frac{\partial^2 \psi}{\partial x^2} - I_1\frac{\partial^2 u}{\partial t^2} - I_2\frac{\partial^2 \psi}{\partial t^2} = 0 \tag{5.19a}$$

$$A_{55}\left(\frac{\partial^2 w}{\partial x^2} + \frac{\partial \psi}{\partial x}\right) - k_w w + N_x^T\frac{\partial^2 w}{\partial x^2} + k_p\frac{\partial^2 w}{\partial x^2} - I_1\frac{\partial^2 w}{\partial t^2} = 0 \tag{5.19b}$$

$$B_{11}\frac{\partial^2 u}{\partial x^2} + D_{11}\frac{\partial^2 \psi}{\partial x^2} - A_{55}\left(\frac{\partial w}{\partial x} + \psi\right) - I_2\frac{\partial^2 u}{\partial t^2} - I_3\frac{\partial^2 \psi}{\partial t^2} = 0 \tag{5.19c}$$

The governing equation of the FG-GOEAM beam is solved using an analytical technique. As shown by citation [3], it is anticipated that the segments inside the displacement domain would be explicitly delineated.

$$u = U.e^{i(\beta x - \omega t)}$$
$$w = W.e^{i(\beta x - \omega t)} \tag{5.20a}$$
$$\psi = \Psi.e^{i(\beta x - \omega t)}$$

The wave amplitudes are represented by the variables U, W, and Ψ. The variable β represents the wave number, while ω denotes the circular frequency of dispersed waves. Replacing for u, w, and ψ from equation (5.20) in equations (5.19a–c), the subsequent equation is achieved [3]:

$$\left([K]_{3\times3} - \omega^2 [M]_{3\times3}\right)[\Delta] = 0 \tag{5.20b}$$

in which

$$[\Delta] = \begin{bmatrix} U \\ W \\ \Psi \end{bmatrix}$$

The stiffness matrix, designated as K, and the mass matrix, denoted as M, are included in Appendix 5.1. The matrices include the elements as described. The eigenvalue issue discussed above is solved for the variable ω, thereby allowing for the identification of the wave frequency. It is essential to set the value of the determinant in question equal to zero.

$$\left| [K]_{3\times 3} - \omega^2 [M]_{3\times 3} \right| = 0 \tag{5.20c}$$

Furthermore, the determination of the phase velocity may be attained by dividing the wave frequency by the wave number, as seen below:

$$c_p = \frac{\omega}{\beta} \tag{5.20d}$$

The governing equation was solved in this part by using the information that was previously supplied. The following graphics provide a concise representation of the research results, which may be used to assess the influence of different variables on the phenomenon of wave propagation. This section largely focuses on the impact on GOri content, folding degree, distribution patterns, and temperature on the wave propagation and phase velocity of auxetic metamaterial beams. With a certain cross-sectional area $A = 0.1 \ m^2$ and a ratio of slenderness $L = 20h$, the object may be analyzed. In the following discussion, copper is selected as the matrix material for nano-composites, characterized by its Young's modulus $E_{Cu} = 65.79 \ Gpa$, Poisson's ratio $v_{Cu} = 0.387$, mass density $\rho_{Cu} = 8.8 \ g/cm^3$, and CTE $\alpha_{Cu} = 16.51 \times 10^{-6} K^{-1}$ at $T = 300K$. The inherent properties of graphene in its pristine condition are defined by certain material factors, such as its electrical conductivity, thermal conductivity, and mechanical strength $E_{GR} = 929.57 \ Gpa$, $v_{GR} = 0.22$, $\alpha_{GR} = -3.98 \times 10^{-6} K^{-1}$, $\rho_{GR} = 1.8 \ g/cm^3$ at $T = 300K$, as well as its geometric attributes, namely length $l_{GR} = 83.76$ Å and thickness $t_{GR} = 3.4$ Å. In order to facilitate comprehension, the dimensionless form of the Winkler-Pasternak foundation coefficient is shown herewith.

$$K_w = k_w \frac{L^4}{\zeta}, \qquad K_p = k_p \frac{L^2}{\zeta}, \qquad \zeta = \frac{E_c h^3}{12(1 - v_c^2)}$$

Figures 5.2a–5.2d illustrate the impact of different levels of GOri content on the relationship between wave frequency and wave number for various distribution patterns at a given value of $T = 300$ K, $K_w = K_p = 50$, $H_{GR} = 100\%$. Research has shown that the frequency of waves exhibits a gradual increase from zero for low wave numbers, reaching its peak at the highest wave number. The data suggests that there is a positive correlation between the percentage of GOri concentration in the auxetic beam and the observed wave frequency. Compared to previous GOri dispersion designs, it can be seen that the beam $X - W_{GR}$ exhibits a greater wave frequency (Figure 5.2a).

This phenomenon may be attributed to the presence of a higher concentration of layers with positive Poisson's ratio in close proximity to the central surface of the beam. After careful examination of Figures 5.2c and 5.2d under magnification, it is seen that when the graphene range is below 1%, the wave frequencies of the $U - W_{GR}$ and $O - W_{GR}$ beams exhibit a noticeable resemblance. The cause of this phenomenon may be ascribed to the heightened dispersion of GOri particles in close proximity to the focal point of the beam. Consequently, the differentiation between the two patterns might provide challenges in cases when the range of graphene is limited. Based on the data shown in Figures 5.2a–5.2d, it can be seen that the transformation of GOri/copper nano-composites into auxetic metamaterials occurs when the concentration of GOri exceeds 1.5%. As a consequence, there is a heightened degree of negative Poisson's ratio (NPR) and an augmentation in Young's modulus, hence enhancing the stiffness of the beam $X - W_{GR}$ and $A - W_{GR}$. This phenomenon is shown by a substantial increase in the frequency of the beams. While the increase in stiffness is negligible for the $U - W_{GR}$ beam, it is evident that the $X - W_{GR}$, $A - W_{GR}$, and $O - W_{GR}$ beams exhibit higher beam stiffness in comparison.

Figures 5.3a–5.3d illustrate the variation in phase velocity with respect to the longitudinal wave number across varied amounts and distribution patterns of GOri. The phase velocity experiences a significant drop as the longitudinal wave number grows, mostly owing to the nonlinearity, dispersion, and dissipation properties of the medium. The graph illustrates a positive correlation between the quantity of GOri

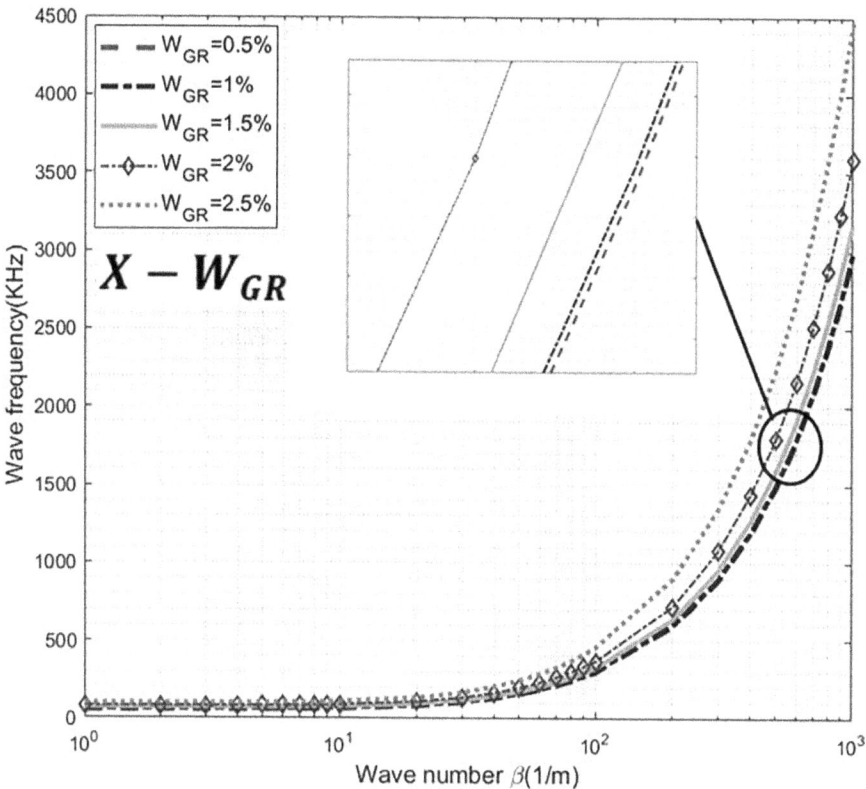

FIGURE 5.2A Wave frequency variation vs. number of waves for various GOri content $(X - W_{GR}, H_{GR} = 100\%, T = 300K, K_w = 50, K_p = 50, L = 20h, N_L = 10)$.

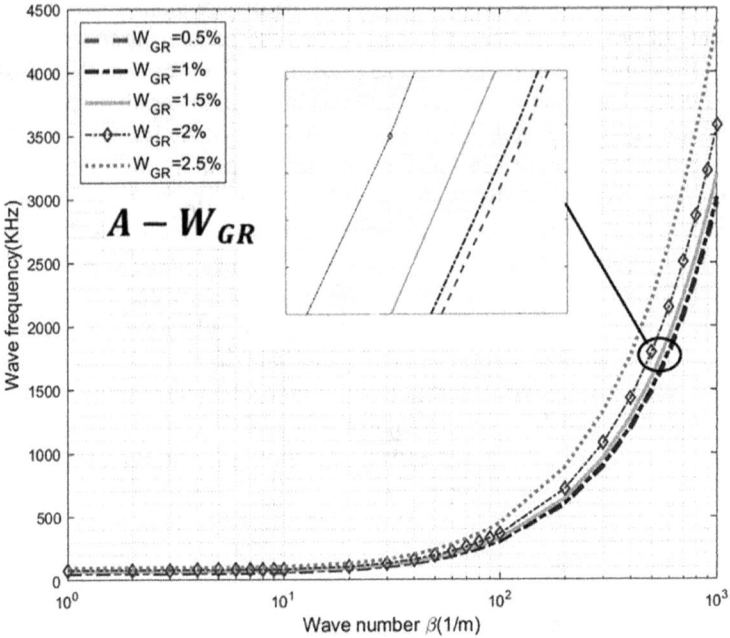

FIGURE 5.2B Wave frequency variation vs. number of waves for various GOri content $(A - W_{GR}, H_{GR} = 100\%, T = 300K, K_w = 50, K_p = 50, L = 20h, N_L = 10)$.

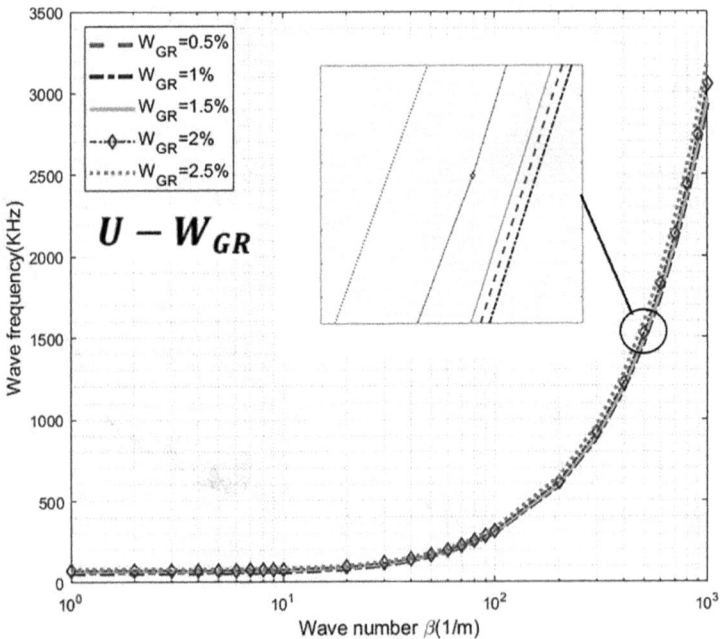

FIGURE 5.2C Wave frequency variation vs. number of waves for various GOri content $(U - W_{GR}, H_{GR} = 100\%, T = 300K, K_w = 50, K_p = 50, L = 20h, N_L = 10)$.

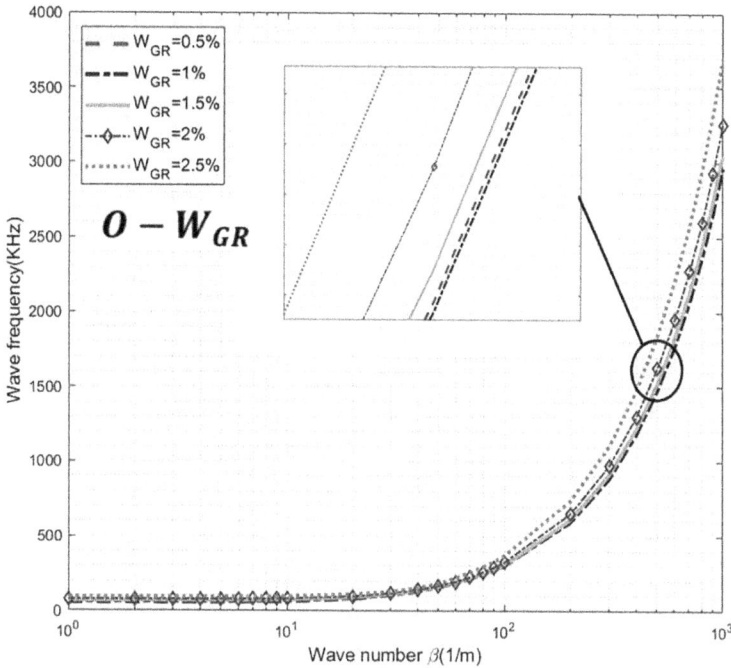

FIGURE 5.2D Wave frequency variation vs. number of waves for various GOri content $(O - W_{GR}, H_{GR} = 100\%, T = 300K, K_w = 50, K_p = 50, L = 20h, N_L = 10)$.

content and the phase velocity, similar to the behavior shown by wave frequency. This implies that when the GOri content rises, there is a corresponding increase in the phase velocity.

Upon conducting a comprehensive examination of Figures 5.3a–5.3d, the phase velocity of beams composed of $X - W_{GR}$, $A - W_{GR}$, and $O - W_{GR}$ is greater than that of the beam $U - W_{GR}$ owing to increased stiffness.

Furthermore, the impact of the GOri content on the phase velocity and wave frequency may be seen in Figures 5.4a and 5.4b, which depict the different disturbance patterns. As previously indicated, the increase in wave frequency and phase velocity has been attributed to the growth in GOri content.

Table 5.1 provides the precise frequency values for various GOri composition and dispersion patterns. Based on the examination of Figure 5.4b, it can be seen that there is a marginal reduction in wave frequency for both the $U - W_{GR}$ and $O - W_{GR}$ beams when the GOri scope ascends by a magnitude of less than 1%. The primary cause of this decrease may be primarily ascribed to the decrease in Poisson's ratio and the increase in Young's modulus that arise from the elevated GOri concentration.

The influence of varying temperatures on wave frequency and wave number for distinct distribution patterns is clearly seen in Figures 5.5a–5.5d. The frequency of a

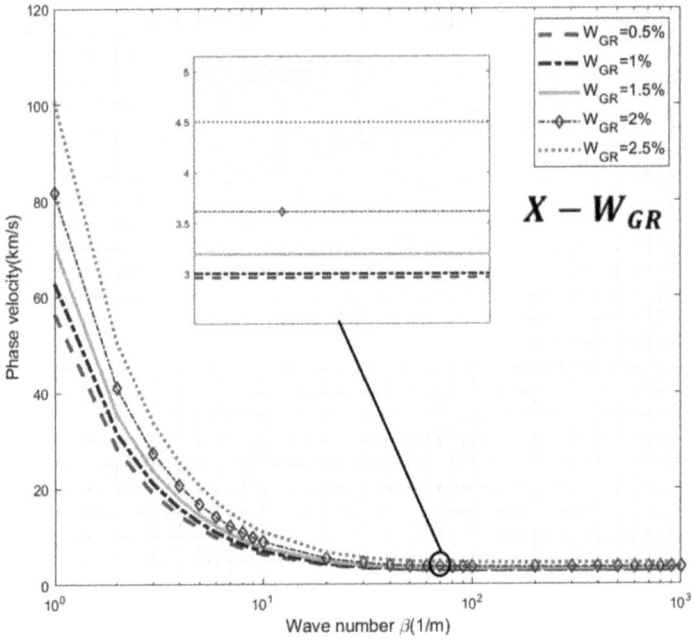

FIGURE 5.3A Phase velocity vs. number of waves variation various GOri content $(X - W_{GR}, H_{GR} = 100\%, T = 300K, K_w = 50, K_p = 50, L = 20h, N_L = 10)$.

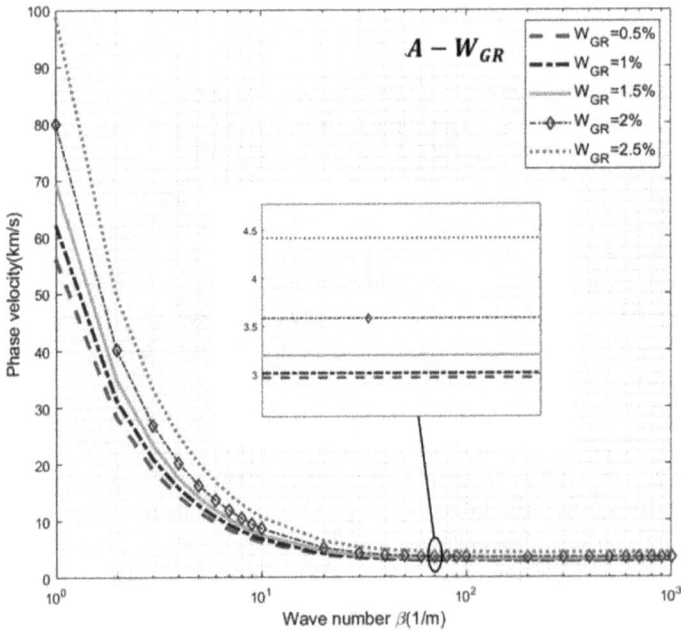

FIGURE 5.3B Phase velocity vs. number of waves variation various GOri content $(A - W_{GR}, H_{GR} = 100\%, T = 300K, K_w = 50, K_p = 50, L = 20h, N_L = 10)$.

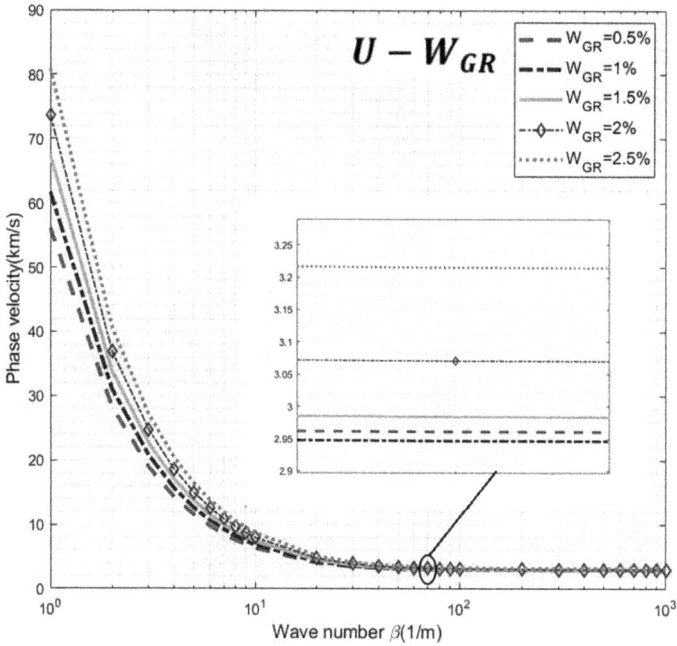

FIGURE 5.3C Phase velocity vs. number of waves variation various GOri content $(U - W_{GR}, H_{GR} = 100\%, T = 300K, K_w = 50, K_p = 50, L = 20h, N_L = 10)$.

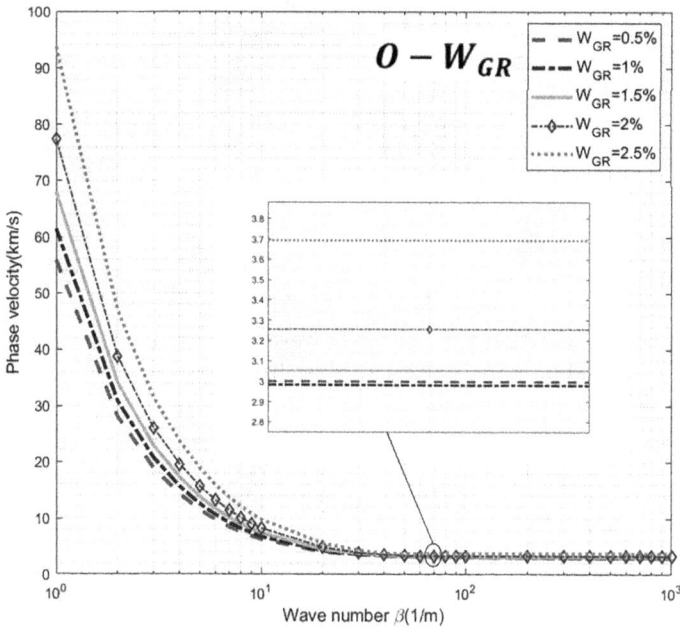

FIGURE 5.3D Phase velocity vs. number of waves variation various GOri content $(O - W_{GR}, H_{GR} = 100\%, T = 300K, K_w = 50, K_p = 50, L = 20h, N_L = 10)$.

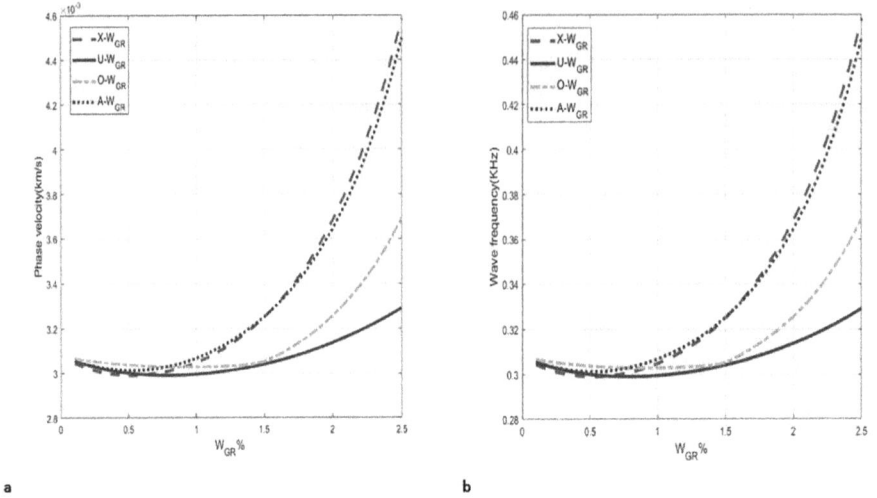

a   b

FIGURE 5.4 Coupled effects, the phase velocity (a) and frequency of waves (b) changes according to GOri content and distribution ($H_{GR} = 100\%, T = 300K, K_w = 50, K_p = 50$, $L = 20h$, $N_L = 10, \beta = 100$).

TABLE 5.1

Wave Frequency Variation for Various Methods of Distribution Using GOri Content ($H_{GR} = 100\%, T = 300K, k_w = 50, k_p = 50, \beta = 100, L = 20h, N_L = 10$)

Distribution pattern	Wave frequency (KHz)				
	$W_{GR} = 0.5\%$	$W_{GR} = 1\%$	$W_{GR} = 1.5\%$	$W_{GR} = 2\%$	$W_{GR} = 2.5\%$
$X - W_{GR}$	299.0577	304.7979	324.9113	368.3487	458.2572
$A - W_{GR}$	301.2139	306.7850	325.1925	364.4752	449.7022
$O - W_{GR}$	303.7661	302.8647	305.2635	325.5575	369.3238
$U - W_{GR}$	300.0591	299.5055	304.0975	313.6969	329.1350

wave has a positive correlation with greater wave numbers, indicating that it begins at a lower value and progressively rises. It has been noted that an increase in temperature results in a little decrease in wave frequency. Research findings have shown that the wave frequency of beams $X - W_{GR}$ and $A - W_{GR}$, namely those denoted as patterns A and B in Figure 5.7a and 5.7b, respectively, exhibit a lesser degree of susceptibility to temperature variations compared to other patterns.

Figures 5.6a–5.6d depict the variations in phase velocity with respect to the longitudinal wave number across different temperatures and distribution patterns. When the longitudinal wave number increases, there is a notable drop in the phase velocity. It has been noted that a rise in temperature is associated with a marginal reduction in phase velocity. The observed occurrence of a decrease in phase velocity due

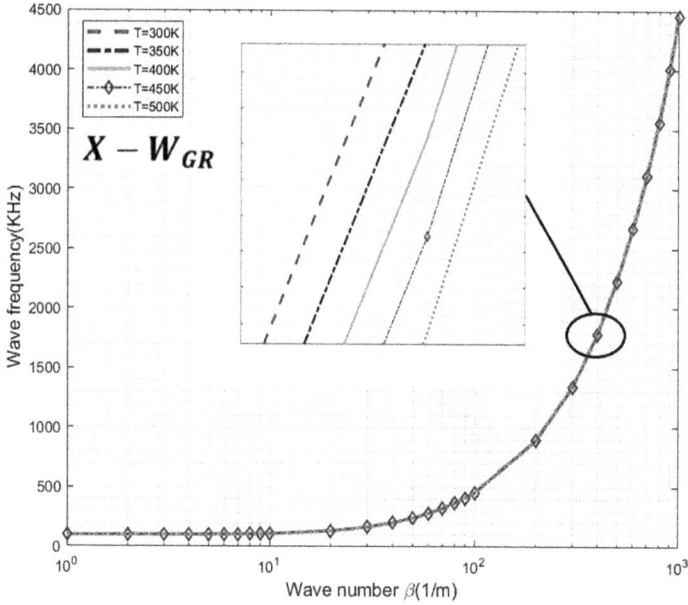

FIGURE 5.5A Wave frequency variation vs. number of waves for various temperatures $(X - W_{GR}, W_{GR} = 2.5\%, T = 300K, K_w = 50, K_p = 50, L = 20h, N_L = 10)$.

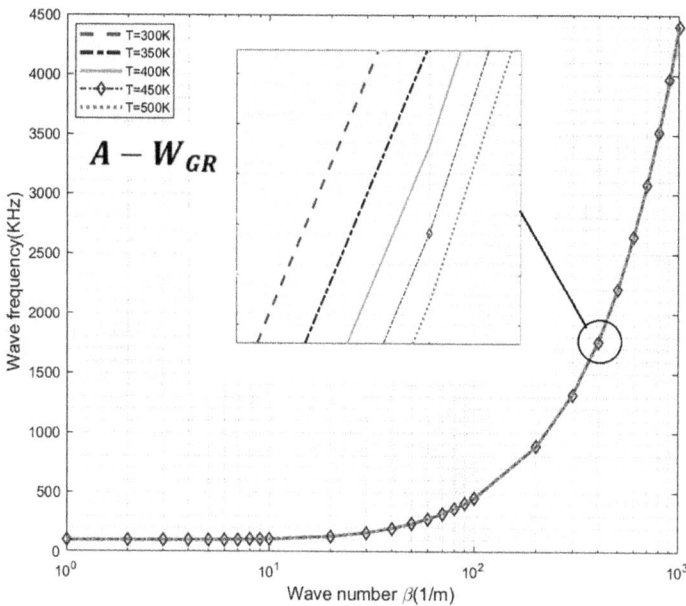

FIGURE 5.5B Wave frequency variation vs. number of waves for various temperatures $(A - W_{GR}, W_{GR} = 2.5\%, T = 300K, K_w = 50, K_p = 50, L = 20h, N_L = 10)$.

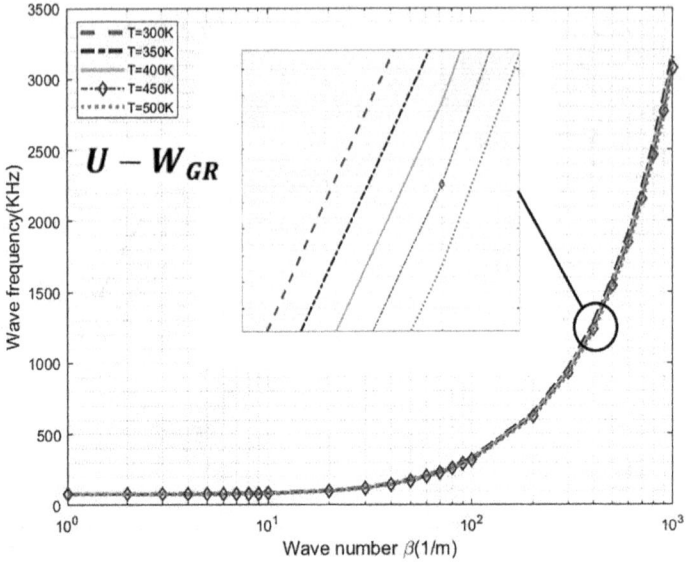

FIGURE 5.5C Wave frequency variation vs. number of waves for various temperatures $(U - W_{GR}, W_{GR} = 2.5\%, T = 300K, K_w = 50, K_p = 50, L = 20h, N_L = 10)$.

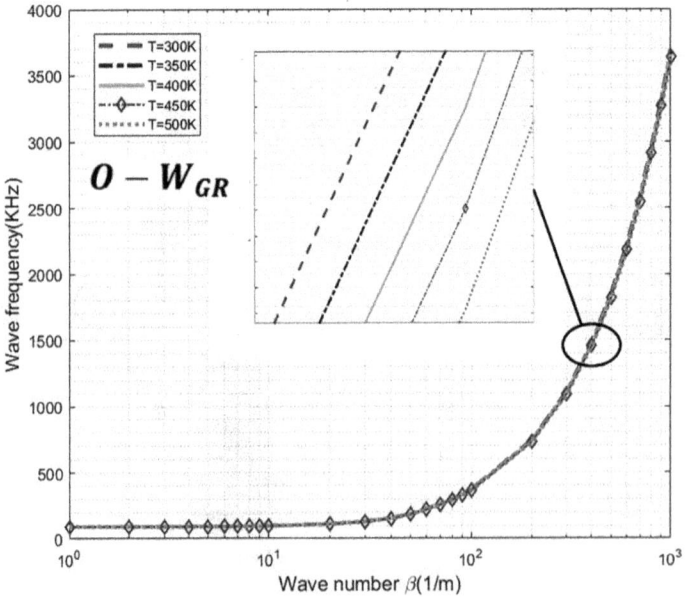

FIGURE 5.5D Wave frequency variation vs. number of waves for various temperatures $(O - W_{GR}, W_{GR} = 2.5\%, T = 300K, K_w = 50, K_p = 50, L = 20h, N_L = 10)$.

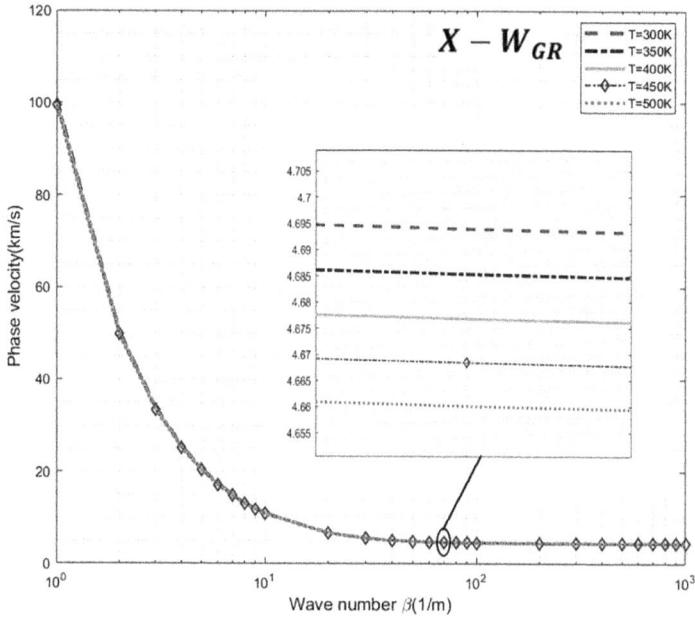

FIGURE 5.6A Phase velocity vs. number of waves variation based on various temperatures $(X - W_{GR}, W_{GR} = 2.5\%, T = 300K, K_w = 50, K_p = 50, L = 20h, N_L = 10)$.

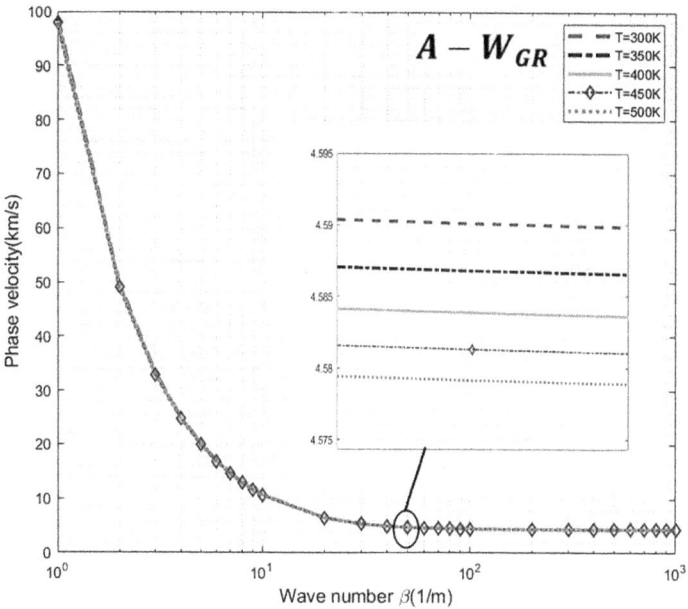

FIGURE 5.6B Phase velocity vs. number of waves variation based on various temperatures $(A - W_{GR}, W_{GR} = 2.5\%, T = 300K, K_w = 50, K_p = 50, L = 20h, N_L = 10)$.

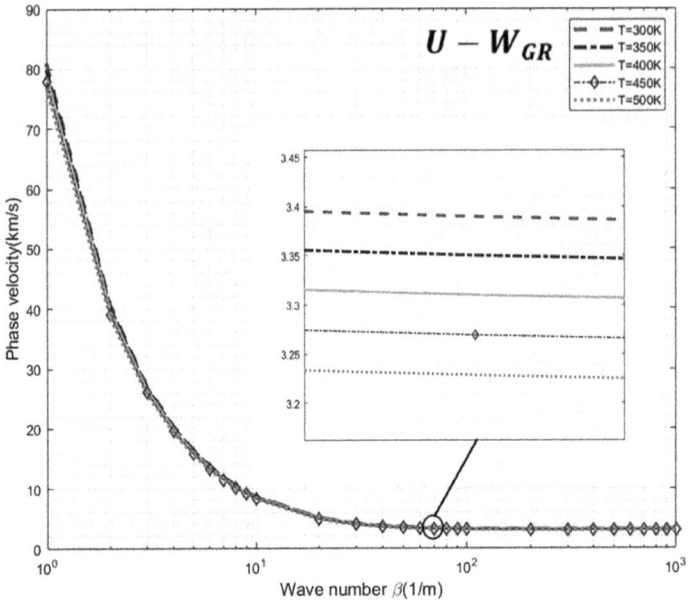

FIGURE 5.6C Phase velocity vs. number of waves variation based on various temperatures $(U - W_{GR}, W_{GR} = 2.5\%, T = 300K, K_w = 50, K_p = 50, L = 20h, N_L = 10)$.

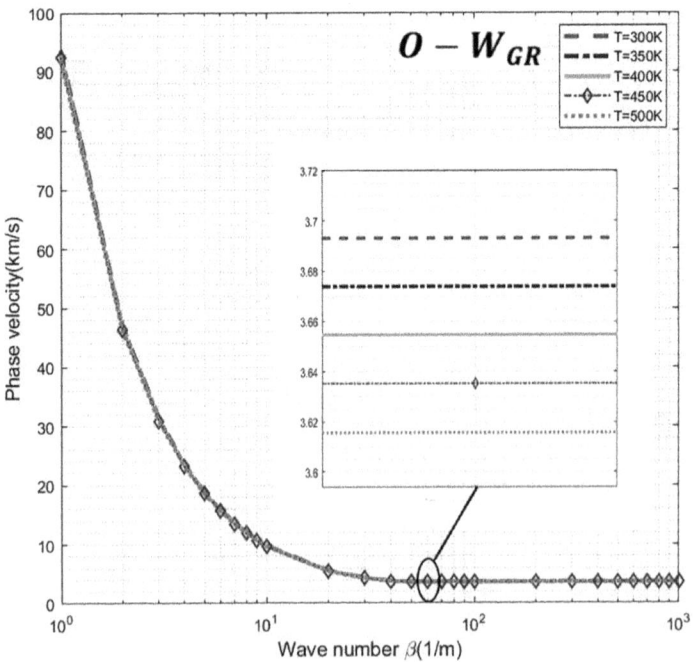

FIGURE 5.6D Phase velocity vs. number of waves variation based on various temperatures $(O - W_{GR}, W_{GR} = 2.5\%, T = 300K, K_w = 50, K_p = 50, L = 20h, N_L = 10)$.

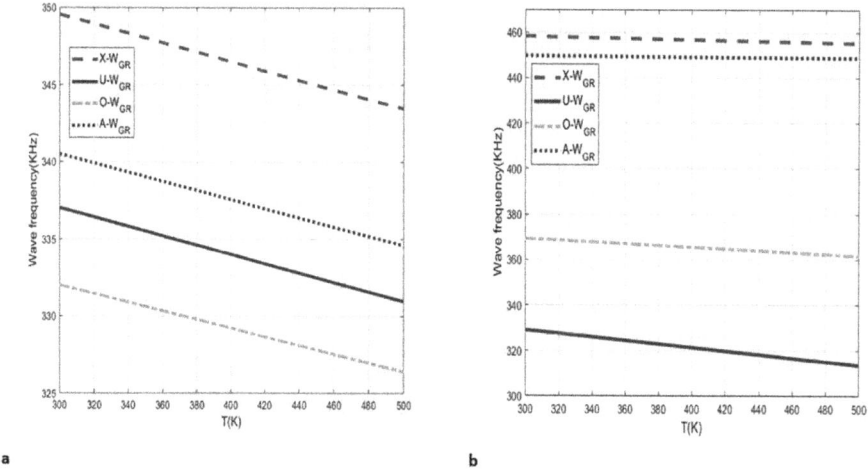

FIGURE 5.7 The variation in wave frequency at different temperatures includes the NPR effect, (a) without NPR ($W_{GR} = 0.5\%, H_{GR} = 10\%$), (b) with NPR effect ($W_{GR} = 2.5\%, H_{GR} = 100\%$).

to changes in temperature has been appropriately acknowledged. According to the graph, it can be seen that changes in temperature have a negligible effect on the phase velocity characteristic.

Figures 5.7a and 5.7b show the impact of different temperatures on wave frequency, including the effect of the negative Poisson's ratio (NPR). The attainability of NPR features may be attributed to the high concentration of GOri and the folding degree, which contribute to the better mechanical capabilities of NPR. These qualities are less susceptible to variations in temperature, as seen in Figure 5.7a. The lack of NPR may be seen in Figure 5.7b, as shown by the relatively low levels of GOri content and folding degree. This phenomenon leads to a significant decrease in wave frequency as the temperature rises.

Figures 5.8a–5.8d demonstrate the influence of different degrees of GOri folding on the correlation between wave frequency and wave number for varied distribution patterns at $W_{GR} = 2.5\%$, $T = 300$ K, $K_w = K_p = 50$. The increase in the degree of GOri folding leads to a notable drop in the Young's modulus of the GOri/copper nano-composites, thus leading to a reduction in beam stiffness and a subsequent fall in wave frequency. Moreover, the impact of the degree of folding on wave frequency for the pattern shown in Figure 5.8c exhibits greater efficacy compared to the other patterns, an anticipated outcome attributed to its somewhat lower stiffness in relation to the other designs.

Figures 5.9a–5.9d illustrate the variations in phase velocity with longitudinal wave number for different degrees of GOri folding and distribution patterns. As the quantity of longitudinal waves grows, there is a notable decrease in the phase velocity. The observed behavior of phase velocity exhibits a similar pattern to that of wave frequency. The phase velocity exhibits an increase when the GOri folding degree value decreases at low wave numbers. Nevertheless, it can be seen that the phase

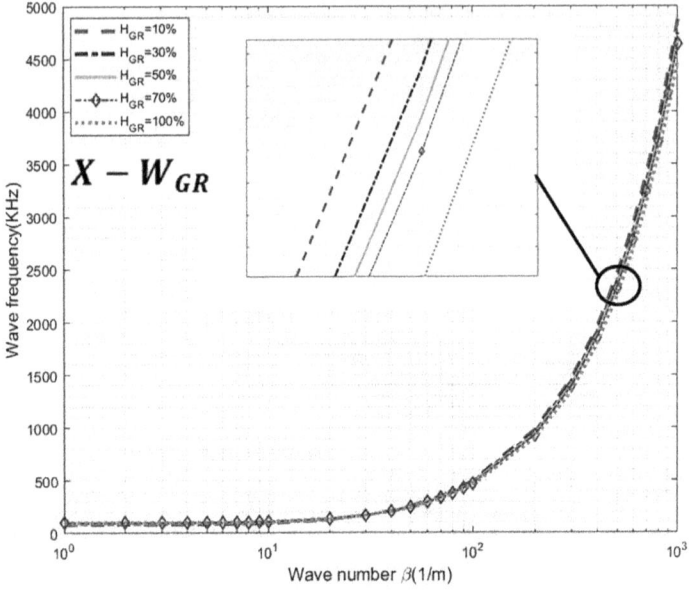

FIGURE 5.8A Wave frequency variation vs. number of waves for various graphene origami folding degree ($X - W_{GR}, W_{GR} = 2.5\%, H_{GR} = 100\%, K_w = 50, K_p = 50, L = 20h, N_L = 10$).

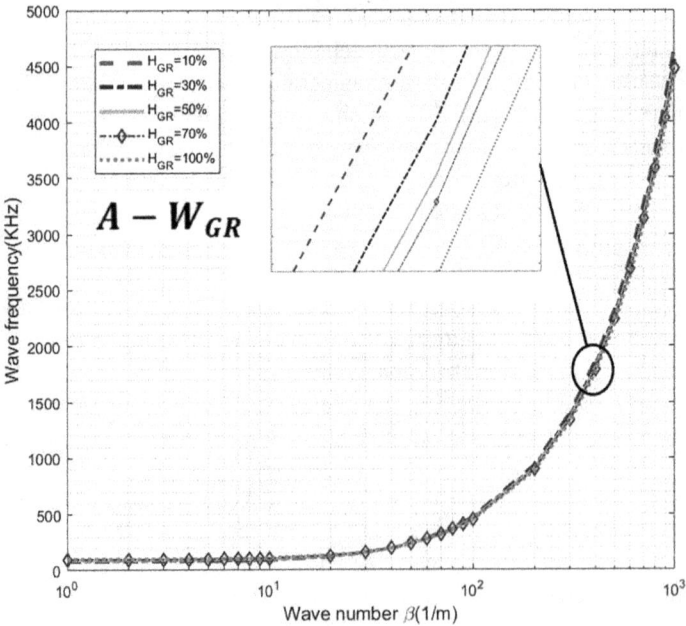

FIGURE 5.8B Wave frequency variation vs. number of waves for various graphene origami folding degree ($A - W_{GR}, W_{GR} = 2.5\%, H_{GR} = 100\%, K_w = 50, K_p = 50, L = 20h, N_L = 10$) .

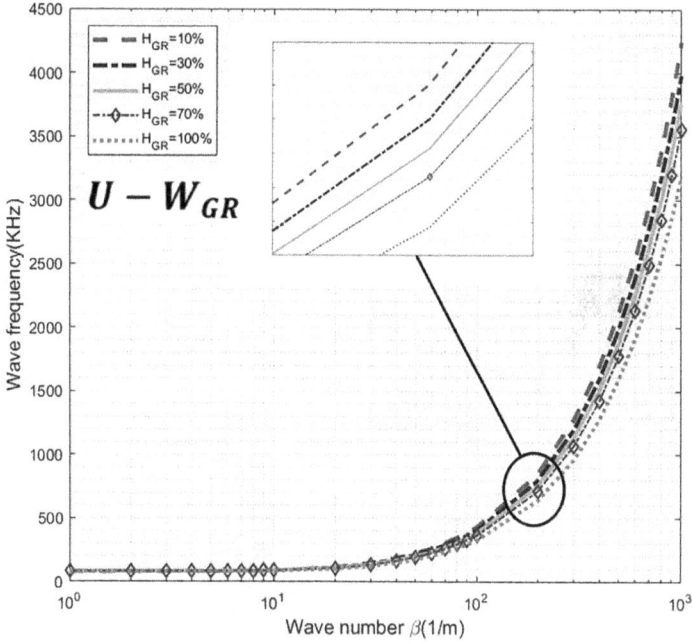

FIGURE 5.8C Wave frequency variation vs. number of waves for various graphene origami folding degree $(U - W_{GR}, W_{GR} = 2.5\%, H_{GR} = 100\%, K_w = 50, K_p = 50, L = 20h, N_L = 10)$.

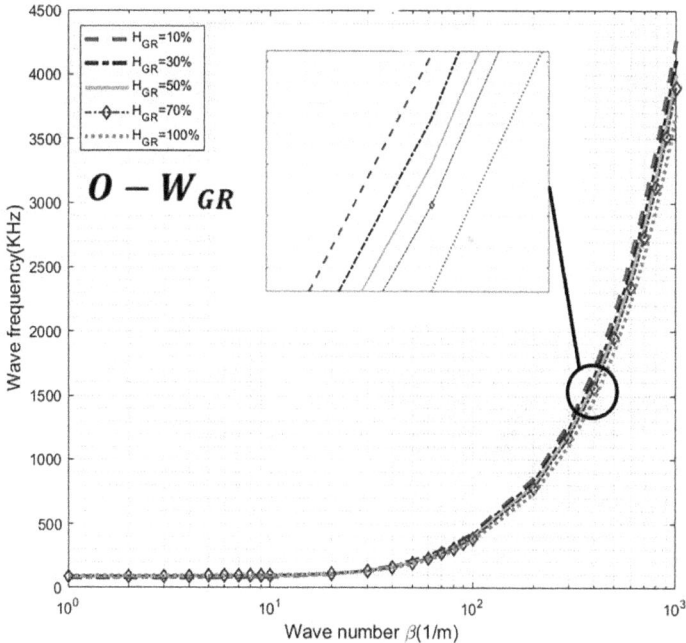

FIGURE 5.8D Wave frequency variation vs. number of waves for various graphene origami folding degree $(O - W_{GR}, W_{GR} = 2.5\%, H_{GR} = 100\%, K_w = 50, K_p = 50, L = 20h, N_L = 10)$.

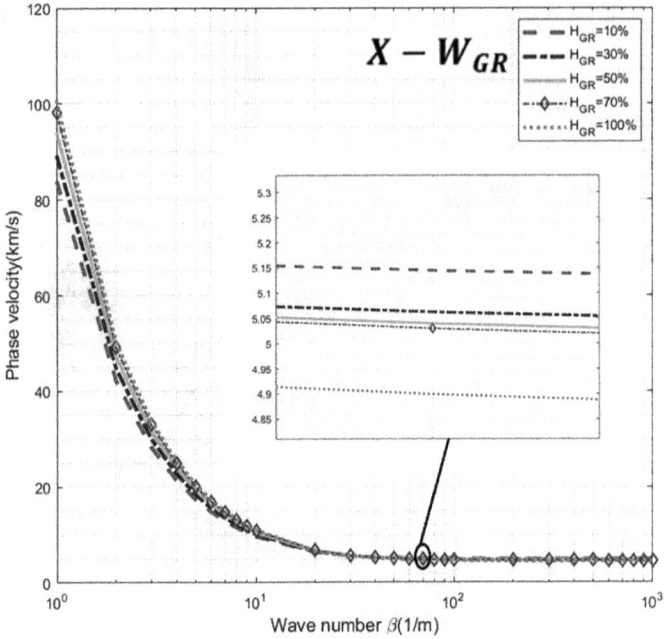

FIGURE 5.9A   Phase velocity vs. number of waves variation based on various graphene origami folding degree ($X - W_{GR}, W_{GR} = 2.5\%, H_{GR} = 100\%, K_w = 50, K_p = 50, L = 20h, N_L = 10$).

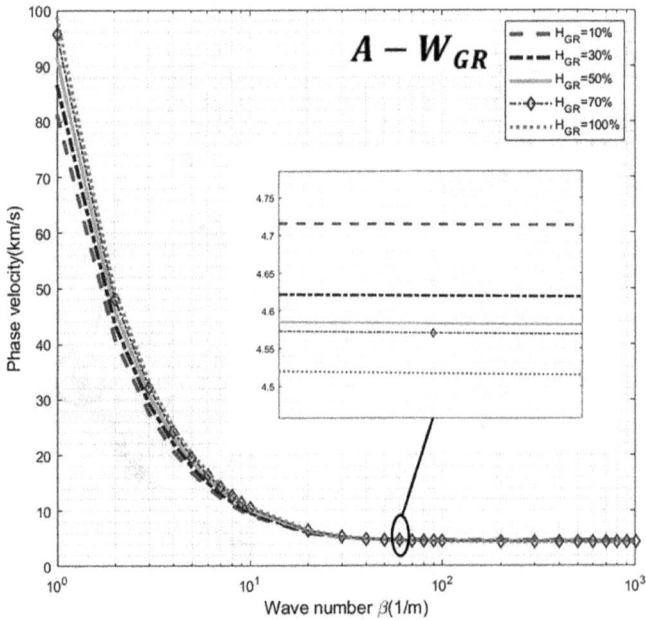

FIGURE 5.9B   Phase velocity vs. number of waves variation based on various graphene origami folding degree ($A - W_{GR}, W_{GR} = 2.5\%, H_{GR} = 100\%, K_w = 50, K_p = 50, L = 20h, N_L = 10$).

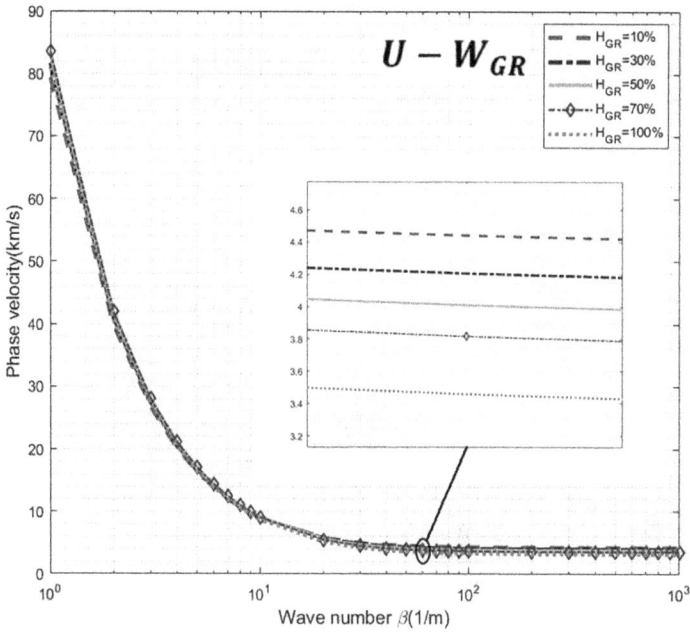

FIGURE 5.9C Phase velocity vs. number of waves variation based on various graphene origami folding degree $(U - W_{GR}, W_{GR} = 2.5\%, H_{GR} = 100\%, K_w = 50, K_p = 50, L = 20h, N_L = 10)$.

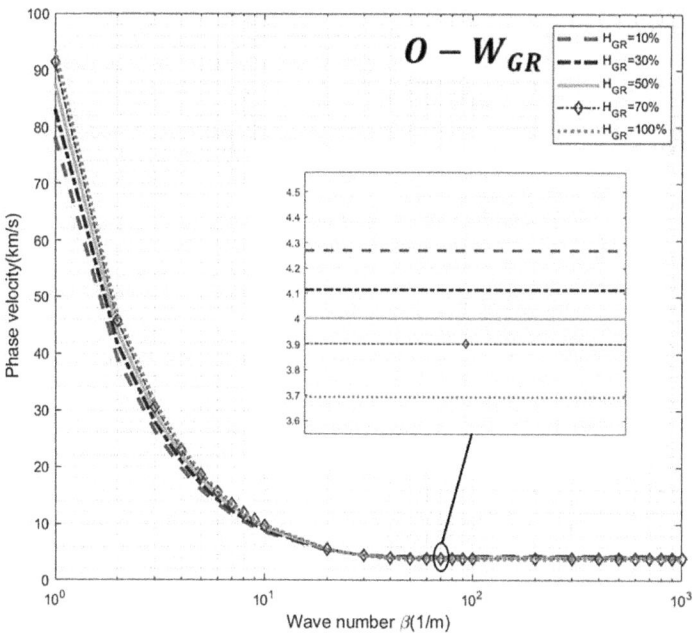

FIGURE 5.9D Phase velocity vs. number of waves variation based on various graphene origami folding degree $(O - W_{GR}, W_{GR} = 2.5\%, H_{GR} = 100\%, K_w = 50, K_p = 50, L = 20h, N_L = 10)$.

TABLE 5.2

Phase Velocity Values Corresponding to Varying Degrees of Origami Folding and Wave Numbers for $X - W_{GR}$ **beam** ($W_{GR} = 2.5\%, T = 300K, k_w = 50, k_p = 50,$ $L = 20h, N_l = 10$)

	Phase velocity (km/s)				
Wave number	$H_{GR} = 10\%$	$H_{GR} = 30\%$	$H_{GR} = 50\%$	$H_{GR} = 70\%$	$H_{GR} = 100\%$
1	84.0407	89.4078	94.2767	98.3405	100.8259
5	17.4937	18.4995	19.4310	20.2134	20.6644
10	9.7176	10.1301	10.5383	10.8875	11.0479
100	4.9365	4.8227	4.7717	4.7372	4.5825
1000	4.8653	4.7399	4.6785	4.6348	4.4710

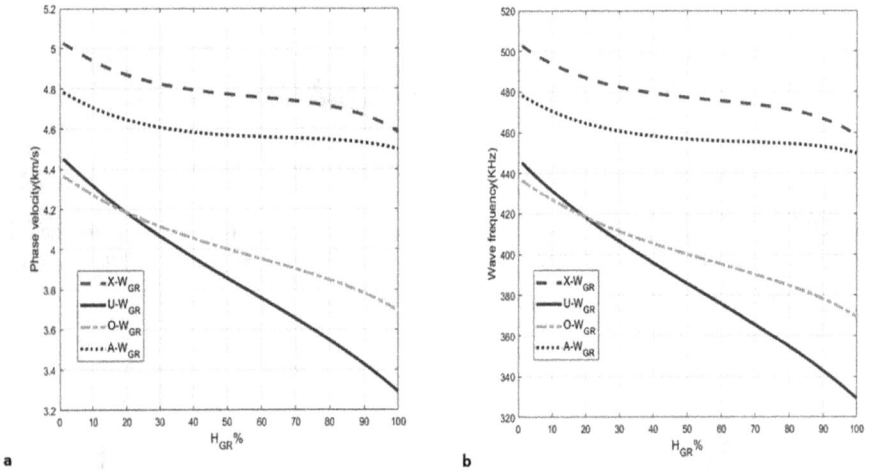

FIGURE 5.10 Variation of phase velocity (a) and wave frequency (b) versus graphene origami folding degree rises with respect to distribution pattern.

velocity exhibits a reduction when the longitudinal wave number grows in relation to the GOri folding degree value.

It is noteworthy to observe that the phase velocity of the beam in Figure 5.9c is lower when compared to the phase velocities of other GOri dispersion schemes. The primary factor contributing to this phenomenon is the decreased level of stiffness.

Table 5.2 presents the specific phase velocity values associated with different levels of origami folding and wave numbers for a $X - W_{GR}$ beam. The phase velocity exhibits an upward trend with a rise in the folding degree of GOri at low wave numbers, but it has a downward trend as the longitudinal wave number grows.

The impact of GOri folding on the wave frequency of diverse distribution patterns is seen in Figures 5.10a and 5.10b. As stated before, an increase in the

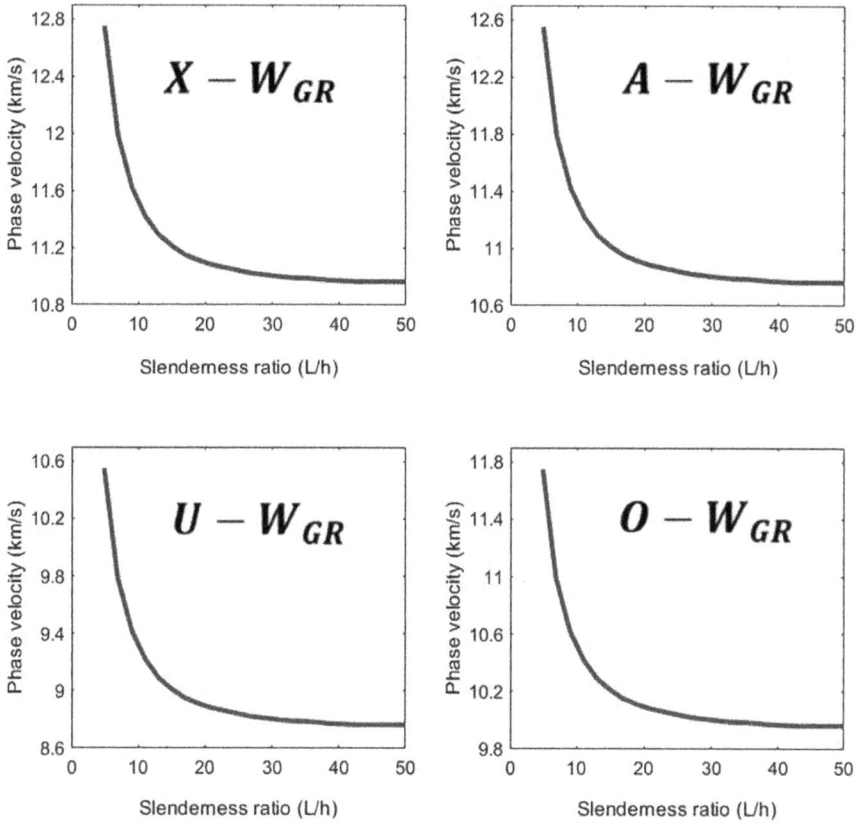

FIGURE 5.11   Phase velocity against slenderness ratio variances for various distribution patterns $(W_{GR} = 2.5\%, H_{GR} = 100\%, T = 300K, K_w = 50, K_p = 50, N_L = 10, \beta = 10)$.

degree of GOri folding is associated with a reduction in wave frequency. The identification of this inverse link was achieved by an analysis of the association between the increasing level of GOri folding and the frequency of waves.

The graphical representation in Figure 5.11 illustrates the relationship between variations in slenderness ratio and the corresponding change in phase velocity for different distribution patterns. It is evident that increasing the slenderness ratio leads to a decrease in phase velocity, whereas increasing the slenderness ratio results in a constant phase velocity that remains unchanged.

The association between wave frequency fluctuation and the Winkler coefficient for different distribution patterns is seen in Figure 5.12. An observable phenomenon of a marginal rise in wave frequency occurs with an increase in the Winkler coefficient. Nevertheless, the Winkler coefficient does not have a substantial influence

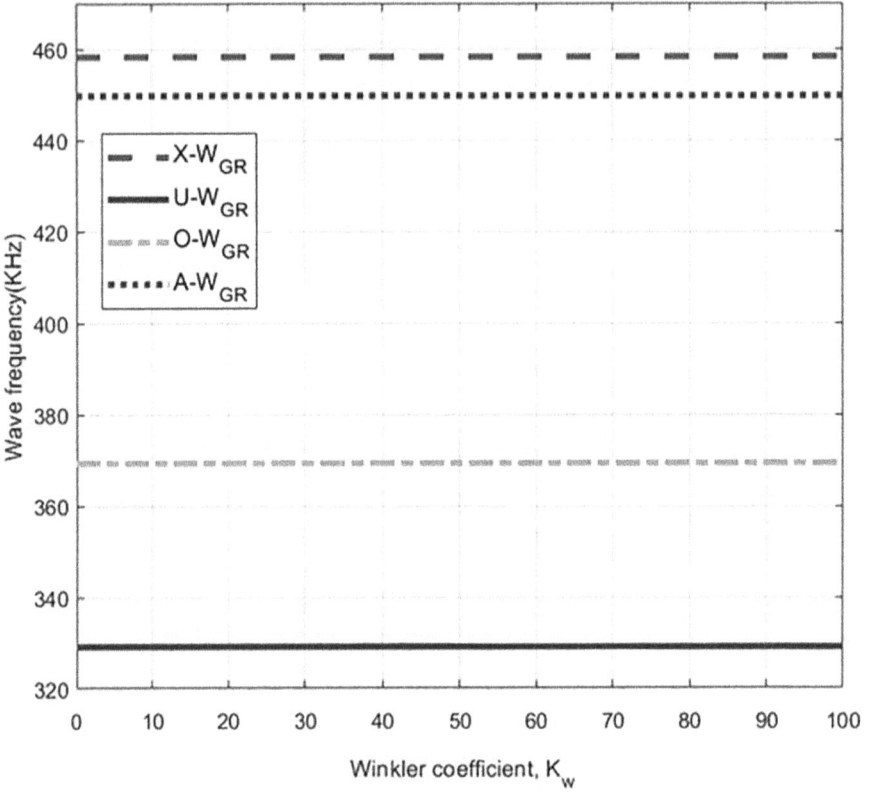

FIGURE 5.12 The relationship between the variation in the frequency of waves and the Winkler coefficient using several sorts of distribution patterns $(H_{GR} = 100\%, W_{GR} = 2.5\%, T = 300K, K_p = 30, \beta = 100, L = 20h, N_L = 10)$.

on wave frequency. Opting for a lower wave number facilitates a more conspicuous observation of the ascending trajectory of curves. In order to provide further clarification, it should be noted that augmenting the values of the elastic foundation leads to a more rigid structure, accompanied by an elevation in both wave frequency and phase velocity.

Figure 5.13 presents an analysis of the influence of the Pasternak coefficient on wave frequency, taking into account different distribution patterns. Based on the findings, a modest positive correlation is shown between the Pasternak coefficient and wave frequency. The wave frequency of the beam exhibits a linear increment as a result of including the Pasternak coefficient term into the matrix component using a linear formulation.

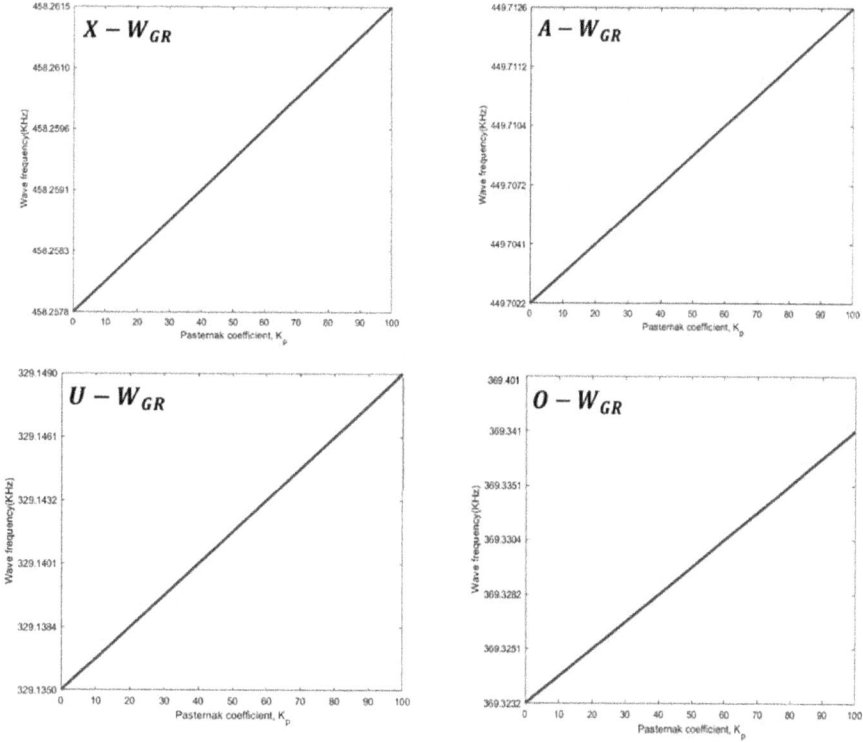

FIGURE 5.13 The relationship between the variation in the frequency of waves and the Pasternak coefficient using several sorts of distribution patterns ($H_{GR} = 100\%, W_{GR} = 2.5\%$, $T = 300K, K_w = 30, \beta = 100, L = 20h, N_L = 10$).

5.3 AUXETIC PLATES: WAVE PROPAGATION ANALYSIS

A sandwich composite plate with intelligent properties is expected to be shown in Figure 5.14. The smart piezo nanocomposite layers are regarded as the constituent materials for the top and bottom layers of the smart layers. The micro-electro-mechanical model is used for the purpose of estimating the electro-mechanical characteristics of the intelligent nanocomposite plate.

However, the effective qualities of electro-mechanical systems may be expressed in the following manner:

$$C_{11} = \frac{C^p_{11} C^m_{11}}{\eta C^m_{11} + (1 - \eta) C^p_{11}} \tag{5.21}$$

$$C_{12} = C_{11} \left[\frac{\eta C^p_{12}}{C^p_{11}} + \frac{(1 - \eta) C^m_{12}}{C^m_{11}} \right]$$

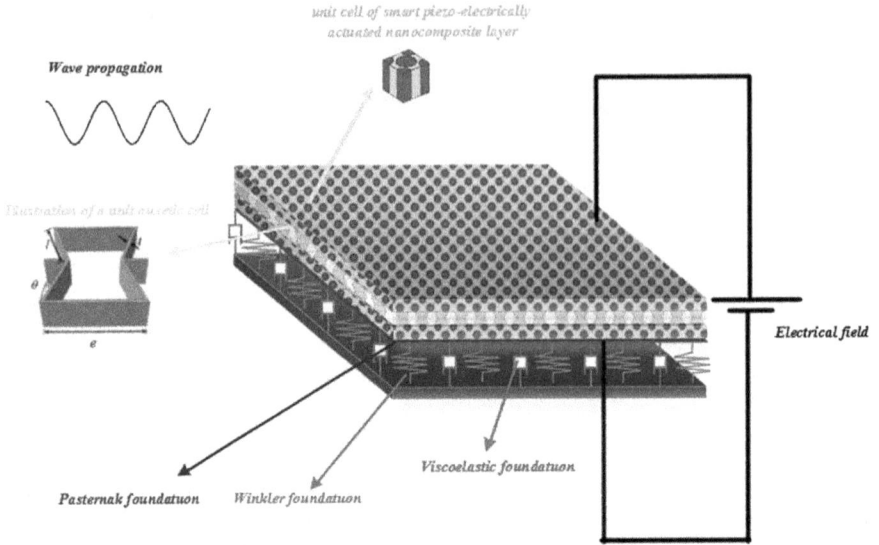

FIGURE 5.14 Schematic view of the smart sandwich plate with an auxetic core based on the Visco-Pasternak foundation.

$$C_{13} = C_{11}\left[\frac{\eta C_{13}^{p}}{C_{11}^{p}} + \frac{(1-\eta)C_{13}^{m}}{C_{11}^{m}}\right]$$

$$C_{22} = \eta C_{22}^{p} + (1-\eta)C_{22}^{m} + \frac{C_{12}^{2}}{C_{11}} - \frac{\eta\left(C_{12}^{p}\right)^{2}}{C_{11}^{p}} - \frac{(1-\eta)\left(C_{12}^{m}\right)^{2}}{C_{11}^{m}}$$

$$C_{23} = \eta C_{23}^{p} + (1-\eta)C_{23}^{m} + \frac{C_{12}C_{13}}{C_{11}} - \frac{\eta C_{12}^{p}C_{13}^{p}}{C_{11}^{p}} - \frac{(1-\eta)C_{12}^{m}C_{13}^{m}}{C_{11}^{m}}$$

$$C_{33} = \eta C_{33}^{p} + (1-\eta)C_{33}^{m} + \frac{C_{13}^{2}}{C_{11}} - \frac{\eta\left(C_{13}^{p}\right)^{2}}{C_{11}^{p}} - \frac{(1-\eta)\left(C_{13}^{m}\right)^{2}}{C_{11}^{m}}$$

$$C_{44} = \eta C_{44}^{p} + (1-\eta)C_{44}^{m}$$

$$C_{55} = \frac{A}{B^{2} + AC}$$

$$C_{66} = \frac{C_{66}^{p}C_{66}^{m}}{\eta C_{66}^{m} + (1-\eta)C_{66}^{p}}$$

$$e_{31} = C_{11}\left[\frac{\eta e_{31}^{p}}{C_{11}^{p}} + \frac{(1-\eta)e_{31}^{m}}{C_{11}^{m}}\right]$$

$$e_{32} = \eta e_{32}^p + (1-\eta) e_{32}^m + \frac{C_{12} e_{31}}{C_{11}} - \frac{\eta C_{12}^p e_{31}^p}{C_{11}^p} - \frac{(1-\eta) C_{12}^m e_{31}^m}{C_{11}^m}$$

$$e_{33} = \eta e_{33}^p + (1-\eta) e_{33}^m + \frac{C_{13} e_{31}}{C_{11}} - \frac{\eta C_{13}^p e_{31}^p}{C_{11}^p} - \frac{(1-\eta) C_{13}^m e_{31}^m}{C_{11}^m}$$

$$e_{24} = \eta e_{24}^p + (1-\eta) e_{24}^m$$

$$e_{15} = \frac{B}{B^2 + AC}$$

$$s_{11} = \frac{C}{B^2 + AC}$$

$$s_{22} = \eta s_{22}^p + (1-\eta) s_{22}^m$$

$$s_{33} = \eta s_{33}^p + (1-\eta) s_{33}^m + \frac{e_{31}^2}{C_{11}} - \frac{\eta \left(e_{31}^p\right)^2}{C_{11}^p} - \frac{(1-\eta)\left(e_{31}^m\right)^2}{C_{11}^m}$$

$$A = \left[\frac{\eta C_{55}^p}{\left(e_{15}^p\right)^2 + C_{55}^p s_{11}^p} + \frac{(1-\eta) C_{55}^m}{\left(e_{15}^m\right)^2 + C_{55}^m s_{11}^m} \right]$$

$$B = \left[\frac{\eta e_{15}^p}{\left(e_{15}^p\right)^2 + C_{55}^p s_{11}^p} + \frac{(1-\eta) e_{15}^m}{\left(e_{15}^m\right)^2 + C_{55}^m s_{11}^m} \right]$$

$$C = \left[\frac{\eta s_{11}^p}{\left(e_{15}^p\right)^2 + C_{55}^p s_{11}^p} + \frac{(1-\eta) s_{11}^m}{\left(e_{15}^m\right)^2 + C_{55}^m s_{11}^m} \right]$$

$$\rho = \eta \rho^p + (1-\eta) \rho^m$$

The variable η represents the volume percentage of the smart nanofiller inside the matrix. The effective mechanical characteristics of the auxetic core may be derived in the following manner:

$$E_{11}^a = E_{Al} (t/l)^3 \frac{\cos\theta}{(e/l + \sin\theta)\sin^2\theta} \tag{5.22}$$

$$E_{22}^a = E_{Al} (t/l)^3 \frac{(e/l + \sin\theta)}{\cos^3\theta}$$

$$G_{12}^a = E_{Al} (t/l)^3 \frac{(e/l + \sin\theta)}{(e/l)^2 (1 + 2e/l)\cos\theta}$$

$$v_{12}^a = \frac{\cos^2\theta}{(e/l + \sin\theta)\sin\theta}$$

$$\rho_a = \rho_{Al} \frac{(t/l)(e/l + 2)}{2\cos\theta (e/l + \sin\theta)}$$

The equations (5.22) provide the designated measures of the effective mechanical characteristics of the auxetic core. Furthermore, the newer formulae include geometrical characteristics such as the thickness-to-length ratio, the length of the vertical cell rib-to-length ratio, and the angle. The electro-elastic constitutive model of a smart composite rectangular plate is expressed in the following manner. Specifically:

$$\sigma_{ij} = \left[C_{ijkl}\varepsilon_{kl} - e_{mij}E_m \right] \tag{5.23-a}$$

$$D_i = \left[e_{ikl}\varepsilon_{kl} + s_{im}E_m \right] \tag{5.233-b}$$

The stress, strain, and elastic tensors are denoted as σ_{ij}, ε_{kl}, and C_{ijkl}, respectively. In addition, the electrical displacement and electrical field are represented by the symbols D_i and E_m, respectively. In addition, the dielectric constants and piezoelectric constants are denoted as s_{im} and e_{mij}, respectively. Furthermore, the correlation between the stress and strain shown by a smart composite rectangular plate may be elucidated in the following manner:

$$
\begin{Bmatrix} \sigma_{xx} \\ \sigma_{yy} \\ \sigma_{yz} \\ \sigma_{zx} \\ \sigma_{xy} \end{Bmatrix} =
\begin{pmatrix} C_{11} & C_{12} & 0 & 0 & 0 \\ C_{12} & C_{11} & 0 & 0 & 0 \\ 0 & 0 & C_{55} & 0 & 0 \\ 0 & 0 & 0 & C_{55} & 0 \\ 0 & 0 & 0 & 0 & C_{66} \end{pmatrix}
\begin{Bmatrix} \varepsilon_{xx} \\ \varepsilon_{yy} \\ \varepsilon_{yz} \\ \varepsilon_{zx} \\ \varepsilon_{xy} \end{Bmatrix} -
\begin{pmatrix} 0 & 0 & e_{31} \\ 0 & 0 & e_{31} \\ 0 & e_{15} & 0 \\ e_{15} & 0 & 0 \\ 0 & 0 & 0 \end{pmatrix}
\begin{Bmatrix} E_x \\ E_y \\ E_z \end{Bmatrix}
$$

$$\tag{5.24}$$

$$
\begin{Bmatrix} D_x \\ D_y \\ D_z \end{Bmatrix} =
\begin{pmatrix} 0 & 0 & 0 & 0 & e_{15} & 0 \\ 0 & 0 & 0 & e_{15} & 0 & 0 \\ e_{31} & e_{31} & 0 & 0 & 0 & 0 \end{pmatrix}
\begin{Bmatrix} \varepsilon_{xx} \\ \varepsilon_{yy} \\ \varepsilon_{zz} \\ \varepsilon_{yz} \\ \varepsilon_{zx} \\ \varepsilon_{xy} \end{Bmatrix} -
\begin{pmatrix} S_{11} & 0 & 0 \\ 0 & S_{11} & 0 \\ 0 & 0 & S_{33} \end{pmatrix}
\begin{Bmatrix} E_x \\ E_y \\ E_z \end{Bmatrix}
$$

In accordance with the revised higher-order shear deformation theory (HSDT), the displacement field of the rectangular plate composed of smart composites may be expressed in the following manner:

$$U(x,y,z,t) = u_0(x,y,t) - z\frac{\partial w_b}{\partial x} - f(z)\frac{\partial w_s}{\partial x} \tag{5.25a}$$

$$V(x,y,z,t) = v_0(x,y,t) - z\frac{\partial w_b}{\partial y} - f(z)\frac{\partial w_s}{\partial y} \tag{5.25b}$$

$$W(x,y,z,t) = w_b(x,y,t) + w_s(x,y,t) \tag{5.25c}$$

The longitudinal and transverse displacements are indicated by the functions $u_0(x,y,t)$ and $v_0(x,y,t)$, respectively. Additionally, the bending and shear deflections are represented by the functions $w_b(x,y,t)$ and $w_s(x,y,t)$. Moreover, the present work utilizes form functions as $f(z) = z - \dfrac{\sin(\zeta z)}{\zeta}$ and $f(z) = \dfrac{\zeta e^z}{\zeta^2 + 1} \sin(\zeta z) + \pi \dfrac{\cos(\zeta z)}{\zeta}$ [2].

Furthermore, it is assumed that the smart composite rectangular plate is subjected to an electrical potential. This potential is calculated to equal the sum of linear and harmonic terms throughout the whole thickness of the smart plate. In order to adhere to Maxwell's law, the electrical potential is estimated in the following manner:

$$\Phi(x,y,z,t) = -\cos(\zeta z)\phi(x,z,t) + \frac{2z}{h}V \tag{5.26}$$

in which $\zeta = \dfrac{\pi}{h}$ and the specific electric voltage being referred to is denoted by the symbol V. The strains of the smart composite rectangular plate may be determined using the following calculations:

$$\begin{Bmatrix} \varepsilon_{xx} \\ \varepsilon_{yy} \\ \varepsilon_{xy} \end{Bmatrix} = \begin{Bmatrix} \varepsilon_{xx}^0 \\ \varepsilon_{yy}^0 \\ \varepsilon_{xy}^0 \end{Bmatrix} + z \begin{Bmatrix} k_{xx}^b \\ k_{yy}^b \\ k_{xy}^b \end{Bmatrix} + f(z) \begin{Bmatrix} k_{xx}^s \\ k_{yy}^s \\ k_{xy}^s \end{Bmatrix}, \qquad \begin{Bmatrix} \gamma_{xz} \\ \gamma_{yz} \end{Bmatrix} = \begin{Bmatrix} \gamma_{xz}^0 \\ \gamma_{yz}^0 \end{Bmatrix} \tag{5.27}$$

where

$$\begin{Bmatrix} \varepsilon_{xx}^0 \\ \varepsilon_{yy}^0 \\ \varepsilon_{xy}^0 \end{Bmatrix} = \begin{Bmatrix} \dfrac{\partial u_0}{\partial x} \\ \dfrac{\partial v_0}{\partial x} \\ \dfrac{\partial u_0}{\partial y} + \dfrac{\partial v_0}{\partial x} \end{Bmatrix}, \begin{Bmatrix} k_{xx}^b \\ k_{yy}^b \\ k_{xy}^b \end{Bmatrix} = \begin{Bmatrix} -\dfrac{\partial^2 w_b}{\partial x^2} \\ -\dfrac{\partial^2 w_b}{\partial y^2} \\ -\dfrac{\partial^2 w_b}{\partial x \partial y} \end{Bmatrix}, \begin{Bmatrix} k_{xx}^s \\ k_{yy}^s \\ k_{xy}^s \end{Bmatrix} = \begin{Bmatrix} -\dfrac{\partial^2 w_s}{\partial x^2} \\ -\dfrac{\partial^2 w_s}{\partial y^2} \\ -\dfrac{\partial^2 w_s}{\partial x \partial y} \end{Bmatrix}, \begin{Bmatrix} \gamma_{xz}^0 \\ \gamma_{yz}^0 \end{Bmatrix} = \begin{Bmatrix} \dfrac{\partial w_s}{\partial x} \\ \dfrac{\partial w_s}{\partial y} \end{Bmatrix} \tag{5.28}$$

that

$$g(z) = 1 - \frac{df(z)}{dz} \tag{5.29}$$

The expression for the electric field is given by:

$$E_x = -\frac{\partial \Phi}{\partial x} = \cos(\zeta z)\frac{\partial \phi}{\partial x} \tag{5.30a}$$

$$E_y = -\frac{\partial \Phi}{\partial y} = \cos(\zeta z)\frac{\partial \phi}{\partial y} \tag{5.30b}$$

$$E_z = -\frac{\partial \Phi}{\partial z} = -\zeta \sin(\zeta z)\phi - \frac{2V}{h} \tag{5.30c}$$

By applying Hamilton's concept to the smart composite rectangular plate, we get the following equation:

$$\int_0^t \delta(\Pi_S - \Pi_K + \Pi_W)dt = 0 \tag{5.31}$$

where Π_S, Π_W, and Π_K represent the strain energy, external work, and kinetic energy, respectively. The expression of strain energy may be formulated as follows:

$$\delta \Pi_S = \int_V (\sigma_{ij}\delta\varepsilon_{ij})dV = \int_V \left(\begin{array}{c} \sigma_{xx}\delta\varepsilon_{xx} + \sigma_{yy}\delta\varepsilon_{yy} + \sigma_{xy}\delta\varepsilon_{xy} + \sigma_{xz}\delta\varepsilon_{xz} + \sigma_{yz}\delta\varepsilon_{yz} \\ -D_x\delta E_x - D_y\delta E_y - D_z\delta E_z \end{array} \right)dV \tag{5.32}$$

By substituting equations (5.28) and (5.30) into equation (5.32), the expression for the strain energy may be reformulated as follows:

$$\delta \Pi_S = \int_A \left(\begin{array}{c} N_{xx}\dfrac{\partial \delta u_0}{\partial x} - M_{xx}^b \dfrac{\partial^2 w_b}{\partial x^2} - M_{xx}^s \dfrac{\partial^2 w_s}{\partial x^2} + N_{yy}\dfrac{\partial \delta u_0}{\partial y} - M_{yy}^b \dfrac{\partial^2 w_b}{\partial y^2} - M_{yy}^s \dfrac{\partial^2 w_s}{\partial y^2} \\ + N_{xy}\left(\dfrac{\partial u_0}{\partial y} + \dfrac{\partial v_0}{\partial x}\right) - 2M_{xy}^b \dfrac{\partial^2 w_b}{\partial x \partial y} - 2M_{xy}^s \dfrac{\partial^2 w_s}{\partial x \partial y} + Q_{xz}\dfrac{\partial w_s}{\partial x} + Q_{yz}\dfrac{\partial w_s}{\partial y} \end{array} \right)dA$$

$$+ \int_{-h/2}^{h/2}\int_A \left(-D_x \cos(\zeta z)\delta(\dfrac{\partial \phi}{\partial x}) - D_y \cos(\zeta z)\delta(\dfrac{\partial \phi}{\partial y}) + D_z \sin(\zeta z)\delta\phi \right)dAdz \tag{5.33}$$

The forces and moments may be acquired using the following methods:

$$\left(N_i, M_i^b, M_i^s \right) = \int_{-h/2}^{h/2} \left(1, z, f(z) \right)\sigma_i dz, \quad i = (xx, yy, xy) \tag{5.34a}$$

$$Q_j = \int_{-h/2}^{h/2} g(z)\sigma_j dz, \quad j = (xz, yz) \tag{5.34b}$$

The first variation of external work would be expressed as:

$$\delta \Pi_W = \int_A \left(\begin{array}{c} N_x^E \dfrac{\partial^2 \left(w_b + w_s \right)}{\partial x^2} + N_y^E \dfrac{\partial^2 \left(w_b + w_s \right)}{\partial y^2} \\ -k_w \left(w_b + w_s \right) + k_p \left(\dfrac{\partial^2 \left(w_b + w_s \right)}{\partial x^2} + \dfrac{\partial^2 \left(w_b + w_s \right)}{\partial y^2} \right) \\ -c_d \dfrac{\partial \left(w_b + w_s \right)}{\partial t} \end{array} \right)\delta\left(w_b + w_s \right)dA \tag{5.35a}$$

The variables N_x^E and N_y^E denote the external electrical loads acting in the x and y directions, respectively, that are applied to the smart composite rectangular plate. In addition, the coefficients associated with Winkler, Pasternak, and damping founda-tions are represented by k_w, k_p, and c_d, respectively. Additionally, the initial form of the equation for kinetic energy may be expressed as:

$$\delta \Pi_K = \int_V (\dot{U}\delta\dot{U} + \dot{V}\delta\dot{V} + \dot{W}\delta\dot{W})\rho(z)dV =$$

$$\int_A \left(\begin{array}{l} I_0\left(\dfrac{\partial u_0}{\partial t}\dfrac{\partial\delta u_0}{\partial t} + \dfrac{\partial v_0}{\partial t}\dfrac{\partial\delta v_0}{\partial t} + \dfrac{\partial(w_b+w_s)}{\partial t}\dfrac{\partial\delta(w_b+w_s)}{\partial t}\right) \\[3mm] -I_1\left(\dfrac{\partial u_0}{\partial t}\dfrac{\partial^2\delta w_b}{\partial x\partial t} + \dfrac{\partial^2 w_b}{\partial x\partial t}\dfrac{\partial\delta u_0}{\partial t} + \dfrac{\partial v_0}{\partial t}\dfrac{\partial^2\delta w_b}{\partial y\partial t} + \dfrac{\partial^2 w_b}{\partial y\partial t}\dfrac{\partial\delta v_0}{\partial t}\right) \\[3mm] -J_1\left(\dfrac{\partial u_0}{\partial t}\dfrac{\partial^2\delta w_s}{\partial x\partial t} + \dfrac{\partial^2 w_s}{\partial x\partial t}\dfrac{\partial\delta u_0}{\partial t} + \dfrac{\partial v_0}{\partial t}\dfrac{\partial^2\delta w_s}{\partial y\partial t} + \dfrac{\partial^2 w_s}{\partial y\partial t}\dfrac{\partial\delta v_0}{\partial t}\right) \\[3mm] +I_2\left(\dfrac{\partial^2 w_b}{\partial x\partial t}\dfrac{\partial^2\delta w_b}{\partial x\partial t} + \dfrac{\partial^2 w_b}{\partial y\partial t}\dfrac{\partial^2\delta w_b}{\partial y\partial t}\right) + K_2\left(\dfrac{\partial^2 w_s}{\partial x\partial t}\dfrac{\partial^2\delta w_s}{\partial x\partial t} + \dfrac{\partial^2 w_s}{\partial y\partial t}\dfrac{\partial^2\delta w_s}{\partial y\partial t}\right) \\[3mm] +J_2\left(\dfrac{\partial^2 w_b}{\partial x\partial t}\dfrac{\partial^2\delta w_s}{\partial x\partial t} + \dfrac{\partial^2 w_s}{\partial x\partial t}\dfrac{\partial^2\delta w_b}{\partial x\partial t} + \dfrac{\partial^2 w_b}{\partial y\partial t}\dfrac{\partial^2\delta w_s}{\partial y\partial t} + \dfrac{\partial^2 w_s}{\partial y\partial t}\dfrac{\partial^2\delta w_b}{\partial y\partial t}\right) \end{array}\right) dA \tag{5.35b}$$

The mass inertias could be obtained as:

$$(I_0,I_1,J_1,I_2,J_2,K_2) = \int_{-h/2}^{h/2} (1,z,f(z),z^2,zf(z),f^2(z))\rho(z)dz \tag{5.36}$$

By substituting equations (5.33), (5.35a), and (5.35b) into equation (5.31), the equations governing the motion of the smart composite rectangular plate may be formulated as:

$$\frac{\partial N_{xx}}{\partial x} + \frac{\partial N_{xy}}{\partial y} = I_0\frac{\partial^2 u_0}{\partial t^2} - I_1\frac{\partial^3 w_b}{\partial x\partial t^2} - J_1\frac{\partial^3 w_s}{\partial x\partial t^2} \tag{5.37a}$$

$$\frac{\partial N_{xy}}{\partial x} + \frac{\partial N_{yy}}{\partial y} = I_0\frac{\partial^2 v_0}{\partial t^2} - I_1\frac{\partial^3 w_b}{\partial y\partial t^2} - J_1\frac{\partial^3 w_s}{\partial y\partial t^2} \tag{5.37b}$$

$$\frac{\partial^2 M_{xx}^b}{\partial x^2} + 2\frac{\partial^2 M_{xy}^b}{\partial x\partial y} + \frac{\partial^2 M_{yy}^b}{\partial y^2} + (k_p - N^E)\nabla^2(w_b+w_s) - k_w\left(w_b+w_s\right) - c_d\frac{\partial\left(w_b+w_s\right)}{\partial t}$$

$$= I_0\frac{\partial^2(w_b+w_s)}{\partial t^2} + I_1(\frac{\partial^3 u_0}{\partial x\partial t^2} + \frac{\partial^3 v_0}{\partial y\partial t^2}) - I_2\nabla^2\frac{\partial^2 w_b}{\partial t^2} - J_2\nabla^2\frac{\partial^2 w_s}{\partial t^2} \tag{5.37c}$$

$$\frac{\partial^2 M_{xx}^s}{\partial x^2} + 2\frac{\partial^2 M_{xy}^s}{\partial x \partial y} + \frac{\partial^2 M_{yy}^s}{\partial y^2} + \frac{\partial Q_{xz}}{\partial x} + \frac{\partial Q_{yz}}{\partial y} + (k_p - N^E)\nabla^2(w_b + w_s)$$

$$-k_w(w_b + w_s) - c_d \frac{\partial(w_b + w_s)}{\partial t} = I_0 \frac{\partial^2(w_b + w_s)}{\partial t^2} + J_1(\frac{\partial^3 u_0}{\partial x \partial t^2} + \frac{\partial^3 v_0}{\partial y \partial t^2}) \quad (5.37\text{d})$$

$$-J_2\nabla^2 \frac{\partial^2 w_b}{\partial t^2} - K_2\nabla^2 \frac{\partial^2 w_s}{\partial t^2}$$

$$\int_A \left(\cos(\zeta z)\frac{\partial D_x}{\partial x} + \cos(\zeta z)\frac{\partial D_y}{\partial y} + \zeta\sin(\zeta z)D_z \right) dA = 0 \quad (5.37\text{e})$$

In accordance with the constitutive model, the force and moment are determined by using the displacement fields in the following manner:

$$\begin{Bmatrix} N_{xx} \\ N_{yy} \\ N_{xy} \end{Bmatrix} = \begin{pmatrix} \begin{bmatrix} A_{11} & A_{12} & 0 \\ A_{21} & A_{22} & 0 \\ 0 & 0 & A_{66} \end{bmatrix} \begin{Bmatrix} \dfrac{\partial u_0}{\partial x} \\ \dfrac{\partial v_0}{\partial y} \\ \dfrac{\partial u_0}{\partial y} + \dfrac{\partial v_0}{\partial x} \end{Bmatrix} + \begin{bmatrix} B_{11} & B_{12} & 0 \\ B_{21} & B_{22} & 0 \\ 0 & 0 & B_{66} \end{bmatrix} \begin{Bmatrix} -\dfrac{\partial^2 w_b}{\partial x^2} \\ -\dfrac{\partial^2 w_b}{\partial y^2} \\ -2\dfrac{\partial^2 w_b}{\partial x \partial y} \end{Bmatrix} \\[3em] + \begin{bmatrix} B_{11}^s & B_{12}^s & 0 \\ B_{21}^s & B_{22}^s & 0 \\ 0 & 0 & B_{66}^s \end{bmatrix} \begin{Bmatrix} -\dfrac{\partial^2 w_s}{\partial x^2} \\ -\dfrac{\partial^2 w_s}{\partial y^2} \\ -2\dfrac{\partial^2 w_s}{\partial x \partial y} \end{Bmatrix} + \begin{Bmatrix} A_{31}^e \\ A_{31}^e \\ 0 \end{Bmatrix}\phi - \begin{Bmatrix} N_x^E \\ N_y^E \\ 0 \end{Bmatrix} \end{pmatrix} \quad (5.38\text{a})$$

$$\begin{Bmatrix} M_{xx}^b \\ M_{yy}^b \\ M_{xy}^b \end{Bmatrix} = \begin{pmatrix} \begin{bmatrix} B_{11} & B_{12} & 0 \\ B_{21} & B_{22} & 0 \\ 0 & 0 & B_{66} \end{bmatrix} \begin{Bmatrix} \dfrac{\partial u_0}{\partial x} \\ \dfrac{\partial v_0}{\partial y} \\ \dfrac{\partial u_0}{\partial y} + \dfrac{\partial v_0}{\partial x} \end{Bmatrix} + \begin{bmatrix} D_{11} & D_{12} & 0 \\ D_{21} & D_{22} & 0 \\ 0 & 0 & D_{66} \end{bmatrix} \begin{Bmatrix} -\dfrac{\partial^2 w_b}{\partial x^2} \\ -\dfrac{\partial^2 w_b}{\partial y^2} \\ -2\dfrac{\partial^2 w_b}{\partial x \partial y} \end{Bmatrix} \\[3em] + \begin{bmatrix} D_{11}^s & D_{12}^s & 0 \\ D_{21}^s & D_{22}^s & 0 \\ 0 & 0 & D_{66}^s \end{bmatrix} \begin{Bmatrix} -\dfrac{\partial^2 w_s}{\partial x^2} \\ -\dfrac{\partial^2 w_s}{\partial y^2} \\ -2\dfrac{\partial^2 w_s}{\partial x \partial y} \end{Bmatrix} + \begin{Bmatrix} E_{31}^e \\ E_{31}^e \\ 0 \end{Bmatrix}\phi - \begin{Bmatrix} M_{bx}^E \\ M_{by}^E \\ 0 \end{Bmatrix} \end{pmatrix} \quad (5.38\text{b})$$

$$
\begin{aligned}
\begin{Bmatrix} M_{xx}^s \\ M_{yy}^s \\ M_{xy}^s \end{Bmatrix} =
& \begin{pmatrix} B_{11}^s & B_{12}^s & 0 \\ B_{21}^s & B_{22}^s & 0 \\ 0 & 0 & B_{66}^s \end{pmatrix}
\begin{Bmatrix} \dfrac{\partial u_0}{\partial x} \\[2mm] \dfrac{\partial v_0}{\partial y} \\[2mm] \dfrac{\partial u_0}{\partial y}+\dfrac{\partial v_0}{\partial x} \end{Bmatrix}
+ \begin{pmatrix} D_{11}^s & D_{12}^s & 0 \\ D_{21}^s & D_{22}^s & 0 \\ 0 & 0 & D_{66}^s \end{pmatrix}
\begin{Bmatrix} -\dfrac{\partial^2 w_b}{\partial x^2} \\[2mm] -\dfrac{\partial^2 w_b}{\partial y^2} \\[2mm] -2\dfrac{\partial^2 w_b}{\partial x \partial y} \end{Bmatrix} \\[4mm]
& + \begin{pmatrix} H_{11}^s & H_{12}^s & 0 \\ H_{21}^s & H_{22}^s & 0 \\ 0 & 0 & H_{66}^s \end{pmatrix}
\begin{Bmatrix} -\dfrac{\partial^2 w_s}{\partial x^2} \\[2mm] -\dfrac{\partial^2 w_s}{\partial y^2} \\[2mm] -2\dfrac{\partial^2 w_s}{\partial x \partial y} \end{Bmatrix}
+ \begin{Bmatrix} F_{31}^e \\ F_{31}^e \\ 0 \end{Bmatrix}\phi - \begin{Bmatrix} M_{sx}^E \\ M_{sy}^E \\ 0 \end{Bmatrix}
\end{aligned} \tag{5.38c}
$$

$$
\begin{Bmatrix} Q_{xz} \\ Q_{yz} \end{Bmatrix} = \begin{pmatrix} A_{44}^s & 0 \\ 0 & A_{55}^s \end{pmatrix}
\begin{Bmatrix} \dfrac{\partial w_s}{\partial x} \\[2mm] \dfrac{\partial w_s}{\partial y} \end{Bmatrix}
- A_{15}^e \begin{Bmatrix} \dfrac{\partial \phi}{\partial x} \\[2mm] \dfrac{\partial \phi}{\partial y} \end{Bmatrix} \tag{5.38d}
$$

$$
\int_A \left(\begin{Bmatrix} D_x \\ D_y \end{Bmatrix} \cos(\zeta z) \right) dA = E_{15}^e \begin{Bmatrix} \dfrac{\partial w_s}{\partial x} \\[2mm] \dfrac{\partial w_s}{\partial y} \end{Bmatrix}
- F_{11}^e \begin{Bmatrix} \dfrac{\partial \phi}{\partial x} \\[2mm] \dfrac{\partial \phi}{\partial y} \end{Bmatrix} \tag{5.38e}
$$

$$
\int_A (D_z \zeta \sin(\zeta z)) dA = A_{31}^e \left(\frac{\partial u_0}{\partial x}+\frac{\partial v_0}{\partial y} \right) - E_{31}^e \nabla^2 w_b - F_{31}^e \nabla^2 w_s - F_{33}^e \phi \tag{5.38f}
$$

Additionally, the cross-sectional rigidities may be derived in the following manner:

$$
\begin{pmatrix} A_{11} & B_{11} & D_{11} & B_{11}^s & D_{11}^s & H_{11}^s \\ A_{12} & B_{12} & D_{12} & B_{12}^s & D_{12}^s & H_{12}^s \\ A_{66} & B_{66} & D_{66} & B_{66}^s & D_{66}^s & H_{66}^s \end{pmatrix} =
$$
$$
\left(1, z, z^2, f(z), zf(z), f^2(z) \right) \left(2 \int_{-h/2}^{-h/2+h_{SNC}} \begin{Bmatrix} c_{11}^{SNC} \\ c_{12}^{SNC} \\ c_{66}^{SNC} \end{Bmatrix} dz + \int_{-h/2+h_{SNC}}^{-h/2+h_{SNC}+h_{Auxetic}} \begin{Bmatrix} c_{11}^{Auxetic} \\ c_{12}^{Auxetic} \\ c_{66}^{Auxetic} \end{Bmatrix} dz \right) \tag{5.39a}
$$

$$\left(A_{31}^e, E_{31}^e, F_{31}^e\right) = 2 \int\limits_{-h/2}^{-h/2+h_{SNC}} \left(1, z, f(z)\right) e_{31}\zeta \sin(\zeta z)dz$$

$$\left(A_{15}^e, E_{15}^e\right) = 2 \int\limits_{-h/2}^{-h/2+h_{SNC}} \left(1, g(z)\right) e_{31}\cos(\zeta z)dz \tag{5.39b}$$

$$\left(F_{11}^e, F_{33}^e\right) = 2 \int\limits_{-h/2}^{-h/2+h_{SNC}} \left(s_{11}\cos^2(\zeta z), s_{33}\zeta^2 \sin^2(\zeta z)\right)dz$$

$$\left(A_{44}^s, A_{55}^s\right) = 2 \int\limits_{-h/2}^{-h/2+h_{SNC}} \left(c_{44}g^2(z)\right)dz$$

Furthermore, the external pressures and moments, which are derived from external electrical work, may be expressed as follows:

$$N_x^E = N_y^E = -2 \int\limits_{-h/2}^{-h/2+h_{SNC}} \left(e_{31}\frac{2V_0}{h_{SNC}}\right)dz \tag{5.40a}$$

$$M_{bx}^E = M_{by}^E = -2 \int\limits_{-h/2}^{-h/2+h_{SNC}} \left(e_{31}\frac{2V_0}{h_{SNC}}\right)dz \tag{5.40b}$$

$$M_{sx}^E = M_{sy}^E = -2 \int\limits_{-h/2}^{-h/2+h_{SNC}} \left(e_{31}\frac{2V_0}{h_{SNC}}f(z)\right)dz \tag{5.40c}$$

The governing equations of the smart rectangular plate may be determined by inserting equations (5.38a–f) into equations (5.37a–e).

$$\begin{aligned} &\left(A_{11}\frac{\partial^2 u_0}{\partial x^2} + \left(A_{12} + A_{66}\right)\frac{\partial^2 v_0}{\partial x \partial y} + A_{66}\frac{\partial^2 u_0}{\partial y^2} - B_{11}\frac{\partial^3 w_b}{\partial x^3} - \left(B_{12} + 2B_{66}\right)\frac{\partial^3 w_b}{\partial x \partial y^2}\right.\\ &\left.\qquad - B_{11}^s \frac{\partial^3 w_s}{\partial x^3} - \left(B_{12}^s + 2B_{66}^s\right)\frac{\partial^3 w_s}{\partial x \partial y^2} + A_{31}^e \frac{\partial \phi}{\partial x}\right)\\ &+ \left(-I_0\frac{\partial^2 u_0}{\partial t^2} + I_1\frac{\partial^3 w_b}{\partial x \partial t^2} + J_1\frac{\partial^3 w_s}{\partial x \partial t^2}\right) = 0 \end{aligned} \tag{5.41}$$

$$\begin{aligned} &\left(A_{22}\frac{\partial^2 v_0}{\partial y^2} + \left(A_{12} + A_{66}\right)\frac{\partial^2 u_0}{\partial x \partial y} + A_{66}\frac{\partial^2 v_0}{\partial x^2} - B_{22}\frac{\partial^3 w_b}{\partial y^3} - \left(B_{12} + 2B_{66}\right)\frac{\partial^3 w_b}{\partial x^2 \partial y}\right.\\ &\left.\qquad - B_{22}^s \frac{\partial^3 w_s}{\partial y^3} - \left(B_{12}^s + 2B_{66}^s\right)\frac{\partial^3 w_s}{\partial x^2 \partial y} + A_{31}^e \frac{\partial \phi}{\partial y}\right)\\ &+ \left(-I_0\frac{\partial^2 v_0}{\partial t^2} + I_1\frac{\partial^3 w_b}{\partial y \partial t^2} + J_1\frac{\partial^3 w_s}{\partial y \partial t^2}\right) = 0 \end{aligned} \tag{5.42}$$

$$
+\left\{
\begin{array}{c}
\left(
\begin{array}{c}
B_{11}\dfrac{\partial^3 u_0}{\partial x^3}+\left(B_{12}+2B_{66}\right)\dfrac{\partial^3 u_0}{\partial x\partial y^2}+B_{22}\dfrac{\partial^3 v_0}{\partial y^3}+\left(B_{12}+2B_{66}\right)\dfrac{\partial^3 v_0}{\partial x^2\partial y} \\[4mm]
-D_{11}\dfrac{\partial^4 w_b}{\partial x^4}-2\left(D_{12}+2D_{66}\right)\dfrac{\partial^4 w_b}{\partial x^2\partial y^2}-D_{22}\dfrac{\partial^4 w_b}{\partial y^4}+E_{31}^e\left(\dfrac{\partial^2\phi}{\partial x^2}+\dfrac{\partial^2\phi}{\partial y^2}\right) \\[4mm]
-D_{11}^s\dfrac{\partial^4 w_s}{\partial x^4}-2\left(D_{12}^s+2D_{66}^s\right)\dfrac{\partial^4 w_s}{\partial x^2\partial y^2}-D_{22}^s\dfrac{\partial^4 w_s}{\partial y^4}
\end{array}
\right) \\[18mm]
-I_0\dfrac{\partial^2\left(w_b+w_s\right)}{\partial t^2}-I_1\left(\dfrac{\partial^3 u_0}{\partial x\partial t^2}+\dfrac{\partial^3 v_0}{\partial y\partial t^2}\right)+I_2\left(\dfrac{\partial^2}{\partial x^2}+\dfrac{\partial^2}{\partial y^2}\right)\dfrac{\partial^2 w_b}{\partial t^2} \\[5mm]
+J_2\left(\dfrac{\partial^2}{\partial x^2}+\dfrac{\partial^2}{\partial y^2}\right)\dfrac{\partial^2 w_s}{\partial t^2}-N^E\left(\dfrac{\partial^2}{\partial x^2}+\dfrac{\partial^2}{\partial y^2}\right)\left(w_b+w_s\right) \\[5mm]
-k_w\left(w_b+w_s\right)+k_p\left(\dfrac{\partial^2\left(w_b+w_s\right)}{\partial x^2}+\dfrac{\partial^2\left(w_b+w_s\right)}{\partial y^2}\right)-c_d\dfrac{\partial\left(w_b+w_s\right)}{\partial t}
\end{array}
\right\}=0
$$

(5.43)

$$
+\left\{
\begin{array}{c}
\left(
\begin{array}{c}
B_{11}^s\dfrac{\partial^3 u_0}{\partial x^3}+\left(B_{12}^s+2B_{66}^s\right)\dfrac{\partial^3 u_0}{\partial x\partial y^2}+B_{22}^s\dfrac{\partial^3 v_0}{\partial y^3}+\left(B_{12}^s+2B_{66}^s\right)\dfrac{\partial^3 v_0}{\partial x^2\partial y} \\[4mm]
-D_{11}^s\dfrac{\partial^4 w_b}{\partial x^4}-2\left(D_{12}^s+2D_{66}^s\right)\dfrac{\partial^4 w_b}{\partial x^2\partial y^2}-D_{22}^s\dfrac{\partial^4 w_b}{\partial y^4}+\left(F_{31}^e-E_{15}^e\right)\left(\dfrac{\partial^2\phi}{\partial x^2}+\dfrac{\partial^2\phi}{\partial y^2}\right) \\[4mm]
-H_{11}^s\dfrac{\partial^4 w_s}{\partial x^4}-2\left(H_{12}^s+2H_{66}^s\right)\dfrac{\partial^4 w_s}{\partial x^2\partial y^2}-H_{22}^s\dfrac{\partial^4 w_s}{\partial y^4}+A_{44}^s\left(\dfrac{\partial^2 w_s}{\partial x^2}+\dfrac{\partial^2 w_s}{\partial y^2}\right)
\end{array}
\right) \\[18mm]
-I_0\dfrac{\partial^2\left(w_b+w_s\right)}{\partial t^2}-J_1\left(\dfrac{\partial^3 u_0}{\partial x\partial t^2}+\dfrac{\partial^3 v_0}{\partial y\partial t^2}\right)+J_2\nabla^2\dfrac{\partial^2 w_b}{\partial t^2} \\[5mm]
+K_2\left(\dfrac{\partial^2}{\partial x^2}+\dfrac{\partial^2}{\partial y^2}\right)\dfrac{\partial^2 w_s}{\partial t^2}-N^E\left(\dfrac{\partial^2}{\partial x^2}+\dfrac{\partial^2}{\partial y^2}\right)\left(w_b+w_s\right) \\[5mm]
-k_w\left(w_b+w_s\right)+k_p\left(\dfrac{\partial^2\left(w_b+w_s\right)}{\partial x^2}+\dfrac{\partial^2\left(w_b+w_s\right)}{\partial y^2}\right)-c_d\dfrac{\partial\left(w_b+w_s\right)}{\partial t}
\end{array}
\right\}=0
$$

(5.44)

$$
\left\{
\begin{array}{c}
A_{31}^e\left(\dfrac{\partial u_0}{\partial x}+\dfrac{\partial v_0}{\partial y}\right)-E_{31}^e\left(\dfrac{\partial^2 w_b}{\partial x^2}+\dfrac{\partial^2 w_b}{\partial y^2}\right) \\[5mm]
-\left(F_{31}^e-E_{15}^e\right)\left(\dfrac{\partial^2 w_s}{\partial x^2}+\dfrac{\partial^2 w_s}{\partial y^2}\right)+F_{11}^e\left(\dfrac{\partial^2}{\partial x^2}+\dfrac{\partial^2}{\partial y^2}\right)\phi-F_{33}^e\phi
\end{array}
\right\}=0
$$

(5.45)

In this section, the solution to the governing equations is provided using the analytical technique, namely in the form of an exponential function.

$$
\begin{Bmatrix} u_0(x,y,t) \\ v_0(x,y,t) \\ w_b(x,y,t) \\ w_s(x,y,t) \\ \phi(x,y,t) \end{Bmatrix} = \begin{Bmatrix} U\exp\left[i\left(k_1 x + k_2 y - \omega t\right)\right] \\ V\exp\left[i\left(k_1 x + k_2 y - \omega t\right)\right] \\ W_b\exp\left[i\left(k_1 x + k_2 y - \omega t\right)\right] \\ W_s\exp\left[i\left(k_1 x + k_2 y - \omega t\right)\right] \\ \Phi\exp\left[i\left(k_1 x + k_2 y - \omega t\right)\right] \end{Bmatrix}
\tag{5.46}
$$

The wave numbers, angular frequency, are represented by the symbols k_1, k_2, and ω, respectively. The substitution of exponential functions with the ultimate governing equations yields the ensuing equation called the eigenvalue issue.

$$
\left([\mathbf{K}] + i[\mathbf{C}]\omega - \omega^2[\mathbf{M}]\right)\{d\} = 0
\tag{5.47}
$$

The d can be defined as:

$$
\{d\} = \{U, V, W_b, W_s, \Phi\}^T
\tag{5.48}
$$

The determination of the angular wave frequency of the smart composite plate involves the process of setting the determinant of equation (5.47) to zero, as shown:

$$
\left\| [\mathbf{K}] + i[\mathbf{C}]\omega - \omega^2[\mathbf{M}] \right\| = 0
\tag{5.49}
$$

The angular frequency of the smart composite plate may be determined by solving equation (5.49) in the specified format.

$$
\omega_0 = M_0(k), \quad \omega_1 = M_1(k), \quad \omega_2 = M_2(k), \quad \omega_3 = M_3(k)
\tag{5.50a}
$$

Furthermore, the phase velocity may be seen as comparable to:

$$
c_i = \frac{M_i}{k}, \quad i = 0,1,2,3
\tag{5.50b}
$$

The geometrical parameters of the auxetic cell are yet undefined $\theta = 55$, $e/l = 4$, and $t/l = 0.0138571$. Additionally, it is expected that the substance of the auxetic core is aluminum, as indicated by $E_{Al} = 70$ GPa $\rho_{Al} = 2705$ kg / m^3, and $v_{Al} = 0.3$. Table 5.3 presents the electro-mechanical properties of the nanofiller and matrix materials used in the nanocomposite piezoelectric layers.

Figure 5.15 illustrates the fluctuations in phase velocity of a smart sandwich plate as a function of wave number (k), with respect to various geometrical parameters (e/l). As seen in this figure, there is a noticeable rise in the phase velocity when the ratio of e to l is raised from 1 to 4. It is evident that the phase velocity increases as the wave number increases from 0 to 100.

TABLE 5.3

Material Properties $BaTiO_3$ and $CoFe_2O_4$ [3]

Properties		
$c_{11} = c_{22}$ (GPa)	166	286
c_{33}	162	269.5
$c_{13} = c_{23}$	78	170.5
c_{12}	77	173
c_{55}	43	45.3
c_{66}	44.5	56.5
e_{31} (Cm^{-2})	−4.4	0
e_{33}	18.6	0
e_{15}	11.6	0
s_{11} ($10^{-9}C^2m^{-2}N^{-1}$)	11.2	0.08
s_{33}	12.6	0.093
ρ (kgm^{-3})	5800	5300

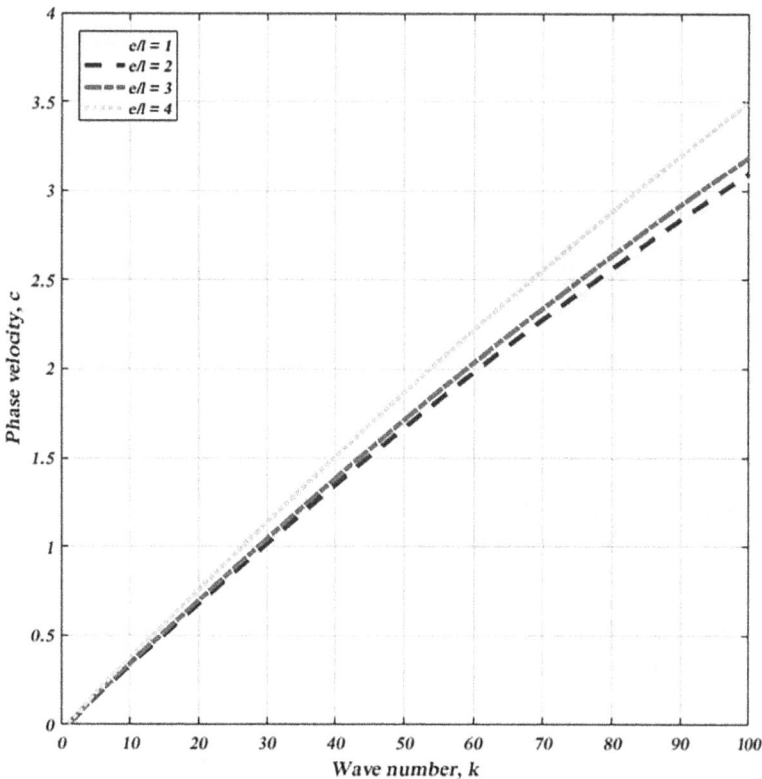

FIGURE 5.15 Variations of phase velocity of a smart sandwich plate versus wave number (k) for various geometrical parameters (e/l).

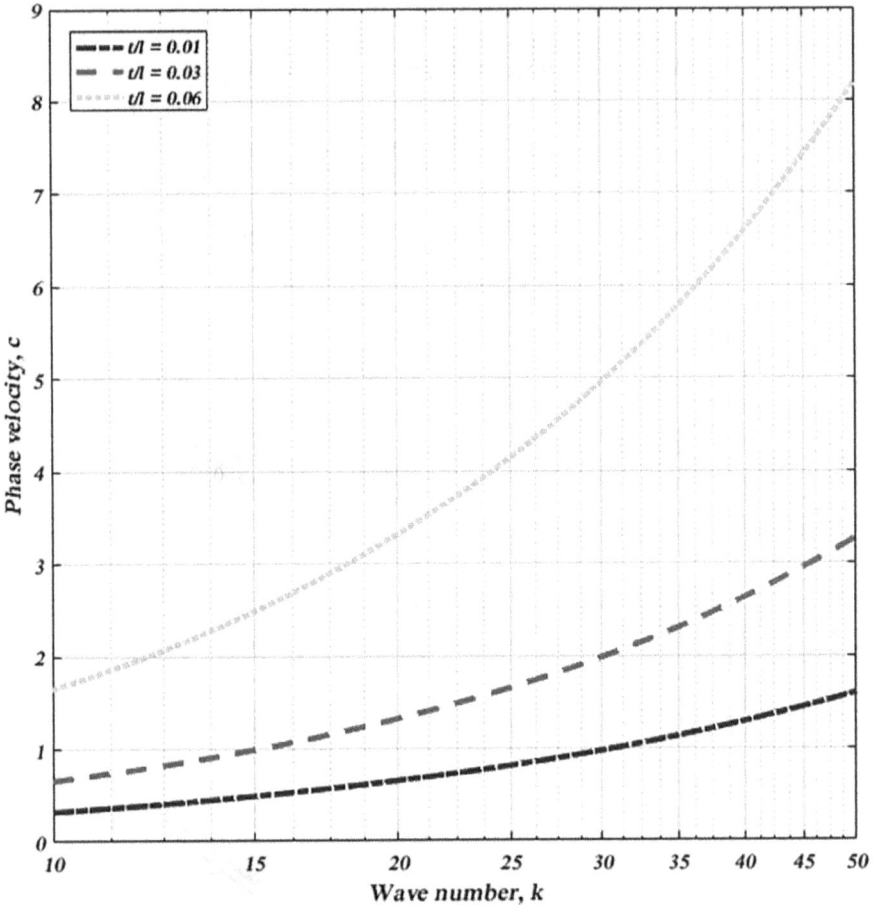

FIGURE 5.16 Variations of phase velocity of a smart sandwich plate versus wave number (k) for various geometrical parameters (t/l).

Figure 5.16 illustrates the fluctuations in phase velocity of a smart sandwich plate as a function of wave number (k), with respect to various geometric parameters (t/l). As seen in this figure, the phase velocity demonstrates an increase when the geometrical parameter (t/l) exceeds its previous range, namely from 0.01 to 0.06. The observation indicates that there was a noticeable rise in the phase velocity as the wave number climbed from 10 to 50.

Figure 5.17 illustrates the changes in phase velocity of a smart sandwich plate with respect to wave number (k) for various geometrical parameters (θ). As seen in this figure, the alteration of the geometrical parameter (θ) from −15 to −45 results in an increase in the phase velocity. It is evident that the phase velocity exhibits an increase when the wave number is incremented from 1 to 100.

Figure 5.18 presents the phase velocity fluctuations of a smart sandwich plate as a function of wave number (k) for various thicknesses of smart nanocomposite. As

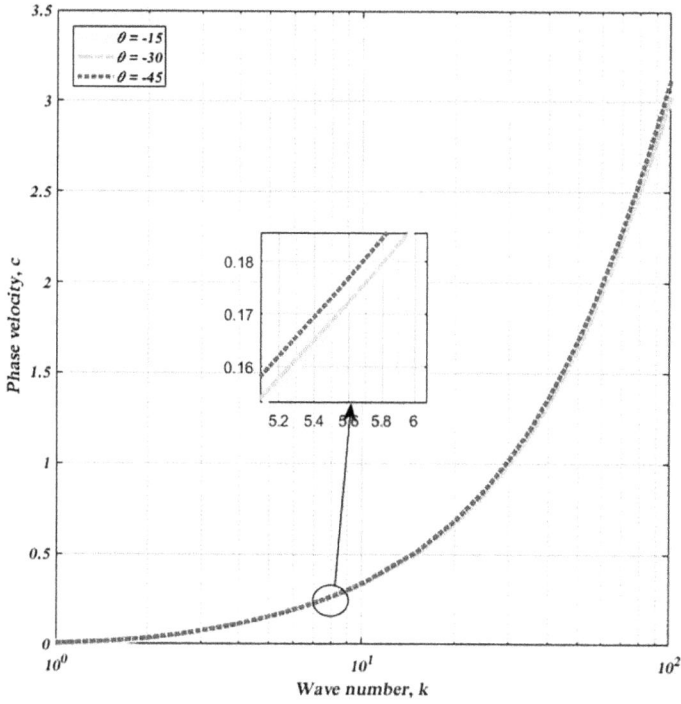

FIGURE 5.17 Variations of phase velocity of a smart sandwich plate versus wave number (k) for various geometrical parameters (θ).

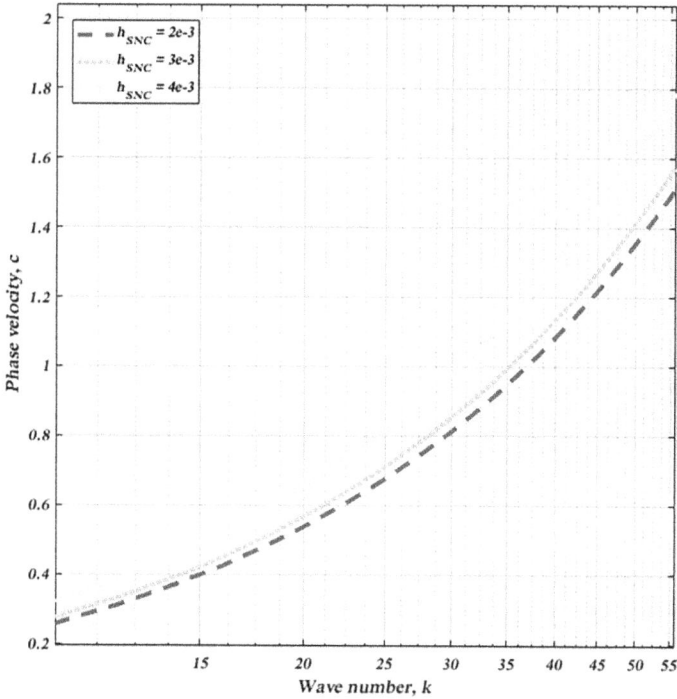

FIGURE 5.18 Variations of phase velocity of a smart sandwich plate versus wave number (k) for various smart nanocomposite thickness (h_{SNC}).

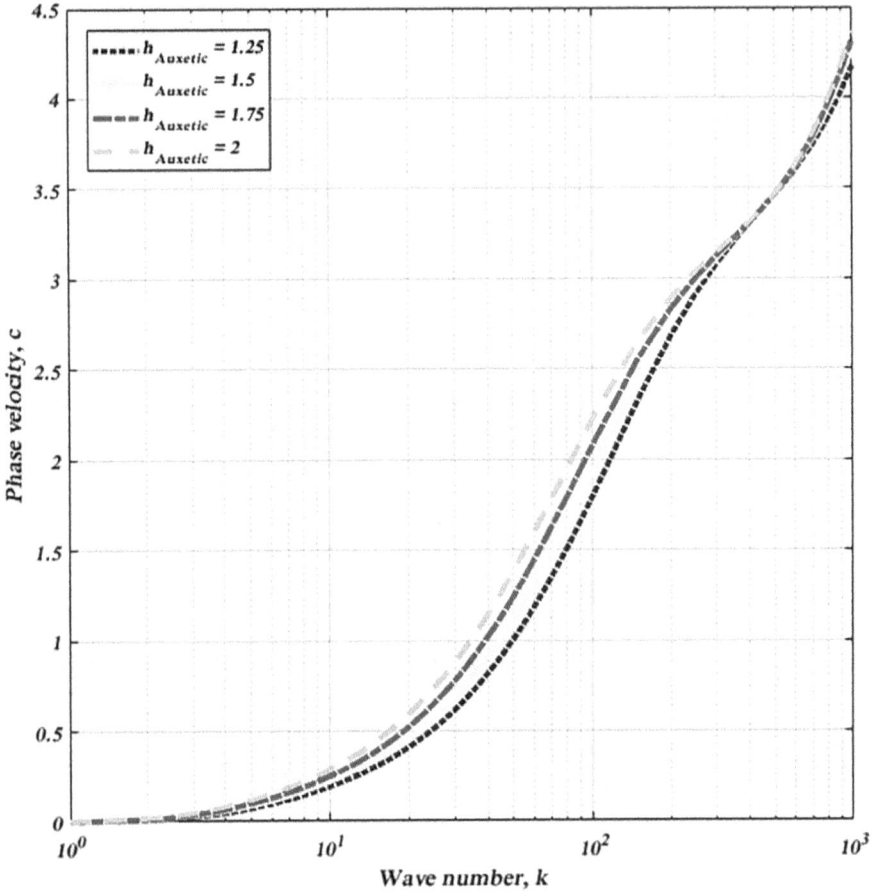

FIGURE 5.19 Variations of phase velocity of a smart sandwich plate versus wave number (k) for various auxetic core thicknesses ($h_{Auxetic}$).

seen in this figure, the phase velocity exhibits an increasing trend when the thickness of the smart nanocomposite is augmented from 2 to 4, while maintaining a constant value of k=10:55.

Figure 5.19 illustrates the fluctuations in phase velocity of a smart sandwich plate as a function of wave number (k), with consideration to varying auxetic core thickness. As seen in this figure, the phase velocity demonstrates an increase when the thickness of the auxetic core is augmented from 1.25 to 2. Additionally, it is noteworthy that the phase velocity exhibits an increasing trend when the wave number is incremented from 1 to 1000.

Figure 5.20 presents the observed phase velocity fluctuations of a smart sandwich plate in relation to the wave number (k) and the Winkler foundation coefficient (k_w). As seen in this figure, it is evident that the phase velocity exhibits an increasing trend

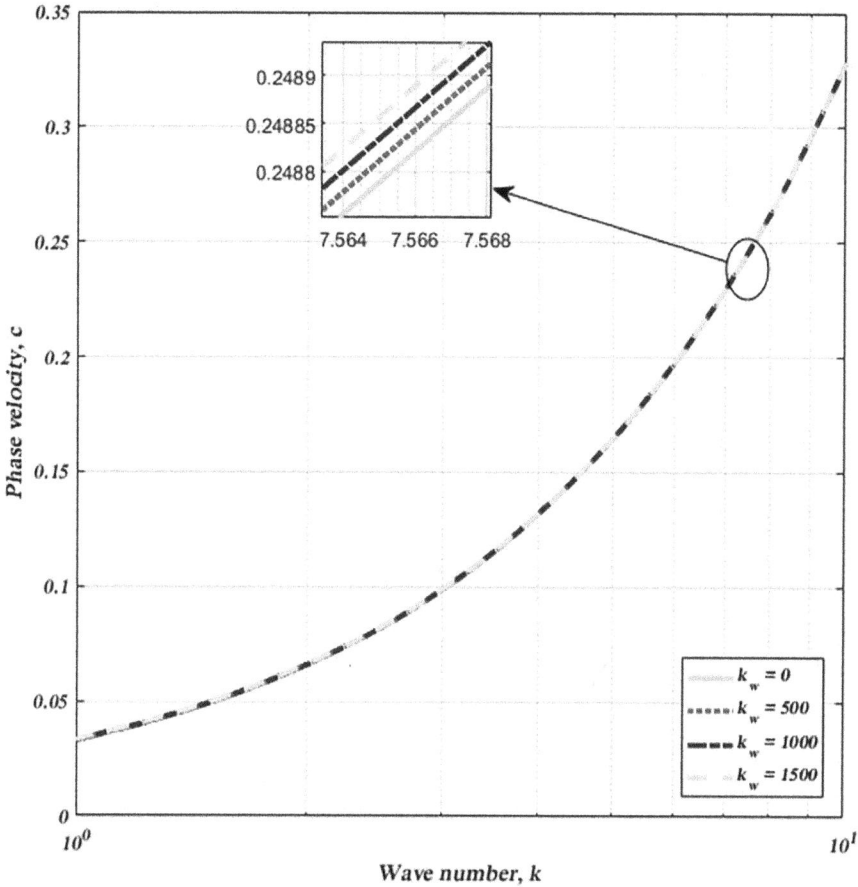

FIGURE 5.20 Variations of phase velocity of a smart sandwich plate versus wave number (k) for various Winkler foundation coefficient (k_w).

when the wave number (k) is incremented from 1 to 10. The relationship between the value of phase velocity and the Winkler foundation coefficient (k_w) is clearly shown by the observed increase in phase velocity when k_w is varied from 0 to 1500.

Figure 5.21 illustrates the changes in phase velocity of a smart sandwich plate with respect to wave number (k) for various values of the Pasternak foundation coefficient (k_P). As seen in this figure, an increase in the Pasternak foundation coefficient (k_P) from 100 to 500 results in a rise in the phase velocity. It is evident that the phase velocity exhibits an increase when the wave number transitions from 0 to 10.

Figure 5.22 illustrates the phase velocity fluctuations of a smart sandwich plate with respect to the wave number (k), considering various values of the viscoelastic damping foundation coefficient (c_d). As seen in this figure, the phase velocity experiences a decrease when the viscoelastic damping foundation coefficient is incremented from 20 to 30, 40, and 50 within the range of $k = 1:10$. It is evident that by increasing the wave number, the phase velocity exhibits an increase.

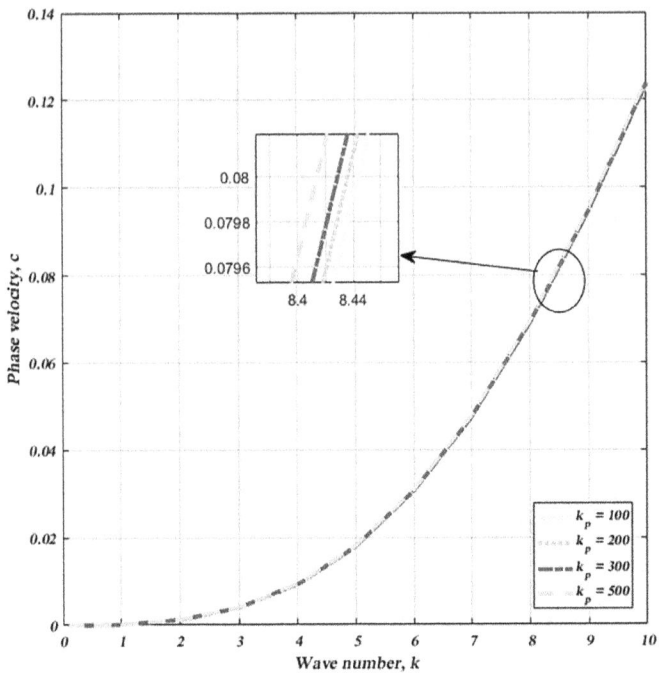

FIGURE 5.21 Variations of phase velocity of a smart sandwich plate versus wave number (k) for various Pasternak foundation coefficient (k_P).

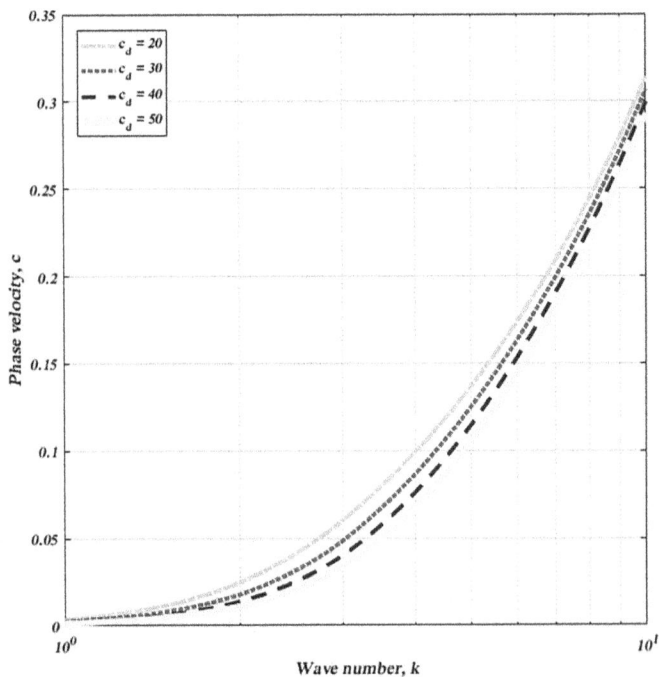

FIGURE 5.22 Variations of phase velocity of a smart sandwich plate versus wave number (k) for various viscoelastic damping foundation coefficient (c_d).

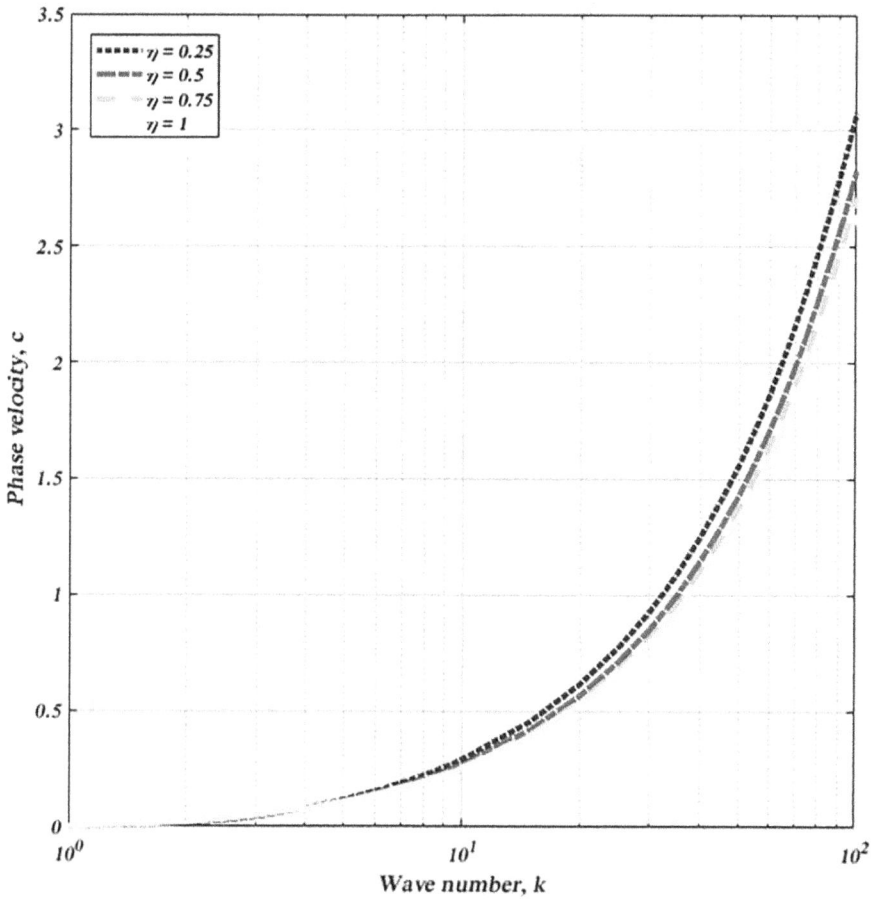

FIGURE 5.23 Variations of phase velocity of a smart sandwich plate versus wave number (k) for various volume percent of the smart nanofiller in the matrix (η).

Figure 5.23 presents the phase velocity changes of a sandwich plate with smart properties, specifically in relation to the wave number (k), across varying volume percentages of the smart nanofiller included inside the matrix (η). The figure illustrates a decrease in phase velocity as the volume percent of the smart nanofiller in the matrix (η) increases from 0.25 to 1, within the range of $k = 1:10$. However, it should be noted that for values of k greater than 1, the phase velocity experiences an increase as the wave number (k) is augmented.

Figure 5.24 displays the observed alterations in phase velocity of a smart sandwich plate as a function of wave number (k) for various external voltage (V_0) values. As seen in this figure, the phase velocity demonstrates an increase when the wave number (k) is incremented from 1 to 100. Moreover, it can be seen that

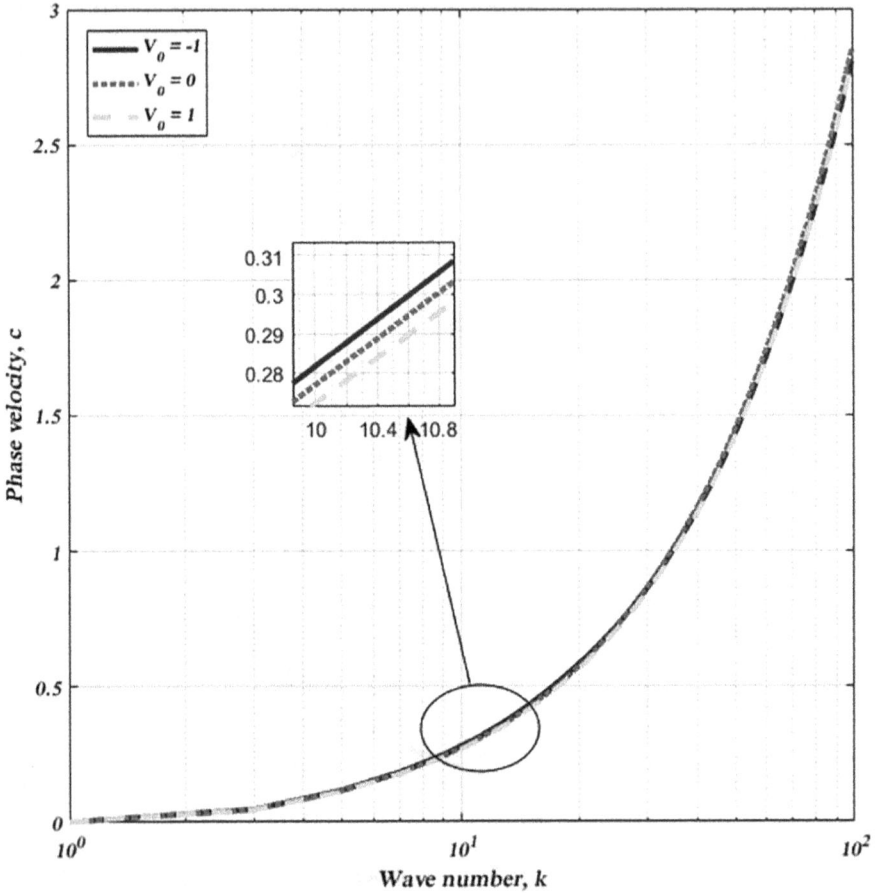

FIGURE 5.24 Variations of phase velocity of a smart sandwich plate versus wave number (k) for various external voltage (V_0).

the phase velocity decreases when the external voltage (V_0) is increased from $V_0 = -1$ to $V_0 = 1$.

5.4 AUXETIC SHELLS: WAVE PROPAGATION ANALYSIS

This section presents an investigation of the wave propagation of cylindrical sandwich shells composed of auxetic material. Figure 5.25 illustrates the geometric representation, form, and intricate details of the auxetic structure. The mechanical parameters of the auxetic material may be represented as follows at the first stage [38,39]:

$$\left|\begin{aligned}
E_1^{(c)} &= E^c \frac{\lambda_3^3(\lambda_1 - \sin\alpha)}{\cos\alpha[1 + (\tan^2\alpha + \lambda_1\sec^2\alpha)\lambda_3^2]} \\[2mm]
E_2^{(c)} &= E^c \frac{\lambda_3^3}{\cos\alpha(\tan^2\alpha + \lambda_3^2)(\lambda_1 - \sin\alpha)} \\[2mm]
G_{12}^{(C)} &= G^C \frac{\lambda_3^3}{\lambda_1(1 + 2\lambda_1)\cos\alpha} \\[2mm]
G_{23}^{(C)} &= G^C \frac{\lambda_3 \cos\alpha}{\lambda_1 - \sin\alpha} \\[2mm]
G_{13}^{(C)} &= G^C \frac{\lambda_3^3}{2\cos\alpha}[\frac{\lambda_1 - \sin\alpha}{1 + 2\lambda_1} + \frac{\lambda_1 + 2\sin^2\alpha}{2(\lambda_1 - \sin\alpha)}] \\[2mm]
v_{12}^{(c)} &= -\frac{\sin\alpha(1 - \lambda_3^2)(\lambda_1 - \sin\alpha)}{\cos^2\alpha[1 + (\tan^2 + \sec^2\alpha\lambda_1)\lambda_3^2}] \\[2mm]
v_{21}^{(c)} &= \frac{-\sin\alpha(1 - \lambda_3^2)}{(\tan^2\alpha + \lambda_3^2)(\lambda_1 - \sin\alpha)} \\[2mm]
\rho^c &= \rho \frac{\lambda_3(\lambda_1 + 2)}{2\cos\alpha \ (\lambda_1 - \sin\alpha)}
\end{aligned}\right. \qquad (5.51)$$

The symbols d, l, and t represent the horizontal rib length, vertical inclined rib length, and rib thickness of the auxetic cell, respectively. The symbols $E_1^{(c)}$, $E_2^{(c)}$, and $G_{12}^{(C)}$ $G_{23}^{(C)}$ signify the effective Young's modulus and effective shear modulus of the auxetic core, while $v_{12}^{(c)}$ $v_{21}^{(c)}$ represents the effective auxetic Poisson's ratio.

Furthermore, the terms Young's modulus, shear modulus, and Poisson's ratio refer to the mechanical properties of auxetic materials v^c, E^c, G^c. In the context of density, the variable ρ represents the density of the auxetic material, while ρ^c denotes the effective density of the auxetic core. The parameters λ_1, λ_3 correspond to the auxetic rib-length ratio, which may be determined using the following formula:

$$\lambda_1 = d/l \qquad \lambda_3 = t/l$$

This part pertains to the topic of wave propagation. In the context of wave propagation, a deformable shear solution will be used. By using the first-order shear deformable shell theory, one may establish the kinematic equations that describe the behavior of the shell.

$$\begin{aligned}
u_x(x,\varphi,z,t) &= u(x,\varphi,t) + z\phi_x(x,\varphi,t) \\
u_\phi(x,\varphi,z,t) &= v(x,\varphi,t) + z\phi_\varphi(x,\varphi,t) \\
u_z(x,\varphi,z,t) &= w(x,\varphi,t)
\end{aligned} \qquad (5.52)$$

The variables v, u, and w are used to denote circumferential, axial, and lateral displacements, correspondingly. Furthermore, the variables ϕ_x and ϕ_φ

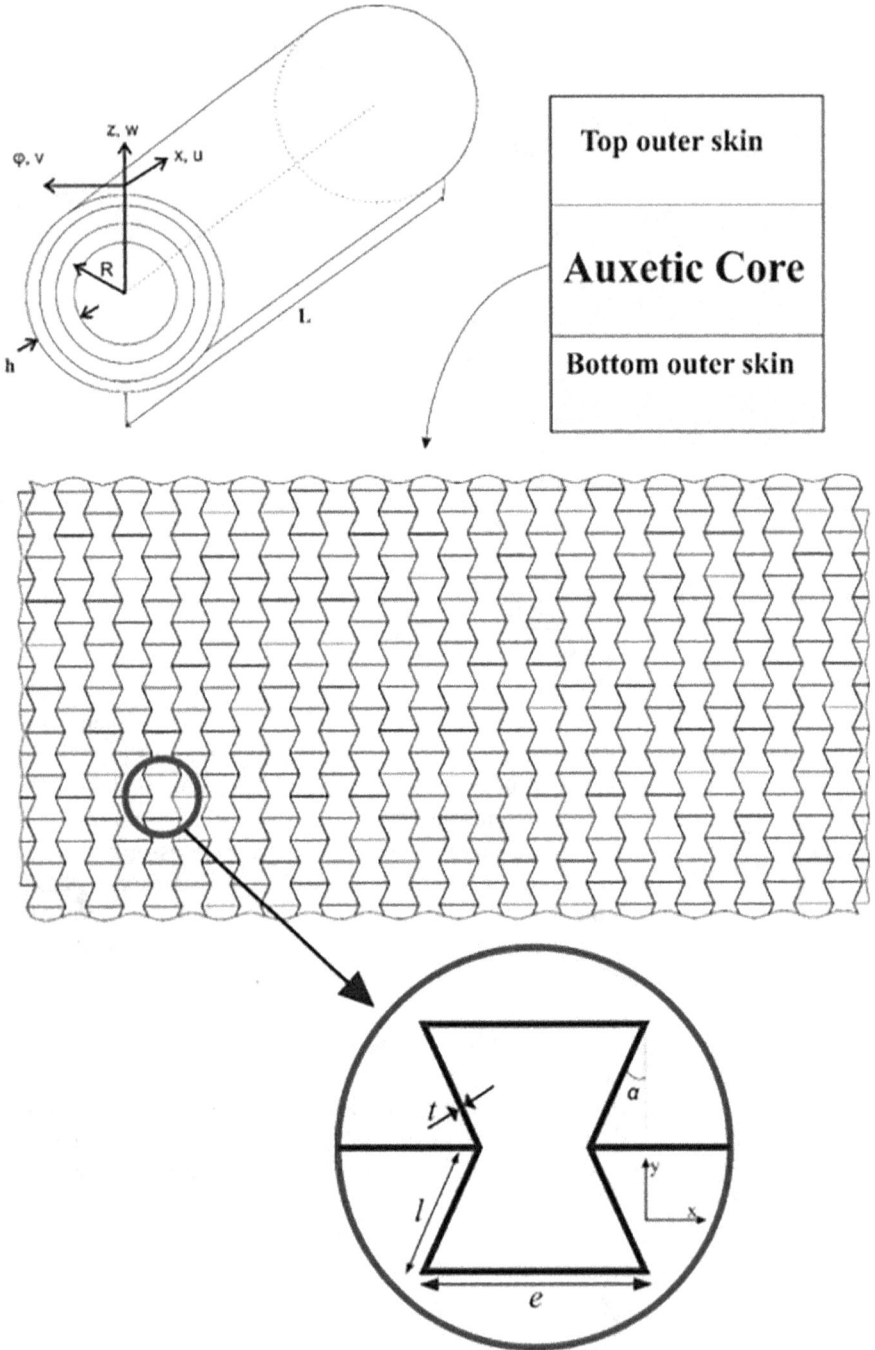

FIGURE 5.25   Cell geometry of the auxetic sandwich cylindrical shell.

represent the sequential rotation components along the axial and circumferential axes, respectively. The non-zero components of shell strain are expressed as follows:

$$\varepsilon_{xx} = \frac{\partial u}{\partial x} + z\frac{\partial \theta_x}{\partial x}$$ (5.53a)

$$\varepsilon_{\varphi\varphi} = \frac{1}{R}\left(\left(\frac{\partial V}{\partial \varphi} + z\frac{\partial \theta_\varphi}{\partial \varphi} + w\right)\right)$$ (5.53b)

$$\varepsilon_{x\varphi} = \frac{1}{R}\frac{\partial u}{\partial \varphi} + \frac{\partial v}{\partial x} + \frac{z}{R}\frac{\partial \theta_x}{\partial \varphi} + z\frac{\partial \theta_\varphi}{\partial x}$$ (5.53c)

$$\varepsilon_{xz} = \theta_x + \frac{\partial w}{\partial x}$$ (5.53d)

$$\varepsilon_{\varphi z} = \theta_\varphi + \frac{1}{R}\frac{\partial w}{\partial \varphi} - \frac{v}{R}$$ (5.53e)

The propagation of waves may be deduced by the use of the Hamilton principle. According to this theory, the total energy at any given instant must equal to zero. The Hamilton equation has the following form when using the Euler-Lagrange equations and utilizing the first-order shear deformable shell theory:

$$\int_0^t \delta(U - T - V)dt = 0$$ (5.54)

The terms U, T, V refer to the concepts of strain, kinetic energy, and work done by an external load, respectively. The expression for energy variation is derived based on the loadings of radial, axial, and circumferential nature N_r, N_x, N_φ:

$$\delta U = \int_{-\frac{h}{2}}^{\frac{h}{2}} \int_0^{2\pi} \int_0^L \sigma_{ij}\delta\varepsilon_{ij} Rd\forall$$ (5.55a)

$$\delta T = \int_{-\frac{h}{2}}^{\frac{h}{2}} \int_0^{2\pi} \int_0^L \rho(z)\left[\left(\frac{\partial \delta_x}{\partial t}\right)^2 + \left(\frac{\partial \delta u_\varphi}{\partial t}\right)^2 + \left(\frac{\partial \delta u_z}{\partial t}\right)^2\right]Rd\forall$$ (5.55b)

$$\delta V = \int_{-\frac{h}{2}}^{\frac{h}{2}} \int_0^{2\pi} \int_0^L \left(N_r + N_x\frac{\partial^2 w}{\partial x^2} + \frac{N_\varphi}{R^2}\frac{\partial^2 w}{\partial \varphi^2}\right)\delta w R\forall$$ (5.55c)

This chapter posits the absence of any external load acting on the cylindrical sandwich shell, hence resulting in zero energy generated by an external load.

The entire volume of the shell $d\forall$ is comparable to $dxd\varphi dz$. The equation of motion, which is represented by the location in equation (5.54), is reformulated according to reference [39].

$$\frac{\partial N_{xx}}{\partial x} + \frac{1}{R}\frac{\partial N_{x\varphi}}{\partial \varphi} = I_0\frac{\partial^2 u}{\partial t^2} + I_0\frac{\partial^2 \theta_x}{\partial t^2}$$

$$\frac{\partial N_x}{\partial x} + \frac{1}{R}\frac{\partial N_{\varphi\varphi}}{\partial \varphi} + \frac{Q_{z\varphi}}{R} = I_0\frac{\partial^2 v}{\partial t^2} + I_1\frac{\partial^2 \theta_\varphi}{\partial t^2}$$

$$\frac{\partial Q_{xz}}{\partial x} + \frac{1}{R}\frac{\partial Q_{z\varphi}}{\partial \varphi} - \frac{N_{\varphi\varphi}}{R} = I_0\frac{\partial^2 w}{\partial t^2} \qquad (5.56)$$

$$\frac{\partial M_{xx}}{\partial x} + \frac{1}{R}\frac{\partial M_{x\varphi}}{\partial \varphi} - Q_{xz} = I_1\frac{\partial^2 u}{\partial t^2} + I_2\frac{\partial^2 \theta_x}{\partial t^2}$$

$$\frac{\partial M_{x\varphi}}{\partial x} + \frac{1}{R}\frac{\partial M_{\varphi\varphi}}{\partial \varphi} - Q_{\varphi z} = I_1\frac{\partial^2 v}{\partial t^2} + I_2\frac{\partial^2 \theta_\varphi}{\partial t^2}$$

It is important to acknowledge that the link between stress and strain may be represented as follows:

$$\sigma_{ij} = C_{ijkl}\varepsilon_{kl} \qquad (5.57)$$

The characteristics of Cauchy stress, strain, and elasticity tensors are as σ_{ij}, ε_{kl}, C_{ijkl}, in sequential order. The relationships may be increased in the following manner:

$$\begin{Bmatrix} N_{xx}N_{\varphi\varphi}N_{x\varphi} \\ M_{xx}M_{\varphi\varphi}M_{x\varphi} \end{Bmatrix} = \int_{-\frac{h}{2}}^{\frac{h}{2}} (\sigma_{xx},\sigma_{\varphi\varphi},\sigma_{x\varphi})\begin{Bmatrix} 1 \\ z \end{Bmatrix}dz \qquad (5.58)$$

The determination of the elasticity tensor components may be achieved using the following method:

$$[Q_{xz},Q_{z\varphi}] = k_s\int_{-\frac{h}{2}}^{\frac{h}{2}} [\sigma_{xz},\sigma_{zx}]dz \qquad (5.59)$$

where k_s is the shear correction factor in equation (5.59). Moreover, for the mass moments of inertia, the mathematical relationship becomes:

$$[I_0,I_1,I_2] = \int_{-\frac{h}{2}}^{\frac{h}{2}} [1,z,z^2]\rho(z)dz \qquad (5.60)$$

By incorporating the aforementioned equations into the governing equations of the shell, one may derive the following equations:

$$N_{xx} = A_{11}\frac{\partial u}{\partial x} + B_{11}\frac{\partial \theta_x}{\partial x} + \frac{A_{12}}{R}(\frac{\partial v}{\partial \varphi} + w) + \frac{B_{12}}{R}\frac{\partial \theta_\varphi}{\partial \varphi} \tag{5.61a}$$

$$N_{x\varphi} = \left(A_{66}(\frac{1}{R}\frac{\partial u}{\partial \varphi} + \frac{\partial v}{\partial x})\right) + \left(B_{66}(\frac{1}{R}\frac{\partial \theta_x}{\partial \varphi} + \frac{\partial \theta_\varphi}{\partial x})\right) \tag{5.61b}$$

$$M_{x\varphi} = \left(B_{66}(\frac{1}{R}\frac{\partial u}{\partial \varphi} + \frac{\partial v}{\partial x})\right) + \left(D_{66}(\frac{1}{R}\frac{\partial \theta_x}{\partial \varphi} + \frac{\partial \theta_\varphi}{\partial x})\right) \tag{5.61c}$$

$$M_{x\varphi} = \left(B_{66}(\frac{1}{R}\frac{\partial u}{\partial \varphi} + \frac{\partial v}{\partial x})\right) + \left(D_{66}(\frac{1}{R}\frac{\partial \theta_x}{\partial \varphi} + \frac{\partial \theta_\varphi}{\partial x})\right) \tag{5.61d}$$

$$M_{xx} = B_{11}\frac{\partial u}{\partial x} + D_{11}\frac{\partial \theta_x}{\partial x} + \frac{B_{12}}{R}(\frac{\partial v}{\partial \varphi} + w) + \frac{D_{12}}{R}\frac{\partial \theta_\varphi}{\partial \varphi} \tag{5.61e}$$

$$N_{\varphi\varphi} = A_{12}\frac{\partial u}{\partial x} + B_{12}\frac{\partial \theta_x}{\partial x} + \frac{A_{11}}{R}(\frac{\partial v}{\partial \varphi} + w) + \frac{B_{11}}{R}\frac{\partial \theta_\varphi}{\partial \varphi} \tag{5.61f}$$

$$M_{\varphi\varphi} = B_{12}\frac{\partial u}{\partial x} + D_{12}\frac{\partial \theta_x}{\partial x} + \frac{B_{11}}{R}(\frac{\partial v}{\partial \varphi} + w) + \frac{D_{11}}{R}\frac{\partial \theta_\varphi}{\partial \varphi} \tag{5.61g}$$

$$Q_{xz} = A^s_{55}\left(\theta_x + \frac{\partial w}{\partial x}\right) \tag{5.61h}$$

$$Q_{\varphi z} = A^s_{55}\left(\theta_\varphi + \frac{1}{R}\frac{\partial w}{\partial x} - \frac{v}{R}\right) \tag{5.61i}$$

The following are items used in equations (5.61a–i) [39]:

$$[A_{11}, A_{12}, A_{66}] = \int_{-\frac{h}{2}}^{\frac{h}{2}} Q_{11}.dz$$

$$[B_{11}, B_{12}, B_{66}] = \int_{-\frac{h}{2}}^{\frac{h}{2}} Q_{12}.z.dz$$

$$\tag{5.62}$$

$$[D_{11}, D_{12}, D_{66}] = \int_{-\frac{h}{2}}^{\frac{h}{2}} Q_{66}.z^2.dz$$

$$A^s_{55} = k_s \int_{-\frac{h}{2}}^{\frac{h}{2}} \frac{E_{11}}{2(1+v_{12})}dz$$

The expression of mass inertia to get mass matrix can be rewritten as:

$$
\begin{cases}
I_0 = \displaystyle\int_{-\frac{h}{2}}^{-\frac{h}{2}+h_b} \rho_a(z)dz + \int_{-\frac{h}{2}+h_b}^{-\frac{h}{2}+h_b+h_a} \rho_c(z)dz + \int_{-\frac{h}{2}+h_b+h_a}^{-\frac{h}{2}+h_b+h_a+h_t} \rho_a(z)dz \\[4mm]
I_1 = \displaystyle\int_{-\frac{h}{2}}^{-\frac{h}{2}+h_b} z\rho_a(z)dz + \int_{-\frac{h}{2}+h_b}^{-\frac{h}{2}+h_b+h_a} z\rho_c(z)dz + \int_{-\frac{h}{2}+h_b+h_a}^{-\frac{h}{2}+h_b+h_a+h_t} z\rho_a(z)dz \\[4mm]
I_2 = \displaystyle\int_{-\frac{h}{2}}^{-\frac{h}{2}+h_b} z^2\rho_a(z)dz + \int_{-\frac{h}{2}+h_b}^{-\frac{h}{2}+h_b+h_a} z^2\rho_c(z)dz + \int_{-\frac{h}{2}+h_b+h_a}^{-\frac{h}{2}+h_b+h_a+h_t} z^2\rho_a(z)dz
\end{cases}
\tag{5.63}
$$

To determine the governing equation of motion, it is necessary to express the terms indicated above in the prescribed formulae in the following manner:

$$
\begin{Bmatrix} A_{11} \\ A_{12} \\ A_{66} \end{Bmatrix} = \int_{-\frac{h}{2}}^{-\frac{h}{2}+h_b} \begin{Bmatrix} Q^a_{11} \\ Q^a_{12} \\ Q^a_{66} \end{Bmatrix} dz + \int_{-\frac{h}{2}+h_b}^{-\frac{h}{2}+h_b+h_a} \begin{Bmatrix} Q^c_{11} \\ Q^c_{12} \\ Q^c_{66} \end{Bmatrix} dz + \int_{-\frac{h}{2}+h_b+h_a}^{-\frac{h}{2}+h_b+h_a+h_t} \begin{Bmatrix} Q^a_{11} \\ Q^a_{12} \\ Q^a_{66} \end{Bmatrix} dz
$$

$$
\begin{Bmatrix} B_{11} \\ B_{12} \\ B_{66} \end{Bmatrix} = \int_{-\frac{h}{2}}^{-\frac{h}{2}+h_b} \begin{Bmatrix} Q^a_{11} \\ Q^a_{12} \\ Q^a_{66} \end{Bmatrix} z\,dz + \int_{-\frac{h}{2}+h_b}^{-\frac{h}{2}+h_b+h_a} \begin{Bmatrix} Q^c_{11} \\ Q^c_{12} \\ Q^c_{66} \end{Bmatrix} z\,dz + \int_{-\frac{h}{2}+h_b+h_a}^{-\frac{h}{2}+h_b+h_a+h_t} \begin{Bmatrix} Q^a_{11} \\ Q^a_{12} \\ Q^a_{66} \end{Bmatrix} z\,dz
\tag{5.64}
$$

$$
\begin{Bmatrix} D_{11} \\ D_{12} \\ D_{66} \end{Bmatrix} = \int_{-\frac{h}{2}}^{-\frac{h}{2}+h_b} \begin{Bmatrix} Q^a_{11} \\ Q^a_{12} \\ Q^a_{66} \end{Bmatrix} z^2\,dz + \int_{-\frac{h}{2}+h_b}^{-\frac{h}{2}+h_b+h_a} \begin{Bmatrix} Q^c_{11} \\ Q^c_{12} \\ Q^c_{66} \end{Bmatrix} z^2\,dz + \int_{-\frac{h}{2}+h_b+h_a}^{-\frac{h}{2}+h_b+h_a+h_t} \begin{Bmatrix} Q^a_{11} \\ Q^a_{12} \\ Q^a_{66} \end{Bmatrix} z^2\,dz
$$

In the aforementioned equations, the variable represents auxetic materials, whereas the variable a represents aluminum materials. Additionally, the aforementioned term has the suffix t to indicate the uppermost layer, b to denote the lowermost layer, and a to signify the intermediate layer.

In order to analytically solve the governing equations of a cylindrical sandwich shell, it is postulated that an exponential solution function may be used, which is expressed as follows:

$$
\begin{aligned}
u &= U.e^{i(\beta x + n\varphi - \omega x)} \\
v &= V.e^{i(\beta x + n\varphi - \omega x)} \\
w &= W.e^{i(\beta x + n\varphi - \omega x)} \\
\theta_x &= \Theta_x.e^{i(\beta x + n\varphi - \omega x)} \\
\theta_\varphi &= \Theta_\varphi.e^{i(\beta x + n\varphi - \omega x)}
\end{aligned}
\tag{5.65}
$$

where U, V, W are the displacement amplitudes refer to the magnitudes of displacement, whereas the rotation amplitudes Θ_x, Θ_φ pertain to the magnitudes of rotation. Similarly, the term ω denotes the frequency, whereas the terms β, n refer to certain wave characteristics. The mathematical motion equation for a cylindrical sandwich shell may be derived by using eigenvalue analysis and including the elements acquired from the aforementioned formulae $(u, v, w, \theta_x, \theta_\varphi)$.

$$\left([K]_{5\times5} - \omega^2 [M]_{5\times5}\right)([\Delta]) = 0 \tag{5.66}$$

in which

$$[\Delta] = \begin{bmatrix} U \\ V \\ W \\ \Theta_x \\ \Theta_\varphi \end{bmatrix}$$

The stiffness and mass matrices are defined as matrices denoted by K and M, respectively. The specific values of these matrices may be found in Appendix 5.2.

In order to get a solution for equation (5.66), it is necessary for the determinants of the matrices on the left side of the aforementioned eigenvalue equation to be equal to zero.

$$\left|\left([K]_{5\times5} - \omega^2 [M]_{5\times5}\right)\right| = 0 \tag{5.67}$$

By substituting the values K, M into the aforementioned equation, one may determine the frequency and explore the wave propagation in cylindrical sandwich shells composed of auxetic materials. Furthermore, assuming the values of β, n, k are equivalent, the calculation of phase velocity may be performed as follows:

$$c_p = \frac{\omega}{k} \tag{5.68}$$

This section examines the investigation of wave frequency and phase velocity in cylindrical sandwich auxetic shells, as well as the impact of auxetic cells on these properties [39]. Within the given set of calculations, it is postulated that the variance in auxetic inclination angle ranges from −75 to 75 degrees. It is important to acknowledge that variations in rib-length ratios may have an influence on the propagation of waves, a phenomenon that has been thoroughly examined in the diagrams. The range in auxetic rib-length ratios λ_1 is assumed to be 1, 2.5, 3.5, and 5.5. Similarly, the assumed value of λ_3 is 0.013857. The variation of the radius-to-thickness ratio ranges from 20 to 100. The influence of the radius-to-thickness ratio on the required parameters is explored and shown in graph form in the subsequent paragraph. The characteristics of aluminum, which serve as the substructure of the sandwich shell structure used in the numerical case studies, and the properties of the auxetic core are $E^c = 69$ GPa and $G^c = 26$ GPa. The value of Poisson's ratio is 0.3.

In addition, it should be noted that the density of auxetic material fabricated from aluminum is 2700 (kg / m^3). The assumed values for the shear correction factor and height of the shell layers are 1 and 3mm, respectively. This study aims to address the governing equation for a sandwich shell, which has three distinct layers. Specifically, the top layer is composed of aluminum, the inner sandwich layer consists of auxetic material, and the bottom layer is also made of aluminum. To do this, the Euler-Lagrange equation is used as a means of solution.

The implication of the structural arrangement of the studied structures is also evident in this essay. As an illustration, consider a structure with a 1–1–1 configuration, denoting three layers of equal length. Conversely, a structure denoted as 2–1–2 would imply a configuration with two layers of equal length sandwiching a central layer. The structural arrangement consists of five components, with two components in the top layer and two components in the lower layer being constructed from aluminum. Additionally, a single component inside the middle layer is composed of auxetic material. The aforementioned structural arrangement is subject to the same regulation.

In this section, the governing equation was successfully solved, taking into account all the relevant points stated in earlier sections. The findings are shown in a series of graphs in order to analyze the impact of various factors on the propagation of waves [39]. The graphical representation in Figure 5.26 illustrates the relationship

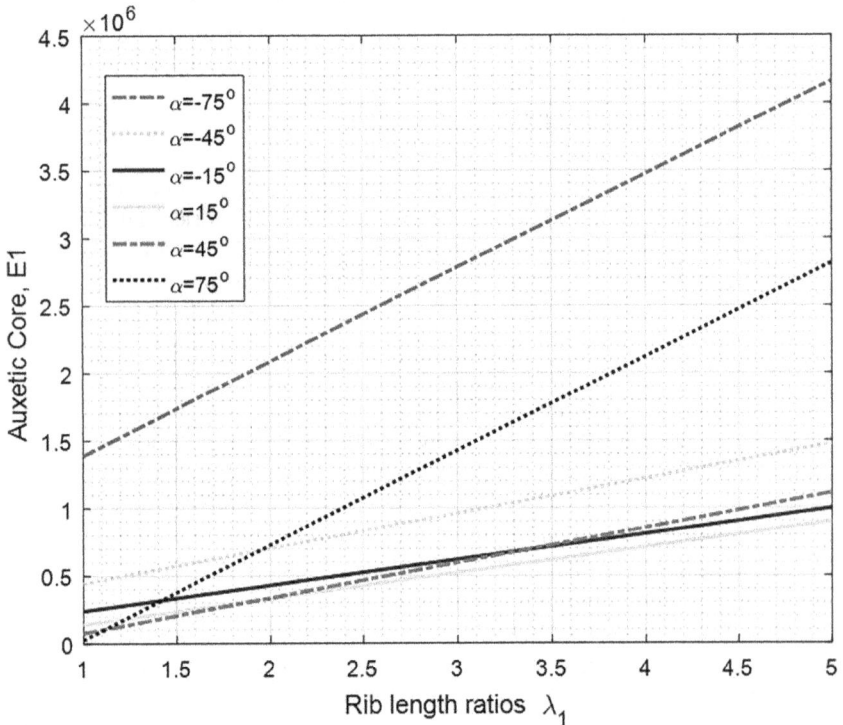

FIGURE 5.26 Auxetic core elastic modulus (E_1) vs. rib-length ratios with different auxetic inclination angle.

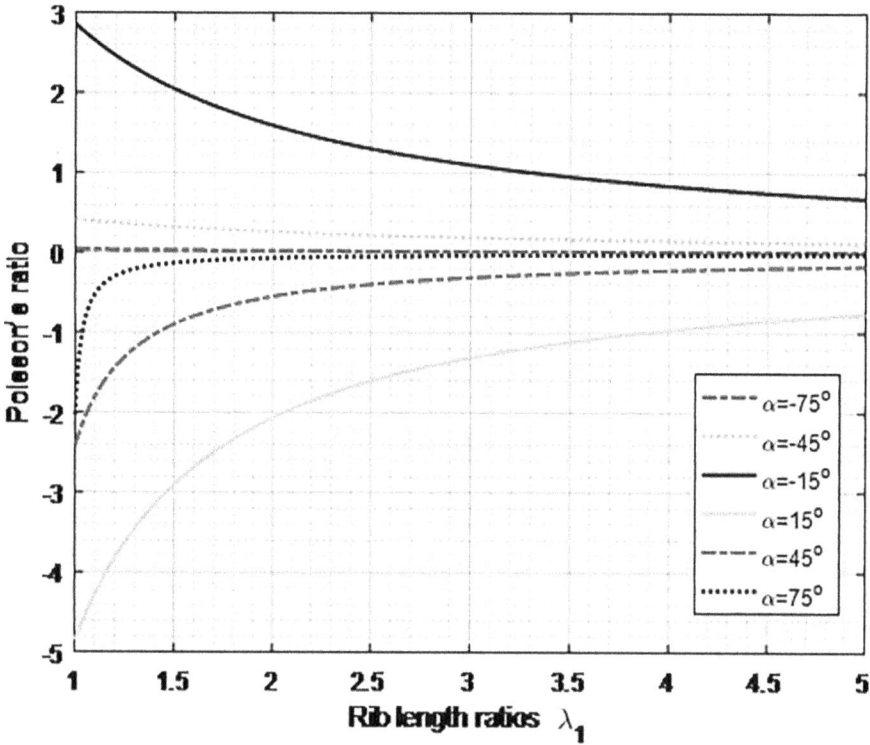

FIGURE 5.27   Auxetic inclination angle variation and rib-length ratios effect on the Poisson's ratio.

between the Young's modulus of cylindrical shells with an auxetic core and the ratio of rib length. The graph illustrates the relationship between the ratio of elastic modulus to unit rib length and the auxetic inclination angles. In a general sense, it can be seen that there is a positive correlation between the rise in λ_1 and the corresponding increase in the value of the elastic module for all auxetic inclination angles [39].

Figure 5.27 demonstrates the impact of varying auxetic inclination angles on the relationship between Poisson's ratio and rib-length ratio. As shown, it is seen that the Poisson's ratio for negative inclination angles is positive. Furthermore, it is noted that for greater absolute values of negative angles, the Poisson's ratio exhibits a more significant magnitude. Furthermore, it has been observed that in cases where the inclination angles are positive, Poisson's ratio exhibits a negative value. Additionally, it has been noted that as the inclination angle grows, the magnitude of the negative value of Poisson's ratio also increases [39].

The distribution of the elastic modulus (E1) of auxetic core materials in relation to the cell inclination angle is shown in Figure 5.28. Based on the graphical representation, it can be shown that a rise in inclination angle leads to a subsequent decline in the elastic modulus, followed by a gradual recovery. As obvious, the pace of auxetic

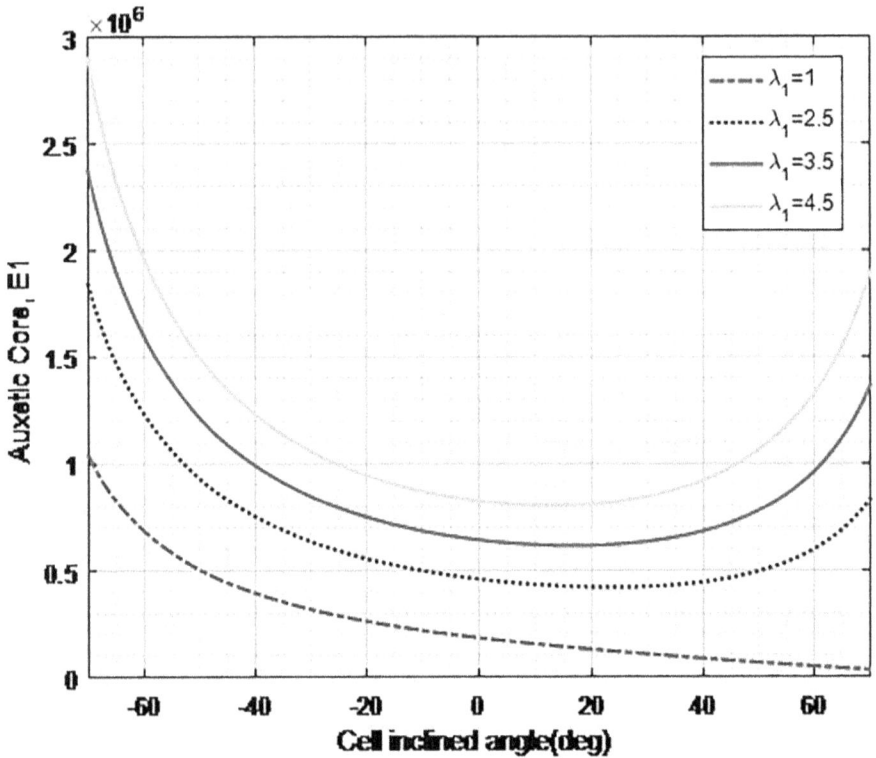

FIGURE 5.28 Auxetic core elastic modulus (*EI*) vs. cell angle (*deg*) with different rib-length ratios.

core elastic modulus changes rely on rib-length ratios. In other words, an increase in the quantity of λ_3 resulted in a corresponding increase in the value of the elastic modulus [39].

Based on the visual representation provided in Figure 5.29, it can be observed that the Poisson's ratio exhibits a negative value for positive cell inclined angles, while conversely, a positive value is observed for negative cell inclined angles. It is evident that the parabolic trajectory of each entity signifies a decrease in Poisson's ratio as the quantity of [39] λ_3 increases.

Figure 5.30 illustrates the variations in density with respect to the inclination angle of the cell, considering different ratios of rib length. As clarified by the diagram, an increase in cell inclined angles caused the amount of density to rise. The presented figure illustrates that the range of density variation is dependent on the quantity of λ_1 [39].

Figure 5.31 presents the variation of phase velocity with respect to circumferential wave number for different ratios of core-to-face sheet thickness. The diagram illustrates a notable trend wherein the phase velocity of a shell containing an auxetic core experiences a significant decrease as the wave number increases. It is important

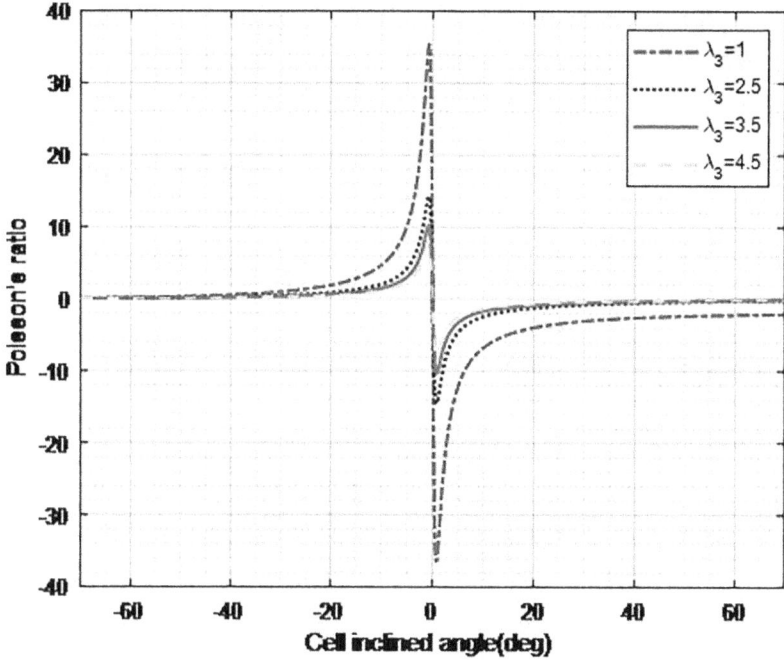

FIGURE 5.29 Poisson's ratio vs. cell inclined angle with different rib-length ratios.

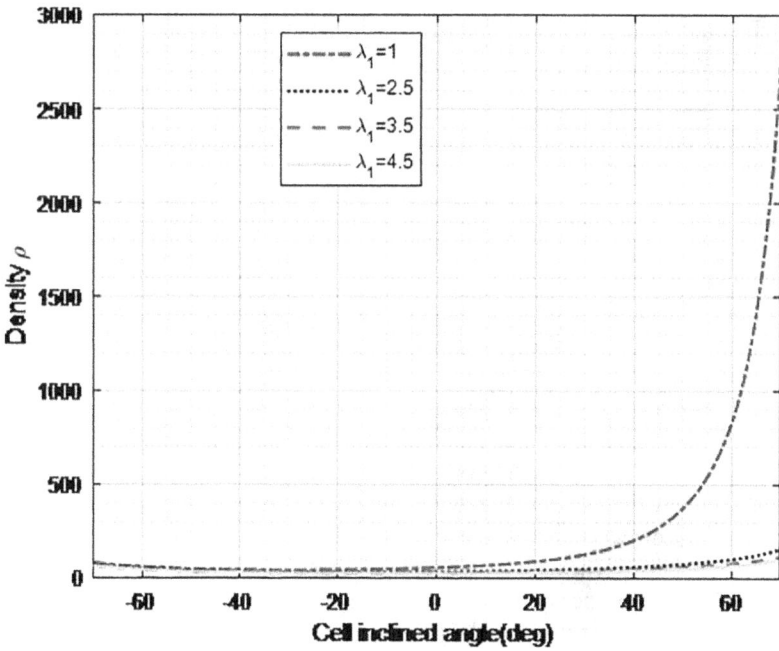

FIGURE 5.30 Density vs. cell inclined (*deg*) with different rib-length ratios.

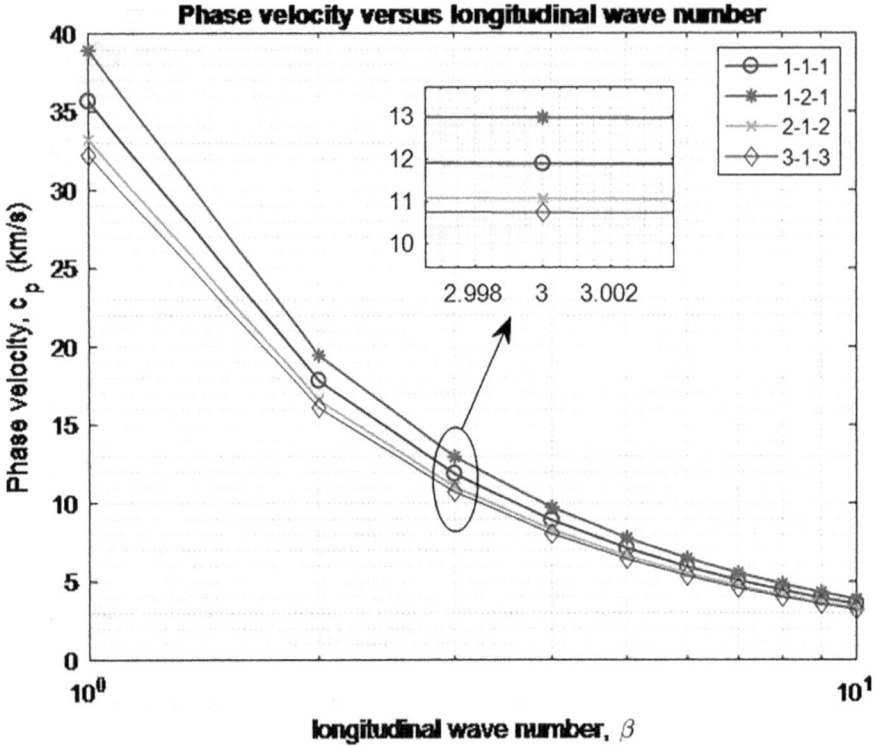

FIGURE 5.31 Variation of phase velocity versus circumferential wave number for various core-to-face sheet thickness ratios $(\frac{R}{h} = 20, n = 1, \beta = 10, h = 3e - 3)$.

to observe that higher phase velocity is associated with greater core-to-face sheet thickness ratios, wherein the top and bottom layers contain a higher proportion of aluminum. Table 5.4 presents a comprehensive depiction of graph results with greater elaboration [39].

Figure 5.32 presents the relationship between wave frequency and circumferential wave number for various cell angles. Based on the depicted trend in the diagram, there is a positive correlation between wave frequency and wave number. According to the given presentation, it can be observed that there is a consistent rise in wave frequency across all angles. Additionally, it is noted that the graphs corresponding to smaller degrees exhibit a more pronounced incline [39].

Figure 5.33 depicts the relationship between phase velocity and circumferential wave number for each cell angle. According to the diagrams, during the initial phase, velocity experiences a rapid decrease, while in subsequent phases, it gradually diminishes. It is evident that there exist varying phase velocities for each cell angle. Specifically, as the wave frequency increases within a given cell angle, the corresponding phase velocity decreases [39].

TABLE 5.4
Various Core-to-Face Sheet Thickness Ratios Effect on Phase Velocity Versus Circumferential Wave Number

n	1.1.1	1.2.1	2.1.2	3.1.3
1	35.66944256	38.92316441	33.20171336	32.20898978
2	17.83162774	19.46158221	16.60074799	16.08780546
3	12.9743911	12.9743911	11.06697224	10.73633003
4	8.981864525	10.73633003	8.300428674	8.052247364
5	7.133902096	7.784632876	6.640449612	6.44180321
6	5.945527895	6.486681855	5.533754498	5.368166491
7	5.096610833	5.565645217	4.739887383	4.601247772
8	4.468716711	4.86513281	4.150184441	4.026168907
9	3.963269715	4.324591873	3.689078986	3.578776686
10	3.566946655	3.892316389	3.320152163	3.220899017

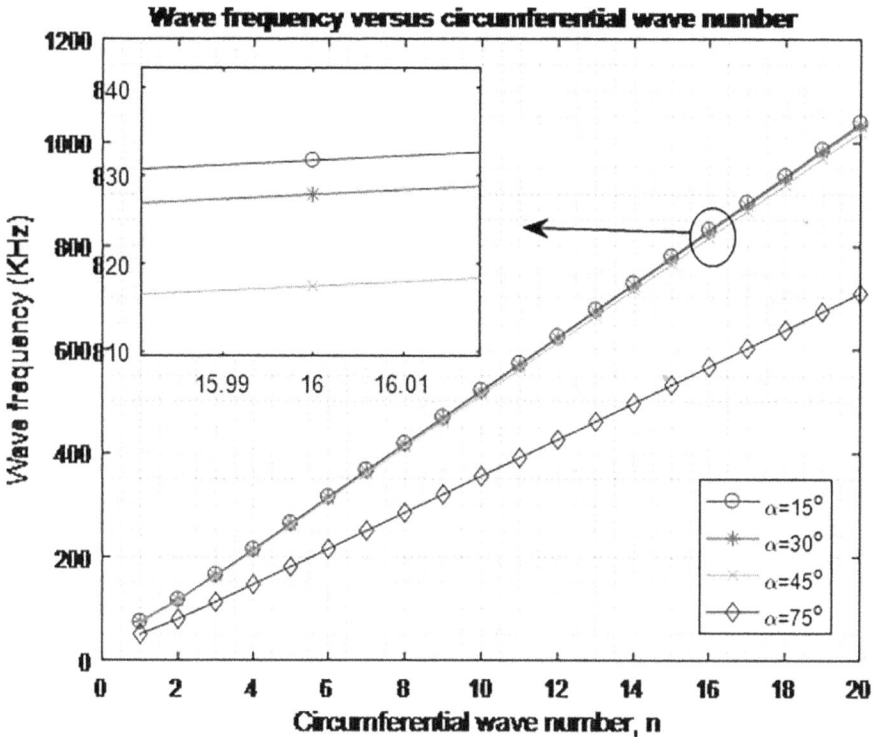

FIGURE 5.32 Wave frequency variation versus circumferential wave number for different cell angle ($\frac{R}{h} = 20$, $\beta = 10$, $h = 3e - 3$).

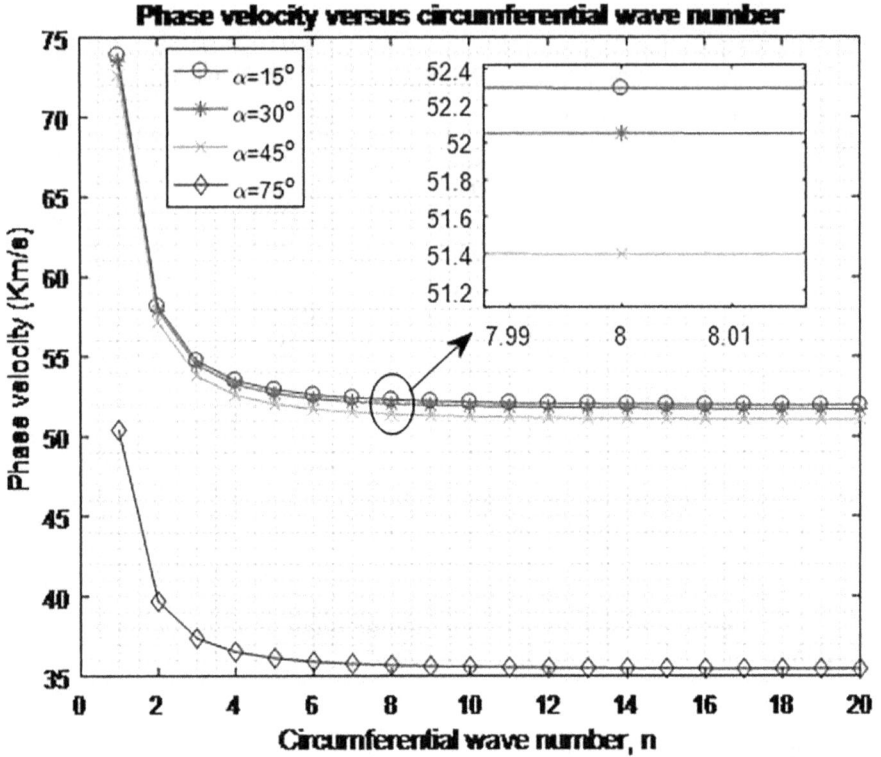

FIGURE 5.33 Phase velocity variation versus circumferential wave number for different cell angle $(\dfrac{R}{h} = 20, \beta = 10, h = 3e - 3)$.

The wave frequency variation per unit of circumferential wave number for various core-to-face sheet thickness ratios is described in Figure 5.34. According to the outline, increasing in wave number caused an increase in wave frequency. It is evident that the greater thickness of the isotropic aluminum layers promotes a higher wave frequency and has a more significant effect on wave frequency.

In Figure 5.35, different ratios of auxetic rib lengths are shown as effects on phase velocity variation versus circumferential wave number. The graph explains that phase velocity decreases significantly with increasing circumferential wave number; also, this indicates that the amount of λ_1 can reduce the phase velocity sufficiently. On the other hand, the smaller amount of λ_1 has smaller values of phase velocity [39].

Wave frequency changes versus alterations in circumferential wave number, as evident by Figure 5.36, for different auxetic rib-length ratios. Concurring to the graph, with increasing wave number, wave frequency also increases. Clearly, the greater value of λ_1 caused more wave frequency promotion.

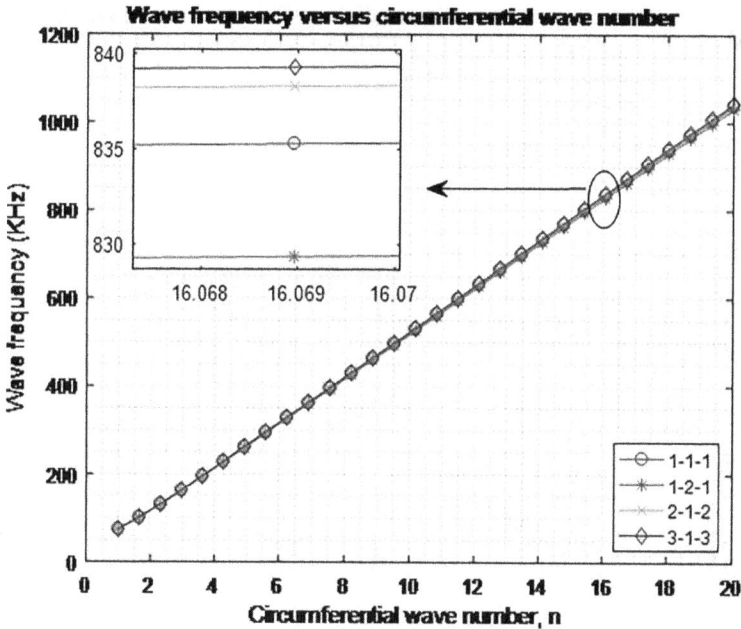

FIGURE 5.34 Variation of wave frequency versus circumferential wave number for various core-to-face sheet thickness ratios. ($\frac{R}{h} = 20$, $n = 1$, $\beta = 10$, $h = 3e-3$).

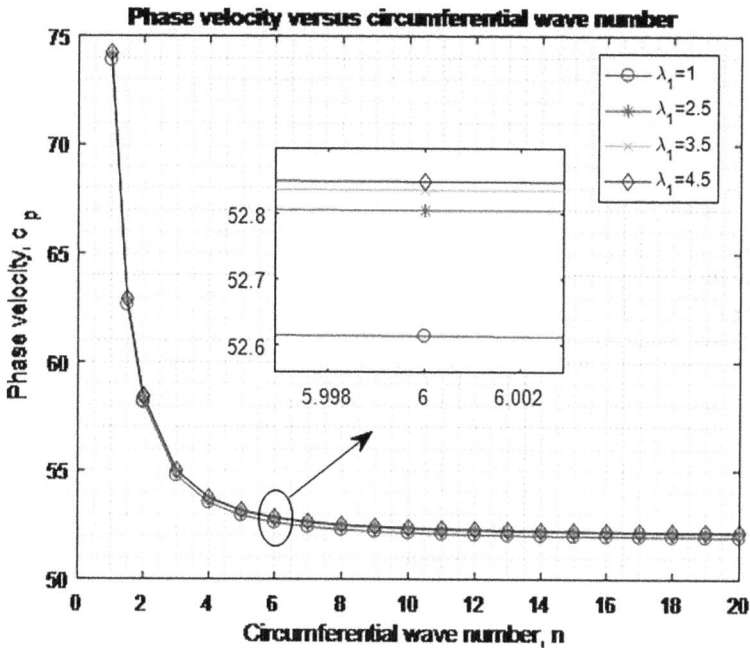

FIGURE 5.35 Phase velocity variation versus circumferential wave number for different auxetic rib-length ratios ($\frac{R}{h} = 20$, $\beta = 10$, $h = 3e-3$).

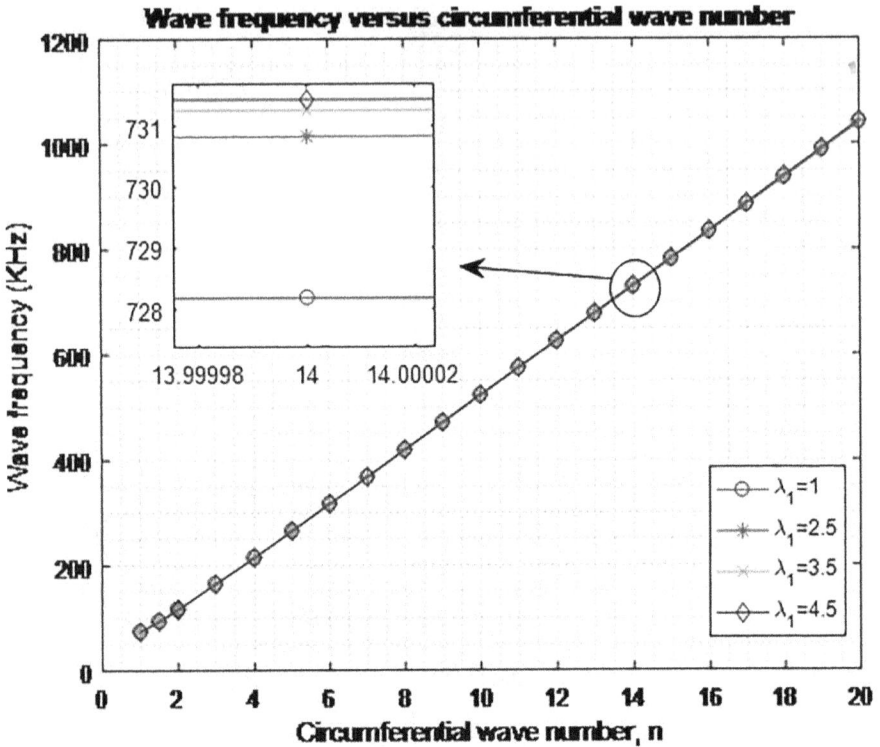

FIGURE 5.36 Wave frequency variation versus circumferential wave number for different auxetic rib-length ratios ($\frac{R}{h} = 20$, $\beta = 10$, $h = 3e - 3$).

In Figure 5.37, it is observed how radius-to-thickness ratio influences phase velocity variation. The phase velocity decreases continuously per unit circumferential wave number. Apparently, the larger radius-to-thickness ratio has the lower phase velocity value [39].

The wave frequency variation versus circumferential wave number for different radius-to-thickness ratios is described in Figure 5.38. The graph exhibits promotion of wave frequency of shell with auxetic core per unit of circumferential wave number depending on radius-to-thickness ratio; with a smaller radius-to-thickness ratio, a higher value of wave frequency occurs [39].

Figure 5.39 describes the phase velocity changing range in term of longitudinal wave number and the effect of various core-to-face sheet thickness ratios. As it is explicit, phase velocity has rapid reduction versus longitudinal wave number. Also, the influence of core-to-face sheet thickness ratio in 3–1–3 induced larger phase velocity than 1–1–1 [39].

Figure 5.40 displays the wave frequency promotion versus longitudinal wave number. As is clear, the larger the amount of aluminum in the sandwich construction

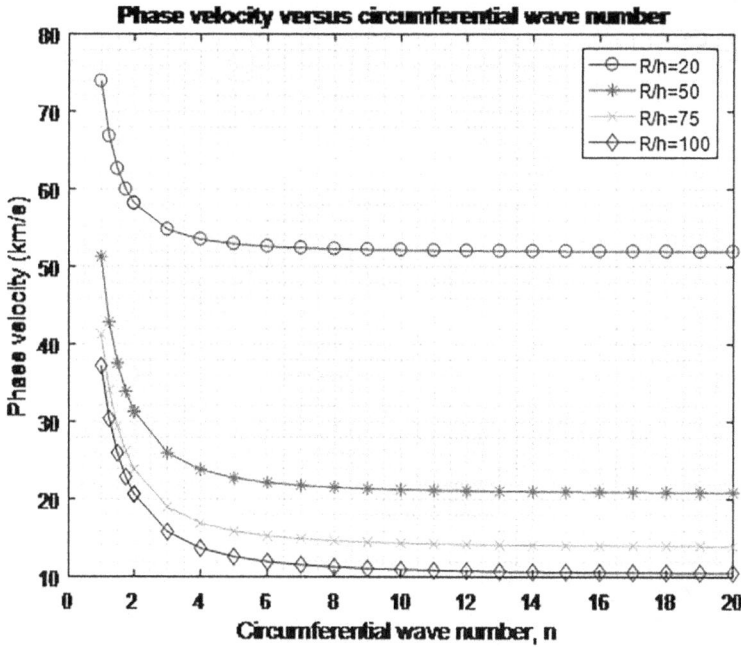

FIGURE 5.37 Phase velocity variation versus circumferential wave number for different radius-to-thickness ratio $(n = 1, \beta = 10, h = 3e - 3)$.

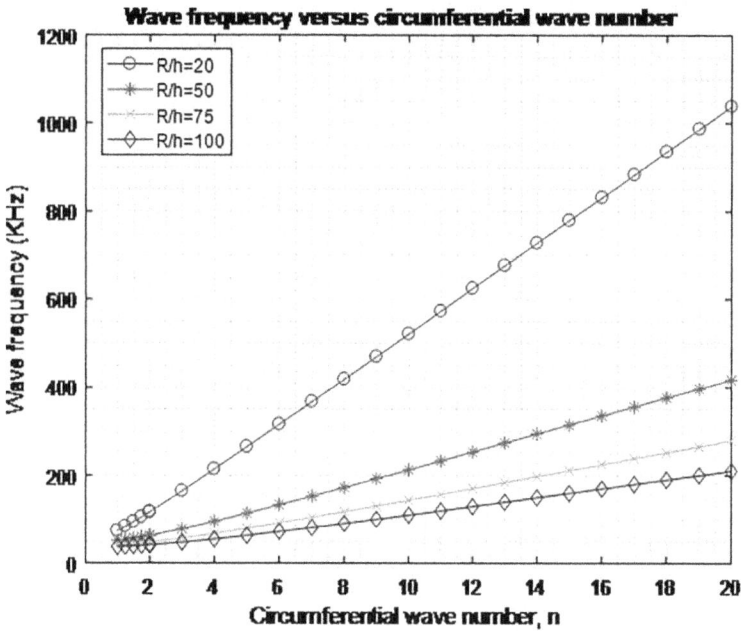

FIGURE 5.38 Wave frequency variation versus circumferential wave number for different radius-to-thickness ratio $(n = 1, \beta = 10, h = 3e - 3)$.

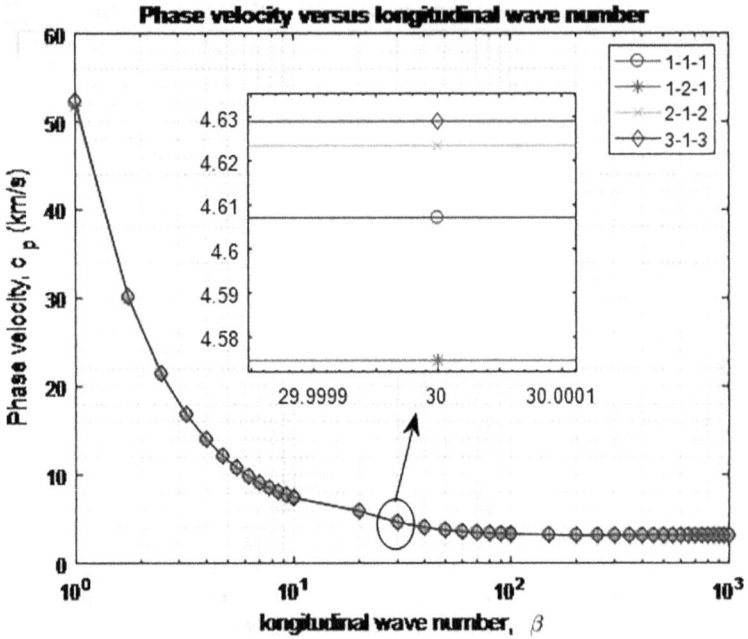

FIGURE 5.39   Phase velocity variation of versus longitudinal wave number for various core-to-face sheet thickness ratio.

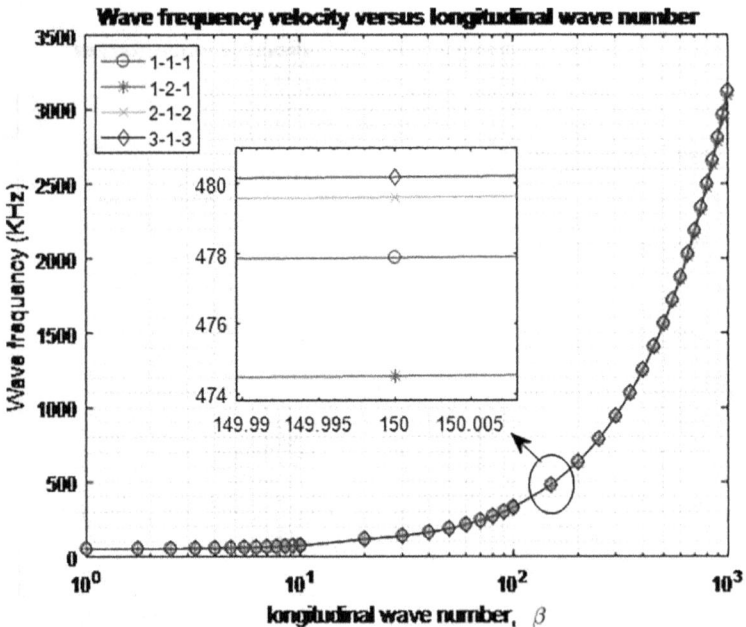

FIGURE 5.40   Wave frequency variation of versus longitudinal wave number for various core-to-face sheet thickness.

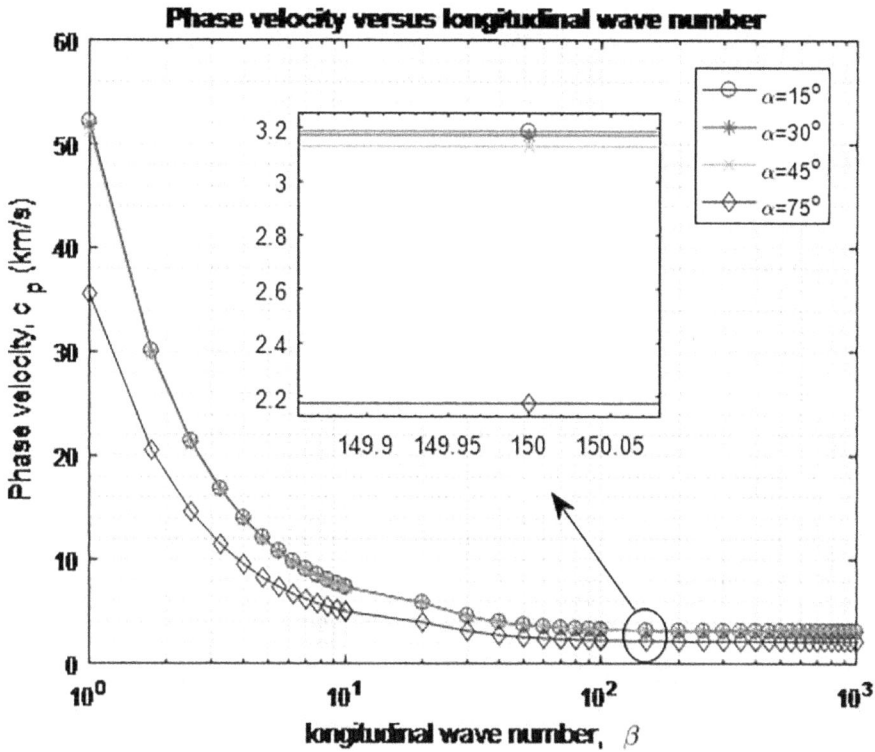

FIGURE 5.41 Phase velocity variation versus longitudinal wave number for various cell angle.

created the more substantial wave frequency. Moreover, as indicated in the picture, the wave frequency exhibits an upward slope during longitudinal wave number growing [39].

Figure 5.41 depicts the link between phase velocity variation and longitudinal wave number as a function of cell angles. There is no question that all curves follow the same pattern in which phase velocity reduces as longitudinal wave number grows. The greater phase velocity occurs in a lower cell angle plainly [39].

To study the impact of cell angles on wave frequency fluctuation during the longitudinal wave number changes, Figure 5.42 is depicted. It is clear that wave frequency grows progressively in terms of longitudinal wave number. The effect of cell angle created a change in the value of wave frequency. Besides, for the same longitudinal wave number, graphs with the lower cell angles have greater wave frequency [39].

Figure 5.43 depicts the phase velocity variation versus longitudinal wave number for various auxetic rib-length ratios. By rising of longitudinal wave number, phase velocity drops considerably. According to the graph, phase velocity rises by climbing λ_1.

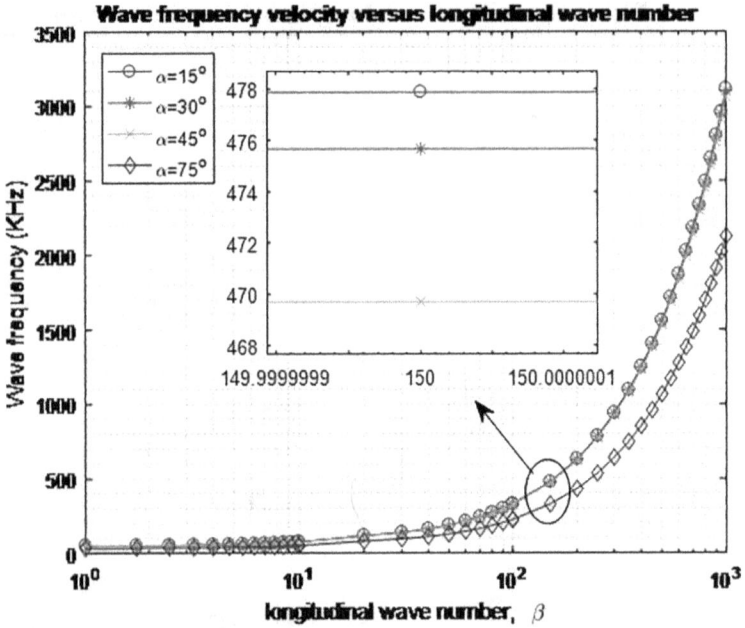

FIGURE 5.42 Wave frequency variation versus longitudinal wave number for various cell angle.

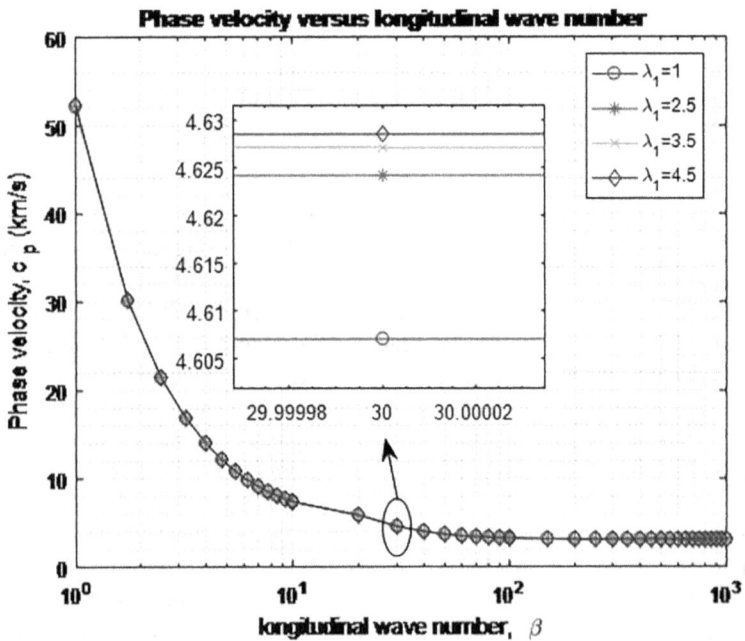

FIGURE 5.43 Phase velocity variation versus longitudinal wave number for different auxetic rib-length ratios.

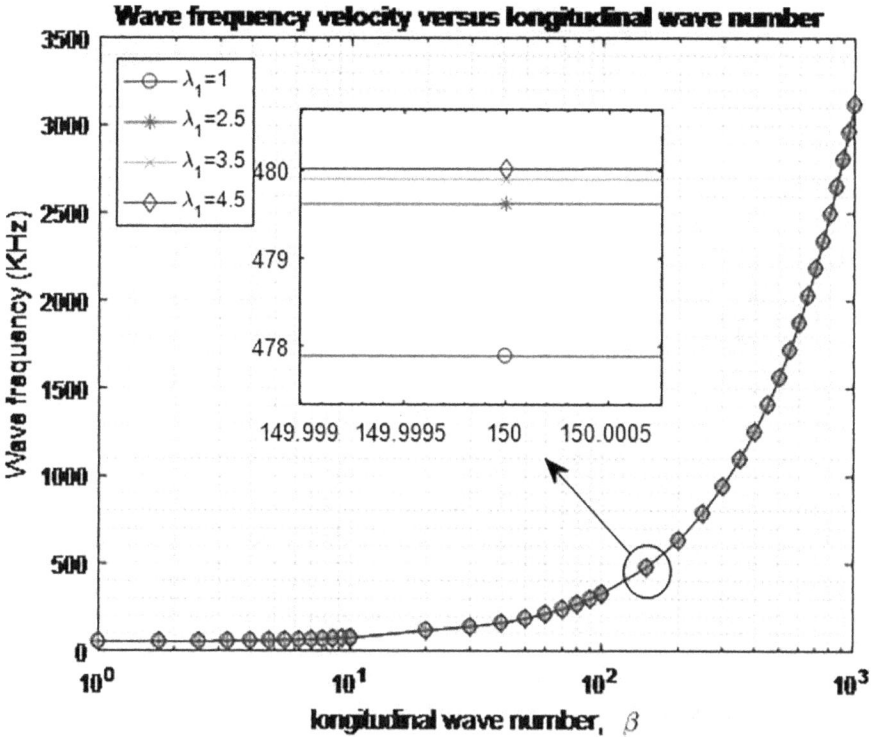

FIGURE 5.44 Wave frequency variation of versus longitudinal wave number for different auxetic rib-length ratios.

The impact of varied auxetic rib-length ratios on the wave frequency modification versus longitudinal wave number is examined by Figure 5.44. According to the graph's trend, a rise in longitudinal wave number resulted in fast increase in wave frequency. The expanded curve illustrates the effect of rib-length ratios causing variances in the value of wave frequency. Furthermore, for a fixed longitudinal wave number value, increased wave frequency occurs with larger levels of λ_1 [39].

Figure 5.45 demonstrates the change in phase velocity as a function of longitudinal wave number for varied ratios of radius to thickness. As longitudinal wave number grows, the wave velocity drops exponentially. As shown in previous research, it has been observed that the wave velocity drops when the radius-to-thickness ratio is increased [39].

The influence of varying radius-to-thickness ratio on wave frequency fluctuation in terms of longitudinal wave number is investigated in Figure 5.46. Based on the graph, it can be inferred that a decrease in the radius-to-thickness ratio has a positive impact on the relationship between wave frequency and longitudinal wave number, as shown by previous research [39].

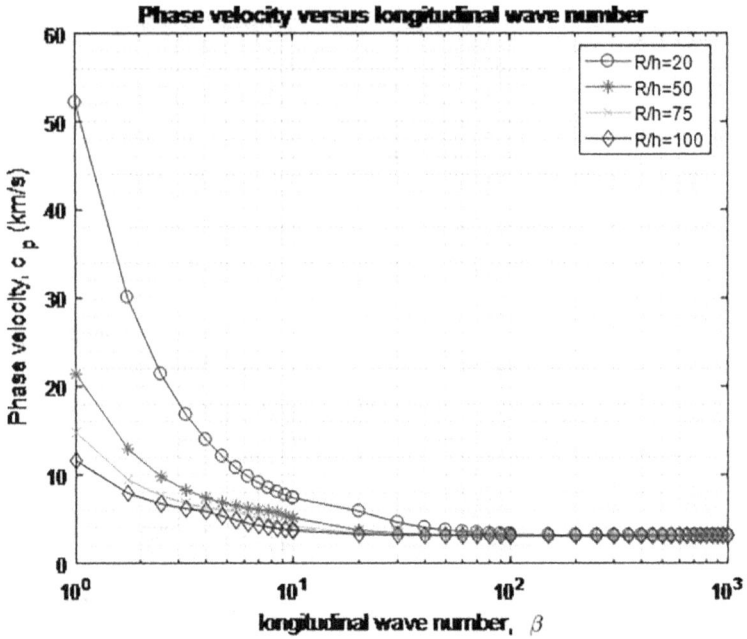

FIGURE 5.45 Variation of phase velocity versus longitudinal wave number for different radius-to-thickness ratio.

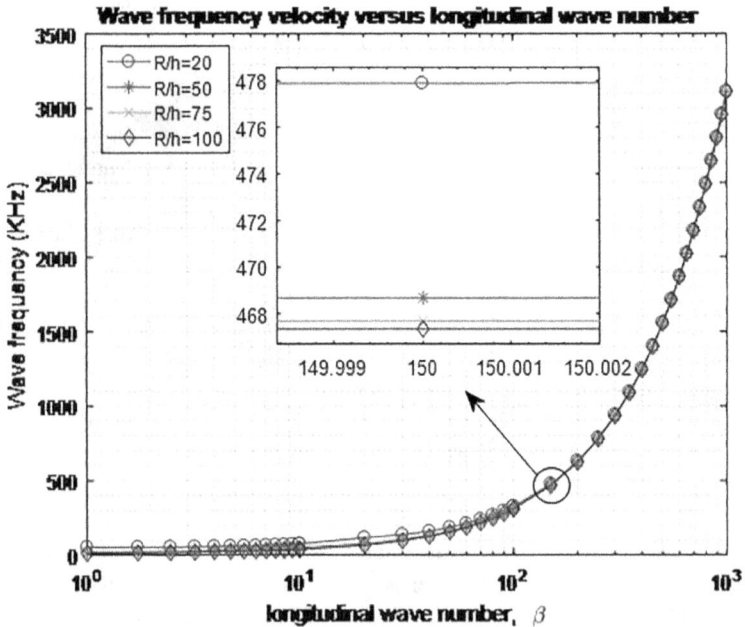

FIGURE 5.46 Wave frequency variation versus longitudinal wave number for different radius-to-thickness ratio.

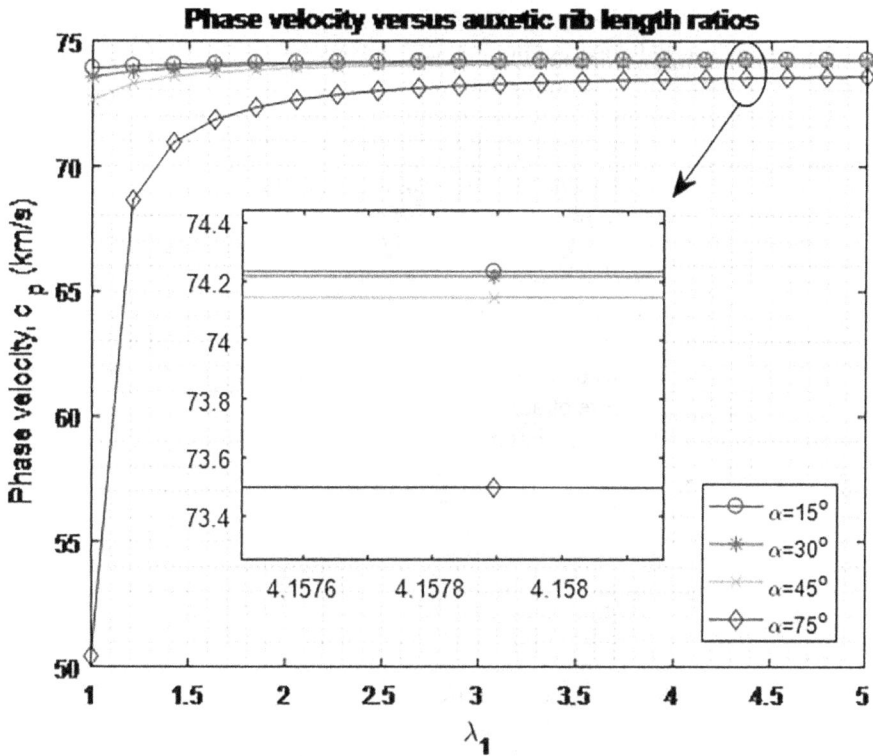

FIGURE 5.47 Phase velocity variation versus rib-length ratios for different cell angles.

To study the impact of cell angles on the fluctuation of phase velocity against rib-length ratios, Figure 5.47 is presented. The graphic demonstrates that with large angle values (75 degrees), phase velocity fluctuates fast and then has a constant pattern. As is obvious, the phase velocity changes process is slower at smaller angles [39].

5.5 SUMMARY

This chapter examines the wave propagation characteristics of beams, plates, and sandwich shells that have an auxetic core. Hamilton's concept was used during the investigation of the governing equation of the system. Ultimately, the numerical findings collected exhibit significant consequences that may be succinctly described as follows [18–39]:

- The trajectory of auxetic beams in FG-GOEAM beams is greatly influenced by the composition and quantities of elements used in the nanocomposite design. The integration of auxetic features into the structure has the potential to enhance the wave frequency while concurrently reducing the phase velocity.

- Increasing the concentration of GOri leads to the development of mechanical metamaterials that exhibit a negative Poisson's ratio (NPR). This, in turn, results in an elevation of the wave frequency and phase velocity of the FG-GOEAM beam in auxetic beams.
- Increasing the level of GOri folding leads to the formation of mechanical metamaterials that demonstrate an NPR, resulting in a decrease in the wave frequency and phase velocity of the FG-GOEAM beam.
- The $X - W_{GR}$ and $A - W_{GR}$ beams have a greater wave frequency and phase velocity in comparison to the other beams. The observed phenomenon may be attributed to the heightened dispersion of graphene oxide inside the surface layers of the fiber-reinforced polymer beams. This higher distribution of graphene oxide leads to an augmentation in the stiffness of the beams, particularly in the case of auxetic beams.
- The auxetic metamaterial has notable characteristics that enable it to maintain a steady wave frequency and phase velocity, even in the presence of temperature variations inside auxetic beams.
- It is worth noting that auxetic beams have a greater phase velocity and wave frequency when the proportion of non-metamaterials in the structure is increased. Therefore, it is crucial to use auxetic metamaterials in order to efficiently mitigate wave propagation and decrease vibration.
- An augmentation in the Winkler and Pasternak parameters leads to a marginal augmentation in the wave frequency and phase velocity in auxetic beams. The phase velocity in auxetic plates is reduced by increasing the external voltage (V_0), the volume percentage of the smart nanofiller in the matrix (η), and the viscoelastic damping foundation coefficient (c_d).
- The phase velocity in auxetic plates exhibits a rise due to several factors, including the expansion of the auxetic core thickness, the Winkler foundation coefficient, the Pasternak foundation coefficient, the geometrical parameters, the thickness of the smart nanocomposite, and the wave number.
- The findings indicate that auxetic shells exhibit a negative correlation between the rib-length ratio of auxetic cores and their density.
- The proposed structure, composed of an auxetic core, exhibits a positive or potentially negative Poisson's ratio, as determined by the inclination angle of the auxetic cells.
- The configuration and quantity of materials used in the sandwich construction of the cylindrical shell significantly impact wave propagation. The incorporation of an auxetic core inside the structure has the potential to decrease the phase velocity while simultaneously increasing the wave frequency.
- The reduced radius-to-thickness ratio was a consequence of the thicker shell, which in turn had a notable impact on both phase velocity and wave frequency.
- Both the phase velocity and wave frequency in auxetic shells are significantly influenced by the ratio of horizontal to vertical length of the auxetic cell and the cell inclination angle.

- In auxetic shells, it has been observed that for a given cell dimension of rib-length ratio, the phase velocity is greater for smaller cell angles compared to larger angles. Additionally, an increase in longitudinal wave number leads to an increase in wave frequency and a decrease in phase velocity in auxetic shells. Furthermore, it has been found that higher rib-length ratio and smaller radius-to-thickness ratio contribute to an increase in both wave frequency and phase velocity.
- The presence of a greater proportion of non-auxetic materials, such as aluminum as discussed in this study, inside auxetic shells resulted in an increase in phase velocity and wave frequency. Consequently, the use of auxetic materials proves to be more efficient in reducing wave propagation and minimizing vibrations.

REFERENCES

1. Ebrahimi, F. & Dabbagh, A., 2020. *Mechanics of Nanocomposites: Homogenization and Analysis*. 1st ed. Boca Raton, FL: CRC Press.
2. Ebrahimi, F. & Dabbagh, A., 2022. *Mechanics of Multiscale Hybrid Nanocomposites*. Duxford: Elsevier. ISBN: 9780128196144
3. Ebrahimi, F. & Dabbagh, A., 2019. *Wave Propagation Analysis of Smart Nanostructures*. Boca Raton, FL: CRC Press.
4. Ebrahimi, F. & Barati, M. R., 2016f. Wave propagation analysis of quasi-3D FG nanobeams in thermal environment based on nonlocal strain gradient theory. *Applied Physics A*, 122(9), p. 843.
5. Ebrahimi, F. & Barati, M. R., 2017g. Flexural wave propagation analysis of embedded S-FGM nanobeams under longitudinal magnetic field based on nonlocal strain gradient theory. *Arabian Journal for Science and Engineering*, 42(5), pp. 1715–1726.
6. Ebrahimi, F., Barati, M. R. & Dabbagh, A., 2016. A nonlocal strain gradient theory for wave propagation analysis in temperature-dependent inhomogeneous nanoplates. *International Journal of Engineering Science*, 107, pp. 169–182.
7. Ebrahimi, F., Barati, M. R. & Dabbagh, A., 2016d. Wave dispersion characteristics of axially loaded magneto-electro-elastic nanobeams. *Applied Physics A*, 122(11), p. 949.
8. Ebrahimi, F., Barati, M. R. & Haghi, P., 2017. Thermal effects on wave propagation characteristics of rotating strain gradient temperature-dependent functionally graded nanoscale beams. *Journal of Thermal Stresses*, 40(5), pp. 535–547.
9. Ebrahimi, F., Barati, M. R. & Haghi, P., 2016c. Nonlocal thermo-elastic wave propagation in temperature-dependent embedded small-scaled nonhomogeneous beams. *The European Physical Journal Plus*, 131(11), p. 383.
10. Ebrahimi, F., Barati, M. R. & Haghi, P., 2017a. Wave propagation analysis of size-dependent rotating inhomogeneous nanobeams based on nonlocal elasticity theory. *Journal of Vibration and Control*, p. 1077546317711537.
11. Ebrahimi, F. & Dabbagh, A., 2017a. Nonlocal strain gradient based wave dispersion behavior of smart rotating magneto-electro-elastic nanoplates. *Materials Research Express*, 4(2), p. 025003.
12. Ebrahimi, F. & Dabbagh, A., 2017d. Wave propagation analysis of embedded nanoplates based on a nonlocal strain gradient-based surface piezoelectricity theory. *The European Physical Journal Plus*, 132(11), p. 449.

13. Ebrahimi, F. & Dabbagh, A., 2018a. Viscoelastic wave propagation analysis of axially motivated double-layered graphene sheets via nonlocal strain gradient theory. *Waves in Random and Complex Media*, pp. 1–20.

14. Ebrahimi, F. & Dabbagh, A., 2018b. Wave dispersion characteristics of nonlocal strain gradient double-layered graphene sheets in hygro-thermal environments. *Structural Engineering and Mechanics*, 65(6), pp. 645–656.

15. Ebrahimi, F. & Dabbagh, A., 2018c. Thermo-magnetic field effects on the wave propagation behavior of smart magnetostrictive sandwich nanoplates. *The European Physical Journal Plus*, 133(3), p. 97.

16. Ebrahimi, F. & Dabbagh, A., 2018. Effect of humid-thermal environment on wave dispersion characteristics of single-layered graphene sheets. *Applied Physics A*, 124(4), p. 301.

17. Ebrahimi, F., Dabbagh, A. & Barati, M. R., 2016a. Wave propagation analysis of a size-dependent magneto-electro-elastic heterogeneous nanoplate. *The European Physical Journal Plus*, 131(12), p. 433.

18. Ebrahimi, F. & Dabbagh, A. Thermo-mechanical wave dispersion analysis of nonlocal strain gradient single-layered graphene sheet rested on elastic medium. *Microsystem Technologies*, 25(2019): 587–597.

19. Ebrahimi, F. & Haghi, P., 2018. Wave dispersion analysis of rotating heterogeneous nanobeams in thermal environment. *Advances in Nano Research*, 6(1), pp. 21–37.

20. Ebrahimi, F. & Haghi, P., 2017. Wave propagation analysis of rotating thermoelastically-actuated nanobeams based on nonlocal strain gradient theory. *Acta Mechanica Solida Sinica*, 30(6), pp. 647–657.

21. Ebrahimi, F., Dehghan, M. & Seyfi, A., 2019. Eringen's nonlocal elasticity theory for wave propagation analysis of magneto-electro-elastic nanotubes. *Advances in Nano Research*, 7(1), p. 1.

22. Ebrahimi, F., Seyfi, A. & Dabbagh, A., 2019. Dispersion of waves in FG porous nanoscale plates based on NSGT in thermal environment. *Advances in Nano Research*, 7(5), pp. 325–335.

23. Ebrahimi, F., Seyfi, A. & Dabbagh, A., 2019. A novel porosity-dependent homogenization procedure for wave dispersion in nonlocal strain gradient inhomogeneous nanobeams. *The European Physical Journal Plus*, 134(5), p. 226.

24. Eringen, A. C., 1984. Plane waves in nonlocal micropolar elasticity. *International Journal of Engineering Science*, 22(8–10), pp. 1113–1121.

25. Ebrahimi, F., Seyfi, A., Dabbagh, A. & Tornabene, F., 2019. Wave dispersion characteristics of porous graphene platelet-reinforced composite shells. *Structural Engineering and Mechanics, An Int'l Journal*, 71(1), pp. 99–107.

26. Ebrahimi, F. & Dabbagh, A., 2018. Wave dispersion characteristics of embedded graphene platelets-reinforced composite microplates. *The European Physical Journal Plus*, 133, pp. 1–13.

27. Mahinzare, M., Rastgoo, A. & Ebrahimi, F., 2023. Magnetic field effects on wave dispersion of piezo-electrically actuated auxetic sandwich shell via GPL reinforcement.

28. Ebrahimi, F., Dabbagh, A. & Rabczuk, T., 2021. On wave dispersion characteristics of magnetostrictive sandwich nanoplates in thermal environments. *European Journal of Mechanics-A/Solids*, 85, p. 104130.

29. Ebrahimi, F., et al., 2019. Hygro-thermal effects on wave dispersion responses of magnetostrictive sandwich nanoplates. *Advances in Nano Research*, 7(3), p. 157.

30. Ebrahimi, F. & Seyfi, A., 2021. A wave propagation study for porous metal foam beams resting on an elastic foundation. *Waves in Random and Complex Media*, pp. 1–15.

31. Ebrahimi, F., Seyfi, A. & Dabbagh, A., 2021. The effects of thermal loadings on wave propagation analysis of multi-scale hybrid composite beams. *Waves in Random and Complex Media*, pp. 1–24.
32. Ebrahimi, F., et al., 2022. Influence of magnetic field on the wave propagation response of functionally graded (FG) beam lying on elastic foundation in thermal environment. *Waves in Random and Complex Media*, 32(5), pp. 2158–2176.
33. Ebrahimi, F. & Dabbagh, A., 2017. On flexural wave propagation responses of smart FG magneto-electro-elastic nanoplates via nonlocal strain gradient theory. *Composite Structures*, 162, pp. 281–293.
34. Ebrahimi, F. & Dabbagh, A., 2017. Wave propagation analysis of smart rotating porous heterogeneous piezo-electric nanobeams. *The European Physical Journal Plus*, 132(4), p. 153.
35. Zhao, S., et al., 2022. Genetic programming-assisted micromechanical models of graphene origami-enabled metal metamaterials. *Acta Materialia*, 228, p. 117791.
36. Zhao, S., et al., 2022. A functionally graded auxetic metamaterial beam with tunable nonlinear free vibration characteristics via graphene origami. *Thin-Walled Structures*, 181, p. 109997.
37. Zhao, S., et al., 2022. Vibrational characteristics of functionally graded graphene origami-enabled auxetic metamaterial beams based on machine learning assisted models. *Aerospace Science and Technology*, 130, p. 107906.
38. Ebrahimi, F. & Seyfi, A., 2021. Wave propagation response of multi-scale hybrid nanocomposite shell by considering aggregation effect of CNTs. *Mechanics Based Design of Structures and Machines*, 49(1), pp. 59–80. doi: 10.1080/15397734.2019.1666722.
39. Ebrahimi, F. & Sepahvand, M., 2022. Wave propagation analysis of cylindrical sandwich shell with auxetic core utilizing first-order shear deformable theory (FSDT). *Mechanics Based Design of Structures and Machines*, pp. 1–25.

APPENDICES

APPENDIX 5.1

$$
M = \begin{bmatrix} m_{11} & 0 & m_{13} \\ 0 & m_{22} & 0 \\ m_{31} & 0 & m_{33} \end{bmatrix}, \quad K = \begin{bmatrix} k_{11} & 0 & k_{13} \\ 0 & k_{22} & k_{23} \\ k_{31} & k_{32} & k_{33} \end{bmatrix}
$$

where:

$$m_{11} = I_1,\, m_{13} = I_2 \qquad k_{11} = -A_{11}\beta^2,\; k_{13} = -B_{11}\beta^2,$$
$$m_{22} = I_1 \qquad k_{22} = -A_{55}\beta^2 - k_w - k_p\beta^2 + N_x^T\beta^2,\; k_{23} = A_{55}\beta i$$
$$m_{31} = I_2,\, m_{33} = I_3 \qquad k_{31} = -B_{11}\beta^2,\; k_{32} = -A_{55}\beta i,\; k_{33} = -D_{11}\beta^2 - A_{55}$$

APPENDIX 5.2

Where:

$$m_{11} = m_{22} = m_{33} = I_0$$
$$m_{14} = m_{25} = I_1$$
$$m_{44} = m_{55} = I_2$$

in which, the mass moments of inertia calculate from equation (5.66).

$$k_{11} = -A_{11}\beta^2 - \frac{A_{66}}{R^2}n^2$$

$$k_{12} = -\beta n\left(\frac{A_{12}+A_{66}}{R}\right)$$

$$k_{13} = \frac{A_{12}}{R}\beta i$$

$$k_{14} = -B_{11}\beta^2 - \frac{B_{66}}{R^2}n^2$$

$$k_{15} = -\beta n\left(\frac{B_{12}+B_{66}}{R}\right)$$

$$k_{22} = -A_{66}\beta^2 - \frac{A_{11}}{R^2}n^2 - \frac{A^S_{55}}{R^2}$$

$$k_{23} = n\left(\frac{A_{11}+A^S_{55}}{R^2}\right)i$$

$$k_{24} = -\beta n\left(\frac{B_{12}+B_{66}}{R}\right)$$

$$k_{25} = -B_{66}\beta^2 - \frac{B_{11}}{R^2}n^2 - \frac{A^S_{55}}{R}$$

$$k_{33} = -A^S_{55}\beta^2 - \left[\frac{A_{11}}{R^2} + n^2\frac{A^S_{55}}{R^2}\right]$$

$$k_{34} = \left(A^S_{55} - \frac{B_{12}}{R}\right)i\beta$$

$$k_{35} = n\left(\frac{A^S_{55}}{R} - \frac{B_{11}}{R^2}\right)i$$

$$k_{44} = -D_{11}\beta^2 - n^2\frac{D_{66}}{R^2} - A^S_{55}$$

$$k_{45} = -\beta n\frac{D_{12}+D_{66}}{R}$$

$$k_{55} = -D_{66}\beta^2 - n^2\frac{D_{11}}{R^2} - A^S_{55}$$

Index

For Product Safety Concerns and Information please contact our EU
representative GPSR@taylorandfrancis.com
Taylor & Francis Verlag GmbH, Kaufingerstraße 24, 80331 München, Germany